Handbook of Food Science and Technology

Edited by Sarah Scott

SYRAWOOD
PUBLISHING HOUSE

New York

Published by Syrawood Publishing House,
750 Third Avenue, 9th Floor,
New York, NY 10017, USA
www.syrawoodpublishinghouse.com

Handbook of Food Science and Technology
Edited by Sarah Scott

International Standard Book Number: 978-1-64740-251-8 (Hardback)

Cataloging-in-Publication Data

Handbook of food science and technology / edited by Sarah Scott.
      p. cm.
Includes bibliographical references and index.
ISBN 978-1-64740-251-8
1. Food. 2. Food industry and trade. 3. Food industry and trade--Appropriate technology.
4. Food--Biotechnology. I. Scott, Sarah.
TP370 .H36 2022
641.3--dc23

PREFACE

Food science is the discipline concerned with the study of food. It can be defined as a subject where other disciplines like biology, engineering, Physical sciences, biochemistry, chemistry and microbiology are used in order to get a detailed understanding of the nature of food and its processing. It also focuses on the improvement of food to ensure its fitness for consumption. Food science applies basic science and engineering to examine the biological, chemical and physical nature of food and its processing. Food technology is concerned with developing new food products. It also focuses on developing better packaging materials, studying shelf-life, and microbiological and chemical testing. This book provides comprehensive insights into the field of food science and technology. It presents researches and studies performed by experts across the globe. This book will prove to be immensely beneficial to students and researchers in this field.

The information shared in this book is based on empirical researches made by veterans in this field of study. The elaborative information provided in this book will help the readers further their scope of knowledge leading to advancements in this field.

Finally, I would like to thank my fellow researchers who gave constructive feedback and my family members who supported me at every step of my research.

Editor

TABLE OF CONTENTS

Microbiological Safety of Kitchen Sponges used in Food Establishments

Tesfaye Wolde[1,2] and Ketema Bacha[2]

[1]Applied Biology Department, Wolkite University, Wolkite, Ethiopia
[2]Biology Department, Jimma University, Jimma, Ethiopia

Correspondence should be addressed to Tesfaye Wolde; tesfalem_atnafu@yahoo.com

Academic Editor: Marie Walsh

Kitchen sponges are among the possible sources of contaminants in food establishments. The main purpose of the current study was, therefore, to assess the microbiological safety of sponges as it has been used in selected food establishments of Jimma town. Accordingly, the microbiological safety of a total of 201 kitchen sponges randomly collected from food establishments was evaluated against the total counts of aerobic mesophilic bacteria (AMB), Enterobacteriaceae, coliforms, and yeast and molds. The mean counts of aerobic mesophilic bacteria ranged from 7.43 to 12.44 log CFU/mm^3. The isolated genera were dominated by *Pseudomonas* (16.9%), *Bacillus* (11.1%), *Micrococcus* (10.6%), *Streptococcus* (7.8%), and *Lactobacillus* (6%) excluding the unidentified Gram positive rods (4.9%) and Gram negative rods (9.9%). The high microbial counts (aerobic mesophilic bacteria, coliforms, Enterobacteriaceae, and yeast and molds) reveal the existence of poor kitchen sponge sanitization practice. Awareness creation training on basic hygienic practices to food handlers and periodic change of kitchen sponges are recommended.

1. Introduction

During the cleaning of kitchen utensils, the prewashing and washing steps are usually carried out using sponges in order to remove food residues. In due course, some parts of the food residues could adhere to the sponges. Food remains, together with humidity retained in sponges, tender a positive environment for growth and survival of bacteria. Report shows that such heavily contaminated sponges could be the main vehicle that significantly contributes to the dissemination of potentially pathogenic bacteria in back of house setting [1]. Research on the significance of bacterial contamination of kitchen environments started 40 years ago. According to the early studies, uncooked material is the major cause of contaminants in kitchen although the area nearby the kitchen could contribute free-living bacteria. Accordingly, numerous studies revealed that sponges can be vital disseminators of pathogens and can transfer bacteria to surfaces and utensils, leading to cross-contamination of food [1, 2].

Earlier studies conducted to evaluate microbial safety of kitchens utensils and its environment [1, 2] showed that bacterial profiles of hand towels, dishcloths, tea towels, steel sinks, and working surfaces are significant and contributed to food contamination. Foodborne illness related to foods prepared in unhygienic kitchen is recurrently associated with *Salmonella* [3–5]. Several other bacterial infections associated with contaminated kitchen environment are *Listeria*, *Campylobacter Bacillus cereus*, *Staphylococcus aureus*, and *Escherichia coli* [4, 6]. *S. aureus* is one of the pathogenic bacteria isolated from 34.3% used synthetic washing sponges [7].

Jimma town, the study site, has been visited by many travelers as the town is a strategic place for the southwest part of the country in terms of transportation and trading. As a result, many visitors are coming to the town for several reasons. Thus, the town and its food establishments are visited by many individuals from different parts of the country traveling to and away from the city. In addition, the numbers of food vendors are currently increasing faster. Nevertheless, there is a scarcity of reports on the microbiological safety of kitchen sponges of food establishments in Ethiopia including Jimma town. This prompted the current study to evaluate the microbial wellbeing of synthetic sponges from food establishments of Jimma town. Assessment was made in terms of prevalence and load of aerobic mesophilic bacteria, coliforms, Enterobacteriaceae, and yeast and molds.

2. Methods

A total of 201 kitchen sponge samples were collected from food establishments (20 restaurants, 101 hotels, 47 cafeterias, and 33 pastry shops) in Jimma town using simple random sampling technique during the months of October 2010 to June 2011. Sponges from the selected food establishments were collected aseptically in germ-free polythene bags discretely and transported to Postgraduate Microbiology Laboratory, Department of Biology, Jimma University. Samples were processed in an hour. Questionnaire was used to obtain first-hand information about personal history/status assessments of the establishment owners, food handlers, food servers, and care being taken during washing steps of kitchen sponges.

Of the collected kitchen sponge samples, an amount of $25 \, mm^3$ was aseptically cut using sterile blade and the pieces were blended separately in $225 \, mm^3$ of sterile peptone water (OXOID). Kitchen sponges were mixed and proper dilutions of peptone water were plated in triplicate on dried surfaces of relevant media for microbial enumeration. Therefore, standard plate counts were counted on Standard Plate Count Agar (SPCA) (OXOID) following incubation at thirty-two degrees Celsius for forty-eight hours; VRBA (OXOID) was used to enumerate coliforms following incubation for forty-eight hours at thirty-two degrees Celsius. Enterobacteriaceae were enumerated on MacConkey agar (OXOID) following incubation at thirty-two degree Celsius for forty-eight hours (pink-to-red purple colonies with/without haloes of precipitation were counted as Enterobacteriaceae. Similarly, yeast and molds were counted on Chloramphenicol Bromophenol blue agar incubated at 27 degrees Celsius for 3 days. Media composition (Gram per 500 ml distilled water) includes yeast extracts (OXOID) 2.5, glucose 10, chloramphenicol 0.05, Bromophenol blue 0.005, and agar 7.5; pH 6.3). Smooth colony-forming units with no extension at margin were considered as yeasts. Furry colony-forming units with elongations at edge were considered as molds. All the microbiological media used for this study were OXOID products.

Following record, fifteen colonies were indiscriminately singled out from enumerable plates of SPCA. Culture characteristics on solid media, cell morphology, and Gram reaction were performed following usual microbiological techniques: lipopolysaccharide test was made to differentiate between Gram negative and Gram positive bacteria using the quick method suggested by Gregersen [8]. Catalase test was done by applying little drops of 3% H_2O_2 on agar plates containing an overnight initiated culture for production of air bubbles. Cytochrome oxidase test was conducted as suggested by Kovacs [9] using newly ready kovac's reagents for revelation of a blue colour on recently activated colonies in 30 seconds— few minutes as a positive reaction.

TABLE 1: Demographic features of proprietors/workers in food enterprise, Jimma, 2011.

Demographic features	Incidence	Percent
Sex		
Women	94	46.7
Men	107	53.3
Age		
19–34	27	13.2
35–50	130	65.0
>50	44	21.8
Educational status		
Illiterate	5	2.5
Literate	196	97.5
Marital status		
Single	33	16.4
Married	168	83.6

3. Results

The median service period of the food establishments was 9 yrs, arraying from 3 months to 43 years (data not shown). Middle age of the establishments' proprietors was 43 years, varying from 23 to 70 yrs. Majorities (97.5%) of the hotel managers were literate (Table 1). More than half (67.1%) of the study participants had received training on food hygiene and safe handling, although 32.9% did not get any training opportunity.

Majority (97.5%) of the food establishments were found using different kinds of detergents for dish washing while 39.3% use warm (65–71°C) water to clean the dishware at least two times per a week. Among the 201 food preparing personnel who were asked about their familiarity and practices related to food cleanliness, about 180 (89.5%) were found familiar with at least one type of foodborne illness (Table 2). The proportion of food handlers who said that foodborne diseases are due to pathogens was 160 (79.6%), with even more numbers (83.6%) associating the vehicle of transmission of foodborne diseases with contaminated food. Moreover, 167 (83%) of the food handlers had serious concerns on the risks of dirty hands in contaminating food. It was observed that all of the food establishments were washing their kitchen utensils using sponges. Mean service time of a kitchen sponge was 8 days. Majority, 181 (91%), of the respondents had no knowledge of the presence of antimicrobial chemicals in kitchen sponges (Table 2).

The average enumerations ($\log CFU/mm^3$) of bacteria which are aerobic mesophiles of kitchen sponge samples were the highest in pastry shop samples (11.8), followed by hotel (10.9), cafeteria (10.6), and restaurant (9.87). With mean counts of high margin (>9 log CFU/mm^3) in all sample sources, the actual figures fall within the ranges (log CFU/mm^3) of 7.43 to 10.4 (restaurant), 7.45 to 11.9 (hotel), 7.45 to 11.7 (cafeteria), and 9.25 to 12.44 (pastry shop) (Table 3). Irrespective of the microbial groups and food establishment

TABLE 2: Food handlers' familiarity with foodborne illness, Jimma, 2011 ($n = 201$).

Familiarity and practice	Incidence	Percent
Had information about foodborne diseases	180	89.6
Causes of foodborne illness		
(i) Microbes	160	79.6
(ii) Substance	3	1.5
(iii) Unhygienic food preparation	50	24.9
(iv) Others	3	1.5
Vehicle for transmission of foodborne disease		
(i) Contaminated food	168	83.6
(ii) Contaminated water	71	35.5
(iii) Vectors like flies and cockroaches	112	55.7
Factors contributing to food contamination		
(i) Unclean hands	167	83.1
(ii) Grimy food handlers	37	18.4
(iii) Dirty utensils	124	61.6
(iv) Unhygienic working environment	63	31.5
(v) Infestation by insects and rats	45	22.6
Dish washing and kitchen cleaning material		
(i) Sponge	201	100
(ii) Cloth	152	75.6
(iii) Towel	38	18.9
Life span of a kitchen sponge (days)		
(i) 3–5	31	15.4
(ii) 6–8	121	60.2
(iii) 9–11	49	24.4

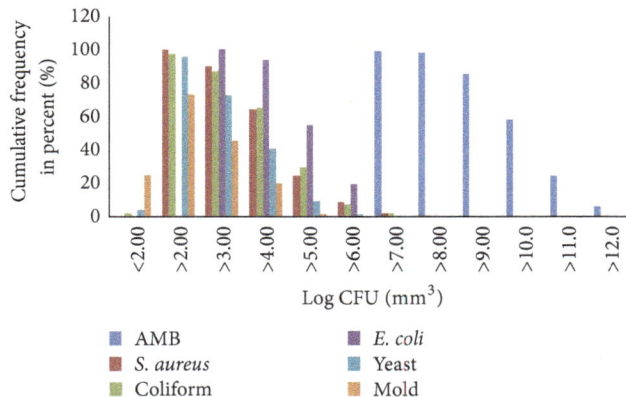

FIGURE 1: Cumulative frequency of microbial groups from kitchen sponges ($n = 201$).

types, there were statistically significant variations in microbial counts within groups (CV > 10%).

A total of 138, 732, 259, and 377 bacterial strains were isolated from kitchen sponge samples collected from restaurants, hotels, pastry shops, and cafeterias, respectively, and characterized with various genera and bacterial groups (Table 4). *Pseudomonas* (16.9%), *Bacillus* (11.1%), *Micrococcus* (10.6%), *Streptococcus* (7.8%), and *Lactobacillus* (6%) were among the dominant bacterial genera besides the unidentified Gram positive rods (4.9%) and Gram negative rods (9.9%). In general, 61.6% of the sponges were Gram positive organisms, with about 38.4% of the total 1506 aerobic mesophilic bacterial grouped as Gram negative. The proportion of *Pseudomonas* spp. (16.9%) among the aerobic mesophilic bacterial flora of kitchen sponges was relatively high.

Out of 201 different kitchen sponges examined, 64.9% had coliform counts of greater than 4.0 LOG CFU $(mm^3)^{-1}$ with 2% of the kitchen sponges' samples containing coliform counts above 7 log CFU/mm^3. Likewise, about 55.6% of the samples had Enterobacteriaceae count greater than 5 log CFU/mm^3. The counts of yeasts were relatively low where 72.8% of the samples had yeast counts greater than log 3 CFU/mm^3 with 4% of the samples having yeast counts below detectable number. About 45.7% of the kitchen sponges

had mold count greater than 3 log CFU/mm^3 with only 24.5% of the samples having counts below detectable level (Figure 1).

4. Discussion

The current report on microbiology of synthetic kitchen sponges worn in food establishments is the first of its kind from Ethiopia, particularly from Jimma town. Unhygienic handling and use of the same sponges for extended period of time contributed to the observed gross contamination with aerobic mesophilic counts as high as 11.8 log cfu/mm^3. Other related studies also indicate that food establishments that lack basic sanitary facilities and utensils used for handling of food are grossly contaminated with microorganisms and could contribute to the occurrence of foodborne diseases [10]. Recently, increase in population mobility for business or leisure purposes and visiting food establishments has become common practice. On the other hand, kitchen sponges are unhygienic with disease causing bacteria and indicator bacteria owing to cross-contamination, inappropriate handling and storage, or improper sanitizing, leading to foodborne illness [11]. Thus, pattern of disease development data are required to update community health authorities about the nature and extent of the problem and to keep an eye on trends over time. Result of the current study clearly revealed limitation of experience on hygienic handling of kitchen utensils and poor sanitization strategies tracked to diminish pathogenic bacteria in the kitchen sponges as experienced by the food establishments.

The most extensively used and usual methods of washing of tools for food handling are the three-partition sink or holder system [12]; conversely, a reduced amount of half (46%) of the enterprises had been washing utensils by means of this system. Almost three-fourth (76%) of the food serving institutions, particularly those who prepare food in their own kitchen and dish up meals, and 38.5% of the establishments portioning both meals and drinks had been using three compartments and sponge to clean utensils. Even if the washing practices in Jimma were superior to the practices

TABLE 3: Microbial count (log CFU/mm^3) of different microbial groups detected in *kitchen sponge samples*, Jimma, 2011.

Microbial group	Source of kitchen sponge																			
	Restaurant					Hotel					Pastry					Cafeteria				
	Avg	SD	CV%	Min	Max	Avg	SD	CV%	Min	Max	Avg	SD	CV%	Min	Max	Avg	SD	CV%	Min	Max
AMB	9.87	9.99	101.2	7.43	10.4	10.9	11.2	102.8	7.45	11.9	11.8	11.9	100.8	9.25	12.4	10.6	10.9	102.8	7.45	11.7
Coliform	3.41	3.67	107.6	<2.0	5.3	6.08	6.37	104.8	3.11	6.93	4.94	5.24	106	2.20	5.95	5.81	6.26	107.7	2.00	6.91
Yeast	4.04	4.32	106.9	<2.0	4.87	4.67	5.17	110.7	<2.0	5.95	5.59	6.47	115.7	2.00	6.78	4.78	5.23	109.4	<2.0	5.95
EB	4.90	4.92	100.4	3.99	5.84	6.42	7.08	110.3	1.45	7.93	6.00	6.16	102.7	3.07	6.79	6.04	6.36	105.3	3.49	6.93
Molds	3.11	3.49	112.2	<2.0	4.00	3.88	4.23	109	<2.0	5.00	3.34	3.55	106.3	<2.0	4.11	3.83	4.25	110.9	<2.0	5.00

AMB = aerobic mesophilic bacteria; EB = Enterobacteriaceae.

TABLE 4: Frequency distribution (%) of dominant bacteria in kitchen sponges collected from food establishments of Jimma town, 2011.

Food establishment	Number of isolates	Pseudomonas	Enterobacteriaceae	Bacillus	Micrococcus	Gram negative coccus	Streptococcus	Lactobacillus	Other Gram positive rods
Restaurant	138	6	5	23	23	6	3	7	22
Hotel	732	139	93	81	69	76	61	43	29
Pastry shop	259	43	29	26	25	28	23	14	10
Cafeteria	377	67	47	38	43	39	31	27	13
Total	1506	255	174	168	160	149	118	91	74
%	100	16.9	11.5	11.1	10.6	9.9	7.8	6.0	4.9

elsewhere [13], they are still below meeting the sanitary standard and necessitate extra advance.

From this study, food handlers had good knowledge or awareness of the basis of foodborne disease transmission mechanisms, aggravating factors for food contamination, and as a whole foodborne disease. Though they have good acquaintance, they did not keep up with the poor sanitary conditions of the facilities where they have been working, predominantly with the neatness and maintenance of the kitchens and food utensils and the reduced storage of organized foods and food utensils. The observed discrepancy between comprehension and practice could be accounted to reluctance to practice what they know due to negligence, lack of commitment, and attitudinal change. Equally important, they might also not be operational and/or supplied with the essential materials that would allow them to uphold the sanitary excellence of their property. In fact, all food handlers have duty to keep high degree of personal hygiene and be preinformed about hygienic and nondangerous food handling; failure to do so could contribute to contamination of food sources with pathogens [14]. In this report the food handlers' hygiene and food management practices were found below standard and not satisfactory. Such condition facilitates the spread of foodborne diseases if corrective hygienic procedures are not set in place.

The detection of high counts of coliforms, Enterobacteriaceae, and aerobic mesophilic bacteria in kitchen sponge samples was an indication of underprivileged sanitary eminence of utensils used for food handling. Outbreaks of food poisoning recurrently happen as a result of unacceptable food preparation in which cross-contamination in combination with insufficient storage or cooking was concerned in many occasions [15]. Kitchen sponges and dishcloths could lead to cross-contamination in kitchens since they can transfer microorganisms to surfaces where microorganisms can survive for hours or days and contaminate food persisting in these disease vehicles [16].

In the current report, diverse microorganisms including probably pathogenic microbes were encountered in kitchen sponges. Among these microbes are *Pseudomonas* sp., *Bacillus* sp., and *Streptococcus* sp. As reported earlier [16], microorganisms, together with pathogenic species, frequently exist in all areas of the home surroundings. Accordingly, wet sites including kitchen sink areas, toilets, and nappy buckets are most commonly associated with heavy contamination and the occurrence of potentially harmful species. Other wet sites, such as dishcloths and similar cleaning materials, were also found to be frequently and heavily contaminated by microbes of different arrays. As to Keeratipibul and his colleagues [17], the presence of coliforms is unacceptable because it reveals poor sanitary conditions. Contamination in sponges may come from leftovers, inadequate hygienic practices during food preparation, cross-contamination due to contaminated surfaces, and storage in places. These results imply that, in kitchen environment, although raw food is most likely the main source of contamination, the sink, waste trap, and the surrounding areas can also act as reservoirs of different arrays of microorganisms which harbor and promote the concern of free-living bacteria and fungi populations.

Among 201 kitchen sponges evaluated, 98.7% of them had aerobic mesophilic bacterial count greater than $8 \log \text{CFU/mm}^3$, a value comparable to the $7.5 \log \text{CFU/mm}^3$ count reported elsewhere [18] but much lower than the $10 \log \text{CFU/mm}^3$ reported by Erdogrul and Erbilir [13] from Turkey. The significantly high count could be accounted to failure to use sanitizing chemicals to clean kitchen sponges, extended use of the same sponge for cleaning, and underprivileged hygienic conditions of the food serving establishments as observed during data collection.

Results of this study indicated that the counts of coliform ranged between 1.3 and $7.93 \log \text{CFU/mm}^3$ with an average count of $6.25 \log \text{CFU/mm}^3$. Furthermore, 94.7% of the restaurants, 98% of the hotels, and 100% each of the pastry shops and cafeterias kitchen sponge samples had coliform count greater than detectable level. In a related study [16], it was reported that the coliform load of kitchen sponges was within the range of below detectable level to $8.82 \log \text{CFU/mm}^3$. This indicates that significantly high proportion of synthetic kitchen sponges in many food serving establishments at Jimma town was contaminated with microorganisms of fecal origin because of poor hygienic practices. Coliform should be eliminated during washing steps using hot water. Presence of such high counts among the kitchen sponge samples could indicate either inappropriateness of the heating temperature used during washing or postwashing contamination because of inadequate storage conditions being practiced in the food establishments accompanied by proliferation of those wash-surviving microorganisms during storage.

The Enterobacteriaceae count of the present study ranged between 3.08 and $6.9 \log \text{CFU/mm}^3$ with mean number $6.04 \log \text{CFU/mm}^3$. Of the total kitchen sponge samples evaluated, 94.7% of restaurants samples, 96% of hotels, and 100% each of pastry shops and cafeterias samples had Enterobacteriaceae count greater than 10^3CFU/mm^3. In addition, among the total 1506 characterized aerobic mesophilic bacteria, 11.6% were Enterobacteriaceae. If not necessarily at all, many of the Enterobacteriaceae could be potentially pathogenic as indicated by the presence high counts of indicator coliforms as discussed above. The members of Enterobacteriaceae isolated in a related study conducted by Scott et al. [19] included *Klebsiella*, *Enterobacter*, *Proteus*, and *Citrobacter*. Speirs et al. [1] also reported almost similar isolates from domestic kitchens.

About 72.8% and 45.7% of the kitchen sponge samples had yeast and mold counts greater than 10^3CFU/mm^3 with only 4% and 24.5% of the samples with yeast and mold count below detectable level, respectively. In general, the mean counts of aerobic mesophilic bacteria, Enterobacteriaceae, coliforms, and yeast and molds of kitchen sponge samples were significantly different among the different food establishments. Relatively, high counts of these microbial groups were obtained in samples from pastry shops. This might indicate the prolonged usage of kitchen sponge when compared to other food establishment types. In fact, the hygienic practice and sanitary conditions of kitchen environment among pastry shops were better than kitchens of other food establishment types as observed during sample collection.

In conclusion, high microbial load of any sort is clear indicator for below standard handling practice in a given setting and calls for regular monitoring of the practice by owners and staff working at the same food establishment besides timely supervision by concerned government bodies to ensure safety of customers.

Competing Interests

The authors affirm that there are no competing interests.

Acknowledgments

The authors would like to thank Jimma University for financial support. The study participants of all food establishments deserve special thanks for genuine provision of information and cooperation during sample collection.

References

[1] J. P. Speirs, A. Anderton, and J. G. Anderson, "A study of the microbial content of the domestic kitchen," *International Journal of Environmental Health Research*, vol. 5, no. 2, pp. 109–122, 1995.

[2] J. E. Finch, J. Prince, and M. Hawksworth, "A bacteriological survey of the domestic environment," *Journal of Applied Bacteriology*, vol. 45, no. 3, pp. 357–364, 1978.

[3] J. T. Holah and R. H. Thorpe, "Cleanability in relation to bacterial retention on unused and abraded domestic sink materials," *Journal of Applied Bacteriology*, vol. 69, no. 4, pp. 599–608, 1990.

[4] J. Dufrenne, W. Ritmeester, E. Deffgou-van Asch, F. Van Leuden, and R. de Jonge, "Quantification of the contamination of chicken and chicken products in the netherlands with Salmonella and Campylobacter," *Journal of Food Protection*, vol. 64, no. 4, pp. 538–541, 2001.

[5] H. D. Kusumaningrum, G. Riboldi, W. C. Hazeleger, and R. R. Beumer, "Survival of foodborne pathogens on stainless steel surfaces and cross-contamination to foods," *International Journal of Food Microbiology*, vol. 85, no. 3, pp. 227–236, 2003.

[6] T. Regnath, M. Kreutzberger, S. Illing, R. Oehme, and O. Liesenfeld, "Prevalence of *Pseudomonas aeruginosa* in households of patients with cystic fibrosis," *International Journal of Hygiene and Environmental Health*, vol. 207, no. 6, pp. 585–588, 2004.

[7] W. Tesfaye, B. Ketema, A. Melese, and S. Henok, "Prevalence and antibiotics resistance patterns of *S. aureus* from kitchen sponge's," *International Journal of Research Studies in Biosciences*, vol. 3, pp. 63–71, 2015.

[8] T. Gregersen, "Rapid method for distinction of gram-negative from gram-positive bacteria," *European Journal of Applied Microbiology and Biotechnology*, vol. 5, no. 2, pp. 123–127, 1978.

[9] N. Kovacs, "Identification of *Pseudomonas pyocyanea* by the oxidase reaction," *Nature*, vol. 178, no. 4535, p. 703, 1956.

[10] A. Kumie and K. Zeru, "Sanitary conditions of food establishments in Mekelle town, Tigray, North Ethiopia," *Ethiopian Journal of Health Development*, vol. 21, no. 1, pp. 3–11, 2007.

[11] M. S. Myint, *Epidemiology of Salmonella contamination of poultry products; knowledge gaps in the farm to store products [Ph.D. dissertation]*, Faculty of the Graduate School of the University of Maryland, College Park, Md, USA, 2004.

[12] T. Gebre-Emanual, "Public food service establishment hygiene," in *Food Hygiene: Principles and Methods of Foodborne Diseases Control*, Department of Community Health, Faculty of Medicine, Addis Ababa University (AAU), Addis Ababa, Ethiopia, 1997.

[13] O. Erdogrul and F. Erbilir, "Microorganisms in kitchen sponges," *International Journal of Food Safety*, vol. 6, pp. 17–22, 2003.

[14] A. Kumie, K. Genete, H. Worku, E. Kebede, F. Ayele, and H. Mulugeta, "Sanitary conditions of food and drink establishments in Zeway," *The Ethiopian Journal of Health Development*, vol. 16, pp. 95–104, 2002.

[15] J. A. Salvato, *Environmental Engineering and Sanitation*, Wiley-Interscience Publication, John Wiley & Sons, New York, NY, USA, 4th edition, 1992.

[16] S. Olsen, L. MacKinnon, J. Goulding, N. Bean, and L. Slutsker, "Surveillance for food borne disease outbreaks, 1993–1997," *Surveillance Summaries*, vol. 49, pp. 1–62, 2000.

[17] S. Keeratipibul, P. Techaruwichit, and Y. Chaturongkasumrit, "Contamination sources of coliforms in two different types of frozen ready-to-eat shrimps," *Food Control*, vol. 20, no. 3, pp. 289–293, 2009.

[18] M. Sharma, J. Eastridge, and C. Mudd, "Effective household disinfection methods of kitchen sponges," *Food Control*, vol. 20, no. 3, pp. 310–313, 2009.

[19] E. Scott, S. Bloomfield, and C. Baelow, "Investigation of microbial contamination in home," *Journal of Hygiene*, vol. 89, no. 2, pp. 279–293, 1982.

Brilliant Blue Dyes in Daily Food: How Could Purinergic System be Affected?

Leonardo Gomes Braga Ferreira,[1] **Robson Xavier Faria,**[2]
Natiele Carla da Silva Ferreira,[3] **and Rômulo José Soares-Bezerra**[3]

[1]*Laboratory of Inflammation, Oswaldo Cruz Foundation, Av. Brazil, 4365 Rio de Janeiro, RJ, Brazil*
[2]*Laboratory of Toxoplasmosis, Oswaldo Cruz Foundation, Av. Brazil, 4365 Rio de Janeiro, RJ, Brazil*
[3]*Laboratory of Cellular Communication, Oswaldo Cruz Foundation, Av. Brazil, 4365 Rio de Janeiro, RJ, Brazil*

Correspondence should be addressed to Robson Xavier Faria; robson.xavier@gmail.com

Academic Editor: Rosana G. Moreira

Dyes were first obtained from the extraction of plant sources in the Neolithic period to produce dyed clothes. At the beginning of the 19th century, synthetic dyes were produced to color clothes on a large scale. Other applications for synthetic dyes include the pharmaceutical and food industries, which are important interference factors in our lives and health. Herein, we analyzed the possible implications of some dyes that are already described as antagonists of purinergic receptors, including special Brilliant Blue G and its derivative FD&C Blue No. 1. Purinergic receptor family is widely expressed in the body and is critical to relate to much cellular homeostasis maintenance as well as inflammation and cell death. In this review, we discuss previous studies and show purinergic signaling as an important issue to be aware of in food additives development and their correlations with the physiological functions.

1. Introduction

The purinergic receptor superfamily has ionotropic and metabotropic receptors. These receptors are widely expressed in the body and shows distinct pharmacological properties and activation pathways [1, 2]. The G protein-coupled P2Y receptor subtypes are activated by adenosine triphosphate (ATP), uridine triphosphate (UTP), and metabolites, such as adenosine diphosphate (ADP) and uridine diphosphate (UDP). There are eight mammalian subtypes: P2Y1R, P2Y2R, P2Y4R, P2Y6R, P2Y11R, P2Y12R, P2Y13R, and P2Y14R. Depending on the type of G protein coupled to the P2Y receptors, its activation triggers different signaling cascades. In general, these events lead to phospholipase C recruitment, inositol 3-phosphate formation, and intracellular Ca^{2+} release from intracellular stores, as well as modulating adenylyl cyclase-related signaling [3, 4]. On the other hand, the mammalian ATP-gated ion channels, namely, P2X, are composed of the following 7 subtypes: P2X1R, P2X2R, P2X3R, P2X4R, P2X5R, P2X6R, and P2X7R [4]. Following their activation, the P2X receptors lead to rapid mobilization of monovalent and divalent cations, such as K^+, Na^+, and Ca^{2+}, which depolarize plasma membrane and trigger several intracellular events. Nevertheless, because of the lack of selective agonists, synthetic ATP analogues were designed, such as 3'-O-(4-benzoyl)benzoyl-ATP and adenosine 5'-[γ-thio]triphosphate, for pharmacological experiments. Conversely, P2YR subtypes have distinct agonist preferences. There are receptors preferentially activated by ATP, such as P2Y11; those preferentially activated by ADP, such as P2Y1R, P2Y12R, P2Y13R; those preferentially activated by UDP, such as P2Y6R; and those preferentially activated by UTP, such as P2Y2R and P2Y4R. Yet, antagonists development has been highly prolific. There are many categories of blockers that have been properly described and used against both P2Y and P2X receptors [1, 5–8]. It is noteworthy that many diseases display purinergic signaling involvement, in which several researchers have focused on new pharmacological strategies targeting P2 receptors [5]. Indeed, Gum and colleagues have stated in 2012 some challenges in development of orally

Brilliant Blue Dyes in Daily Food: How Could Purinergic System...

9

Trypan blue

Brilliant Blue G

Brilliant Blue FCF

Reactive blue 2

Phenol red

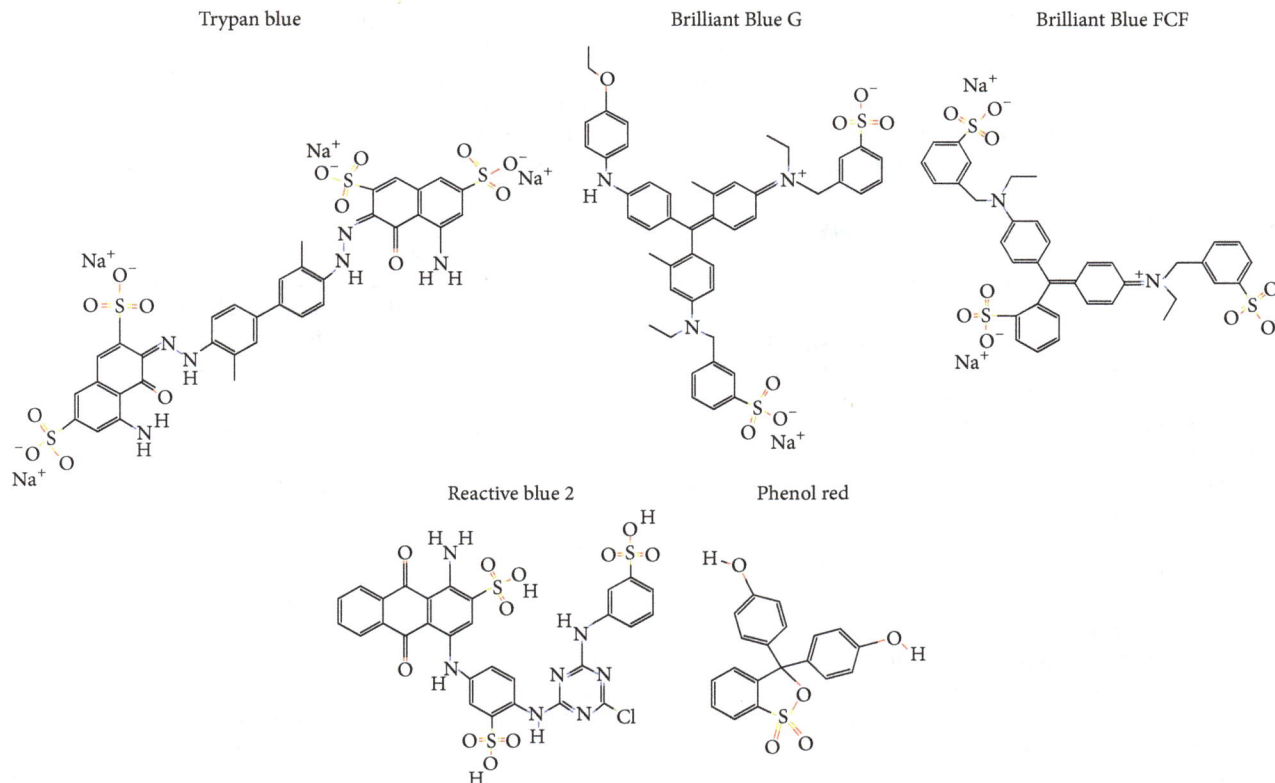

FIGURE 1: Structure of the dyes, which are P2 receptors antagonists.

purinergic receptors specific drugs and achieving suitable bioavailability [9].

Curiously, many available P2 receptor antagonists have biological and industrial uses. Brilliant Blue G (BBG) [10], reactive Blue 2 (RB-2) [11], phenolphthalein (Phenol red) [12], and trypan blue [13], shown in Figure 1, are widely used. Among these first compounds mentioned as purinergic inhibitors, they lack pharmacological selectivity, which, in turn, might inhibit one or more subtypes of P2 receptors and unexpectedly modulate some organ/tissue functions (via P2 receptor inhibition). Nevertheless, only BBG (P2X7 antagonist) was assessed *in vivo* (and in humans) and it has utility in the clothing and food industries. Therefore, as P2X7 has critical roles under both physiological and pathological circumstances as inflammation, infection, and tissue injury [5], this review comes up with warnings about such compound ordinary uses and purinergic signaling.

2. Industrial Uses of Brilliant Blue G and Its Derivative

The Brilliant Blue dye family has several members in which BBG and FD&C Blue No. 1 have wide uses in health sciences and industrial issues. The synthetic dye Brilliant Blue G, also known as Coomassie Brilliant Blue, was first synthesized from coal tar dye. It has a reasonable stability when exposed to light, heat, and acidic conditions, whereas it has low oxidative stability. On the other hand, FD&C Blue No. 1 (also known as

the Brilliant Blue FCF or E133 in the European numbering system) (Figure 1) is one of the most common dyes used in food and cosmetic preparations and medicines. FD&C Blue No. 1 was approved in various countries to be used as a food additive in dairy products, candies, cereals, cheese, toppings, jellies, liquors, and soft drinks. This dye is also used in cosmetics such as shampoos, nail polishes, lip gloss, and lip sticks and in the textile sector [14]. The uses of this dye are justified due to its high cost-benefits as blue is not a color currently found in secretions in the body [15]. It is noteworthy that FD&C Blue No. 1 is also found in green shaded food and drinks as a mixture with yellow dyes. It has a low gastrointestinal absorption, and the amount absorbed is highly excreted by biliary vesicles and urinary routes [16]. However, FD&C Blue No. 1 utilization was not allowed in some countries, such as Germany, Austria, France, Belgium, Norway, Sweden, and Switzerland before European Union foundation. In other countries, such as the United States, its use is unconditional; in Canada, use is limited to 100 ppm; in England, it can be used in some food; in the European Union, it is allowed for any use [17], and in Brazil, consumption is allowed up to 100 g [18]. Prado and Godoy analyzed the concentrations of different dyes by HPLC, including FD&C Blue No. 1 in different types of food in Brazil. Chocolates and candies contained an amount of the dyes within the limits, whereas in gumballs, the FD&C Blue No. 1 content was above the authorized concentration [18]. This fact may be more complicated when one considers alimentary imported products, whose specifications do not conform to the current Brazilian legislation [18]. FD&C Blue No. 1 has been related to skin irritations and bronchial constriction, especially when combined with other dyes [19]. In this sense, FD&C Blue No. 1 may exhibit another impairment to health, such as high genotoxic properties due to DNA base intercalation.

3. Brilliant Blue Dyes *In Vitro* Assays

FD&C Blue No. 1 has been linked to intracellular enzyme modulation, such as protein tyrosine phosphatases (PTPases). PTPases, along with protein tyrosine kinases, regulate the cellular phosphorylation levels [20]. Shrestha and colleagues studied the action of some food color additives on the activity of these phosphatases *in vitro*. FD&C Blue No. 1 inhibited PTP1B function (IC_{50} = 91 μM) while the enzymes TC-PTP and YPTP1 were less sensitive than PTP1B with IC_{50} values greater than 120 μM [21]. Furthermore, FD&C Blue No. 1 has been suggested to inhibit mitochondrial respiration *in vitro*, which was shown to aggravate sepsis by enteral tube feedings [22].

At the bench, Brilliant Blue dyes, more specifically BBG, have been applied in electrophoresis studies, especially for protein visualization. The advantage of these triphenylmethane dyes is that they give strong colored bands when they combine with a cellulose acetate membrane. BBG has been extensively used in several biochemical analyses and protein quantification in protein-dye binding assays [23–26]. Moreover, alternative colorimetric enzyme activity assays were proposed based on BBG properties [27–30]. Similarly, Brilliant Blue R is usually applied in protein analyzers, mainly in electrophoresis and in microscopy [31–33]. In light microscopy procedures, BBG was used to visualize intracellular organelles in hepatocytes and fibroblast cell cultures [34, 35]. BBG is an ingredient of the Bradford assay, which biochemically measures intracellular protein [36]. Additionally, this dye was also used to purify virus-like particles subjected to sucrose density ultracentrifugation. This technique allows for the identification of rubella virus-like particles [37].

Nevertheless, as mentioned above, BBG is a known noncompetitive antagonist for P2X7R [10, 38]. There are many studies showing that BBG has a higher selectivity and potency to rat P2X7R with an IC_{50} of 12 nM [10], but BBG also inhibited human P2X7R at a nanomolar range (IC_{50} = 265 nM) [10]. BBG was also shown to be efficient in blocking P2X7R-induced cytotoxicity in retinal cells [39], microglial cells [40], and astrocyte cells [41]. However, there are divergences concerning the applicability of BBG in human cells. Eschke and colleagues demonstrated a reduction in activity in human macrophage cells following BBG exposure [42].

Alternatively, FD&C Blue No. 1 and its analogue FD&C Green No. 3 have been shown to selectively inhibit pannexin-1 in oocytes-transfected system. Interestingly, both inhibitors showed the same IC_{50} in an oocytes heterologous system (IC_{50} of 0.27 μM) and were ineffective as a P2X7R inhibitor [43]. The pannexin-1 is widely expressed in the organism and it has been connected to the P2X7R-induced pore-forming phenomenon, whereas this point remains controversial [44, 45]. Nevertheless, this relationship should not be ignored,

once this appears to be upon cell type dependence, as in astrocyte. Indeed, Garre et al. showed that BBG inhibited ethidium uptake in astrocyte similarly to $_{10}$Panx1, a mimetic inhibitory peptide of pannexin-1 [46]. However, supposing that pannexin-1 is part of the P2X7-pore phenomenon, which part of this complex is inhibited should be carefully assessed and one cannot affirm that BBG is specifically blocking pannexin-1. Furthermore, there are several indications that pannexin-1 is associated with ATP release and the inflammasome assembling, indicating that it could be related to inflammatory diseases, such as colitis and Crohn's disease [43, 47–52]. In 2011, Jo and Bean observed that BBG was a sodium channel blocker in mouse N1E-115 neuroblastoma cells, whereas FD&C Blue No. 1 had little effect on sodium channel [53]. Table 1 shows major Brilliant Blues dye effects *in vitro* assays.

4. Brilliant Blue Dyes *In Vivo* Assays

BBG should also be carefully studied in this context once it is currently used clinically in ophthalmic procedures [54, 55]. Surely, it is reasonable to consider that inflamed (mostly chronic) tissues would show P2X7R hyperactivity (directly or indirectly by pannexin-1) and that the ingestion of food containing blue dye could relieves or diminishes deleterious processes by P2X7R inhibition [43, 51, 56]. In addition, Mennel and colleagues studied some vital dyes to visualize anatomical structures during vitreoretinal surgery in 2008. BBG had no effects on the morphology and functionality of the optical tissues. Additionally, the effect of BBG in the retinal pigment epithelium (RPE) was analyzed *in vitro*. The transepithelial resistance (TER) was used to measure the barrier integrity after 3 days of BBG treatment. In the fluid-filled eye model, BBG at 0.25 mg/mL did not affect the outer blood retinal barrier. The concentration of 2.4 mg/mL in the fluid-filled eye and air-filled eye models showed a decrease after 1.5 h, which was no longer observed after 24 h. In addition, there were no structural alterations of the RPE cells after BBG treatment. Clinically, BBG did not stain epiretinal membranes, and it represents an appropriate candidate for the future, as BBG has a high affinity for the internal limiting membrane [57]. In vitreoretinal surgical procedures, Höing and collaborators used BBG as a macular surgery stain. Patients received it during vitrectomy for macular holes or epiretinal membranes. The authors analyzed several parameters, such as best corrected visual acuity and intraocular pressure, perimeter, fundus photography, and optical coherence tomography. At the end, they concluded that BBG sufficiently and selectively stained the internal limiting membrane (ILM). Additionally, there was no retinal toxicity or side effects associated with the BBG, and the safety of long-term ingestion should be evaluated in a larger patient series and in a longer follow-up [58]. Based on these preliminary studies and molecular structural similarities among BBG and other Brilliant Blue food dyes, Franke and collaborators investigated purinergic precipitates in microglial activation in the rat nucleus accumbens (NAc) *in vivo* based on extracellular ATP release after brain injury. Using confocal image analysis, they observed increases of P2X1, P2X2, and P2X4 and P2Y1, P2Y2, P2Y4, P2Y6, and P2Y12 subtypes in the region of the lesion, mainly related

TABLE 1: The main studies of the Brilliant Blue dyes effects *in vitro* and *in vivo* assays.

			Brilliant Blue FCF	Ref.
In vitro assays	Phosphatases activities modulation	PTP1B function	$IC_{50} = 91\,\mu M$	[21]
		YPTP1 function	$IC_{50} > 120\,\mu M$	
	Selective pannexin-1 activity inhibition		$IC_{50} = 0.27\,\mu M$	[43]
	Mitochondrial respiration inhibition			[22]
In vivo assays	Food and Drug Administration (FDA) oral absorption limits		12 mg/kg/day	[22]
	Higher gastrointestinal permeability in sepsis		Systemic absorption enteral feedings after tracheostomy for obstructive apnea	[22]
			Systemic absorption chronic renal failure	[22]
			It leads to death	[85]
	No evident alterations (even in carcinogenicity studies) in the animal models used			[86, 87]
	Increase in important hepatic health-indicative enzymes		Alanine transaminase, aspartate transaminase, alkaline phosphatase, and bilirubin	[88]
			Brilliant Blue R	
In vitro assays	Biochemical analyses and protein quantitation in protein-dye binding assays			[23–26]
	Electrophoresis studies microscopy			[31–33]
	Polyacrylamide gels in electrophoresis			[30]
			Brilliant Blue G	
In vitro assays	Biochemical analyses and protein quantitation			[23–26]
	The administration of BBG reduced the agglomeration of pathogenic prion protein (PrPres) *in vitro*			[73]
	Electrophoresis studies			[31–33]
	Microscopy studies			[34, 35]
	Polyacrylamide gels in electrophoresis			[30]
	Bradford assay			[36]
	Virus-like particles purification			[37]
	$IC_{50} = 12\,nM$ (rat)			[10]
	$IC_{50} = 265\,nM$ (human)			[10]
	P2X7R noncompetitive antagonist	P2X7-induced calcium influx	$pIC_{50} < 4$ (mouse) $pIC_{50} = 5.09$ (rat) $pIC_{50} < 4$ (human)	[96]
		P2X7-associated pore formation (Yopro-1 uptake)	$pIC_{50} = 6.71$ (BALB/c mouse) $pIC_{50} = 6.34$ (C57BL/6 mouse) $pIC_{50} = 6.24$ (rat) $pIC_{50} = 5.71$ (human)	[96]
	Pannexin-1 antagonist	Oocytes-expressing Pannexin-1-induced ionic currents	$IC_{50} \sim 3\,\mu M$	[97]

TABLE 1: Continued.

	Ophthalmic procedures		[54, 55]
	Visualization of anatomical structures during vitreoretinal surgery		[57]
	Vitreoretinal surgical procedures		[58]
In vivo assays	Deleterious consequences of ATP release following brain injury	Microglial activation in the rat nucleus accumbens (NAc)	[59]
		Inhibition of ATP-induced caspase-3 activity	
	Traumatic spinal cord injury	Local astrocytes and microglia activity inhibition	[60]
		Inhibition of neutrophil infiltration	
	Huntington's disease (HD)	BBG impaired the symptomatology (body weight loss, motor-coordination deficits, and neuronal apoptosis)	[61]
	Seltzer model of neuropathic pain	Acute thermal nociception	[62]
	Freund's adjuvant (CFA) pain model	Moderate effect in inflammatory pain and edema	
	BBG toxicity assays in *Drosophila melanogaster*	No impairment on survival of the insect model ($LC_{50} = 38$ mM). No neurotoxic effects	[63]
	Suppression of P2X7R-induced ganglion cell death		[64]
	Protective neuronal effects upon oxygen-glucose deprivation	The brain damaged area following ischemia was diminished in the animals pretreated with BBG, compared to treatment with vehicle alone	[67]
	Cerebral ischemia/reperfusion (I/R) injury	BBG (10 μg) protected against transient global cerebral I/R injury, augmented survival rates, retarded I/R-induced learning and memory deficits, and suppressed I/R-induced neuronal death, DNA cleavage, glial activation and inflammatory cytokine overexpression in the hippocampus	[68]
	Traumatic brain injury (TBI) induced by posterior cerebral edema	BBG (25 mg/kg) orally administered before the TBI induction diminished the TBI effects	[69]
	Duchenne muscular dystrophy (*mdx/mdx* mouse)	BBG treatment recovered the CD62L in blood lymphocytes and it maintained its ability to migrate to the heart	[71]
		Treatment with BBG impaired the number of degeneration-regeneration cycles in mdx skeletal muscles *in vivo*	[72]
	Agglomeration of pathogenic prion protein (PrPres)	Diminished number of PrPres in infected mouse brain after the administration of BBG	[73]
	Chronic renal dysfunctions renal injury induced by hypertension	BBG suppressed the diuresis pressure threshold in F344 rats	[78]
	Amyotrophic lateral sclerosis (ALS)	The administration of BBG inhibited the progression of ALS	[79]

to P2X7R immunoreactivity, which colocalized with active caspase 3, but not with the antiapoptotic marker pAkt. P2R agonists augmented the immunoreactivity of P2R and P2X7R. BzATP administered intra-accumbally increased the caspase-3 activity. In contrast, PPADS and BBG diminished injury and immunoreactivity [59]. According to Mennel et al.'s paper evaluating FD&C Blue No. 1 *in vivo*, Peng et al. studied the effect of BBG on traumatic spinal cord injury. The authors gave 10 or 50 mg/kg BBG per day, immediately after injury and for three consecutive days. This treatment did not promote any effects on behavior, weight, survival, or other physiological parameters, including body temperature, blood pH, blood gases, or blood pressure. The evaluation of the blood-brain barrier permeability of BBG in the setting of a medullary lesion was quantified, and following a 10 mg/kg-dose, they measured 9.94 μM BBG within contused spinal cord tissue. At a dose of 50 mg/kg of BBG, a concentration of 43.59 μM was achieved in the lesion 3 days later [60].

Brilliant Blue Dyes in Daily Food: How Could Purinergic System...

13

As described above, to study neuroinflammation, they produced a lesion in the spinal cord, which led to an instantaneous and irreversible loss of tissue at the contusion point and consequent enlargement of tissue injury over time. Although secondary injury should be potentially avoidable, the patients with acute spinal cord injury (SCI) do not have adequate medication options. After the first lesion, the traumatized tissue releases ATP leads to P2X7R activation. The authors applied BBG systemically in a weight-drop model of thoracic SCI in rats to evaluate the neuroprotective reactions. BBG administered for 15 days after the SCI suppressed spinal cord anatomic damage and improved motor restoration without apparent toxicity. The local BBG application inhibited local activation of astrocytes and microglia and neutrophil infiltration [60].

Díaz-Hernández and collaborators studied the reduction of ATP production in *Drosophila* models of Huntington's disease (HD), which is associated with synaptic disturbances and the elevation of neuronal apoptosis. In this initial paper, they demonstrated the influence of the P2X7R on apoptosis triggered by activation of its receptor *in vitro*. The administration of the BBG in HD mice, *in vivo*, impaired the symptomatology, such as body weight loss, motor-coordination deficits, and neuronal apoptosis [61].

In 2010, Andó et al. compared the analgesic activity of antagonists acting at P2X1R, P2X7R, and P2Y12R and agonists acting at P2Y1R, P2Y2R, P2Y4R, and P2Y6R in neuropathies associated with neurogenic and inflammatory pain. BBG and other drugs were evaluated by mechanical allodynia in the Seltzer model of neuropathic pain, by acute thermal nociception, and by the inflammatory pain and edema induced by complete Freund's adjuvant (CFA). BBG only presented a moderate effect on inflammatory pain [62].

Also in 2010, Okamoto and collaborators analyzed the toxicity of BBG using *Drosophila melanogaster*. They evaluated the long-term toxic effects of continuous and a single exposure of *Drosophila melanogaster* to BBG added to the culture medium. BBG concentrations did not affect the survival of the insect model. Because BBG concentrations higher than 15 mM cannot be evaluated (BBG was insoluble), LC_{50} was estimated to be 38 mM. Additionally, BBG was not neurotoxic [63].

P2X7R is abundantly expressed on microglia; this cell type is capable of modulating long-term potentiation (LTP) of spinal pain. Possibly, this process occurs through regulating the communication between microglia and neurons. Additionally, microglial cells are the main components that activate the spinal LTP, which are the C-fiber induced field potentials generated by tetanic stimulation of the sciatic nerve (TSS). In this context, Chu and colleagues investigated the function of the P2X7R in the LTP evoked by TSS in rats. *In vitro*, BBG and oxidized ATP diminished the TSS stimulation of spinal LTP in spinal cord slices. In turn, *in vivo* BBG administration inhibited the induction of spinal LTP and diminished mechanical allodynia. The intrathecal application of BBG blocked the upregulation of microglial P2X7R. However, IL-1β expression was inhibited in the model of LTP evoked by TSS pretreated with BBG. This result was confirmed by a reduction in the expression of this cytokine after pretreatment with an IL-1 receptor antagonist (IL-1ra) [64].

Isolated retinal ganglion cells (RGCs), glia, and other cell types surrounding the ganglion cells may evoke stimulatory or inhibitory effects through P2X7R activation. Hu and collaborators investigated the participation of the P2X7R in retinal ganglion cell death. BzATP was applied intravitreally in the superior nasal region of rats in the presence or absence of BBG and MRS 2540. Both antagonists suppressed ganglion cell death [65].

In 2011, another paper demonstrated the involvement of the P2X7R in the pain mechanism. They studied the mustard oil-induced glutamatergic-dependent central sensitization, a mechanism present in medullary dorsal horn (MDH) nociceptive neurons of the tooth pulp. In this model, the central sensitization occurs due to an increase of the neuronal excitability of nociceptive pathways with subsequent peripheral tissue injury and inflammation. The expression of P2X7R was assessed in the presynaptic terminal and glial cells, and in the latter, its association in chronic inflammatory and neuropathic pain was evaluated. In this paper, the authors described P2X7R participation in the acute inflammatory pain model evaluated in anesthetized rats [66]. For the study of the nociceptive neurons in the MDH, they measured unitary records of mechanoreceptive fields, the mechanical activation threshold, and responses to noxious stimuli. Mustard oil-(MO-) induced MDH central sensitization was reduced by continuous intrathecal superfusion of both BBG and oxidized ATP. Interestingly, the microglial blocker minocycline also diminished the MO-induced MDH central sensitization [66]. According to the authors, these results confirm that dorsal horn P2X7R is expressed on microglia and that P2X7R may be connected to central sensitization in an acute inflammatory pain model [66].

In 2012, Arbeloa et al. evaluated P2X7R participation in neuronal excitotoxicity. In primary neuronal cultures and in brain slices, electrophysiological and Ca^{2+} assays recorded BzATP inhibition by BBG administration. In addition, the neuronal death promoted by oxygen-glucose deprivation was impaired by BBG treatment. The formation of a middle cerebral artery occlusion in rats was used as a model of transient focal cerebral ischemia. The area of brain damaged after the ischemia was diminished in the animals pretreated with BBG, compared to treatment with vehicle alone [67].

Also in 2012, Chu et al. studied P2X7R action in neuroinflammation observed in cerebral ischemia/reperfusion (I/R) injury. In this context, they used the rat model of transient global cerebral I/R injury. Transient global cerebral I/R was induced using the four-vessel occlusion (4-VO) method after 20 minutes of infusion with BBG, oxidized ATP (oxATP), or A-438079. The high dosage of BBG (10 μg) and A-438079 (3 μg) and low dosage of oxATP protected against transient global cerebral I/R injury in a dosage-dependent manner, augmented survival rates, retarded I/R-induced learning and memory deficits, and suppressed I/R-induced neuronal death, DNA cleavage, glial activation, and inflammatory cytokine overexpression in the hippocampus [68].

In 2012, Kimbler and colleagues assessed the effects of BBG, orally administered, in traumatic brain injury (TBI), which is associated with posterior cerebral edema. Neurosurgical procedures to ameliorate the edema-induced augmented intracranial pressure are questionable, and there are no efficient drugs to treat this symptom. Using P2X7R knockout mice (P2X7$^{-/-}$ mice), the expressions of IL-1β and cerebral edema were impaired. In another strategy to inhibit P2X7R, BBG (25 mg/kg) was administered via the drinking water for one week before the TBI induction or via an intravenous bolus four hours after the TBI. Both strategies diminished the TBI effects. The final BBG concentration in the plasma was quantified after its intravenous administration. They measured a BBG value of 383 μM and 1.73 mM following the application of 50 mg/kg and 100 mg/kg, respectively. These results implicate P2X7R as a therapeutic target to avoid the secondary effects of TBI [69].

As mentioned above, P2X7R activation leads to diverse intracellular signaling, such as CD62L expression pathways in different cell types [70]. In the paper published by Cascabulho and collaborators, they studied Duchenne muscular dystrophy (DMD) using a mouse model (*mdx/mdx* mouse). This model presents most aspects of the disease, such as a low quantity of T cells in damaged muscles, the degeneration of skeletal and cardiac muscles, and chronic inflammation. The authors investigated the migration of T cells to the heart and disturbances in adhesion molecules in the *mdx/mdx* mouse model. They demonstrated a reduction in the positive CD62L expression in blood leukocytes, including T cells, of six-week-old *mdx/mdx* mice. A downregulation and reduction of approximately 40% of T cells expressing CD62L in 12-week-old mice were observed. This diminished CD62L quantity was associated with the suppression of the ability of blood T cells to adhere to cardiac vessels *in vitro* and reach cardiac tissue *in vivo*. BBG treatment recovered the CD62L in blood lymphocytes, and they maintained their ability to migrate to the heart [71]. Another scientific group in 2012 studied the role of the P2X7R in DMD using the *mdx* mouse model. Dystrophies muscles had upregulated P2X7R mRNA and protein expression *in vitro*. They linked the protein quantity to the change in the responsiveness of intracellular Ca^{2+} and extracellular signal-regulated kinase (ERK) phosphorylation. Dystrophies mice exhibited alterations in P2X7R in isolated primary muscle cells and dystrophic muscles *in vivo*. Treatment with BBG impaired the number of degeneration-regeneration cycles in mdx skeletal muscles *in vivo*. Both papers suggested that the disturbance in P2X7R expression and function occurs in dystrophic *mdx* muscle. In addition, treatment with P2X7R antagonists, such as BBG, may reverse this effect [72].

There are reports of neuronal injuries in a large number of neurodegenerative disturbances. These damages, in some cases, are related to P2X7R activation. In prion-related diseases, there are lethal neurodegenerative disturbances that do not have effective therapies. Based on this information, Iwamaru and collaborates stated that the blockade of the P2X7R could impair prion replication conjointly as a therapeutic tool for prion infection. The administration of BBG reduced the agglomeration of pathogenic prion protein (PrPres) *in vitro*. In addition, the brain of infected mice had a diminished number of PrPres after the administration of BBG *in vivo* [73]. It is noteworthy that recently BBG was shown to prevent

neuronal loss in mouse models and modulates amyloid-β aggregation and cytotoxicity in cell-based assays, suggesting a new therapy for AD [74, 75].

In 2013, Kakurai and colleagues investigated the action of the P2X7R in cultured retinal ganglion cells after optic nerve crush (ONC) injury. P2X7R agonists reduced the viability of RGCs *in vitro*, which was reversed with P2X7R antagonist pretreatment. Using rats with ONC injury, BBG suppressed impairments in the vitreous body by approximately 61% of baseline 7 days after the lesion was generated [76].

A relevant function associated with P2X7R activity is the activation of the inflammasome, such as NLRP3. In this context, lupus nephritis (LN) is composed of inflammatory and autoimmune events. Interestingly, NLRP3 may modulate both characteristics. Additionally, the binding between P2X7R and the NLRP3 inflammasome pathway was investigated in the pathogenesis of LN. The authors used the mouse lupus model, MLR/*lpr*, and the parameters analyzed were anti-dsDNA antibody production, survival, activation of the NLRP3/ASC/caspase-1 inflammasome, and the Th17/Treg ratio on renal lesions. The MLR/*lpr* mice presented a substantial upregulation of the P2X7/NLRP3 inflammasome compared to nontreated mice. Following 8 weeks of BBG treatment, the treated mice exhibited P2X7R inhibition and diminished NLRP3/ASC/caspase-1 assembly and IL-1β release. Consequently, there was suppression of the nephritis, circulating anti-dsDNA antibody levels, and an augmented survival of the mice. Moreover, the IL-1β, IL-17, and Th17/Treg ratio levels in the serum were impaired with BBG administration. Similar results to those with BBG were observed after silencing P2X7R using interference RNA *in vivo* [77].

Menzies and collaborators studied chronic renal dysfunctions. This pathology is globally prevalent in approximately 10% of the population. In this work, they reported that the Fischer (F344) rat had a lower glomerular filtration rate (GFR) than the Lewis rat and is more susceptible to renal injury induced by hypertension. They performed kidney activity microarray assays to find candidate genes for impaired blood pressure control using the endeavor enrichment tool. The *P2rx7* and *P2rx4* genes showed seven- and threefold increased expression, respectively, in F344 rats. These purinergic receptors were visualized in the endothelium of the preglomerular vasculature and in the renal tubule, and this expression was similar in both rat species. The BBG treatment of Lewis rats did not affect the blood pressure but elevated renal vascular resistance (possibly due to inhibition of some basal vasodilatory tone). Both parameters were diminished in F344 rats (possibly due to an increase in the vasoconstrictor tone). In addition, BBG suppressed the diuresis pressure threshold in F344 rats [78].

Among the diverse functions attributed to P2X7R, the receptor may induce motor neuron death in brain alterations. Based on amyotrophic lateral sclerosis (ALS), which is the progressive degeneration of motor neurons that is currently without treatment, the relationship between P2X7R and ALS was investigated by Cervetto et al. in 2013. The transgenic mice have mutations causing superexpression of the superoxide dismutase 1 gene, which was utilized to study the influence of the gender, disease course, motor performance, weight loss, and life span. The administration of BBG inhibited the progression of ALS [79].

As an attempt to reduce the toxic action of the trypan blue (TB) dye, a combination containing TB, Brilliant Blue G (BBG), and polyethylene glycol had been used in ARPE retinal pigment epithelial cells. In this paper, the authors inquired the toxic effect promoted by this association. TB and BBG were exposed alone or in combination with ARPE cells. TB in concentrations above 0.075% and BBG in concentrations above 0.1% were toxic to the cells after 30 min incubation. In the case of BBG, in concentrations of 0.1% and higher, it showed a protective effect on cells when incubated by 5 min. Concentration of 0.025% BBG was able to preserve against TB-induced damage at 5 min and 30 min incubation [80].

Also in 2013, Henrich, as principal investigator of University Hospital, Basel, Switzerland, patented the clinical trial entitled "Intraoperative Utility of Brilliant Blue G (BBG) and Indocyanine Green (Icg) Assisted Chromovitrectomy". Although Icg is the dye more indicated to color the ILM in a surgical extraction, in comparison to BBG, it is not approved to intravitreal use, because there are description of ocular toxicity caused by its utilization. In this clinical trial, the medicines intend to find hypothetical alterations in Icg and BBG dyes to obtain improved intraoperative dye utility associated with a safety profile (ClinicalTrials. NCT01485575) [81, 82].

In 2014, based on limited information about the relationship between dyes and the causes of retinal damage after ERM and ILM peeling, Giansanti and colleagues evaluated the toxicities of blue dyes on retinal pigment epithelium and retinal ganglion cells with light exposure (with and without halogen and xenon light exposure) [83]. Indeed, the researchers treated the human retinal pigment epithelium line ARPE-19 and the rat retinal ganglion cells RGC5 with 0.5% vital dyes for 5 min and exposed them to light. Cell viability was estimated by proliferation assay using WST-1 reagent for 12, 24, or 120 h after the light exposition. Time-lapse video microscopy recorded the morphological aspects of the apoptosis for 72 h. They did not observe toxicity to both cell lines, when they were exposed to light. ARPE-19 cells present moderate toxicity to BBG after xenon illumination, in contrast; RGC5 cell line did not exhibit toxicity to BBG.

Balaiya and collaborates studied the light-induced decomposition of vital dyes, in the case of the BBG, that occurring during the chromovitrectomy [84]. Phototoxic effects of the BBG have not yet been described, but the phototoxicity depends on the light source, the intensity of illumination, the distance of the light source from the surface of the retina, and the duration of exposure. Then, they investigated the BBG toxicity in human retinal pigment epithelium (HRPE) cells stimulated to metal halide surgical endoilluminator (SE) in different distances of illumination. BBG was used in the concentrations of 0.25 and 0.5 mg/mL on ARPE-19 cell line and illuminated with SE for 1, 5, and 15 min. The surgical distance of illumination was hypothesized varying distances (1 and 2.5 cm) of the illumination used. Cell viability and the distance of illumination of 1 cm during the times of 1 (about 90%), 5 (about 60%), and 15 min (about 35%) of exposure was similar

TABLE 2: Calculus 1.

Male	Female
0.1 mg Brilliant Blue FCF, 1 L	0.1 mg Brilliant Blue FCF, 1 L
X mg Brilliant Blue FCF, 4.2 L	Z mg Brilliant Blue FCF, 3.9 L
X = 0.42 mg Brilliant Blue FCF	Z = 0.39 mg Brilliant Blue FCF

to BBG treatment with doses of 0.25 mg/mL and 0.5 mg/mL. In contrast, the distance of 2.5 cm exhibited augment in the viability with 0.25 mg/mL BBG 1 min (98.85%–3.3%), 5 min (95.31%–7.12%), and 15 min (62.07%–3.0%). In consequence, they concluded that BBG in the concentration of 0.25 mg/mL during vitreoretinal surgery is not toxic to HRPE under focal illumination (1 cm) or diffuse illumination (2.5 cm).

As a potential food dye, FD&C Blue No. 1 in vivo assays were carried out according to the Food and Drug Administration (FDA) oral absorption and oral-intake limits, which in healthy animals was 12 mg/kg/day. However, in critical situations, such as during sepsis, there is a higher gastrointestinal permeability due to enterocyte death and a loss of barrier function at intercellular gaps. In this context, systemic absorption of FD&C Blue No. 1 from enteral feedings after tracheostomy for obstructive apnea and chronic renal failure have been described [22] and, in some cases, the systemic absorption of FD&C Blue No. 1 leads to death [85]. However, possibly because of its low oral absorption, only a few studies have documented toxicity related to FD&C Blue No. 1 in different tissues, as there are no evident alterations (even in carcinogenicity studies) in the animal models used [86, 87]. In addition, Abd El-Wahab and Moram studied some of the most used colorants and flavors and their toxic effects on essential organs. Increasing in FD&C Blue No. 1 intake altered some parameters, whereas it did not increase mass weight. Additionally, the ingestion of this dye was associated with an increase in important hepatic health-indicative enzymes, such as alanine transaminase, aspartate transaminase, alkaline phosphatase, and the amount of bilirubin [88]. Table 2 shows the main studies of the Brilliant Blue dye effects in vivo assays.

5. Dye Concentrations Used in the Food Industry and the Concentration to Inhibit P2 Receptor Function in Humans

Among the dyes with the capacity to inhibit the P2 receptors listed above, only BBG (or its analogue) is used in the food industry. Therefore, we will focus on this dye to investigate the possible correlations between the concentrations used in food and in scientific experiments.

In beverages, such as Gatorade® Lemon-Lime, the quantity of FD&C Blue No. 1 added in a bottle of 946 mL is 0.1 mg/L. In other types of Gatorade, this concentration is higher: Gatorade Glacier Freeze (1 mg/L), Gatorade Tropical Blend (2 mg/L), and Gatorade Blueberry Pomegranate (10 mg/L).

The Food and Drug Administration (FDA) investigates new food coloring (dye) when it is developed to analyze toxicology and safety studies, estimates human dietary intake, and then publishes literature for this drug and identifies an acceptable daily intake (ADI) level. The value calculated for FD&C Blue No. 1 according to body weight for an adult is 12.5 mg/kg/day. The EU Scientific Committee for Food (SCF) revised the ADI to 10 mg/kg/day in 1984, based on new long-term studies. In theory, an adult weighing 60 kg would tolerate a maximum amount of 720 mg of FD&C Blue No. 1. More recently, the Panel on Food Additives and Nutrient Sources determined that the no observed adverse effect level (NOAEL) of 631 mg/kg/day of the chronic toxicity study in rats could be used to designate another ADI for FD&C Blue No. 1. Applying an uncertainty factor of 100, the Panel supports a new ADI for FD&C Blue No. 1 of 6 mg/kg/day [EFSA Panel on Food Additives and Nutrient Sources added to Food (ANS), 2010]. However, practically, it is difficult to estimate the quantity of dyes consumed without information about the quantity of the dye used in common foods. Additionally, new studies about the impact of a prolonged intake of this dye and cancer and neurological disturbances in humans are necessary.

A number of papers have used BBG to inhibit P2X7R *in vivo*. BBG concentrations administered to the animals of the studies varied, in general, from 1 mg/kg to 100 mg/kg. Although a large number of scientific publications used this dye *in vivo*, among those found for us, only three quantify the final concentration of BBG in the body. In 2009, Peng and colleagues quantified BBG concentrations in the medullary tissue as 9.94 μM after giving a dose of 10 mg/kg and 43.59 μM after a dose of 50 mg/kg [60]. In the same year, Díaz-Hernández and coworkers also quantified BBG in the plasma after 45 mg/kg administration in one- and four-month-old mice. In the former, BBG was measured at 7.08 μM in the plasma and 152.6 nM in the brain; in the 4-month-old mice, the plasma concentration was 0.04 μM and 226.12 nM in the brain [61]. Kimbler and collaborates quantified the concentration in the plasma as 383 μM and 1.73 mM of BBG after the administration of 50 mg/kg and 100 mg/kg, respectively [69].

Taking into consideration the results of the Kimbler and Díaz-Hernández groups, the final BBG concentration obtained is sufficient to functionally block P2X7R *in vivo* [61, 69]. Based on IC_{50} values, *in vitro* concentrations can inhibit rat P2X7R (10 nM), human P2X7R (270 μM), and mouse P2X7R (170 nM). At a micromolar range, this antagonist may reach other P2X receptors, such as rat P2X2R with an IC_{50} of 1.4 μM, rat P2X4R with IC_{50} of 5 μM, human P2X4R with an IC_{50} of 3.2 μM, and other channels, such as voltage-gated sodium channels with an IC_{50} of 2 μM [43]. In this study, FD&C Blue No. 1 inhibited voltage-gated sodium channel, producing a modest effect compared to BBG [43]. More convincing results were published in cells transfected with pannexin-1, where this dye inhibited the pannexin-1 opening with an IC_{50} value of 0.27 μM [43].

Diverse beverages and foods contain the maximum permitted level of FD&C Blue No. 1 ranging from 100 mg/L to 200 mg/L and 50 to 500 mg/kg, respectively (EFSA Panel on Food Additives and Nutrient Sources added to Food (ANS), 2010). However, studies on the FD&C Blue No. 1 concentration used in drinks and food have indicated

TABLE 3: Calculus 2.

Male	Female
60 mg Brilliant Blue FCF, 1 L	60 mg Brilliant Blue FCF, 1 L
X mg Brilliant Blue FCF, 4.2 L	Z mg Brilliant Blue FCF, 3.9 L
X = 252 mg Brilliant Blue FCF	Z = 234 mg Brilliant Blue FCF

TABLE 4: Calculus 3.

Male	Female
300 mg Brilliant Blue FCF, 1 L	300 mg Brilliant Blue FCF, 1 L
X mg Brilliant Blue FCF, 4.2 L	Z mg Brilliant Blue FCF, 3.9 L
X = 1260 mg Brilliant Blue FCF	Z = 1170 mg Brilliant Blue FCF

a typical use concentration in the industry ranging from 0.1 to 60 mg/L. Moreover, researchers have also reported extreme use levels ranging from 0.9 to 200 mg/L and typical use levels ranging from 0.1 to 300 mg/L for beverages reported and reported typical use levels ranging from 0.2 to 500 mg/L for foods (DyeDiet, 2011; EFSA Panel on Food Additives and Nutrient Sources added to Food (ANS), 2010).

Using the example of an adult male weighing 60 kg and an adult female of 55 kg and the male having 70 mL/kg and the female 65 mL/kg of blood, we estimate a total blood volume of 4.2 L for the male and 3.9 L for the female. Using the reported typical use levels of beverages cited above as a reference, we could predict the quantity in milligrams of the dye for the lower limit as shown in Table 3. In a male weighing 60 kg, the calculation of the mass is 0.007 mg/kg and for a female of 55 kg the calculation is 0.007 mg/kg. In micrograms, this relation remains 7 μg/kg for both genders. The upper limit is shown in Table 4. In a male weighing 60 kg, the calculation of the mass is 4.2 mg/kg and for a female of 55 kg the calculation is 4.25 mg/kg. With regard to the reported typical use levels of foods, the quantity in milligrams of the dye for the lower limit is the same as Calculus 1 (Table 2). In addition, the upper limit was calculated below. In a male weighing 60 kg, the calculation of the mass is 21 mg/kg and for a female of 55 kg the calculation is 21.27 mg/kg.

Evidently, the final intake of blue dyes in the alimentary is variable, depending on the portion of blue dye in each food and quantity of the food absorbed. Our intention is to warn consumers about the possibility of a functional inhibition of the P2 receptors, including P2X7R in particular, caused by food intake. We highlight that FD&C Blue No. 1 concentrations used in the alimentary described above may be superior to IC_{50} values that potently inhibit P2X7R (and other P2X), considering pannexin-1-dependent ATP releasing ($IC_{50} = 0.27$ μM) and the voltage dependent sodium channels. There are diverse pathological situations favoring an augment in the gut and skin absorption and clinical procedures as enteral feeding. In these cases, the quantity of FD&C Blue No. 1 in the tissues increases in comparison to healthy patients. Nevertheless, it is noteworthy that scientific researches of dyes in food should not be limited to this area for cumulative absorption studies. These studies should not ignore the great amount of daily life products and have to account other pathways. In 2013, Lucová et al. [89] showed important assays of cumulative absorption through intact skin, shaven skin, and lingual mucosa. They showed that the skin integrity is relevant to systemic dye absorption, which significantly increases after shaving and epilation procedures. Hemorrhagic shock, nonsteroidal anti-inflammatory drug use, renal failure, inflammatory bowel disease, and cystic fibrosis, among others, are some diseases related to increase in the gut permeability [15]. Once P2X7R, in general, is upregulated under pathological conditions, including the cases cited above, possibly, BBG could achieve plasmatic concentration sufficient to inhibit P2X7R. Additionally, the intravenous administration of 50 mg/kg BBG in a mice model of traumatic brain injury promoted a plasmatic value of 383 μM after 4 hours and the application of 100 mg/kg BBG produced a plasmatic value of 1.73 mM also after 4 hours [60]. This could be unfavorable to proinflammatory functions, while, in contrast, inhibition of the P2X7R *in vivo* would be central to the prevention of spine injury or Crohn's disease.

6. Conclusions

In this work, we discuss the effect of the two Brilliant Blue dyes related to purinergic signaling (BBG and FD&C Blue No. 1) that were widely used in many manufactured products until recently. Our purpose was to elucidate the link between the industry concentration usage to produce dyed food or medicines and its dosage in the human system. These studies have become necessary to elucidate the possible therapeutic or deleterious effects obtained through the ingestion of manufactured products, which could modulate some diseases related to P2R function. Surely, this work does not focus on the abolishment of food dyes but is to shed light on the putative relationship within food dyes and purinergic signaling. A few works have shown that restriction of synthetic food color additives has ameliorated some attention deficit or hyperactivity disorder symptoms [90–92]. Whereas there is no work linking purinergic signaling and deficit or hyperactivity disorder symptoms, it will not be a surprise if soon they appear. Indeed, there are many purinergic receptors in glial cells, which have been studied in these disorders [93–95]. Therefore, scientific researches should assess purinergic signaling and manufactured products (as food dyes) relationship and should not be omitted in several inflammatory and infectious diseases.

Competing Interests

The authors declare no conflict of interests.

Acknowledgments

This work was supported by grants from FAPERJ, CNPq, and the Oswaldo Cruz Institute.

References

[1] C. Coddou, Z. Yan, T. Obsil, J. Pablo Huidobro-Toro, and S. S. Stojilkovic, "Activation and regulation of purinergic P2X receptor channels," *Pharmacological Reviews*, vol. 63, no. 3, pp. 641–683, 2011.

[2] G. Burnstock, "Purine and pyrimidine receptors," *Cellular and Molecular Life Sciences*, vol. 64, no. 12, pp. 1471–1483, 2007.

[3] G. Burnstock and M. Williams, "P2 purinergic receptors: modulation of cell function and therapeutic potential," *Journal of Pharmacology and Experimental Therapeutics*, vol. 295, no. 3, pp. 862–869, 2000.

[4] V. Ralevic and G. Burnstock, "Receptors for purines and pyrimidines," *Pharmacological Reviews*, vol. 50, no. 3, pp. 413–492, 1998.

[5] L. A. Alves, R. J. Soares Bezerra, R. X. Faria, L. G. B. Ferreira, and V. D. S. Frutuoso, "Physiological roles and potential therapeutic applications of the P2X7 receptor in inflammation and pain," *Molecules*, vol. 18, no. 9, pp. 10953–10972, 2013.

[6] G. A. Weisman, D. Ajit, R. Garrad et al., "Neuroprotective roles of the P2Y$_2$ receptor," *Purinergic Signalling*, vol. 8, no. 3, pp. 559–578, 2012.

[7] M. Schuchardt, M. Tölle, J. Prüfer et al., "Uridine adenosine tetraphosphate activation of the purinergic receptor P2Y enhances in vitro vascular calcification," *Kidney International*, vol. 81, no. 3, pp. 256–265, 2012.

[8] B. R. Bianchi, K. J. Lynch, E. Touma et al., "Pharmacological characterization of recombinant human and rat P2X receptor subtypes," *European Journal of Pharmacology*, vol. 376, no. 1-2, pp. 127–138, 1999.

[9] R. J. Gum, B. Wakefield, and M. F. Jarvis, "P2X receptor antagonists for pain management: examination of binding and physicochemical properties," *Purinergic Signalling*, vol. 8, no. 1, pp. 41–56, 2012.

[10] L.-H. Jiang, A. B. Mackenzie, R. A. North, and A. Surprenant, "Brilliant blue G selectively blocks ATP-gated rat P2X$_7$ receptors," *Molecular Pharmacology*, vol. 58, no. 1, pp. 82–88, 2000.

[11] H. Uneyama, C. Uneyama, S. Ebihara, and N. Akaike, "Suramin and reactive blue 2 are antagonists for a newly identified purinoceptor on rat megakaryocyte," *British Journal of Pharmacology*, vol. 111, no. 1, pp. 245–249, 1994.

[12] B. F. King, M. Liu, A. Townsend-Nicholson et al., "Antagonism of ATP responses at P2X receptor subtypes by the pH indicator dye, phenol red," *British Journal of Pharmacology*, vol. 145, no. 3, pp. 313–322, 2005.

[13] R. Bültmann, H. Wittenburg, B. Pause, G. Kurz, P. Nickel, and K. Starke, "P2-purinoceptor antagonists: III. Blockade of P2-purinoceptor subtypes and ecto-nucleotidases by compounds related to suramin," *Naunyn-Schmiedeberg's Archives of Pharmacology*, vol. 354, no. 4, pp. 498–504, 1996.

[14] A. D. Watharkar, R. V. Khandare, A. A. Kamble, A. Y. Mulla, S. P. Govindwar, and J. P. Jadhav, "Phytoremediation potential of *Petunia grandiflora* Juss., an ornamental plant to degrade a disperse, disulfonated triphenylmethane textile dye Brilliant Blue G," *Environmental Science and Pollution Research*, vol. 20, no. 2, pp. 939–949, 2013.

[15] J. P. Maloney, T. A. Ryan, K. J. Brasel et al., "Food dye use in enteral feedings: a review and a call for a moratorium," *Nutrition in Clinical Practice*, vol. 17, no. 3, pp. 169–181, 2002.

[16] J. P. Brown, A. Dorsky, F. E. Enderlin, R. L. Hale, V. A. Wright, and T. M. Parkinson, "Synthesis of 14C-labelled FD & C blue no. 1 (brilliant blue FCF) and its intestinal absorption and metabolic fate in rats," *Food and Cosmetics Toxicology*, vol. 18, no. 1, pp. 1–5, 1980.

[17] E. Insumos, Additives & ingredients. The food coloring, 2013, http://www.insumos.com.br/aditivos_e_ingredientes/materias/119.pdf.

[18] M. A. Prado and H. T. Godoy, "Teores de corantes artificiais em alimentos determinados por cromatografia líquida de alta eficiência," *Química Nova*, vol. 30, no. 2, pp. 268–273, 2007.

[19] C. F. O. de Queiroz, *Studies on the Intercalation of the Dye Brilliant Blue-FCF and Coomassie Blue G-250 in Deoxyribonucleic Acid*, Federal University of Uberlândia, 2007.

[20] B. G. Neel and N. K. Tonks, "Protein tyrosine phosphatases in signal transduction," *Current Opinion in Cell Biology*, vol. 9, no. 2, pp. 193–204, 1997.

[21] S. Shrestha, R. B. Bharat, K.-H. Lee, and H. Cho, "Some of the food color additives are potent inhibitors of human protein tyrosine phosphatases," *Bulletin of the Korean Chemical Society*, vol. 27, no. 10, pp. 1567–1571, 2006.

[22] J. P. Maloney, A. C. Halbower, B. F. Fouty et al., "Systemic absorption of food dye in patients with sepsis," *The New England Journal of Medicine*, vol. 343, no. 14, pp. 1047–1048, 2000.

[23] J. C. Bearden Jr., "Quantitation of submicrogram quantities of protein by an improved protein-dye binding assay," *Biochimica et Biophysica Acta (BBA)—Protein Structure*, vol. 533, no. 2, pp. 525–529, 1978.

[24] S. Serra and L. Morgante, "Method of determination of proteins with Coomassie brilliant blue G 250. I. General characteristics and comparative analysis with the biuret method and Lowry's method," *Bollettino della Societa Italiana di Biologia Sperimentale*, vol. 56, no. 2, pp. 160–165, 1980.

[25] S. Serra and L. Morgante, "Method of determination of proteins with Coomassie brilliant blue G 250. II. Influence of pH and inorganic ions," *Bollettino della Società Italiana di Biologia Sperimentale*, vol. 56, no. 2, pp. 166–170, 1980.

[26] S. Serra, L. Morgante, and R. Di Perri, "Determination of proteins with the Coomassie brilliant blue G 250 method. IV. Use with cerebrospinal fluid proteins and comparative analysis with the biuret and Lowry methods," *Bollettino Della Societa Italiana di Biologia Sperimentale*, vol. 56, no. 5, pp. 463–467, 1980.

[27] R. A. Asryants, I. V. Duszenkova, and N. K. Nagradova, "Determination of Sepharose-bound protein with Coomassie brilliant blue G-250," *Analytical Biochemistry*, vol. 151, no. 2, pp. 571–574, 1985.

[28] G. F. Bickerstaff and H. Zhou, "Protease activity and autodigestion (autolysis) assays using Coomassie blue dye binding," *Analytical Biochemistry*, vol. 210, no. 1, pp. 155–158, 1993.

[29] M. Buroker-Kilgore and K. K. W. Wang, "A Coomassie brilliant blue G-250-based colorimetric assay for measuring activity of calpain and other proteases," *Analytical Biochemistry*, vol. 208, no. 2, pp. 387–392, 1993.

[30] M. Saleemuddin, H. Ahmad, and A. Husain, "A simple, rapid, and sensitive procedure for the assay of endoproteases using Coomassie brilliant blue G-250," *Analytical Biochemistry*, vol. 105, no. 1, pp. 202–206, 1980.

[31] S. W. Paddock, "Incident light microscopy of normal and transformed cultured fibroblasts stained with Coomassie blue R250," *Journal of Microscopy*, vol. 128, pp. 203–205, 1982.

[32] J. Borejdo and C. Flynn, "Electrophoresis in the presence of Coomassie brilliant blue R-250 stains polyacrylamide gels during protein fractionation," *Analytical Biochemistry*, vol. 140, no. 1, pp. 84–86, 1984.

[33] A. M. Saoji, C. Y. Jad, and S. S. Kelkar, "Modified critical conditions for prestaining human serum proteins with Remazol Brilliant Blue and separation by disc electrophoresis," *Clinical Chemistry*, vol. 31, no. 2, article 349, 1985.

[34] Y. Mochizuki and K. Furukawa, "Application of Coomassie brilliant blue staining to cultured hepatocytes," *Cell Biology International Reports*, vol. 11, no. 5, pp. 367–371, 1987.

[35] S. D. J. Pena, "A new technique for the visualization of the cytoskeleton in cultured fibroblasts with Coomassie blue R250," *Cell Biology International Reports*, vol. 4, no. 2, pp. 149–153, 1980.

[36] M. M. Bradford, "A rapid and sensitive method for the quantitation of microgram quantities of protein utilizing the principle of protein-dye binding," *Analytical Biochemistry*, vol. 72, no. 1-2, pp. 248–254, 1976.

[37] A. Giessauf, M. Flaim, M. P. Dierich, and R. Würzner, "A technique for isolation of rubella virus-like particles by sucrose gradient ultracentrifugation using Coomassie brilliant blue G crystals," *Analytical Biochemistry*, vol. 308, no. 2, pp. 232–238, 2002.

[38] S. P. Soltoff, M. K. McMillian, and B. R. Talamo, "Coomassie Brilliant Blue G is a more potent antagonist of P_2 purinergic responses than Reactive Blue 2 (Cibacron Blue 3GA) in rat parotid acinar cells," *Biochemical and Biophysical Research Communications*, vol. 165, no. 3, pp. 1279–1285, 1989.

[39] X. Zhang, M. Zhang, A. M. Laties, and C. H. Mitchell, "Stimulation of $P2X_7$ receptors elevates Ca^{2+} and kills retinal ganglion cells," *Investigative Ophthalmology and Visual Science*, vol. 46, no. 6, pp. 2183–2191, 2005.

[40] T. Suzuki, I. Hide, K. Ido, S. Kohsaka, K. Inoue, and Y. Nakata, "Production and release of neuroprotective tumor necrosis factor by $P2X_7$ receptor-activated microglia," *The Journal of Neuroscience*, vol. 24, no. 1, pp. 1–7, 2004.

[41] M. C. Jacques-Silva, R. Rodnight, G. Lenz et al., "$P2X_7$ receptors stimulate AKT phosphorylation in astrocytes," *British Journal of Pharmacology*, vol. 141, no. 7, pp. 1106–1117, 2004.

[42] D. Eschke, M. Wüst, S. Hauschildt, and K. Nieber, "Pharmacological characterization of the P2X7 receptor on human macrophages using the patch-clamp technique," *Naunyn-Schmiedeberg's Archives of Pharmacology*, vol. 365, no. 2, pp. 168–171, 2002.

[43] J. Wang, D. G. Jackson, and G. Dahl, "The food dye FD&C Blue No. 1 is a selective inhibitor of the ATP release channel Panx1," *Journal of General Physiology*, vol. 141, no. 5, pp. 649–656, 2013.

[44] R. Bruzzone, S. G. Hormuzdi, M. T. Barbe, A. Herb, and H. Monyer, "Pannexins, a family of gap junction proteins expressed in brain," *Proceedings of the National Academy of Sciences of the United States of America*, vol. 100, no. 23, pp. 13644–13649, 2003.

[45] A. V. P. Alberto, R. X. Faria, C. G. C. Couto et al., "Is pannexin the pore associated with the P2X7 receptor?" *Naunyn-Schmiedeberg's Archives of Pharmacology*, vol. 386, no. 9, pp. 775–787, 2013.

[46] J. M. Garre, G. Yang, F. F. Bukauskas, and M. V. L. Bennett, "FGF-1 triggers pannexin-1 hemichannel opening in spinal astrocytes of rodents and promotes inflammatory responses in acute spinal cord slices," *The Journal of Neuroscience*, vol. 36, no. 17, pp. 4785–4801, 2016.

[47] T. Woehrle, L. Yip, A. Elkhal et al., "Pannexin-1 hemichannel-mediated ATP release together with P2X1 and P2X4 receptors regulate T-cell activation at the immune synapse," *Blood*, vol. 116, no. 18, pp. 3475–3484, 2010.

[48] F. B. Chekeni, M. R. Elliott, J. K. Sandilos et al., "Pannexin 1 channels mediate 'find-me' signal release and membrane permeability during apoptosis," *Nature*, vol. 467, no. 7317, pp. 863–867, 2010.

[49] Y.-J. Huang, Y. Maruyama, G. Dvoryanchikov, E. Pereira, N. Chaudhari, and S. D. Roper, "The role of pannexin 1 hemichannels in ATP release and cell-cell communication in mouse taste buds," *Proceedings of the National Academy of Sciences of the United States of America*, vol. 104, no. 15, pp. 6436–6441, 2007.

[50] M. A. Timóteo, I. Carneiro, I. Silva et al., "ATP released via pannexin-1 hemichannels mediates bladder overactivity triggered by urothelial $P2Y_6$ receptors," *Biochemical Pharmacology*, vol. 87, no. 2, pp. 371–379, 2014.

[51] K. A. Sharkey and A. B. A. Kroese, "Consequences of intestinal inflammation on the enteric nervous system: neuronal activation induced by inflammatory mediators," *Anatomical Record*, vol. 262, no. 1, pp. 79–90, 2001.

[52] N. Marina-García, L. Franchi, Y.-G. Kim et al., "Pannexin-1-mediated intracellular delivery of muramyl dipeptide induces caspase-1 activation via cryopyrin/NLRP3 independently of Nod2," *The Journal of Immunology*, vol. 180, no. 6, pp. 4050–4057, 2008.

[53] S. Jo and B. P. Bean, "Inhibition of neuronal voltage-gated sodium channels by Brilliant Blue G," *Molecular Pharmacology*, vol. 80, no. 2, pp. 247–257, 2011.

[54] P. Naithani, N. Vashisht, S. Khanduja, S. Sinha, and S. Garg, "Brilliant blue G-assisted peeling of the internal limiting membrane in macular hole surgery," *Indian Journal of Ophthalmology*, vol. 59, no. 2, pp. 158–160, 2011.

[55] T. Hisatomi, H. Enaida, H. Matsumoto et al., "Staining ability and biocompatibility of brilliant blue G: preclinical study of brilliant blue G as an adjunct for capsular staining," *Archives of Ophthalmology*, vol. 124, no. 4, pp. 514–519, 2006.

[56] B. D. Gulbransen, M. Bashashati, S. A. Hirota et al., "Activation of neuronal P2X7 receptor-pannexin-1 mediates death of enteric neurons during colitis," *Nature Medicine*, vol. 18, no. 4, pp. 600–604, 2012.

[57] S. Mennel, C. Meyer, J. Schmidt, S. Kaempf, and G. Thumann, "Trityl dyes patent blue V and brilliant blue G—clinical relevance and in vitro analysis of the function of the outer blood-retinal barrier," *Developments in Ophthalmology*, vol. 42, pp. 101–114, 2008.

[58] A. Höing, M. Remy, M. Dirisamer et al., "Anfärbung der inneren grenzmembran der netzhaut mit brilliant blau," *Klinische Monatsblätter für Augenheilkunde*, vol. 228, no. 8, pp. 724–728, 2011.

[59] H. Franke, C. Schepper, P. Illes, and U. Krügel, "Involvement of P2X and P2Y receptors in microglial activation in vivo," *Purinergic Signalling*, vol. 3, no. 4, pp. 435–445, 2007.

[60] W. Peng, M. L. Cotrina, X. Han et al., "Systemic administration of an antagonist of the ATP-sensitive receptor P2X7 improves recovery after spinal cord injury," *Proceedings of the National Academy of Sciences of the United States of America*, vol. 106, no. 30, pp. 12489–12493, 2009.

[61] M. Díaz-Hernández, M. Díez-Zaera, J. Sánchez-Nogueiro et al., "Altered P2X7-receptor level and function in mouse models of Huntington's disease and therapeutic efficacy of antagonist administration," *The FASEB Journal*, vol. 23, no. 6, pp. 1893–1906, 2009.

[62] R. D. Andó, B. Méhész, K. Gyires, P. Illes, and B. Sperlágh, "A comparative analysis of the activity of ligands acting at P2X and P2Y receptor subtypes in models of neuropathic, acute and inflammatory pain," *British Journal of Pharmacology*, vol. 159, no. 5, pp. 1106–1117, 2010.

[63] D. N. Okamoto, L. C. G. De Oliveira, A. J. Manzato et al., "Coomassie brilliant blue dye toxicity screen using *Drosophila melanogaster* (Diptera—Drosophilidae)," *Drosophila Information Service*, vol. 93, pp. 40–47, 2010.

[64] Y.-X. Chu, Y. Zhang, Y.-Q. Zhang, and Z.-Q. Zhao, "Involvement of microglial P2X7 receptors and downstream signaling pathways in long-term potentiation of spinal nociceptive responses," *Brain, Behavior, and Immunity*, vol. 24, no. 7, pp. 1176–1189, 2010.

[65] H. Hu, W. Lu, M. Zhang et al., "Stimulation of the $P2X_7$ receptor kills rat retinal ganglion cells in vivo," *Experimental Eye Research*, vol. 91, no. 3, pp. 425–432, 2010.

[66] K. Itoh, C.-Y. Chiang, Z. Li, J.-C. Lee, J. O. Dostrovsky, and B. J. Sessle, "Central sensitization of nociceptive neurons in rat medullary dorsal horn involves purinergic P2X7 receptors," *Neuroscience*, vol. 192, pp. 721–731, 2011.

[67] J. Arbeloa, A. Pérez-Samartín, M. Gottlieb, and C. Matute, "P2X7 receptor blockade prevents ATP excitotoxicity in neurons and reduces brain damage after ischemia," *Neurobiology of Disease*, vol. 45, no. 3, pp. 954–961, 2012.

[68] K. Chu, B. Yin, J. Wang et al., "Inhibition of P2X7 receptor ameliorates transient global cerebral ischemia/reperfusion injury via modulating inflammatory responses in the rat hippocampus," *Journal of Neuroinflammation*, vol. 9, article 69, 2012.

[69] D. E. Kimbler, J. Shields, N. Yanasak, J. R. Vender, and K. M. Dhandapani, "Activation of P2X7 promotes cerebral edema and neurological injury after traumatic brain injury in mice," *PLoS ONE*, vol. 7, no. 7, Article ID e41229, 2012.

[70] C. M. Cascabulho, C. B. Corrêa, V. Cotta-De-Almeida, and A. Henriques-Pons, "Defective T-lymphocyte migration to muscles in dystrophin-deficient mice," *The American Journal of Pathology*, vol. 181, no. 2, pp. 593–604, 2012.

[71] J. S. Wiley, C. E. Gargett, W. Zhang, M. B. Snook, and G. P. Jamieson, "Partial agonists and antagonists reveal a second permeability state of human lymphocyte $P2Z/P2X_7$ channel," *American Journal of Physiology—Cell Physiology*, vol. 275, no. 5, pp. C1224–C1231, 1998.

[72] C. N. J. Young, W. Brutkowski, C.-F. Lien et al., "P2X7 purinoceptor alterations in dystrophic mdx mouse muscles: relationship to pathology and potential target for treatment," *Journal of Cellular and Molecular Medicine*, vol. 16, no. 5, pp. 1026–1037, 2012.

[73] Y. Iwamaru, T. Takenouchi, Y. Murayama et al., "Anti-prion activity of brilliant blue G," *PLoS ONE*, vol. 7, no. 5, Article ID e37896, 2012.

[74] X. Chen, J. Hu, L. Jiang et al., "Brilliant Blue G improves cognition in an animal model of Alzheimer's disease and inhibits amyloid-β-induced loss of filopodia and dendrite spines in hippocampal neurons," *Neuroscience*, vol. 279, pp. 94–101, 2014.

[75] J. A. Irwin, A. Erisir, and I. Kwon, "Oral triphenylmethane food dye analog, brilliant blue G, prevents neuronal loss in APPSwDI/NOS2−/− mouse model," *Current Alzheimer Research*, vol. 13, no. 6, pp. 663–677, 2016.

[76] K. Kakurai, T. Sugiyama, T. Kurimoto, H. Oku, and T. Ikeda, "Involvement of $P2X_7$ receptors in retinal ganglion cell death after optic nerve crush injury in rats," *Neuroscience Letters*, vol. 534, pp. 237–241, 2013.

[77] J. Zhao, H. Wang, C. Dai et al., "P2X7 blockade attenuates murine lupus nephritis by inhibiting activation of the NLRP3/ASC/Caspase 1 pathway," *Arthritis & Rheumatism*, vol. 65, no. 12, pp. 3176–3185, 2013.

[78] R. I. Menzies, R. J. Unwin, R. K. Dash et al., "Effect of $P2X_4$ and $P2X_7$ receptor antagonism on the pressure diuresis relationship in rats," *Frontiers in Physiology*, vol. 4, article 305, 2013.

[79] C. Cervetto, S. Alloisio, D. Frattaroli et al., "The P2X7 receptor as a route for non-exocytotic glutamate release: dependence on the carboxyl tail," *Journal of Neurochemistry*, vol. 124, no. 6, pp. 821–831, 2013.

[80] D. Awad, I. Schrader, M. Bartok, N. Sudumbrekar, A. Mohr, and D. Gabel, "Brilliant Blue G as protective agent against trypan blue toxicity in human retinal pigment epithelial cells in vitro," *Graefe's Archive for Clinical and Experimental Ophthalmology*, vol. 251, no. 7, pp. 1735–1740, 2013.

[81] P. B. Henrich, C. Valmaggia, C. Lang, and P. C. Cattin, "The price for reduced light toxicity: do endoilluminator spectral filters decrease color contrast during Brilliant Blue G-assisted chromovitrectomy?" *Graefe's Archive for Clinical and Experimental Ophthalmology*, vol. 252, no. 3, pp. 367–374, 2014.

[82] P. B. Henrich, C. Valmaggia, C. Lang et al., "Contrast recognizability during brilliant blue G—and heavier-than-water brilliant blue G-assisted chromovitrectomy: a quantitative analysis," *Acta Ophthalmologica*, vol. 91, no. 2, pp. e120–e124, 2013.

[83] F. Giansanti, N. Schiavone, L. Papucci et al., "Safety testing of blue vital dyes using cell culture models," *Journal of Ocular Pharmacology and Therapeutics*, vol. 30, no. 5, pp. 406–412, 2014.

[84] S. Balaiya, K. Koushan, T. McLauchlan, and K. V. Chalam, "Assessment of the effect of distance and duration of illumination on retinal pigment epithelial cells exposed to varying doses of brilliant blue green," *Journal of Ocular Pharmacology and Therapeutics*, vol. 30, no. 8, pp. 625–633, 2014.

[85] M. R. Lucarelli, M. B. Shirk, M. W. Julian, and E. D. Crouser, "Toxicity of food drug and cosmetic blue no. 1 dye in critically ill patients," *Chest*, vol. 125, no. 2, pp. 793–795, 2004.

[86] S. D. Gettings, D. L. Blaszcak, M. T. Roddy, A. S. Curry, and G. N. McEwen Jr., "Evaluation of the cumulative (repeated application) eye irritation and corneal staining potential of FD&C yellow no. 5, FD&C blue no. 1 and FD&C blue no. 1 aluminium lake," *Food and Chemical Toxicology*, vol. 30, no. 12, pp. 1051–1055, 1992.

[87] J. F. Borzelleca, K. Depukat, and J. B. Hallagan, "Lifetime toxicity/carcinogenicity studies of FD & C blue No. 1 (Brilliant blue FCF) in rats and mice," *Food and Chemical Toxicology*, vol. 28, no. 4, pp. 221–234, 1990.

[88] H. M. F. Abd El-Wahab and G. S. E.-D. Moram, "Toxic effects of some synthetic food colorants and/or flavor additives on male rats," *Toxicology and Industrial Health*, vol. 29, no. 2, pp. 224–232, 2012.

[89] M. Lucová, J. Hojerová, S. Pažoureková, and Z. Klimová, "Absorption of triphenylmethane dyes Brilliant Blue and Patent Blue through intact skin, shaven skin and lingual mucosa from daily life products," *Food and Chemical Toxicology*, vol. 52, pp. 19–27, 2013.

[90] J. T. Nigg, K. Lewis, T. Edinger, and M. Falk, "Meta-analysis of attention-deficit/hyperactivity disorder or attention-deficit/hyperactivity disorder symptoms, restriction diet, and synthetic food color additives," *Journal of the American Academy of Child and Adolescent Psychiatry*, vol. 51, no. 1, pp. 86.e8–97.e8, 2012.

[91] D. W. Schab and N.-H. T. Trinh, "Do artificial food colors promote hyperactivity in children with hyperactive syndromes? A meta-analysis of double-blind placebo-controlled trials," *Journal of Developmental and Behavioral Pediatrics*, vol. 25, no. 6, pp. 423–434, 2004.

[92] J. G. Millichap and M. M. Yee, "The diet factor in attention-deficit/hyperactivity disorder," *Pediatrics*, vol. 129, no. 2, pp. 330–337, 2012.

[93] E. B. Johansen, T. Sagvolden, H. Aase, and V. A. Russell, "The dynamic developmental theory of attention-deficit/hyperactivity disorder (ADHD): present status and future perspectives," *Behavioral and Brain Sciences*, vol. 28, no. 3, pp. 451–468, 2005.

[94] R. D. Oades, "Dopamine-serotonin interactions in attention-deficit hyperactivity disorder (ADHD)," *Progress in Brain Research*, vol. 172, pp. 543–565, 2008.

[95] R. D. Oades, M. R. Dauvermann, B. G. Schimmelmann, M. J. Schwarz, and A.-M. Myint, "Attention-deficit hyperactivity disorder (ADHD) and glial integrity: S100B, cytokines and kynurenine metabolism—effects of medication," *Behavioral and Brain Functions*, vol. 6, article 29, 2010.

[96] D. L. Donnelly-Roberts, M. T. Namovic, P. Han, and M. F. Jarvis, "Mammalian P2X7 receptor pharmacology: comparison of recombinant mouse, rat and human P2X7 receptors," *British Journal of Pharmacology*, vol. 157, no. 7, pp. 1203–1214, 2009.

[97] F. Qiu and G. Dahl, "A permeant regulating its permeation pore: inhibition of pannexin 1 channels by ATP," *American Journal of Physiology—Cell Physiology*, vol. 296, no. 2, pp. C250–C255, 2009.

3

Isolation, Characterization, and Quantification of Bacteria from African Sausages Sold in Nairobi County, Kenya

W. H. Karoki ⓘ, D. N. Karanja, L. C. Bebora, and L. W. Njagi

Department of Veterinary Pathology, Microbiology and Parasitology, University of Nairobi, P.O. Box 29053-00625, Nairobi, Kenya

Correspondence should be addressed to W. H. Karoki; henrykaroki18@gmail.com

Academic Editor: Amy Simonne

African sausages are local popular delicacies in Kenya. Demand for these sausages has resulted in this delicacy's vendors being on the increase. However, health risk posed to unsuspecting consumers of African sausages sold in informal, unhygienic make shift road-side kiosks in major cities of Kenya is largely unknown. A descriptive study was designed to isolate, characterize and quantify bacteria from African sausages sold in Nairobi County. A total of hundred (100) African sausages (62 roasted and 38 nonroasted) were conveniently collected from three meat eatery points of Westlands, Kangemi slum, and Pangani estates. Five genera of bacteria, namely, *Staphylococcus* spp. at 50.4%, *Bacillus* spp. at 19.5%, *Streptococcus* spp. 9.8%, *Proteus* spp. 2.4%, and *E. coli* spp. at 1.6%, were isolated from 80 African sausage samples. The total aerobic bacterial count range was between $1.0-9.9 \times 10^1$ and $1.0-9.9 \times 10^7$ log cfu/g with 37 samples having total aerobic bacterial count of between $1.0-9.9 \times 10^4$ and $1.0-9.9 \times 10^7$ log cfu/g. There was no significant difference ($p > 0.05$) in distribution of isolates and total aerobic bacterial count across geographical sites studied among the roasted and nonroasted African sausages. This study has demonstrated presence of bacteria in African sausages which are potentially zoonotic to humans. Comprehensive study is needed to sample more eatery meat points in Nairobi and other areas in order to demonstrate pathogenic attributes of these isolates and establish the respective total aerobic bacterial count. There is also need to establish the sources of bacteria due to high total aerobic bacterial count determined in the current study.

1. Introduction

African sausages, popularly known as Kenyan sausages locally known as "Mutura" in Kikuyu dialect, is a local delicacy for low, middle income earners and beer drinkers. It is a protein rich meat snack comprising goat or cow cleaned intestines stuffed with cooked small pieces of meat and formed into long coils; sometimes blood is added. Processed African sausages are then placed in boiling water or soup for 30 to 40 minutes and then roasted over coals on outdoor grills using low to medium heat and turning frequently to dehydrate the meat and give it the sensational smoky taste. The internal temperature should be at least 160 degrees Fahrenheit [1].

Animal proteins such as meat, meat products, and even blood are regarded as high-risk perishable commodities with respect to pathogen content, natural toxins, and other possible contaminants [2]. Among the bacteria isolated from

animal products in recent studies include *Staphylococcus aureus*, *Streptococcus* sp., *Escherichia coli*, *Campylobacter jejuni*, *Clostridium perfringens*, *Shigella* sp., and *Salmonella* sp.

Consumption of food with such microbial pathogens and toxins is estimated to result in approximately 1.5 billion episodes of diarrhea and over 3 million deaths globally each year [3]. Increase in demand for finger-licking African sausages has resulted in this delicacy's vendors being on the increase, especially in slum areas [4]. These informal businesses mostly operate in unhygienic makeshift and road-side meat points of Nairobi and major cities in Kenya, where they are unregulated and the standardized preparation of these African sausages is disregarded.

In regard to Torok *et al.* [5] study, food-borne disease from African sausages arises from intestinal bacteria or external contamination, which is a result of unhygienic food preparation especially if the vendor fails to adhere to hygiene practices during processing, preparation, handling, and/or storage; it may therefore pose serious health risk to unsuspecting consumers resulting in outbreaks of foodborne illnesses. Results of this study will help the relevant regulatory body in laying down food safety measures for the African sausages.

2. Materials and Methods

2.1. Study Design. A descriptive study design was employed whereby a convenience sampling of retail meat outlets from Westlands, Pangani, and Kangemi was carried out. One hundred (100) African sausages samples (38 nonroasted and 62 roasted) were collected, processed, isolated, characterized, and quantified.

2.2. Study Area. The study was carried out in Nairobi County which is one of the 47 counties of Kenya [6]. It is the smallest yet most populous county, the capital and largest city of Kenya, which has experienced one of the most rapid growths in urban centres with a population of 3,375,000 as at year 2009 census [7]. It has a total area of 696 km^2 with 17 parliamentary constituencies. Nairobi is a cosmopolitan and a multicultural city.

Economically, it can be subdivided into three main categories: (1) the high end or leafy suburbs or upper-class estates, the likes of Muthaiga, Karen, Westlands, among others, (2) the Middle-class estates of Pangani, Buruburu, among others, and (3) low class estates of Mukuru, Mathare, Kangemi slums, among others [8]. Three ready-to-eat vending sites and meat eatery points of Westlands market, Kangemi market, and Pangani estate were conveniently selected on the basis of easy access, perceived sanitation, and relative hygiene levels. The number of vendors in these areas is not known; but they tend to converge around the shopping areas.

2.3. Sample Size Calculation. Prevalence of common meat contaminants in previous studies was used to determine the sample size required to detect the presence of the bacteria. An expected prevalence rate of between 7% was used to estimate the sample size in this study since similar studies [9–12] reported a prevalence rate of between 3 and 14%. Using the above information, the sample size was calculated using the formula: $n=z^2pq/d^2$ where n is the desired sample size, Z is the standard normal deviate set at 1.96, p is the estimated prevalence, q=1-p, and d is the degree of accuracy set at 0.05 as given by Fisher et al. [13].

2.4. Homogenate Preparation. At the laboratory, one-gram portions of the African sausages (roasted and nonroasted) were obtained aseptically from the vendors, picked separately as they are sold using sterile glass bottles, stored on ice before processing in the laboratory within 24 hours of collection, cut into small pieces on a sterile chopping board using a sterile knife, and blended (homogenized) in 4ml of 0.1% peptone water to obtain 1:5 initial dilution.

2.5. Bacterial Isolation and Characterization. Since the researcher suspected presence of coliforms and other fastidious organisms, the homogenates of the African sausages

were streaked on general purpose enriched medium (blood agar) and selective and differential medium for members of family Enterobacteriaceae (MacConkey agar) (Oxoid Ltd. Thermo Scientific, UK) and incubated aerobically at 37°C for 24 hours. The isolated bacteria were identified based on colony morphology, Gram staining reaction, and biochemical characteristics using established standardized methods according to Bergey's Manual of determinative bacteriology [14].

2.6. Quantification of Total Aerobic Bacterial Count of the African Sausages. For the determination of bacterial load (total bacterial count), method given by Miles and Misra [15] was used. Serial dilutions of 10^{-1} to 10^{-10} were prepared from the African sausage homogenate stock solution that was prepared earlier. Nutrient agar plate was divided into four quadrants, and each quadrant served as one plate. Using a 25 μl calibrated dropper (equivalent to 1/40th of an ml), one drop from each dilution tube was placed per quadrant; each dilution was done in quadruplicate.

The drop was then allowed to dry and the plate incubated aerobically at 37°C for 24 hours [16], after which the number of colonies that grew per drop was counted using Quebec Dark Field colony counter taking the average count for the quadruplicate drops of each dilution. The concentration of the original bacterial suspension was then calculated and expressed as colony forming units per millilitre (cfu/ml), using the formula, a x 40 x 10^y, where a is the average number of colonies in the 4 drops of one dilution tube/diluted suspension, 40 is the number of drops that make one millilitre (the drop being equivalent to 1.40th of a ml), and 10^y is the dilution factor of the respective dilution tube/diluted suspension. This is then multiplied by 5, the initial dilution at homogenization stage.

2.7. Statistical Analysis. The experiment was done in triplicate. The means for the prevalence of bacterial isolates across the three geographical sites and the bacterial counts were compared by a Paired-Samples T Test. Differences were considered statistically significant when P<0.05. Statistical analysis was conducted using the software SPSS 13.0 (SPSS, Chicago, Illinois, USA)

3. Results

3.1. Isolation and Characterization. Figure 1 gives results on the prevalence of isolated and characterized bacterial isolates. Five genera of bacteria (123 isolates) were isolated and characterized from 80/100 (80%) roasted and nonroasted African sausages. They were *Staphylococcus, Bacillus* spp., *Streptococcus* spp., *Proteus* spp., and *Escherichia coli*. *Staphylococci* spp. were the most predominant bacteria in all the sausage samples collected with a prevalence of 50.4% (62/123), *Bacillus* spp. at 19.5% (24/123), *Streptococcus* spp. 9.8% (12/123), and *Proteus* spp. 2.4% (3/123) while *E. coli* was isolated at 1.6% (2/123). With respect to roasted African sausages, *Staphylococcus* spp. accounted for 53.6% (15/28) of the isolates in Kangemi, 52.2% (12/23) in Pangani, and 38.1% (8/21) in Westlands. *Bacillus* spp. organisms were isolated at

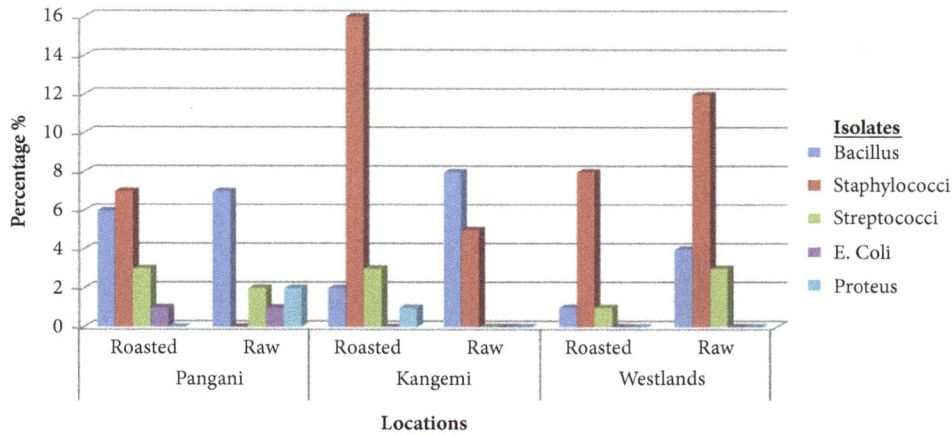

FIGURE 1: Prevalence of the five genera of bacteria isolated from African sausages sampled from Pangani, Kangemi, and Westlands estates, Nairobi County, Kenya: n=100.

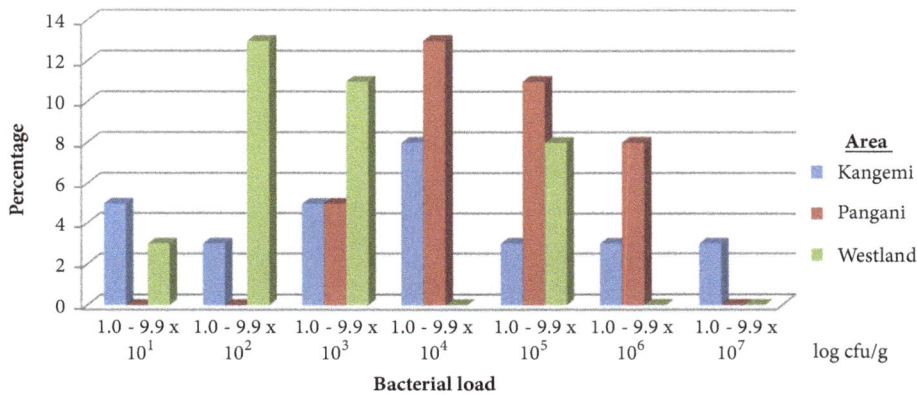

FIGURE 2: Percent total aerobic bacterial counts of the sampled roasted African sausages sold in the three study sites: n=100.

7.1% (2/28) in Kangemi, 26.1% (6/23) in Pangani, and 4.8% (1/21) in Westlands; Streptococcus 10.7% (3/28) in Kangemi, 13% (3/23) in Pangani, and 4.8% 3/21) in Westlands; Proteus 3.6% (1/28) in Kangemi and 0% both in Pangani and in Westlands; *E. coli*, 4.3% (1/23) in Pangani and 0% both in Kangemi and in Westlands.

With respect to nonroasted African sausages, *Staphylococcus* spp. accounted for 50% (6/12) of the isolates in Kangemi, 45.5% (10/22) in Pangani, and 64.7% (11/17) in Westlands. *Bacillus* spp. organisms were isolated at 41.7% (5/12) in Kangemi, 31.8% (7/22) in Pangani, and 17.6% (3/17) in Westlands; *Streptococcus* spp. 0% in Kangemi, 9.1% (2/22) in Pangani, and 17.6% (3/17) in Westlands; *Proteus* spp., 9.1% (2/22) in Pangani and 0% both in Kangemi and in Westlands areas; E. coli, 4.5% (1/22) in Pangani and 0% both in Kangemi and in Westlands areas. Table 1 gives the mean distribution of bacterial isolates. There was no significant difference (p>0.05) in distribution of isolates across the geographical areas under study.

3.2. Total Aerobic Bacterial Count from Roasted African Sausage. The results are as given in Figure 2. 22/62 (35.5%) roasted African sausage samples had a total aerobic bacterial

count of between 1.0 and 9.9x10^1 log cfu/g, 11/62 (17.7%) samples had a total aerobic bacterial count of between 1.0 and 9.9 x10^2 log cfu/g, 12/62 (19%) samples had a total aerobic bacterial count of between 1.0 and 9.9 x10^3 log cfu/g, 9/62 (14.5%) samples had a total aerobic bacterial count of between 1.0 and 9.9 x10^4 log cfu/g, 5/62 (8%) samples had a total aerobic bacterial count of between 1.0 and 9.9 x10^5 log cfu/g, 2/62 (3%) samples had a total aerobic bacterial count of between 1.0 and 9.9 x10^6 log cfu/g, and 1/62 (1.6%) sample had a total aerobic bacterial count of between 1.0 and 9.9 x10^7 log cfu/g.

With respect to individual study sites, 10/24 (41.7%) of samples from Kangemi, 3/19 (15.8%) from Pangani, and 9/20 (45%) from Westlands area had a total aerobic bacterial count of between 1.0 and 9.9 x10^1 log cfu/g. 1/24 (4.17%) of samples from Kangemi, 3/19 (15.8%) from Pangani, and 7/20 (35%) from Westlands area had a bacterial load of between 1.0 and 9.9 x10^2 log cfu/g. 4/24 (16.7%) of samples from Kangemi, 7/19 (36.8%) from Pangani, and 1/20 (5%) from Westlands area had a bacterial load of between 1.0 and 9.9 x10$^{3\,log}$ cfu/g. 3/24 (12.5%) of samples from Kangemi, 4/19 (21%) from Pangani, and 2/20 (10%) from Westlands area had a bacterial load of

TABLE 1: Evaluation of the mean distribution of bacterial isolates from African Sausages across the three geographical areas using paired sample t-test (P values evaluated at 95% confidence limits).

		Paired Differences							
					95% Confidence Interval of the Difference				
		Mean	Std. Deviation	Std. Error Mean	Lower	Upper	t	df	Sig. (2-tailed)
Pair 1	Kangemi - Pangani	-1.333	4.590	1.874	-6.150	3.483	-.712	5	.509
Pair 2	Kangemi - Westlands	.333	2.160	.882	-1.934	2.600	.378	5	.721
Pair 3	Pangani - Westlands	1.667	6.282	2.565	-4.926	8.259	.650	5	.544

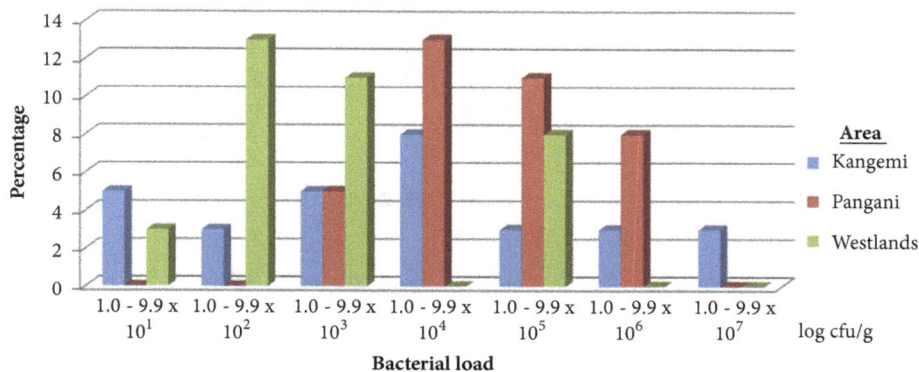

FIGURE 3: Percent total aerobic bacterial counts of the sampled nonroasted African sausages sold in the three study sites: n=100.

between 1.0 and 9.9 x10^4 log cfu/g. 3/24 (12.5%) of samples from Kangemi, 2/19 (10.5%) from Pangani, and 0% from Westlands area had a bacterial load of between 1.0 and 9.9 x10^5 log cfu/g. 2/24 (8.3%) of samples from Kangemi, 0% from Pangani and Westlands area had a bacterial load of between 1.0 and 9.9 x10^6 log cfu/g. 1/24 (4.17%) of samples from Kangemi, 0% from Pangani and Westlands area had a bacterial load of between 1.0 and 9.9 x10^7 cfu/g. Table 2 gives the mean total aerobic bacterial count from roasted African sausages across the three geographical areas. There was no significant difference ($p \geq 0.05$) in mean total aerobic bacterial count across the areas under study.

3.3. Total Aerobic Bacterial Count from Nonroasted African Sausages. The results are as given in Figure 3. 3/38 (7.9%) nonroasted African sausage samples had a total aerobic bacterial count of between 1.0 and 9.9 x10^1 log cfu/g, 6/38 (15.79%) samples had a total aerobic bacterial count of between 1.0 and 9.9 x10^2 log cfu/g, 8/38 (21%) samples had a total aerobic bacterial count of between 1.0 and 9.9 x10^3 log cfu/g, 8/38 (21%) samples had a total aerobic bacterial count of between 1.0 and 9.9 x10^4 log cfu/g, 8/38 (21%) samples had a total aerobic bacterial count of between 1.0 and 9.9 x10^5 log cfu/g, 4/38 (10.5%) samples had a total aerobic bacterial count of between 1.0 and 9.9 x10^6 log cfu/g, and 1/38 (2.6%) sample had a total aerobic bacterial count of between 1.0 and 9.9 x10^7 log cfu/g. With respect to individual study sites, 2/11 (18.8%) of samples from Kangemi, 0% from Pangani, and 1/13 (7.7%) from Westlands area had a total aerobic bacterial count of between 1.0 and 9.9 x10^1 log cfu/g. 1/11 (9%) of samples from Kangemi, 0% from Pangani, and 5/13 (38.5%) from Westlands area had a total aerobic bacterial count of between 1.0 and 9.9 x10^2 log cfu/g. 2/11 (18%) of samples from Kangemi, 2/14 (14.3%) from Pangani, and 4/13 (30.8%) from Westlands area had a total aerobic bacterial count of between 1.0 and 9.9 x10^3 log cfu/g. 3/11 (27.3%) of samples from Kangemi, 5/14 (35.7%) from Pangani, and 0% from Westlands area had a total aerobic bacterial count of between 1.0 and 9.9x10^4 log cfu/g. 1/11 (9%) of samples from Kangemi, 4/14 (28.6%) from Pangani, and 3/13 (23%) from Westlands area had a total aerobic bacterial

count of between 1.0 and 9.9 x10^5 log cfu/g. 1/11 (9%) of samples from Kangemi, 3/14 (21.4)% from Pangani, and 0% from Westlands area had a total aerobic bacterial count of between 1.0 and 9.9 x10^6 log cfu/g. 1/11 (9%) of samples from Kangemi and 0% from Pangani and Westlands area had a total aerobic bacterial count of between 1.0 and 9.9 x10^7 log cfu/g. Table 3 gives the mean total aerobic bacterial count from nonroasted African sausages across the three geographical areas. There was no significant difference ($p \geq 0.05$) in mean total aerobic bacterial count across the areas under study.

4. Discussion

The data obtained on isolation and characterization of bacteria in the present study where *Staphylococcus* spp. and *Bacillus* spp. were the predominant isolates concurs on some aspects with reports by Oluwafemi and Simisaye [17] and Okonko *et al.* [18] working on beef sausages. However, differences are noted whereby in the present study *Streptococci* spp., *Proteus* spp., and *E. coli* were isolated while in the latter, *Enterobacter, Pseudomonas,* and *Klebsiella* species were isolated from beef sausages and seafood, respectively.

The current study showed *Staphylococcus* species prevalence of 50.4%. This was lower than the Staphylococcus species recovery at 58.6% from hotels, restaurants, and cafes, report by Berynestad and Granums [19]. A study by Yusuf *et al.* [20], on percent occurrence of bacteria isolated from the "balangu" meat product in relation to all the retail outlets, reported a lower prevalence of *Staphylococcus aureus* at 12.5%. A study by Aycicek *et al.* [21] reported that processed foods were found to be more prone to *Staphylococcus* species contamination. This may have been attributed to contamination from aerial spores carried in the air, throat, hands, and nail of food handling persons [22].

The results of the roasted African sausages in current study contrast the findings by Orogu and Oshilim [22] who reported a higher prevalence (30%) of *Bacillus* from suya meat. However, similar prevalence was obtained by a study by Matos *et al.* [23] working with dry smoked sausages. Presence of *Bacillus* contamination in some of the samples examined

TABLE 2: Evaluation of the mean total aerobic bacterial count from roasted African sausages across the three geographical areas using paired sample t-test (P values evaluated at 95% confidence limits).

		Paired Differences						
		Mean	Std. Deviation	Std. Error Mean	95% Confidence Interval of the Difference		t	df	Sig. (2-tailed)
					Lower	Upper			
Pair 1	Kangemi - Pangani	147885.000	694148.169	159248.512	-186683.708	482453.708	.929	18	.365
Pair 2	Kangemi - Westlands	175895.500	658535.387	147252.989	-132308.548	484099.548	1.195	19	.247
Pair 3	Pangani - Westlands	26479.211	60777.503	13943.315	-2814.608	55773.029	1.899	18	.074

TABLE 3: Evaluation of the mean total aerobic bacterial count from nonroasted African sausages across the three geographical areas using paired sample t-test (*P* values evaluated at 95% confidence limits).

		Paired Differences							
		Mean	Std. Deviation	Std. Error Mean	95% Confidence Interval of the Difference		t	df	Sig. (2-tailed)
					Lower	Upper			
Pair 1	Kangemi - Pangani	-32490.000	1839501.213	554630.484	-1268283.730	1203303.730	-.059	10	.954
Pair 2	Kangemi - Westlands	504456.364	1407287.946	424313.281	-440972.543	1449885.270	1.189	10	.262
Pair 3	Pangani - Westlands	457091.538	851423.040	236142.264	-57418.255	971601.332	1.936	12	.077

in this study might have resulted from contamination from vendor's skin or the environment.

Similar finding on the prevalence of *Streptococcus* spp. was reported by Onuora *et al.* [24] working on grilled beef. The prevalence of *E. coli* reported in present study was slightly lower than that reported by Syne *et al.* [25] and Onuora *et al.* [24]. *E. coli* presence in African sausages has the potential to cause diarrhea.

The incidence of *E. coli* obtained in this study is a cause for public health concern as this bacterium has been implicated in cases of gastroenteritis [26]. The presence of *Proteus* isolates was remarkably higher in a study by Gwinda *et al.* [27] from beef meat, compared to what was found in the current study. The presence of *Proteus* organisms in the meat samples can obviously be attributed to unhygienic food processing.

Staphylococcus spp., *Bacillus* spp., *Streptococcus* spp., and *E. coli* are known to produce potent enterotoxins and the ingestion of food containing these toxins can cause a sudden onset of illness within three to four hours, with nausea, vomiting, and diarrhea as the major symptoms [28]. There was no significant difference ($p > 0.05$) in distribution of these organisms across the three geographical sites studied. In the present study, it was observed that there was no significant difference ($p > 0.05$) in the total aerobic bacterial count across the three geographical sites studied. Total aerobic bacterial count of between 1.0-9.9×10^2 and $1.0 \times 9.9 \ 10^4$ log cfu/g was reported in most 54/100 (54%) of African sausage samples.

Similar studies by Oluwafemi and Simisaye [17] reported a total aerobic bacterial count level of between 1.3×10^4 and 4.0×10^8 log cfu/g in beef sausage samples. In related studies, Inyang *et al.* [29] reported a total viable count of between 3.7×10^5 to 2.4×10^6 log cfu/g while total viable count by Onuora *et al.* [24] reported a plate count of between 0.9×10^4 log cfu/g and 1.5×10^4 log cfu/g.

The difference in the total bacterial counts may be attributed to the samples used, unhygienic method of transportation, handling, processing, unhygienic environment, and practices such as dirty cutting boards and knifes or utensils. Cheesbrough [28] noted that insects also contribute to contamination by mechanical transfer of microorganisms to food products since they are left uncovered and exposed to dust.

There was no significant difference ($p > 0.05$) in total aerobic bacterial count across the three geographical sites under study. This study therefore concludes that roasted and nonroasted African sausages sold in meat outlets in Nairobi County are contaminated with *Staphylococcus*, *Bacillus*, *Streptococcus*, *Proteus*, and *E. coli* organisms and poses food safety risks to the consumers. The presence of these organisms in ready-to-eat African sausages is a pointer that these African sausages were either processed under poor hygienic and sanitary conditions and insufficient processing or could have been from the animal intestines. Food safety enforcement authority therefore need to scale up inspection of establishments where African sausage vendors are.

Conflicts of Interest

The authors declare that they have no conflicts of interest.

References

[1] M. Wiens, *Kenyan Food Overview: 20 of Kenya's Best Dishes*, 2011, http://migrationology.com/2011/06/kenyan-food-overview-20-of-kenyas-best/dishes/.

[2] A. H. M. Yousuf, M. K. Ahmed, S. Yeasmin, N. Ahsan, M. M. Rahman, and M. M. Islam, "Prevalence of microbial load in shrimp, Penaeus monodon and prawn, Macrobrachium rosenbergii from Bangladesh," *World Journal of Agricultural Science*, vol. 4, pp. 852–855, 2008.

[3] World Health Organization, *Foodborne Hazards in Basic Food Safety for Health Workers*, 2007, http://www.who.int/foodsafety/publications/capacity/en/2.pdf.

[4] O. L. Musa and T. M. Okande, "Effect of health education intervention or food safety practice among food vendors in ilorin," *Journal of Medicine*, vol. 5, pp. 120–124, 2002.

[5] T. J. Torok, R. V. Tanze, R. P. Wise, J. R. Livengood, R. Sokolow, and S. Manvans, "A large community outbreak of salmonellosis caused by intentional contamination of restaurant salad bars," *Journal of the American Medical Association*, vol. 278, no. 8, pp. 389–395, 1997.

[6] Nairobi City County, "Maps". Retrieved from: http://www.nairobi.go.ke/home/maps/, 2017.

[7] Kenya Population and Housing Census (KPHC), "National data archive," http://statistics.knbs.or.ke/nada/index.php/catalog/55/study-description, 2009.

[8] Nairobi Mitaa, "A review of hoods in Nairobi". Retrieved from: http://mitaayanairobi.blogspot.co.ke/2011?m=1, 2011.

[9] J. Miyoko, S. G. Dilma, A. R. Christiane et al., "Occurrence of Salmonella spp and Escherichia coli O157:H7 in raw meat marketed in São Paulo city, Brazil, and evaluation of its cold tolerance in ground beef," *Revista do Instituto Adolfo Lutz*, vol. 63, no. 2, pp. 238–242, 2004.

[10] C. A. Magwira, B. A. Gashe, and E. K. Collison, "Prevalence and antibiotic resistance profiles of Escherichia coli O157:H7 in beef products from retail outlets in Gaborone, Botswana," *Journal of Food Protection*, vol. 68, no. 2, pp. 403–406, 2005.

[11] J. S. Weese, B. P. Avery, J. Rousseau, and R. J. Reid-Smith, "Detection and enumeration of Clostridium difficile spores in retail beef and pork," *Applied and Environmental Microbiology*, vol. 75, no. 15, pp. 5009–5011, 2009.

[12] R. M. Kabwanga, M. M. Kakubua, C. K. Mukeng et al., "Bacteriological assessment of smoked game meat in Lubumbashi, D.R.C," *Biotechnologie, Agronomie, Société et Environnement*, vol. 17, no. 3, pp. 441–449, 2013.

[13] A. A. Fisher, J. E. Laing, J. E. Stoeckel, and J. W. Townsend, *Handbook for Family Planning Operations Research Design*, Population Council, New York, NY, USA, 2nd edition, 1991, https://www.popcouncil.org/.

[14] J. G. Holt, N. R. Kreig, P. H. A. Sneath, and J. T. Staley, *Bergey's Manual of Determinative Bacteriology*, The William and Wilkins Co., Baltimore, Md, USA, 9th edition, 1994.

[15] A. A. Miles and S. S. Misra, "The estimation of the bactericidal power of the blood," *Journal of Hygiene*, vol. 38, no. 6, pp. 732–749, 1938.

[16] B. Jeršek, *Microbiological Examination of Food*, University of Ljubljana, 2017.

[17] F. Oluwafemi and M. T. Simisaye, "Extent of microbial contamination of sausages sold in two Nigerian cities," in *Proceedings of the Book of Abstract of the 29th Annual Conference & General Meeting (Abeokuta 2005) on Microbes as Agents of Sustainable Development, Organized by Nigerian Society for Microbiology (NSM)*, p. 28, University of Agriculture, Abeokuta, Nigeria, 2005.

[18] I. O. Okonko, E. Donbraye, and S. O. I. Babatunde, "Microbiological quality seafood processors and water used in two different sea processing plants in Nigeria," *Electronic Journal of Environmental, Agricultural and Food Chemistry* , vol. 8, no. 8, pp. 621–629, 2009.

[19] M. S. Brynestad and P. E. Granum, "Clostridium perfringens and foodborne infections," *International Journal of Food Microbiology*, vol. 74, no. 3, pp. 195–202, 2002.

[20] M. A. Yusuf, T. H. A. T. Abdul Hamid, and I. Hussain, "Isolation and Identification of Bacteria Associated with Balangu (Roasted Meat Product) Sold in Bauchi. Nigeria," *Journal of Pharmacy*, vol. 2, no. 6, pp. 38–48, 2012.

[21] H. Aycicek, S. Cakiroglu, and T. H. Stevenson, "Incidence of Staphylococcus aureus in ready-to-eat meals from military cafeterias in Ankara, Turkey," *Food Control*, vol. 16, no. 6, pp. 531–534, 2005.

[22] M. Hatakka, K. J. Björkroth, K. Asplund, N. Mäki-Petäys, and H. J. Korkeala, "Genotypes and enterotoxicity of Staphylococcus aureus isolated from the hands and nasal cavities of flight-catering employees," *Journal of Food Protection*, vol. 63, no. 11, pp. 1487–1491, 2000.

[23] J. S. T. Matos, B. B. Jensen, S. F. H. A. Barreto, and O. Hojberg, "Spoilage and pathogenic bacteria isolated from two types of portuguese dry smoked sausages after shelf-life period in modified atmosphere Package," *Journal of Food*, vol. 5, no. 3, pp. 165–174, 2006.

[24] S. Onuorah, I. Obika, F. Odibo, and M. Orji, "An Assessment of the Bacteriological Quality of Tsire-Suya (Grilled Beef) sold in Awka, Nigeria," *American Journal of Life Sciences Researches*, vol. 3, no. 4, pp. 287–292, 2015.

[25] S. M. Syne, A. Ramsubhag, and A. A. Adesiyun, "Microbiological hazard analysis of ready-to-eat meats processed at a food plant in Trinidad, West Indies," *Infection Ecology & Epidemiology*, vol. 3, no. 1, Article ID 20450, 2013.

[26] W. A. Volk, *Essential of Medical Microbiology*, J. B. Lippincott Company, Philadelphia, Pa, USA, 1982.

[27] M. Gwida, H. Hotzel, L. Geue, and H. Tomaso, "Occurrence of *Enterobacteriaceae* in Raw Meat and in Human Samples from Egyptian Retail Sellers," *International Scholarly Research Notices*, vol. 2014, Article ID 565671, 6 pages, 2014.

[28] M. Cheesbrough, *District Laboratory Practice in Tropical Countries*, Cambridge University Press, Cambridge, UK, 2nd edition, 2006.

[29] C. U. Inyang, A. M. Igyor, and E. N. Uma, ""Bacterial Quality of a smoked Meat Product (Suya)," *Nigerian Food Journal*, vol. 23, no. 1, pp. 239–242, 2006.

4

Vegetable Contamination by the Fecal Bacteria of Poultry Manure: Case Study of Gardening Sites in Southern Benin

Séraphin C. Atidégla,[1] Joël Huat,[2] Euloge K. Agbossou,[1] Hervé Saint-Macary,[3] and Romain Glèlè Kakai[1]

[1]*Faculté des Sciences Agronomiques, Université d'Abomey-Calavi, 01 BP 526 Cotonou, Benin*
[2]*CIRAD UPR HortSys, 34498 Montpellier Cedex 05, France*
[3]*CIRAD, UPR Recyclage et Risque, 34398 Montpellier Cedex 05, France*

Correspondence should be addressed to Séraphin C. Atidégla; atideglaser@gmail.com

Academic Editor: Marie Walsh

A study was conducted in southern Benin to assess the contamination of vegetables by fecal coliforms, *Escherichia coli*, and fecal streptococci as one consequence of the intensification of vegetable cropping through fertilization with poultry manure. For this purpose, on-farm trials were conducted in 2009 and 2010 at Yodo-Condji and Ayi-Guinnou with three replications and four fertilization treatments including poultry manure and three vegetable crops (leafy eggplant, tomato, and carrot). Sampling, laboratory analyses, and counts of fecal bacteria in the samples were performed in different cropping seasons. Whatever the fertilization treatment, the logs of mean fecal bacteria count per g of fresh vegetables were variable but higher than AFNOR criteria. The counts ranged from 8 to 10 fecal coliforms, from 5 to 8 fecal streptococci, and from 2 to 6 *Escherichia coli*, whereas AFNOR criteria are, respectively, 0, 1, and 0. The long traditional use of poultry manure and its use during the study helped obtain this high population of fecal pathogens. Results confirmed that the contamination of vegetables by fecal bacteria is mainly due to the use of poultry manure. The use of properly composted poultry manure with innovative cropping techniques should help reduce the number and incidence of pathogens.

1. Introduction

In Sub-Saharan Africa, the products of urban agriculture are considered to be one response to the shortage of foodstuffs [1]. In addition to the contributions of urban agriculture to urban food security, nutrition, and local economies, farming also affects urban water management, sanitation, and health services [2]. In that context, urban production of vegetables is increasing rapidly but, in both Africa and Asia, faces many constraints, especially land pressure, access to water, and low soil fertility [3–5]. In Benin, West Africa, the same problems have been identified in periurban and urban gardening areas, where irrigated vegetable production developed rapidly after 1990, coinciding with the drastic drop in fish resources in the Atlantic Ocean and in the rivers. The main vegetable crops are leafy vegetables, eggplant (*Solanum macrocarpon* L.), carrots (*Daucus carota* L.), tomatoes (*Lycopersicon esculentum* Mill.), peppers (*Capsicum frutescens*), and onions (*Allium cepa*).

To satisfy the increasing demand for vegetables, despite the poverty of coastal soils and land pressure, farmers tend to intensify production by using mineral and organic fertilizers and pesticides. Today, animal manure (60% poultry manure and 40% cattle manure) is frequently used as fertilizer in the study area, Grand-Popo in Benin. Animal manures have been used as effective fertilizers for centuries [6, 7]. Brooks et al. [8] investigated potential microbial runoff associated with the application of poultry litter on the soil. Several other studies pointed to pollution and health risks caused by lack of knowledge and bad practices in the management of livestock manure and chemical fertilizers [1, 3, 9].

Excessive use of fertilizer at each agricultural campaign has been reported in both Africa and Asia, particularly the use of poultry manure at rates of 20 to 50 t·ha^{-1} and the use of mineral fertilizers, such as urea and NPK (10-20-20 nitrogen, phosphorus, and potassium fertilizer), at rates of 1.2 to 2 t·ha^{-1} [6, 9]. Unfortunately, the intensive use of organic matter like cow dung and poultry manure and other animal feces are a significant environmental risk to soils, waters, and crops, including fecal contamination [6].

Many infection outbreaks have been associated with water or food directly or indirectly contaminated by animal manure [10, 11] by identifying *Escherichia coli* and fecal coliforms, which are indicators of fecal pollution [12, 13]. One such example was a major waterborne outbreak of *Escherichia coli* O157:H7 (ECO157) infections with bloody diarrhea and abdominal cramps which lasted from 15 December 1989 to 20 January 1990 in Missouri [14]. Griffin et al. [15] reported the occurrence of many outbreaks of *Escherichia coli* O157:H7 (ECO157) infections in communities, nursing homes, a day care center, and a kindergarten. They mainly took the form of gastrointestinal diseases, bloody diarrhea, hemolytic uremic syndrome, or thrombotic thrombocytopenic purpura. Contaminated manure can contact the product directly when used as a soil fertilizer or indirectly via infiltration of irrigation water or infiltration of water used to wash the product. Ibenyassine et al. [16] and Steele et al. [17] reported that contaminated irrigation water and surface runoff water may be major sources of pathogenic microorganisms that contaminate fruits and vegetables in fields. Animal feces including poultry manure which contain large numbers of bacteria can contaminate croplands and hence agricultural products. Fecal bacteria, including *Escherichia coli*, are responsible for serious outbreaks of diarrhea, particularly in children. Some of these microorganisms, including fecal coliforms, *Escherichia coli*, and fecal streptococci, are life-threatening. Gastrointestinal diseases are ranked as the second most important health problem after malaria for most communities in Accra, Ghana, especially in high-density, low-income areas. Elderly and immune-depressed patients are also exposed to the risk of gastrointestinal problems. Other infectious diseases including hepatitis, typhoid and paratyphoid fever, meningitis, and skin diseases can also be caused by fecal contamination [18–20].

According to the literature, few studies have dealt with the link between the presence of pathogens in freshly harvested vegetables and the wide use of organic material such as poultry manure as fertilizer in market gardening in tropical developing countries [1, 21]. One risk of the intensive use of such organic waste is fecal contamination of the vegetables [3]. Today, poor practices used for the management of livestock manure and chemical fertilizers remain the same as those reported in 2009-2010, since the farmers in the study area have not adopted alternative practices.

In the present study, this issue was addressed by assessing the microbiological quality of vegetables cultivated in market gardening areas in the coastal area of southern Benin. Life-threatening pathogens (fecal coliforms, *Escherichia coli*, and fecal streptococci) were counted both in the poultry manure and in the fresh vegetables after harvest. Both irrigation water and the cultivated soil were analyzed. The main reason why this study focuses on fecal coliforms, *Escherichia coli*, and fecal streptococci is that *Escherichia coli* is an ideal indicator of hygiene in microbiological analyses of raw foods like fresh vegetables. The origin of these fecal pathogens was identified with the aim of recommending ways to reduce these risks.

2. Materials and Methods

2.1. Study Area. The study was conducted in 2009 and 2010 in the coastal area of southern Benin at market gardening sites in Yodo-Condji (01°46′33″N-06°10′10″E, district of Grand-Popo) and Ayi-Guinnou (01°44′36″N-06°15′48″E, district of Agoué).

The climate is subequatorial characterized by little variation in temperature (annual average: 27.4°C) and a bimodal rainfall pattern (annual average rainfall: 882 mm). Marine sandy soils, very seeping and porous, make up the two first soils layers (0–18 cm and 18–40 cm) with slightly basic pH between 7.3 and 7.5 [6] at the two sites. The land has been cultivated continuously for several decades without fallow.

Farmers have easy access to groundwater for crop irrigation (through spraying water method) but are limited by the low fertility of the coastal sandy soils. To satisfy growing urban demand and to improve crop productivity, they have adopted intensive practices, such as application of chemical fertilizers combined with high rates of poultry manure, which is available locally. The length of the crop cycle of these vegetables varies from 1.5 to 3.5 months enabling farmers to successively cultivate four vegetable crops per year.

2.2. Experimental Design. At each site, the experiment was conducted in a split-plot design with two factors (4 fertilization treatments and 3 vegetable crops) and three replications during four successive vegetable growing periods. Each plot measured 2 m^2 (2 m × 1 m).

The four growing periods were from 5 May 2009 to 2 September 2009 (period 1) during the long rainy season, from 10 September 2009 to 12 January 2010 (period 2) during the short rainy season and the long dry season, from 20 January 2010 to 13 May 2010 (period 3) during the dry season and the long rainy season, and from 21 May 2010 to 15 August 2010 (period 4) during the long rainy season and the short dry season.

The main factor analyzed was fertilization including the chemical fertilizers and poultry manure applied during each growing period, using four different treatments (Table 1). The poultry manure was composed of chicken feces and wood shavings and came from a local chicken farm.

The second factor was the vegetable crop: tomato (*Lycopersicon esculentum* M.), traditional eggplant (*Solanum macrocarpon* L.), or carrot (*Daucus carota* L.). The vegetables were harvested the same day and only once.

2.3. Sampling and Analyses. A total of 164 samples from the two sites were analyzed during the study. At harvest time, for each plot and each crop, five individual samples collected from the four corners of the plot and one sample from the middle were pooled to make a composite 100 g sample of fresh

TABLE 1: Fertilization modalities at each growing period: application of poultry manure and mineral fertilizers on the three tested vegetable crops (eggplant, tomato, and carrot). Amendment was applied one week before sowing; top-dressing was applied twice: 2 weeks and 4 weeks after sowing (coastal area of Benin, 2009-2010, 288 plots).

Fertilization modalities	Amendment (t·ha^{-1})	Total top-dressing (t·ha^{-1})	Total per growing period (t·ha^{-1})
T0, control	0	0	0
T1, farmer practice 1			
NPK	0.4	0.8	1.2
Urea	0.4	0.8	1.2
T2, farmer practice 2			
NPK	0.4	0.8	1.2
Urea	0.4	0.8	1.2
Poultry manure	10	10	20
T3, poultry manure only	25	15	40

vegetables. The samples were immediately placed in sterile bags. A total of 96 composite samples of fresh vegetables were analyzed for the period of 2009-2010. Eight samples of poultry manure were collected before each sowing date. One kg of poultry manure was collected using the procedure described above. All samples of poultry manure were immediately sealed in sterile bags at 25°C and transported to the laboratory.

Analyses of the different samples (poultry manure and vegetables) were performed by the "Water and Food Quality Control Laboratory of the Ministry of Health" in Cotonou. At the time, this laboratory was the only one in Cotonou to perform microbiological analyses of both liquid and solid foods. During the study period, a second laboratory, which was only qualified for microbiology analyses in water, had problems with its specific equipment. But as soils and poultry manure are solid and the target of the present study was fecal pathogens, samples of soils and poultry manures were also sent to that qualified laboratory. In the absence of evidence from other organizations (FAO, CEE, and IRD) or other European recommendations that mention levels of fecal pathogens recorded in soils and manure, AFNOR food and water criteria are used in the paper. To our knowledge, the purpose of the AFNOR criteria is to identify the concentration of these pathogens in the substances like vegetables or water, consumed by human beings. For this reason, we compare our data concerning soils and poultry manure with AFNOR criteria and analytical techniques.

(i) Vegetables and Poultry Manure. Note the following:

(a) Fecal coliforms per g were identified by colimetry using the V-08-60 Rapid'E. coli medium (24 h at 44°C).

(b) Escherichia coli per g were identified by colimetry using the V-08-053 Rapid'E. coli medium (24 h at 44°C).

(c) Fecal streptococci were identified by NFT 90416 and Bartley and Slanetz medium (24 h–48 h at 37°C).

The number of fecal microorganisms per gram of composite sample of fresh vegetables or of poultry manure or of soils was used as the unit of measure.

2.4. Data Analysis. Analysis of variance on repeated measures [22] was performed to test the effects of "growing period," "fertilization treatment," "vegetable crop," and "site" on the populations of microorganisms of soils, while traditional analysis of variance (nonrepeated measures) was performed to test the effects of the above-cited factors (except "growing period") on the populations of fecal microorganisms on the vegetables. In the two models, site was considered as random factor, while the other factors were fixed. These statistical analyses were conducted using SAS software version 9.2.

To stabilize the variances, each of the three variables (fecal coliforms, Escherichia coli, and fecal streptococci) was log-transformed according to the following relation: $y = \ln(x+1)$, where x is the number of bacteria observed for each variable and y is the result of the transformation.

The adjusted means of the three variables were compiled with the corresponding coefficients of variation. Student-Newman-Keuls (SNK) tests distinguished means by highlighting the different groups of homogenous treatments.

Factors able to explain the variability of the number of fecal bacteria in the study area were identified.

3. Results and Discussion

3.1. Changes in Populations of Fecal Bacteria in the Poultry Manure as a Function of the Growing Period. High temporal variability of the number of fecal bacteria in the poultry manure at the two sites was observed (Figure 1). Log-transformed fecal coliform counts increased from 5 May 2009 to 2 September 2009 (period 1) and from 10 September 2009 to 12 January 2010 (period 2). Conversely, log-transformed counts of Escherichia coli decreased from 4 in period 1 to slightly less than 4 in period 2 and were absent from 20 January 2010 to 13 May 2010 (period 3) and from 21 May 2010 to 15 August 2010 (period 4). The number of streptococci decreased from more than 6 in period 1 to slightly less than 6 in period 2 and then increased to 8 in period 3 before falling to slightly less than 4 in period 4.

The high temperatures (30°C to 35°C) between December 2009 and January 2010 when the poultry manure was stored outside the henhouse probably helped reduce populations

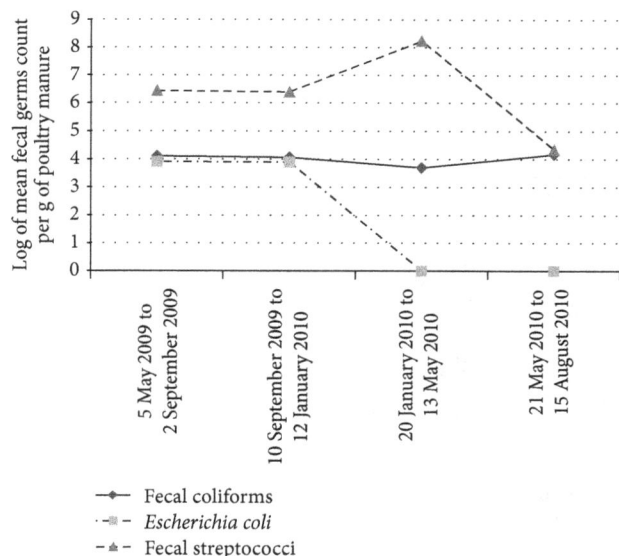

FIGURE 1: Changes in populations of the three fecal microorganisms in poultry manure over the four growing periods (coastal area of Benin, 2009-2010).

of fecal coliforms by more than half (growing period 2 to growing period 3). A similar situation was observed for *Escherichia coli* in irrigation water and soils at both sites, but here we rather witnessed their disappearance from period 1 (harvest period) to period 3. However, concerning the poultry manure, the untimely actions of the driving rain during the trial could explain the variations that occurred in period 4 when the poultry manure was stored outside the henhouse for two months (from April 2010 to May 2010). Rainfall amounted to 147 mm in eight days from 1 April 2010 to 12 April 2010 versus 16 days for the same volume during the same period in 2009. Driving rains could thus also have been responsible for the reduction in (or the absence of) microorganisms at some dates, in particular *Escherichia coli* and fecal streptococci in the poultry manure taken from the henhouse on 10 April 2010 and analyzed on 13 May 2010. This observation is in accordance with the results of Hutchison et al. [23], who contrasted the effect of driving rain on the survival of fecal coliforms, causing their destruction and washing them out, with the effect of drizzle. Like Jamieson et al. [24], our results showed that bacterial survival was optimal in cold wet conditions. But it is possible that competition among microorganisms also affects the survival of fecal bacteria in the soil, in line with the influence of predation [25]. In the study area, predation among bacteria could also be responsible for the disappearance of fecal pathogens.

Based on the results observed through Figure 1, it is possible to conclude that, because there was no organic matter in the soil, as soon as the poultry manure was applied to the soil, the fecal microorganisms remained on the surface and moved directly towards the vegetables. Franz et al. [26] reported that survival of *Escherichia coli* was optimal in soils rich in organic matter and under flooding. According to these authors, water holding capacity, which depends on

soil texture and organic matter content, is known to have an impact on fecal bacteria. Analyses of poultry manure in growing period 4 (the wettest period) revealed the influence of humidity. As soils with a high humus ratio have the highest water holding capacity, they provide a favorable environment for the survival of fecal pathogens. In conclusion, in this study area, a large proportion of the fecal bacteria supplied by the poultry manure did not survive due to unfavorable abiotic conditions (temperature, pH, and organic matter content).

3.2. High Level of Contamination of Harvested Vegetables. Individually, site, fertilization treatment, and type of vegetable had no significant effect on the number of fecal microorganisms (Figures 2(a), 2(b), and 2(c)). Log-transformed fecal coliform counts on vegetables were close to 10, that is, higher than those of *Escherichia coli* and fecal streptococci (each close to six) (Figure 3).

The number of populations of microorganisms varied considerably among the four growing periods and among the three vegetables crops (Figures 2(a), 2(b), and 2(c)). From 5 May 2009 to 13 May 2010 (three successive growing periods), fecal coliforms increased four times on carrot (2.8 to 10.7) and almost two times on eggplant (4.8 to 8.5). In the same periods, fecal coliforms on tomato increased almost four times (from 2.3 to 8.6) but decreased by half on both vegetables in period 4 (Figures 2(a), 2(b), and 2(c)). On tomato (Figure 2(c)), fecal coliforms increased from two in period 1 to nine in period 3 before decreasing to four in period 4. The highest rate of contamination by fecal coliforms (<11) on carrot and by *Escherichia coli* on eggplant (7) occurred in period 3 and in period 2 on carrot (<9). The number of fecal coliforms on the vegetables reached its peak in periods 2 and 3, whereas the number of *Escherichia coli* (7) was highest in period 3 and that of fecal streptococci (>8) was highest in period 2. Fecal coliforms were lowest (around 2) in period 1. *Escherichia coli* were absent during some growing periods and fecal streptococci were absent in period 4 (Figure 2(b)).

In the vegetable samples, the mean number of fecal bacteria per gram was considerably higher than the levels recommended by the French Standards Association [27] for fresh vegetables (Table 2): 8 to 10 times higher than the standard for fecal coliforms, 2 to 6 times higher than the standard for *Escherichia coli*, and 5 to 8 times higher than the standard for fecal streptococci.

Although the original source of contamination of produce has seldom been identified, manure from farm animals has long been suspected of being a leading vehicle of pathogen transmission. Concerning the different results and the fact that the major source of contamination of fresh products is microbial pathogens, we agree with Ijabadeniyi et al. [11] who reported that, at the preharvest stage, the sources include feces, irrigation water, inadequately composted manure, soil, air, animals, and human handling.

3.3. Role of Poultry Manure in the Contamination of Vegetables. Considering all the treatments (T0, T1, T2, and T3) together, we found all three types of fecal bacteria in the three vegetable crops (Figures 2(a), 2(b), and 2(c)). As shown by the results of the analyses of variance (Table 3), at $p < 0.05$, only

(a)

(b)

(c)

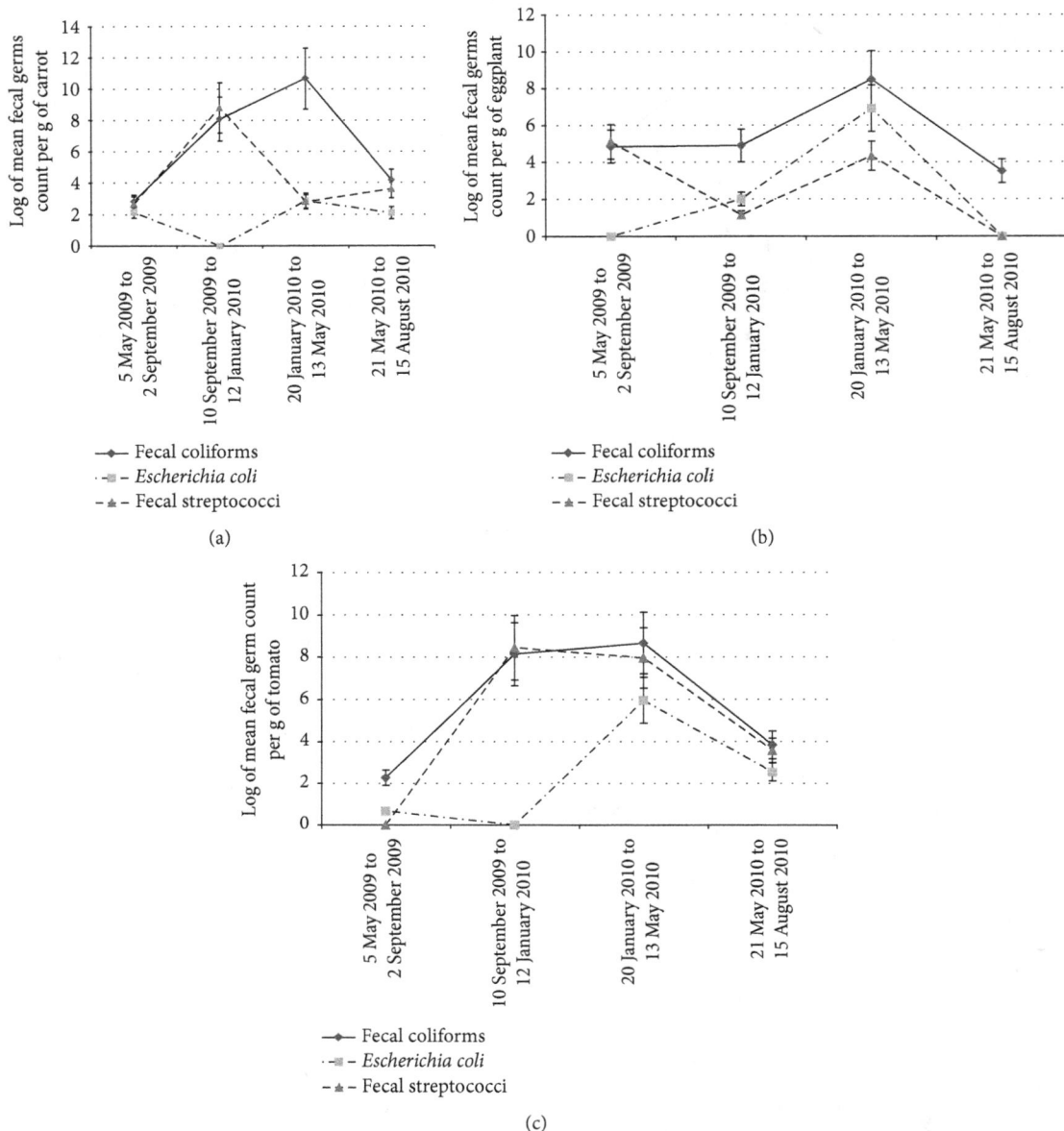

FIGURE 2: (a) Changes in populations of fecal bacteria on carrot per growing period (sites and fertilization treatments considered together (coastal area of Benin, 2009-2010, 288 plots, 32 samples)). (b) Changes in populations of fecal bacteria on eggplant per growing period (sites and fertilization treatments considered together (coastal area of Benin, 2009-2010, 288 plots, 32 samples)). (c) Changes in populations of fecal bacteria on tomato per growing period (sites and fertilization treatments considered together (coastal area of Benin, 2009-2010, 288 plots, 32 samples)).

TABLE 2: Populations of fecal coliforms, *Escherichia coli,* and fecal streptococci on each crop (sites, growing periods, and fertilization treatments considered together) in comparison to AFNOR criteria (coastal area of Benin, 2009-2010, 288 plots, 96 samples of all crops).

Log of mean fecal bacteria counted per g of crop	Carrot		Eggplant		Tomato		AFNOR criteria
	M	SD	M	SD	M	SD	
Fecal coliforms	10	6.09	8	5.21	8	5.46	0
Escherichia coli	2	1.69	6	2.22	5	2.17	0
Fecal streptococci	8	4.26	5	2.55	8	4.85	1

TABLE 3: Analysis of variance onrepeated measures: Fisher values illustrating the effects of growing period, fertilization modality, vegetable crop, and site on the fecal bacteria populations in vegetables (coastal area of Benin, 2009-2010, 288 plots).

Source	Degree of freedom	F_{obs}		
		Fecal coliforms (log)	Escherichia coli (log)	Fecal streptococci (log)
Growing period	3	87.61***	53.15***	10.94***
Growing period * site	3	1.03ns	17.96***	2.25ns
Growing period * vegetable	6	2.82*	5.73**	6.59***
Growing period * fertilization	9	1.37ns	2.21ns	3.32*
Growing period * site * vegetable	6	2.18ns	3.45*	3.55*
Growing period * site * fertilization	9	0.74ns	4.77**	1.07ns
Growing period * vegetable * fertilization	18	1.48ns	1.32ns	1.20ns

Note: *significant at 0.05; **significant at 0.01; ***significant at 0.001; nsnot significant at 0.05.

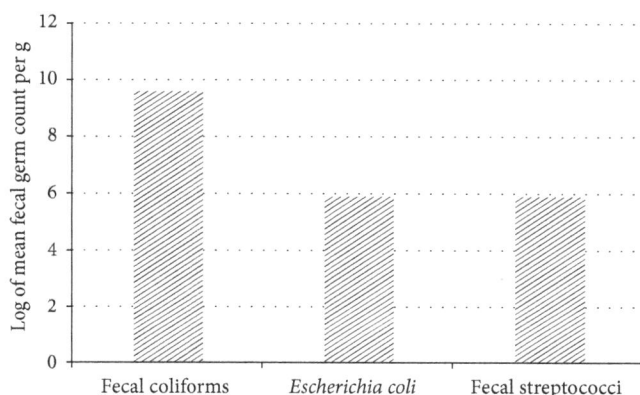

FIGURE 3: Populations of fecal coliforms, Escherichia coli, and fecal streptococci on vegetables (sites, growing periods, fertilization treatments and vegetables considered together (coastal area of Benin, 2009-2010, 288 plots, 96 samples)).

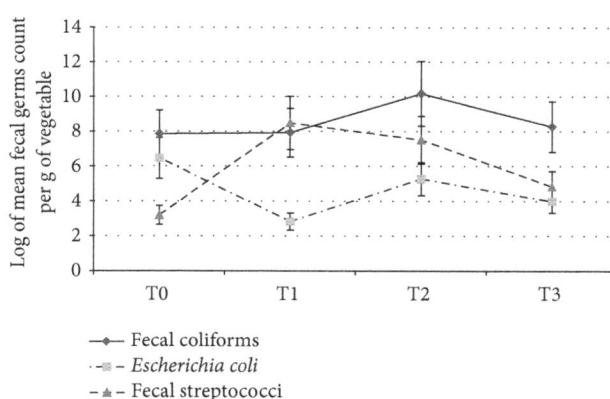

FIGURE 4: Populations of the three fecal bacteria per fertilization treatment: T0, unfertilized control; T1, mineral fertilization (urea + NPK); T2, mineral fertilization + $10 \, t \cdot ha^{-1}$ of poultry manure; and T3, $40 \, t \cdot ha^{-1}$ of poultry manure (coastal area of Benin, 2009-2010, 288 plots). The fertilization, with or without poultry manure, had no significant impact on the high initial contamination of the environment.

the growing period had a significant effect (probability 0.1%) on the concentrations of the three fecal bacteria analyzed. The application of fertilizer had no significant impact on the high initial contamination of the environment. For instance, the control treatment (T0), that is, no poultry manure, resulted in as high level of fecal coliforms as treatments T1 (mineral fertilizer), T2 (mineral fertilizer + poultry manure), and T3 (manure only), on all three vegetables (Figure 4). We recorded log-transformed average counts of eight for fecal coliforms in the plots with the control treatment T0, six for Escherichia coli, and three for fecal streptococci.

On the other hand, treatments T2 and T3 tended to result in more fecal coliforms and fecal streptococci in all the growing periods. This tendency was most marked in period 2 in the short rainy season and in period 3 during the long dry season (Figures 1, 2(a), 2(b), and 2(c)).

The control treatment (T0), under which no poultry manure was applied, resulted in the same high concentration of fecal microorganisms in all three vegetables as treatments T1, T2, and T3. This is likely due to the untimely actions of the flow of irrigation water and rain on the plots under control treatment T0. The long traditional use of poultry manure which probably left fecal bacteria in the soil and water in

the study area before the beginning of the trials is another probable explanation.

On the other hand, the lower number of Escherichia coli in all three vegetables with treatment T3 confirmed that contamination was recent. Even though three months passed between application of the fertilizer and harvest, it is highly probable that Escherichia coli we counted came from the applied poultry manure. So, in the case of Grand-Popo, both traditional organic fertilizers (poultry manure and cow dung) and the poultry manure supplied during the trial can be assumed to be responsible for the contamination. Our results are in agreement with those of Amponsah-Doku et al. [9], who reported that poultry manure, which is the main fertilizer used by 75% of lettuce growers in Accra and Kumasi (Ghana), was responsible for the contamination of the lettuce by numerous fecal coliforms.

Urban areas can also be sources of fecal pollution: storm waters can transport fecal organisms in runoff originating from domestic waste, urban wildlife, or domestic animals. The many heaps of wastes observed in the villages of Ayi-Guinnou and Yodo-Condji and the liquid wastes deposited in the environment along with vegetable leftovers were possible sources of contamination. Moreover, for more than 70% of the local population, river banks and the beach are defecation areas with deleterious consequences for the microbiological quality of the irrigation water, which is also used as drinking water by the local population [1, 28].

4. Conclusion

Application of contaminated poultry manure to the crops in gardening sites may result in contamination of vegetables by the following fecal bacteria: fecal coliforms, *Escherichia coli*, and fecal streptococci. Our study has demonstrated that the use of poultry manure as fertilizer during the agronomic trials has influenced the fecal bacteria counts recorded on the vegetables. The long traditional use of poultry manure which probably left fecal bacteria in the soil and water in the study area before the beginning of the trials is another probable explanation.

This statement of the environment constitutes a genuine danger for the public health: that of the farmers, local populations, and the consumers of the vegetables produced in the area. The combination of market gardening and livestock rising is widely considered to be a good way to increase plant yields. But our results show that it can also have negative impacts on human health. For this reason, to achieve sustainable urban agriculture in Africa, those designing innovative cropping systems need to be mindful of the environment and of public health concerns and such systems should be developed through multidisciplinary research that includes more medical, biological, environmental, and socioeconomic components. But, in the immediate future, the use of properly composted poultry manure should help reduce the number and range of pathogens and hence avoid the application of contaminated manure.

Conflict of Interests

The authors declare that there is no conflict of interests regarding the publication of this paper.

Acknowledgments

The current study was conducted with the financial support of Hydraulic and Water Control Laboratory of the Faculty of Agronomic Science of Abomey-Calavi University through Project BEN/145-UAC/FSA. The authors thank Cécile Fovet-Rabot very much for her comments and corrections made to the paper.

References

[1] H. De Bon, L. Parrot, and P. Moustier, "Sustainable urban agriculture in developing countries. A review," *Agronomy for Sustainable Development*, vol. 30, no. 1, pp. 21–32, 2010.

[2] M. Lydecker and P. Drechsel, "Urban agriculture and sanitation services in Accra, Ghana: the overlooked contribution," *International Journal of Agricultural Sustainability*, vol. 8, no. 1-2, pp. 94–103, 2010.

[3] D. J. Midmore and H. G. P. Jansen, "Supplying vegetables to Asian cities: is there a case for Peri-urban production?" *Food Policy*, vol. 28, no. 1, pp. 13–27, 2003.

[4] O. Cofie, V. R. Veenhuizen, V. de Vreede, and S. Maessen, "Waste Management for Nutrient Recovery: options and challenges for urban agriculture," *Urban Agriculture Magazine*, vol. 23, pp. 3–7, 2010.

[5] P. Drechsel and S. Dongus, "Dynamics and sustainability of urban agriculture: examples from sub-Saharan Africa," *Sustainability Science*, vol. 5, no. 1, pp. 69–78, 2010.

[6] S. Atidégla, *Effets des différentes doses d'engrais minéraux et de la fiente de volaille sur l'accumulation de biocontaminants et polluants (germes fécaux, composés azotés et phosphorés, métaux lourds) dans les eaux, les sols et les légumes de Grand-Popo au Bénin [Ph.D. thesis]*, EDP/FLASH, Université d'Abomey-Calavi (UAC), Bénin, West Africa, 2011.

[7] M. Delgado, C. Rodríguez, J. V. Martín, R. Miralles de Imperial, and F. Alonso, "Environmental assay on the effect of poultry manure application on soil organisms in agroecosystems," *Science of the Total Environment*, vol. 416, pp. 532–535, 2012.

[8] J. P. Brooks, A. Adeli, M. R. Mclaughlin, and D. M. Miles, "The effect of poultry manure application rate and AlCl3 treatment on bacterial fecal indicators in runoff," *Journal of Water and Health*, vol. 10, no. 4, pp. 619–628, 2012.

[9] F. Amponsah-Doku, K. Obiri-Danso, R. C. Abaidoo, L. A. Andoh, P. Drechsel, and F. Kondrasen, "Bacterial contamination of lettuce and associated risk factors at production sites, markets and street food restaurants in urban and peri-urban Kumasi, Ghana," *Scientific Research and Essays*, vol. 5, no. 2, pp. 217–223, 2010.

[10] Centers for Disease Control and Prevention (CDCP), "Outbreaks of Escherichia coli O157 : H7 infections among children associated with farms visits—Pennsylvania and Washington," *Morbidity and Mortality Weekly Report*, vol. 50, no. 15, pp. 293–297, 2001.

[11] O. A. Ijabadeniyi, L. K. Debusho, M. Vanderlinde, and E. M. Buys, "Irrigation water as a potential preharvest source of bacterial contamination of vegetables," *Journal of Food Safety*, vol. 31, no. 4, pp. 452–461, 2011.

[12] T. Garcia-Armisen and P. Servais, "Respective contributions of point and non-point sources of *E. coli* and enterococci in a large urbanized watershed (the Seine river, France)," *Journal of Environmental Management*, vol. 82, no. 4, pp. 512–518, 2007.

[13] M. J. Pantshwa, A. M. van der Walt, S. S. Cilliers, and C. C. Bezuidenhout, "Investigation of faecal pollution and occurrence of antibiotic resistant bacteria in the Mooi river system as a function of a changed environment," 2009, http://www.ewisa.co.za/literature/files/2008_137.pdf.

[14] D. L. Swerdlow, B. A. Woodruff, R. C. Brady et al., "A waterborne outbreak in Missouri of *Escherichia coli* O157:H7 associated with bloody diarrhea and death," *Annals of Internal Medicine*, vol. 117, no. 10, pp. 812–819, 1992.

[15] M. P. Griffin, S. M. Ostroff, R. V. Tauxe et al., "Illnesses associated with *Escherichia coli* O157 : H7 infections: a broad clinical spectrum," *Annals of Internal Medicine*, vol. 109, no. 9, pp. 705–712, 1988.

[16] K. Ibenyassine, R. AitMhand, Y. Karamoko, N. Cohen, and M. M. Ennaji, "Use of repetitive DNA sequences to determine the persistence of enteropathogenic *Escherichia coli* in vegetables and in soil grown in fields treated with contaminated irrigation water," *Letters in Applied Microbiology*, vol. 43, no. 5, pp. 528–533, 2006.

[17] M. Steele, A. Mahdi, and J. Odumeru, "Microbial assessment of irrigation water used for production of fruit and vegetables in Ontario, Canada," *Journal of Food Protection*, vol. 68, no. 7, pp. 1388–1392, 2005.

[18] M. Makoutodé, A. K. Assani, E. M. Ouendo, V. D. Agueh, and P. Diall, "Qualité et mode degestion de l'eau de puits en milieu rural au Bénin: cas de la sous-préfecture de Grand Popo," *Médecine d'Afrique Noire*, vol. 46, pp. 528–534, 1999.

[19] Y. H. B. Nguendo, G. Salem, and J. Thouez, "Risques sanitaires liés aux modes d'assainissement des excreta à Yaoundé, Cameroun," *Natural Science and Sociology*, vol. 16, no. 1, pp. 3–12, 2008.

[20] P. Turgeon, P. Michel, P. Levallois, M. Archambault, and A. Ravel, "Fecal contamination of recreational freshwaters: the effect of time-independent Agroenvironmental factors," *Water Quality, Exposure and Health*, vol. 3, no. 2, pp. 109–118, 2011.

[21] A. Abdulkadir, L. H. Dossa, D. J.-P. Lompo, N. Abdu, and H. van Keulen, "Characterization of urban and peri-urban agroecosystems in three West African cities," *International Journal of Agricultural Sustainability*, vol. 10, no. 4, pp. 289–314, 2012.

[22] M. J. Crowder and D. J. Hand, *Analysis of Repeated Measures*, Chapman and Hall, New York, NY, USA, 1990.

[23] M. L. Hutchison, L. D. Walters, A. Moore, K. M. Crookes, and S. M. Avery, "Effect of length of time before incorporation on survival of pathogenic bacteria present in livestock wastes applied to agricultural soil," *Applied and Environmental Microbiology*, vol. 70, no. 9, pp. 5111–5118, 2004.

[24] R. C. Jamieson, R. J. Gordon, K. E. Sharples, G. W. Stratton, and A. Madani, "Movement and persistence of fecal bacteria in agricultural soils and subsurface drainage water: a review," *Canadian Biosystems Engineering*, vol. 44, pp. 1–9, 2002.

[25] K. R. Reddy, R. Khaleel, and M. R. Overcash, "Behavior and transport of microbial pathogens and indicator organisms in soils treated with organic wastes," *Journal of Environmental Quality*, vol. 10, no. 3, pp. 255–266, 1981.

[26] E. Franz, A. V. Semenov, A. J. Termorshuizen, O. J. de Vos, J. G. Bokhorst, and A. H. C. van Bruggen, "Manure-amended soil characteristics affecting the survival of *E. coli* O157:H7 in 36 Dutch soils," *Environmental Microbiology*, vol. 10, no. 2, pp. 313–327, 2008.

[27] AFNOR (Association Française de Normalisation), *Dictionnaire de L'environnement de l'Association Française de Normalisation: Les Termes Normalisés*, AFNOR (Association Française de Normalisation), Paris, France, 1994.

[28] C. Atidegla and K. Agbossou, "Pollutions chimique et bactériologique des eaux souterraines des exploitations maraîchères irriguées de la commune de Grand-Popo: cas des nitrates et bactéries fécales," *International Journal of Biological and Chemical Sciences*, vol. 4, no. 2, pp. 327–337, 2010.

5

Protein and Metalloprotein Distribution in Different Varieties of Beans (*Phaseolus vulgaris* L.): Effects of Cooking

Aline P. Oliveira, Geyssa Ferreira Andrade, Bianca S. O. Mateó, and Juliana Naozuka

Departamento de Química, Universidade Federal de São Paulo, Diadema, SP, Brazil

Correspondence should be addressed to Juliana Naozuka; jnaozuka@gmail.com

Academic Editor: Salam A. Ibrahim

Beans (*Phaseolus vulgaris* L.) are among the main sources of protein and minerals. The cooking of the grains is imperative, due to reduction of the effect of some toxic and antinutritional substances, as well as increase of protein digestibility. In this study, the effects of cooking on albumins, globulins, prolamins, and glutelins concentration and determination of Fe associated with proteins for different beans varieties and on phaseolin concentration in common and black beans were evaluated. Different extractant solutions (water, NaCl, ethanol, and NaOH) were used for extracting albumins, globulins, prolamins, and glutelins, respectively. For the phaseolin separation NaOH, HCl, and NaCl were used. The total concentration of proteins was determined by Bradford method; Cu and Fe associated with phaseolin and other proteins were obtained by graphite furnace atomic absorption spectrometry and by flame atomic absorption spectrometry, respectively. Cooking promoted a negative effect on (1) the proteins concentrations (17 (glutelin) to 95 (albumin) %) of common beans and (2) phaseolin concentration (90%) for common and black beans. Fe associated with albumin, prolamin, and glutelin was not altered. In Fe and Cu associated with phaseolin there was an increase of 20 and 37% for the common and black varieties, respectively.

1. Introduction

More than 50 varieties of native *Phaseolus vulgaris* L. are present in Latin America, with the common and black bean varieties being the most widely consumed ones in Brazil [1, 2]. In that region, the intake per capita ranges from 1 to 40 kg *per* year [2]. In developed countries, bean consumption is also encouraged because of its health-promoting properties, such as prevention of cardiovascular diseases, obesity, diabetes *mellitus*, and cancer [2, 3].

Beans have great social and economic importance in Brazil being one of the main sources of proteins, plant-derived micronutrients, and minerals for the population [4]. In common beans, the total protein content ranges from 16% to 33%, with high concentrations of aromatic amino acids (lysine, leucine, isoleucine, aspartic acid, and glutamic acid), although low in methionine, cysteine, tryptophan, valine, and threonine [5, 6].

The major protein fractions in *Phaseolus* beans are globulins and albumins, while the minor fractions are prolamins and glutelins [5, 7]. High concentrations of albumins, globulins, and glutelins have been found in raw beans [8]. Salt-soluble globulins, which constitute 34 to 81% of the total protein content, are rich in leucine, lysine, glutamine, and asparagine [5]. The globulin with the highest concentration is phaseolin, whose content corresponds to 40–50% of the total globulins [9, 10].

Phaseolin is a glycoprotein containing neutral sugars, mainly mannose, and consists of three polypeptide subunits with molecular weights between 43 and 53 kDa. Its nutritive value, however, is limited by its low content in sulfur amino acids and its high resistance to enzymatic hydrolysis [3]. Raw phaseolin is highly resistant to in vitro and in vivo digestion, because of glycosylation, which makes its chemical structure rigid and compact. In addition, the low hydrophilic potential of phaseolin limits its accessibility to proteolytic enzymes [2]. Furthermore, the raw albumin and glutelin fractions of the bean proteins show low digestibility; in the case of albumins, this fact is also promoted by a high number of disulfide bridges and the presence of carbohydrates [2].

Low protein digestibility can be improved using thermal treatments, such as domestic cooking [2]. Cooking the bean grains is imperative before intake, since it improves their flavor and palatability and reduces the flatulence factors (raffinose oligosaccharides) and antinutrients (phytic acid and tannins) [3, 4, 11]. However, it is important to consider that cooking can cause considerable changes in the composition of numerous chemical constituents, such as amino acids, vitamins, and minerals. Additionally, studies have shown that cooking may affect the bioavailability of macro- and micronutrients. The digestibility, and hence the absorption of micronutrients such as Fe, is improved by heating processes. With the resultant softening of the food matrix, protein-bound elements are released, thus facilitating their absorption. In addition, heating food alters the inherent factors that inhibit mineral absorption, such as phytates and dietary fiber [8].

In this work, extraction procedures for protein separation and determination of the Fe and Cu concentrations associated with proteins in several *Phaseolus* bean varieties by spectrometric techniques were examined to evaluate the effect of cooking on beans. These studies are essential, because there is a lack of information on the effect of cooking on proteins (albumins, globulins, prolamins, glutelins, and phaseolin) and certain essential elements (Fe and Cu) associated with proteins, especially in the bean cultivars consumed in Brazil (common and black beans).

2. Materials and Methods

2.1. Reagents and Samples. All solutions were prepared from analytical reagent-grade chemicals and high-purity deionized water obtained from a Milli-Q water purification system (Millipore, Belford, MA, USA).

For the sequential extraction, the following reagents (Merck, USA) were used: acetone, chloroform, ethanol, methanol, NaCl, HCl, and NaOH.

The total protein concentration in the extractants was obtained using Bradford's reagent (BioAgency, Brazil), which was diluted five times with deionized water before analysis. The stock solution used to generate a standard curve was prepared dissolving 4.0 mg of ovalbumin (BioAgency, Brazil) in 2.0 mL of deionized water, using Vortex stirring for 2 min. The solution was then diluted ten times with deionized water.

Analytical-grade Titrisol solutions of $1000\,mg\,L^{-1}$ of Cu ($CuCl_2$) and Fe ($FeCl_3$) from Merck were used to prepare the reference analytical solutions to calibrate the instruments.

Two brands of beans comprising seven *Phaseolus* varieties (common, black, rajado, rosinha, bolinha, fradinho, and jalo) were purchased at a local market in São Paulo, each variety weighing 500 g. Six varieties were of the same brand, and the jalo variety was of a different brand. The geographic origin of the varieties was São Paulo (rosinha, rajado, and bolinha) and Minas Gerais (common, black, fradinho, and jalo), according to the producers.

2.2. Instrumentation. A ZEEnit 60 model atomic absorption spectrometer (Analytik Jena AG, Germany) equipped with a transversally heated graphite atomizer, a pyrolytically coated

TABLE 1: Instrumental parameters and heating program for Fe and Cu determination by GF AAS.

Instrumental parameters	Fe[a]	Cu[b]	
λ/nm	248.3	324.8	
Slit/nm	0.8	0.8	
I/mA	4.0	4.0	
Heating program			
Step	T/°C	Ramp/°C s^{-1}	Hold/s
Drying	130	5	20
Pyrolysis	1200[a]/1000[b]	100	15
Atomization	2300	2300	5
Cleaning	2500	1200	2

graphite tube, and a transversal Zeeman-effect background corrector was used for the determination of Cu and Fe. The spectrometer was operated using a hollow cathode lamp. All measurements were based on integrated absorbance values. The instrumental conditions for the spectrometer and the heating program are shown in Table 1. Argon 99.998%, v/v (Air Liquide Brasil, Brazil), was used as the protective and purge gas.

An atomic absorption spectrometer (Model AAS Vario 6, Analytik Jena AG, Jena, Germany), equipped with a hollow cathode lamp of Fe (259 nm, 4 mA, and 0.8 nm) and a deuterium lamp for background correction, was used. The instrumental parameters ($70\,L\,h^{-1}$ acetylene flow, $430\,L\,h^{-1}$ air flow, and 8 mm observation height) were stabilized by a fabricant for the determination of Fe in the water, NaCl, ethanol, and NaOH extractants.

Ultrospec 2100 Pro spectrophotometer (Biochrom Ltd, Cambridge, UK), equipped with a xenon lamp and a wavelength range of 190–900 nm, was used for protein determination at 595 nm.

The samples were dried using a freeze dryer (Thermo Fisher Scientific, USA). Samples milling was carried out in a cryogenic grinder (MA 775 model, Marconi, Brazil). The beans were cooked in an electric pressure cooker (Philips Walita Daily Collection, Brazil).

An orbital shaker (Quimis, Brazil) was used to mix the samples and extractants at a rotation velocity of 250 rpm for 30 min. Phase separation was performed by centrifugation (Spectrafuge 6C Compact model, Labnet International, USA).

2.3. Preliminary Sample Preparation. One part of raw beans was cleaned with deionized water and dried in an oven at 60°C until a constant mass was obtained. The raw beans were ground in the cryogenic grinder, with 5 min of freezing followed by three cycles of grinding for 2 min, with 1 min of freezing between cycles.

The cooking procedure was adapted from the proposed procedure by Carrasco-Castilla et al. [6]. Raw grains (*ca.* 20 g, not ground) were left to soak in deionized water (*ca.* 200 mL) at room temperature for 24 h. The soaking water was discarded and the soaked beans were cooked in deionized water (beans : water = 1 : 4, w/v) for 30 min. The cooked beans

TABLE 2: Characteristic parameters of the analytical calibration curve and LOD obtained for proteins determination by Bradford method.

	Linear range (μg mL^{-1})	R	R^2	Sensibility	Analytical blank	LOD (μg g^{-1})	LOQ (μg g^{-1})
					Water	0.53	1.60
Sequential extraction	2–12	0.9935	0.9871	0.049	NaCl	0.55	1.66
					Ethanol	0.56	1.67
					NaOH	0.54	1.62

and deionized water were mixed and dried in an oven at 60°C until a constant mass was reached. The dried mixture was ground using a porcelain mortar and pestle, previously decontaminated with 10% v/v HNO$_3$.

2.4. Sequential Extraction. The sequential extraction procedure has been described by Naozuka and Oliveira [12]. Around 5.0 g of dried raw and cooked grains (all varieties) were used for the solid-liquid sequential extraction with 10 mL of different extractants: a methanol/chloroform mixture (1:2 v/v), acetone (75% v/v) [13], deionized water, 0.5 mol L^{-1} NaCl, 70% (v/v) ethanol, and 0.5 mol L^{-1} NaOH: the methanol/chloroform was used for defatting, the acetone was to remove phenols, and the water, NaCl, ethanol, and NaOH were used to produce the four protein fractions examined in the study. The extractions were carried out using an orbital shaker at 1520 ×g for 30 minutes. The separation of the solid phase was carried out by centrifugation at 4000 rpm for 10 minutes. Proteins and Fe were determined, respectively, by the Bradford method [14] and flame atomic absorption spectrometry (F AAS) of the supernatants, except for the methanol/chloroform mixture and acetone fractions.

2.5. Phaseolin Separation. Dried and ground raw and cooked grains (common and black varieties, 5.0 g) were subjected to sequential extraction with 10 mL of a methanol/chloroform mixture (1:2 v/v) and 10 mL of acetone (75% v/v). The supernatants of these two extractions were discarded. After that, 10 mL of deionized water and 5 mL of NaOH (1.0 M) were added to the resulting solid. To promote protein precipitation, the supernatant was treated with 1 mol L^{-1} HCl until pH 4.5. A mixture of 2 mol L^{-1} NaCl and 1 mol L^{-1} HCl (1:20 v/v) was added to the precipitated proteins at pH = 2.0. Four milliliters of deionized water was added to the supernatant at 4°C for phaseolin precipitation. The precipitation was completed after leaving the mixture in a freezer for 24 h. For the extraction procedure, the mixture was gently stirred in an orbital shaker at room temperature (350 rpm for 30 min) and phase separation was achieved by centrifugation (1520 ×g for 15 min).

The total concentration of phaseolin and Cu and Fe associated with phaseolin was obtained with the Bradford method [14] and graphite furnace atomic absorption spectrometry (GF AAS), respectively.

2.6. Determination of Total Protein. Protein determination was performed by the Bradford method [14]. Spectrophotometer calibration was performed using analytical reference solutions of 4, 6, 8, 10, 12, 16, and 20 μg of ovalbumin in 1.0 mL of Bradford reagent.

Water, NaCl, ethanol, and NaOH extracts were diluted with deionized water four times (raw and cooked common beans and raw jalo beans) or five times (cooked and raw black, fradinho, rosinha, rajado, and bolinha grains, and cooked jalo grains). The NaOH fractions of rosinha (cooked and raw) and fradinho (raw) were subjected to further dilutions, 10 and 20 times, respectively.

After phaseolin separation, the proteins precipitated with the NaCl/HCl mixture (without phaseolin) and the phaseolin fractions were diluted before analysis. The dilutions were 200 times (raw common and black beans) or 10 times (cooked common and black beans) for the precipitated proteins without phaseolin. The phaseolin fraction was diluted 200 times (raw common beans), 100 times (cooked common and raw black beans), or 10 times (cooked black beans).

2.7. Determination of Fe Associated with Albumins, Globulins, Prolamins, and Glutelins in All Phaseolus Bean Varieties by F AAS. Fe determination was carried out by F AAS for the water, NaCl, ethanol, and NaOH extracts. Instrument calibration was performed using analytical solutions with concentrations ranging from 0.5 to 3.0 mg L^{-1} in 0.1% v/v HNO$_3$. All the supernatants were previously diluted with deionized water four times (raw and cooked common beans and raw jalo beans) or five times (cooked jalo beans, and raw and cooked black, fradinho, rosinha rajado, bolinha, and common beans). For the glutelin fraction, further dilutions were necessary for the cooked rosinha beans (10 times) and raw fradinho and rosinha beans (20 times).

2.8. Determination of Cu and Fe Associated with Phaseolin by GF AAS. Cu and Fe determination was carried out by GF AAS, using the instrument parameters shown in Table 1. A 10 μL aliquot of the analytical solutions or samples (supernatants and precipitated proteins) was introduced into the graphite tube without a chemical modifier and subjected to the heating program described in Table 1. The determination of Cu and Fe in the precipitated proteins (without phaseolin) with NaCl/HCl mixture and in the phaseolin fraction was performed after resuspending the solid in 500 μL of 0.5 mol L^{-1} NaOH.

3. Results and Discussion

3.1. Determination of Albumins, Globulins, Prolamins, and Glutelins in All Varieties of Phaseolus Beans. Protein quantification was carried out with the Bradford method [14]. The characteristic parameters of the analytical calibration curves (linear range, correlation coefficient (R^2), and sensibility) and the limits of detection (LOD) and quantification (LOQ) are shown in Table 2. The LOD was calculated using the standard

TABLE 3: Proteins concentrations in different varieties of *Phaseolus* beans (raw and cooked).

Variety	Condition	Concentration ($mg\,g^{-1}$) $\pm$ standard deviation ($n = 3$)			
		Albumins	Globulins	Prolamins	Glutelins
Common	Raw	1.2 ± 0.1	1.2 ± 0.1	1.3 ± 0.1	14 ± 1
	Cooked	0.99 ± 0.01	0.94 ± 0.01	0.98 ± 0.01	0.74 ± 0.01
Black	Raw	1.3 ± 0.1	1.2 ± 0.1	1.2 ± 0.1	2.8 ± 0.2
	Cooked	0.71 ± 0.01	0.76 ± 0.01	0.64 ± 0.01	2.2 ± 0.2
Rosinha	Raw	0.99 ± 0.01	1.3 ± 0.2	0.87 ± 0.01	6.1 ± 0.1
	Cooked	0.92 ± 0.01	0.81 ± 0.01	0.85 ± 0.01	3.1 ± 0.3
Bolinha	Raw	0.77 ± 0.01	0.73 ± 0.01	0.83 ± 0.01	14 ± 1
	Cooked	0.99 ± 0.01	0.94 ± 0.01	0.97 ± 0.01	2.3 ± 0.1
Rajado	Raw	0.95 ± 0.01	0.83 ± 0.01	0.87 ± 0.01	1.9 ± 0.1
	Cooked	1.1 ± 0.1	0.98 ± 0.01	0.85 ± 0.01	0.72 ± 0.01
Fradinho	Raw	1.0 ± 0.1	1.1 ± 0.1	1.04 ± 0.01	6.3 ± 0.1
	Cooked	1.1 ± 0.1	0.98 ± 0.01	0.95 ± 0.01	4.5 ± 0.1
Jalo	Raw	0.72 ± 0.01	1.1 ± 0.1	0.96 ± 0.01	2.4 ± 0.2
	Cooked	1.1 ± 0.1	0.97 ± 0.01	0.89 ± 0.01	1.7 ± 0.1

deviation of 10 measurements of the analytical blank sample ($3 \times \sigma_{blank}$, where σ is the standard deviation) and the LOQ was calculated as $3 \times$ LOD. For the sequential extraction, the values were obtained in $\mu g\,g^{-1}$, considering a sample mass of 5 g and a final volume of 10 mL.

The separation of lipids and polyphenols was carried out using a mixture of methanol/chloroform or pure acetone [15]. In the absence of lipids and polyphenols, it is possible to separate different protein types by a sequential extraction procedure. The extractants water, NaCl, ethanol, and NaOH were applied sequentially to allow the separation of albumins, globulins, prolamins, and glutelins, respectively [12]. The concentrations of these proteins in the different varieties of *Phaseolus* beans, raw and cooked, are shown in Table 3.

Considering the protein concentrations in Table 3 and applying Student's t-test at a 95% confidence limit, it was possible to confirm that cooking changes the distribution of the different proteins. A decrease was found for the majority of the *Phaseolus* bean varieties, particularly for common beans (albumins, globulins, prolamins, and glutelins). Additionally, a negative effect was observed for glutelins in all the varieties. An increase in protein concentration was found for bolinha (albumins, globulins, and prolamins), rajado (albumins, globulins), and jalo (albumins) beans. No effects were observed in the albumins of fradinho beans; the globulins of fradinho and jalo beans; and the prolamins of rosinha and rajado beans.

Globulins and albumins constitute the major proteins fractions in pulse proteins while prolamins and glutelin exist as minor fractions. The raw albumin and glutelin fractions of beans proteins have low digestibility values. For albumins, this fact is related to high number of disulfide bridges. Thermal treatment induces changes in the protein structure that may inactivate antinutritional factors thus increasing the digestibility and biological values of the beans proteins [7].

TABLE 4: Characteristic parameters of the analytical calibration curve, LOD, and LOQ obtained for Fe determination by F AAS.

λ (mm)	Linear range ($mg\,L^{-1}$)	R^2	LOD ($\mu g\,g^{-1}$)	LOQ ($\mu g\,g^{-1}$)
259	0.25–3.0	0.9961	0.18 (albumins)	0.54 (albumins)
			0.54 (globulins)	1.62 (globulins)
			1.64 (prolamins)	4.92 (prolamins)
			1.68 (glutelins)	5.04 (glutelins)

Previous investigations with common beans showed that the total proteins concentration obtained by the masses sum of all extracts revealed that cooking promoted a sensible decrease in the protein content, mainly in the globulins fraction, when compared to the raw beans [8].

Cooking may therefore promote physical and chemical changes in the proteins, particularly in glutelins, causing variations in the solubility of the proteins. This behavior can be related to changes in the association and dissociation properties of the proteins caused by heating. Protein solubility arises from the thermodynamic equilibrium between protein-protein and protein-solvent interactions and is related to balanced hydrophobic and hydrophilic characteristics of the protein molecules [9, 16–18].

3.2. Determination of Fe Associated with Albumins, Globulins, Prolamins, and Glutelins in All Varieties of Phaseolus Beans. The parameters for Fe determination by F AAS have recently been published by our group [19]. Table 4 lists the characteristics of the analytical calibration curves and the LOD and LOQ values.

The Fe concentration in the different proteins is shown in Table 5. It is worth noting that highly diluted supernatant samples were used for F AAS analysis in order to ensure the absence of chemical interferences during Fe quantification.

TABLE 5: Iron concentration in albumins, globulins, prolamins, and glutelins of *Phaseolus* beans.

Variety	Condition	Concentration (μg g^{-1}) $\pm$ standard deviation ($n = 3$)			
		Albumins	Globulins	Prolamins	Glutelins
Common	Raw	20 ± 1	4 ± 2	3 ± 1	9 ± 1
	Cooked	25 ± 2	12 ± 3	13 ± 2	8 ± 1
Black	Raw	21 ± 1	10 ± 1	10 ± 1	20 ± 6
	Cooked	30 ± 2	10 ± 2	17 ± 1	50 ± 5
Rosinha	Raw	27 ± 2	10 ± 2	17 ± 1	40 ± 5
	Cooked	30 ± 1	16 ± 4	14 ± 2	40 ± 4
Bolinha	Raw	30 ± 1	11 ± 1	13 ± 1	22 ± 6
	Cooked	25 ± 1	8 ± 1	24 ± 1	36 ± 4
Rajado	Raw	23 ± 1	8 ± 3	11 ± 3	27 ± 4
	Cooked	24 ± 1	6 ± 2	9 ± 1	6 ± 2
Fradinho	Raw	21 ± 1	7 ± 2	11 ± 3	24 ± 13
	Cooked	23 ± 2	7 ± 1	10 ± 3	11 ± 4
Jalo	Raw	22 ± 1	4 ± 1	9 ± 2	12 ± 5
	Cooked	26 ± 1	9 ± 2	15 ± 2	34 ± 9

TABLE 6: Concentrations of precipitated proteins without phaseolin and phaseolin in the common and black beans.

	Concentration (mg g^{-1}) $\pm$ standard deviation ($n = 3$)			
	Common		Black	
	Raw	Cooked	Raw	Cooked
Precipitated proteins without phaseolin	5.8 ± 0.1	2.64 ± 0.01	3.31 ± 0.04	2.7 ± 0.1
Phaseolin	0.12 ± 0.01	0.013 ± 0.001	0.10 ± 0.01	0.009 ± 0.001

There was Fe associated with all fractions, both raw and cooked beans. In raw grains, Fe is associated with albumins, prolamins, and glutelins, whereas in cooked grains Fe is linked to albumins and glutelins in most varieties. In addition, variations with respect to the distribution of Fe for the different varieties of beans were observed. The main amino acids in albumins, globulins, prolamins, and glutelins, such as methionine, cysteine, glutamic acid, arginine, aspartic acid, and lysine, are rich in sulfur and charged groups [8, 20–22]. These amino acids have a high affinity toward transition metal ions [23].

Comparing the concentrations in Table 5 and applying Student's t-test at a 95% confidence limit, it is possible to conclude that cooking does not cause significant changes in Fe associated with albumins and prolamins for the rosinha, rajado, and fradinho bean varieties. Globulin levels do not exhibit any significant changes after heating for the rajado, rosinha, black, and fradinho beans. Cooking results in a variation of the Fe concentration when associated with glutelins for the black, bolinha, rajado, and jalo beans, since it is possible to observe an increase (180%) and a decrease (78%) for the jalo and rajado beans, respectively.

For common beans, studies showed that, in raw grains, Fe was associated with albumins, globulins, and glutelins and, after the cooking process, Fe was in the albumins and globulins fractions. The main amino acids constituents of albumins, globulins, prolamins, and glutelins are rich in sulfur and charged groups that present high affinity by metal ions [8]. Heating may result in the denaturation of the proteins. In this scenario, interactions between proteins and essential elements can be established or lost [24]. Softening of the food matrix allows certain elements associated with proteins to be released, facilitating their absorption by the human body [19]. Furthermore, during cooking phytate loses phosphate linkages transforming from inositol hexaphosphate to penta-, tetra- or triphosphate and decreases the inhibitory capacity [25]. Therefore, high temperatures modify the composition of antinutrients (e.g., phytate and dietetic fibers) that inhibit the absorption of essential elements [9].

3.3. Effect of Cooking on Phaseolin Concentration in Common and Black Beans.
The effect of cooking on the phaseolin concentration was studied in the two *Phaseolus* bean varieties most consumed in Brazil: the common and black varieties. With the proposed method, we obtained the characteristic parameters of the analytical calibration curve (linear range (2–20 μg mL^{-1}), correlation coefficient ($R^2 = 0.9917$), and sensibility (0.0106)), as well as the LOD (0.19 μg g^{-1}) and LOQ (0.58 μg g^{-1}) values.

The concentration values of the precipitated proteins without phaseolin and that of the phaseolin fraction are shown in Table 6. Cooking affects negatively the precipitated

TABLE 7: Characteristic parameters of the analytical calibration curve, LOD, and LOQ obtained for Cu and Fe determination by GF AAS.

	Linear range (mg L^{-1})	R^2	Sensibility	LOD (ng g^{-1})	LOQ (ng g^{-1})
Cu	10–80	0.9881	0.0043	0.47	1.41
Fe	10–80	0.9956	0.0055	0.05	0.16

TABLE 8: Cu and Fe concentration associated with precipitated proteins without phaseolin and phaseolin.

	Concentration (μg g^{-1}) ± standard deviation ($n = 3$)			
	Fe		Cu	
	Raw	Cooked	Raw	Cooked
Common beans				
Precipitated proteins without phaseolin	8.9 ± 0.1	17 ± 1	4.2 ± 0.9	2.6 ± 0.4
Phaseolin	3.2 ± 0.1	2.4 ± 0.1	0.72 ± 0.04	0.86 ± 0.03
Black beans				
Precipitated proteins without phaseolin	5.6 ± 0.3	14 ± 1	4.1 ± 0.5	1.8 ± 0.1
Phaseolin	2.9 ± 0.1	2.3 ± 0.1	0.84 ± 0.03	1.2 ± 0.1

proteins (without phaseolin) and the phaseolin fraction for both bean varieties. Regarding the precipitated proteins without phaseolin, the greatest effect is observed for common beans, compared to black beans. The reduction of phaseolin concentration, approximately 90%, is significant for both common and black beans.

Studies have shown that heating promotes a decrease in the total protein concentration, particularly for globulins such as phaseolin [8]. Heating can alter the native conformation of proteins due to disturbances in the noncovalent interactions that stabilize the protein structure, modifying the association and dissociation interactions that occur between amino acids containing opposite charges and/or between protein subunits, altering their isoelectric point and, consequently, their solubility [5, 8, 16]. In the common beans, the secondary structure of the phaseolin is preserved, while its tertiary and quaternary structure suffer alteration, increasing the hydrophilic surfaces that indicate a breakdown of the phaseolin subunit interactions and lead to a higher degree of hydrolysis [2].

3.4. *Effect of Cooking on the Concentrations of Cu and Fe Associated with Phaseolin in Common and Black Beans.* The determination of Cu and Fe associated with phaseolin was carried out by GF AAS. A detector with high sensibility, such as that in GF AAS, is necessary in order to determine Cu and Fe in extracts after the fractionation steps used to isolate phaseolin. The parameters for the determination of Cu and Fe by GF AAS are listed in Table 7.

The concentration of Cu and Fe associated with precipitated proteins (without phaseolin) and that of the phaseolin fraction are shown in Table 8. For common beans, cooking promotes an increase of 91% in the concentration of Fe associated with precipitated proteins (without phaseolin) and a decrease of 24% for Fe associated with phaseolin. The

reverse effect is observed for Cu, an increase of 20% in phaseolin after cooking.

For black beans, similar effects were observed to those with common beans. Cooking promoted an increase of 146% in the concentration of Fe associated with precipitated proteins (without phaseolin) and a reduction of 21% of Fe associated with phaseolin. The heating process reduced the concentration of Cu associated with precipitated proteins by 56% and increased it by 37% in the phaseolin fraction.

The presence of Cu and Fe associated with phaseolin may be a result of the existence of aromatic moieties and acid and basic amino acids, such as arginine, histidine, lysine, aspartate, glutamate, and tyrosine, which are responsible for complexation reactions with essential elements. In previous studies with the Jamapa variety, the chelating activity of phaseolin and other low molecular weight proteins toward Cu was confirmed for proteins of low molecular weight. Additionally, the authors showed that the carboxylic groups have a high affinity toward Cu through electrostatic interactions. Both elements of interest have a similar mechanism of interaction with phaseolin; however, the association of Fe with this globulin is weaker than with Cu since heating decreases the Fe concentration in the phaseolin fraction. It is possible to infer that this effect results from a higher coordination number for Fe chelation than for Cu [6].

Moreover, variations in the Cu and Fe concentration can be promoted through interactions of these elements with the matrix components. Fe bioavailability is modified by the presence of tannins, antinutrients that form insoluble complexes when associated with Fe. The tannins concentration varies with the color of the bean shells, being higher in darker seeds. Therefore, the formation of insoluble complexes with Fe is expected to be greater in black beans, resulting in less Fe available for complexation with the proteins. For the two bean varieties, the increase in Fe concentration in the precipitated proteins (without phaseolin) after cooking arises as the heating process decreases the action of the antinutrient compounds that form insoluble complexes with Fe [8, 17, 25]. For Cu, the reduction in the Cu concentration in the precipitated proteins (without phaseolin) can be explained by the formation of Maillard compounds [12, 26, 27]. The processing of foods rich in protein and carbohydrates promotes the development of Maillard reaction, where the Maillard reaction products behave as anionic polymers, forming stable complexes with metal cations such as copper [28].

4. Conclusion

Cooking *Phaseolus* bean varieties changes their albumin, globulin (including phaseolin), prolamin, and glutelin distribution, the solubility decreases for all four fractions, and the extent to which solubility was affected varied with the type of protein. Additionally, Fe and Cu metalloproteins undergo changes with heating. Chemical reactions can justify these variations. Studies on the effects of cooking are essential, since cooking beans prior to consumption is necessary; the results of such studies would provide nutritional information that will add value to this food, which is widely consumed by the world's population. Finally, it is important to point out that chemical speciation studies are imperative in food sciences to demonstrate the effect of cooking on the distribution of chemical species that are essential for humans.

Competing Interests

No potential conflict of interests was reported by the authors.

Acknowledgments

Aline Pereira de Oliveira acknowledges the Fundação de Amparo à Pesquisa do Estado de São Paulo/FAPESP (2015/01128-6) for the fellowship provided. Juliana Naozuka acknowledges the FAPESP (2015/15510-0 and 2012/11517-1) and Conselho Nacional de Desenvolvimento Científico e Tecnológico (CNPq, 475282/2013-2) for financial support. Geyssa F. Andrade thanks Professor Angerson Nogueira do Nascimento (UNIFESP/Diadema) for initial support in the experiments.

References

[1] P. F. De Lima, C. A. Colombo, A. F. Chiorato, L. F. Yamaguchi, M. J. Kato, and S. A. M. Carbonell, "Occurrence of isoflavonoids in brazilian common bean germplasm (*Phaseolus vulgaris* L.)," *Journal of Agricultural and Food Chemistry*, vol. 62, no. 40, pp. 9699–9704, 2014.

[2] C. A. Montoya, J.-P. Lallès, S. Beebe, and P. Leterme, "Phaseolin diversity as a possible strategy to improve the nutritional value of common beans (*Phaseolus vulgaris*)," *Food Research International*, vol. 43, no. 2, pp. 443–449, 2010.

[3] L. Mojica and E. G. de Mejía, "Characterization and comparison of protein and peptide profiles and their biological activities of improved common bean cultivars (*Phaseolus vulgaris* L.) from Mexico and Brazil," *Plant Foods for Human Nutrition*, vol. 70, no. 2, pp. 105–112, 2015.

[4] P. Brigide, S. G. Canniatt-Brazaca, and M. O. Silva, "Nutritional characteristics of biofortified common beans," *Food Science and Technology (Campinas)*, vol. 34, no. 3, pp. 493–500, 2014.

[5] P. Vidigal Filho, C. Gonçalves-Vidigal, R. Hammerschmidt, and W. W. Kirk, "Characterization and content of total soluble protein and amino acids of traditional common bean (*Phaseolus vulgaris* L.) cultivars collected in Paraná state, Brazil," *Journal of Food Agriculture and Environment*, vol. 9, no. 3-4, pp. 143–147, 2011.

[6] J. Carrasco-Castilla, A. J. Hernández-Álvarez, C. Jiménez-Martínez et al., "Antioxidant and metal chelating activities of *Phaseolus vulgaris* L. var. Jamapa protein isolates, phaseolin and lectin hydrolysates," *Food Chemistry*, vol. 131, no. 4, pp. 1157–1164, 2012.

[7] I. Hayat, A. Ahmad, T. Masud, A. Ahmed, and S. Bashir, "Nutritional and health perspectives of beans (*Phaseolus vulgaris* L.): an overview," *Critical Reviews in Food Science and Nutrition*, vol. 54, no. 5, pp. 580–592, 2014.

[8] J. Naozuka and P. V. Oliveirab, "Cooking effects on iron and proteins content of beans (*Phaseolus vulgaris* L.) by GF AAS and MALDI-TOF MS," *Journal of the Brazilian Chemical Society*, vol. 23, no. 1, pp. 156–162, 2012.

[9] S.-W. Yin, K.-L. huang, C.-H. Tang, X.-Q. Yang, Q.-B. Wen, and J.-R. Qi, "Surface charge and conformational properties of phaseolin, the major globulin in red kidney bean (*Phaseolus vulgaris* L): Effect of pH," *International Journal of Food Science & Technology*, vol. 46, no. 8, pp. 1628–1635, 2011.

[10] C. A. Montoya, P. Leterme, N. F. Victoria et al., "Susceptibility of phaseolin to in vitro proteolysis is highly variable across common bean varieties (*Phaseolus vulgaris*)," *Journal of Agricultural and Food Chemistry*, vol. 56, no. 6, pp. 2183–2191, 2008.

[11] A. P. Oliveira and J. Naozuka, "Chemical speciation of iron in different varieties of beans (*Phaseolus vulgaris* L.): cooking effects," *Journal of the Brazilian Chemical Society*, vol. 26, no. 10, pp. 2144–2149, 2015.

[12] J. Naozuka and P. V. Oliveira, "Cu, Fe, Mn and Zn distribution in protein fractions of Brazil-nut, cupuassu seed and coconut pulp by solid-liquid extraction and electrothermal atomic absorption spectrometry," *Journal of the Brazilian Chemical Society*, vol. 18, no. 8, pp. 1547–1553, 2007.

[13] Q. D. Do, A. E. Angkawijaya, P. L. Tran-Nguyen et al., "Effect of extraction solvent on total phenol content, total flavonoid content, and antioxidant activity of *Limnophila aromatica*," *Journal of Food and Drug Analysis*, vol. 22, no. 3, pp. 296–302, 2014.

[14] M. M. Bradford, "A rapid and sensitive method for the quantitation of microgram quantities of protein utilizing the principle of protein-dye binding," *Analytical Biochemistry*, vol. 72, no. 1-2, pp. 248–254, 1976.

[15] A. N. Nascimento, J. Naozuka, and P. V. Oliveira, "In vitro evaluation of Cu and Fe bioavailability in cashew nuts by off-line coupled SEC–UV and SIMAAS," *Microchemical Journal*, vol. 96, no. 1, pp. 58–63, 2010.

[16] M. Carbonaro, P. Vecchini, and E. Carnovale, "Protein solubility of raw and cooked beans (*Phaseolus vulgaris*): Role of the basic residues," *Journal of Agricultural and Food Chemistry*, vol. 41, no. 8, pp. 1169–1175, 1993.

[17] J. M. Harnly, M. A. Pastor-Corrales, and D. L. Luthria, "Variance in the chemical composition of dry beans determined from UV spectral fingerprints," *Journal of Agricultural and Food Chemistry*, vol. 57, no. 19, pp. 8705–8710, 2009.

[18] A. S. T. Ferreira, J. Naozuka, G. A. R. Kelmer, and P. V. Oliveira, "Effects of the domestic cooking on elemental chemical composition of beans species (*Phaseolus vulgaris* L.)," *Journal of Food Processing*, vol. 2014, Article ID 972508, 6 pages, 2014.

[19] J. Naozuka, S. R. Marana, and P. V. Oliveira, "Water-soluble Cu, Fe, Mn and Zn species in nuts and seeds," *Journal of Food Composition and Analysis*, vol. 23, no. 1, pp. 78–85, 2010.

[20] R. N. Garcia, R. V. Arocena, A. C. Laurena, and E. M. Tecson-Mendoza, "11S and 7S globulins of coconut (*Cocos nucifera* L.): purification and characterization," *Journal of Agricultural and Food Chemistry*, vol. 53, no. 5, pp. 1734–1739, 2005.

[21] F. J. Moreno, J. A. Jenkins, F. A. Mellon et al., "Mass spectrometry and structural characterization of 2S albumin isoforms from Brazil nuts (*Bertholletia excelsa*)," *Biochimica et Biophysica Acta—Proteins and Proteomics*, vol. 1698, no. 2, pp. 175–186, 2004.

[22] S. S. M. Sun, F. W. Leung, and J. C. Tomic, "Brazil nut (Bertholletia excelsa HBK) proteins: fractionation, composition, and identification of a sulfur-rich protein," *Journal of Agricultural and Food Chemistry*, vol. 35, no. 2, pp. 232–235, 1987.

[23] J. S. Garcia, C. S. De Magalhães, and M. A. Z. Arruda, "Trends in metal-binding and metalloprotein analysis," *Talanta*, vol. 69, no. 1, pp. 1–15, 2006.

[24] L. G. Ranilla, M. I. Genovese, and F. M. Lajolo, "Polyphenols and antioxidant capacity of seed coat and cotyledon from Brazilian and Peruvian bean cultivars (*Phaseolus vulgaris* L.)," *Journal of Agricultural and Food Chemistry*, vol. 55, no. 1, pp. 90–98, 2007.

[25] E. Helbig, M. R. Buchweitz, and D. P. Gigante, "Análise dos teores de ácidos cianídrico e fítico em suplemento alimentar: multimistura," *Revista de Nutrição*, vol. 21, no. 3, pp. 323–328, 2008.

[26] C. Barbana and J. I. Boye, "In vitro protein digestibility and physico-chemical properties of flours and protein concentrates from two varieties of lentil (*Lens culinaris*)," *Food & Function*, vol. 4, no. 2, pp. 310–321, 2013.

[27] J. Shibao and D. H. Bastos, "Produtos da reação de Maillard em alimentos: implicações para a saúde," *Revista de Nutrição*, vol. 24, no. 6, pp. 895–904, 2011.

[28] M. Friedman, "Nutritional and toxicological consequences of the food processing," in *Advances in Experimental Medicine and Biology*, vol. 289, pp. 488–491, Springer, Dordrecht, Netherlands, 1991.

6

Effect of Buttermilk on the Physicochemical, Rheological, and Sensory Qualities of Pan and Pita Bread

Amani H. Al-Jahani

Nutrition and Food Science Department, College of Home Economics, Princess Nourah Bint Abdulrahman University, Riyadh, Saudi Arabia

Correspondence should be addressed to Amani H. Al-Jahani; ahaljahani@pnu.edu.sa

Academic Editor: Salam A. Ibrahim

The aim of this study was to evaluate the influence of buttermilk on the physicochemical and sensory attributes of pan and pita breads. Different amounts of buttermilk (30, 60, and 100% of added water) were mixed with other ingredients of pan and pita bread formulations. The doughs and bread were analyzed for rheological, physicochemical, and sensory qualities. The results demonstrated that incorporation of different concentrations of buttermilk in bread formulations progressively enhanced water absorption capacity, dough development time, gelatinization temperature, and peak viscosity, whereas it reduced the dough stability and temperature at peak viscosity. Supplementation of wheat flour with 30% buttermilk significantly ($P \leq 0.05$) enhanced the physical properties of pan bread compared to nonsupplemented control. Incorporation of different percentages of buttermilk in bread formulation concomitantly ($P \leq 0.05$) increased protein, oil, and ash contents and it reduced the carbohydrate contents of both types of bread. Incorporation of 60 and 100% of buttermilk in bread formula showed low scores of all sensory attributes compared to control and 30% buttermilk containing pan and pita bread. In conclusion, supplementation of bread formulas with 30% buttermilk is recommended for improving the nutritional and sensorial qualities of pan and pita bread.

1. Introduction

Wheat is an important and most widely cultivated crop in the world, and wheat-based products provide 20% of all calories consumed by people around the globe. The most important application of wheat is in bread making processes where wheat flour represents the main ingredient. Throughout history, bread is one of the oldest and most staple foods prepared and consumed by humans throughout the world [1]. In recent decades, the global production and consumption of bread have considerably increased due to population explosion and changes in lifestyle and eating habits [2]. In Middle East countries, particularly in Saudi Arabia, the major types of bread prepared and consumed are pan (Samouli) and pita (Mafrood) bread and these bread types are mostly made from white flour [3]. However, during wheat milling processes most of the important nutrients are removed resulting in flour with low protein content and quality due to the deficiency of wheat proteins lacking essential amino acids such as lysine and threonine in wheat protein [4]. In addition, removing

the wheat bran and germ also may result in losses of many nutritious and health promoting compounds such as dietary fiber, minerals, vitamins, and various antioxidant compounds [5]. To overcome such limitations, supplementation of wheat flour with protein rich and nutritious sources has become an increasingly important and challenging research field in recent years. In this regard, several research reports on supplementation or fortification of bread wheat flour with grains and legumes flour have been published [6–12].

Buttermilk, a low-fat milky liquid leftover after the churning of cream, is one of the most important functional dairy products that have excellent health and disease curing potentials and it is now receiving high interest from consumers all over the world. In addition, buttermilk is also considered as an excellent source of nutritional elements such as minerals (potassium, phosphorus, and calcium), vitamin B12, riboflavin, enzymes, and protein [13]. Moreover, buttermilk has a fresh and piquant taste and has applications in a wide variety of foods such as refreshing drinks, low fat yogurt, cheese, ice cream, nutritious bakery products,

TABLE 1: Formulation of pan and pita bread with different concentration of buttermilk.

Ingredients	% of wheat flour	% of added buttermilk			
		0%	30%	60%	100%
Pan bread					
Wheat flour (g)	100	1000	1000	1000	1000
*Water (mL)	60	600	434	256	—
Buttermilk (mL)	—	—	186	384	675
Yeast (g)	3	30	30	30	30
Sugar (g)	5	50	50	50	50
Salt (g)	2	20	20	20	20
Margarine (g)	3	30	30	30	30
Improver (g)	0.01	0.1	0.1	0.1	0.1
Pita bread					
Wheat flour (g)	100	1000	1000	1000	1000
*Water (mL)	60	600	434	256	—
Buttermilk (mL)	—	—	186	384	675
Yeast (g)	3	30	30	30	30
Sugar (g)	5	50	50	50	50
Salt (g)	1	10	10	10	10
Improver (g)	0.01	0.1	0.1	0.1	0.1

*Added water calculated based on the farinograph results of formulated dough.

and confectionaries [14]. Furthermore, buttermilk has several therapeutic potentials such as cholesterol reduction, blood pressure reduction, antiviral effects, and anticancer effects [15–17]. In baking industry, buttermilk could be used to enhance the rheological, nutritional, and organoleptic properties of bread [18, 19]. Due to its high nutritional and health promoting potentials, utilization of buttermilk as a supplement in bread formulations received significant attention in recent years [18, 19]. Therefore, the main aims of the present study are to utilize buttermilk in pan and pita bread formulations and evaluate the influence of buttermilk on the physicochemical and sensory attributes of pan and pita breads.

2. Materials and Methods

2.1. Materials. Hard wheat flour (75% extraction) was obtained from Al-Andalus milling factory, Riyadh, Saudi Arabia. Margarine of hydrated soybean and cotton seed oils was purchased from Goody Middle East Company, Dubai, United Arab Emirates. BakeMate bread improver (containing α-amylase) was brought from Ahmad Abid trading company, Riyadh, Saudi Arabia. Sugar and salt were obtained from local markets in Riyadh, Saudi Arabia. Fresh buttermilk was obtained from Almarai Company, Saudi Arabia. All chemicals used are of analytical grade.

2.2. Preparation of Pan Bread (Loaf). The dough was prepared following the standard straight dough method 10-10A [20] with some modifications. Briefly, the ingredients (Table 1) were mixed on the basis of 1000 g wheat flour (14% moisture content) to the levels of 3% instant yeast, 3% margarine, 2% salt, 5% sugar, and 0.01% improver.

Then, different amounts water and buttermilk (30, 60, and 100% of the added water) were added based on the flour optimum absorption as determined by farinograph. After that, all contents were thoroughly mixed (Tyrone, model TR 202, UK) to the optimum dough development time as determined by farinograph. Then, the dough was formed in ball shape and kept in a fermentation tank for 30 min at 32°C and 85% relative humidity. Thereafter, the dough was freed from gases and divided into three portions of 550 g each and then fermented again for 20 min under the above conditions. Then, the fermented dough was passed through bread forming machine and put into baking pans (22 × 6 × 7 cm) and kept again in the fermentation tank for final fermentation for 30 min at the same condition. After fermentation, the dough was baked in the electric rotary oven (National Co. Ltd., Kyoto, Japan) for 20 min at 225°C. After cooling for 30 min at room temperature, the bread volume was estimated by the seed displacement method. The volume of seeds displaced by the bread was considered as the bread volume. The specific volume of the bread was calculated according to the AACC method [20] by dividing volume (CC) by weight (g).

2.3. Preparation of Pita Bread (Flat). The method of preparation of Arabian bread was used to prepare pita bread as described elsewhere [21]. Briefly, the ingredients (Table 1) were mixed on the basis of 1000 g wheat flour (14% moisture content) to the levels of 3% instant yeast, 5% sugar, 1% salt, and 0.01% improver. Then, different amounts of water and buttermilk (30, 60, and 100% of the added water) were added based on the flour optimum absorption as determined by farinograph. The contents were mixed to form dough and fermented as described for pan bread. Thereafter, the dough

TABLE 2: Farinograph readings of wheat dough fortified with different concentrations of buttermilk (30, 60, and 100% of added water).

Dough samples	% of water absorption corrected to 14%	Arrival time (min)	Dough development time (min)	Dough stability (min)	Departure time (min)	Degree of softening (BU)
Control (Wheat flour + 0% butter milk)	60.0	1.5	2.3	9.7	9.7	23.0
Wheat flour + 30% buttermilk	62.0	1.5	2.9	8.3	9.8	22.0
Wheat flour + 60% buttermilk	64.0	1.5	2.7	8.3	9.8	24.0
Wheat flour + 100% buttermilk	67.0	1.5	2.2	8.1	9.6	24.0

Mean values of triplicates, SD $\leq$ ±5%.

was divided into portions (150 g each) and formed into balls by hand and then fermented again for 20 min under the above-mentioned conditions. Then the flat bread was created by using wooden cylinder having the diameter of 20 cm and thickness of 6 mm and left for final fermentation for 30 min at the same conditions. The bread was baked at 350°C for 2 min and then left to cool for 30 min before analysis.

2.4. Approximate Composition. The approximate composition of bread samples was determined following the official standard method [22]. Moisture content was measured using oven drying method and weight measurements before and after drying (AOAC 935.29). Protein was determined by Kjeldahl method (AOAC 988.05). Oil and ash were estimated by Soxhlet method (AOAC 963.15) and drying methods (AOAC 942.05), respectively. Carbohydrate was calculated by differences.

2.5. Determination of Dough Rheology. The rheological properties of wheat dough supplemented with different concentrations of buttermilk (30, 60, and 100% of added water) followed the standard methods of American Association of Cereal Chemists [20]. Farinogragh and amylograph measurements were carried out according to AACC 54-21 and 54-10 methods, respectively, using Barbender instruments (C.W. Brabender Instrument Inc., South Hackensack, NJ, USA).

2.6. Sensory Attributes. Sensory analysis of bread samples was performed immediately after cooking by 10 semitrained panelists (male, age range 20–35 years old). The panelists were initially trained for sensory evaluation, and the assessment was carried out in three independent sessions. The sensory quality of pan bread was assessed using a 10-point hedonic scale (10 = like extremely, 1 = dislike extremely) and panelists were asked to assess both external and internal properties of the bread such as aroma, taste, crumb texture, crumb color, crumb cell uniformity, and general acceptability, whereas for pita bread a 5-point hedonic scale (5 = like, 1 = dislike) was used for the external (color of the loaf, the shape and degree of consistency, and the degree of cracking and breaking) and internal (color of the pulp, texture of the pulp, the smell and the taste, and the degree of symmetry of the top layer with the

bottom layer) features. All samples were coded with three-digit random numbers and served randomly to the assessors. Sensory analysis was performed in three sessions, and the mean values of the scores of 10 panelists for each sample and session were calculated and used in data analysis.

2.7. Statistical Analysis. Triplicate experiments and measurements were carried out and the data collected were subjected to statistical analysis using SAS program (SAS 8.0 software, SAS Institute, Inc., Cary, NC, USA). Analysis of variance (ANOVA) and Duncan's Multiple Range Test were performed to analyze the effect of treatments on the chemical, rheological, and sensory characteristics of the pan and pita bread. Data were presented as mean, and standard deviation (SD) and statistical significance were accepted at the probability of $P \leq 0.05$.

3. Results and Discussion

3.1. Dough Rheology. The dough rheological properties of wheat flour supplemented with different concentrations of buttermilk were analyzed by farinograph and amylograph. The farinograph results indicated that supplementation of wheat flour with buttermilk significantly affected the water absorption, dough development time, dough stability, and degree of softening (Table 2). Increasing the concentrations of buttermilk in the dough progressively ($P \leq 0.05$) increased the water absorption from 60% in control to 67% in wheat flour supplemented with 100% buttermilk. This could be due to the increase of the protein solubility and content of the dough following the addition of buttermilk whose protein is characterized by its high solubility, hydrophobicity, and absorption capacity. Furthermore, added buttermilk could result in a structural modification in the dough which may allow absorption of more water due to hydrogen bonding. High water absorption capacity of dough represents consistency which is one of the appealing characteristics in bread making. The quantity of added water is considered to be very important for the distribution of the dough materials, their hydration, and the gluten protein network development. Similar to our observations, Hassan et al. [18] reported that replacing water with different concentrations of buttermilk in

TABLE 3: Amylograph readings of wheat dough fortified with different concentrations of buttermilk (30, 60, and 100%).

Dough samples	Gelatinization temperature (°C)	Peak viscosity (BU)	Temperature at peak viscosity (°C)
Control (wheat flour + 0% buttermilk)	60.7	671.0	88.9
Wheat flour + 30% buttermilk	62.0	781.0	86.9
Wheat flour + 60% buttermilk	62.9	848.0	83.9
Wheat flour + 100% buttermilk	63.0	987.0	82.1

Mean values of triplicates, SD ≤ ±5%.

bread formulation substantially increased water absorption of the dough in a concentration dependent manner. In addition, several research reports have demonstrated that supplementation of wheat flour with vegetable and legumes flour, dairy products, or protein isolates significantly increased the water absorption capacity [7, 9, 23, 24]. By contrast, Madenci and Bilgiçli [19] reported that addition of dairy by-products (whey protein concentrate and buttermilk powder) to wheat flour reduced the water absorption of the formed dough. Our results indicated that dough development time was also increased ($P \leq 0.05$) with an increase in buttermilk concentration to 30 and 60% and then decreased again at 100% buttermilk. The enhancement of dough development time upon addition of buttermilk could be due to the differences in the physicochemical properties of the constituents of the buttermilk and those of the wheat flour. Similarly, Hassan et al. [18] and Bilgin et al. [25] stated that the development time of wheat dough was significantly increased upon replacing water with different concentrations of fermented skimmed milk, acid whey, and buttermilk. In addition, Madenci and Bilgiçli [19] reported that incorporation of dairy by-products in wheat dough significantly enhanced the dough development time. Moreover, Mohammed et al. [9] indicated that addition of chickpea flour to wheat flour increased the dough development time. Our findings also showed that incorporation of buttermilk in wheat dough decreased the dough stability compared to control and the reduction was concomitant with an increase in the concentration of buttermilk in the dough. The reduction could be due to the fact that added buttermilk constituents could disrupt the wheat gluten-starch network, competing with wheat flour proteins for water and likely the proteolytic activity in the buttermilk hydrolyze wheat gluten and then decreased its stability. Similarly, Hassan et al. [18] reported a reduction in dough stability at higher supplementation levels of acid whey, buttermilk, and skimmed milk powder in pan bread formulations. In addition, a similar decrease in dough stability was also observed by Sabanis and Tzia [11], Anton et al. [26], Gadallah et al. [27], and Pasha et al. [28] as the percentage level of legume flour in the blend increased. Our results also revealed that the effects of buttermilk fortification on the departure time and degree of softening were minor. Overall, the farinograph analysis demonstrated that incorporation of different concentration of buttermilk in bread formulations positively affected the water absorption capacity and development time of dough, whereas it showed an adverse impact on the dough stability.

The amylograph results (Table 3) showed that fortification of wheat flour with buttermilk significantly ($P \leq 0.05$) affected the gelatinization temperature, peak viscosity, and temperature at peak viscosity of fortified dough compared to control samples. Gelatinization temperature and peak viscosity concomitantly ($P \leq 0.05$) increased with increase in the concentration of buttermilk in the dough, while the temperature at peak viscosity showed concomitant reduction as the buttermilk level increased. The increase in the peak viscosity and gelatinization temperature could be attributed to the added lactose sugar from buttermilk. Similarly, it has been reported that addition of mushroom powder to wheat flour substantially increased peak viscosity from 820 to 1700 BU [29]. Furthermore, increases in peak viscosity and gelatinization temperature were observed in composite flours of wheat, cereals, legumes, or sago flours [30–33]. By contrast, other reports have indicated that addition of legumes protein isolates and milk proteins reduced the peak viscosity of the dough [34, 35]. The difference could be attributed to the variation in the added materials that in the latter is protein isolates which are devoid of sugars or starches whereas in the former the whole materials are added which might contains sugars and starches that increase the viscosity and gelatinization temperature.

3.2. Physical Properties of Pan and Pita Bread. Wheat flour with or without buttermilk supplementation was used for the preparation of two types of bread frequently consumed in Kingdom of Saudi Arabia. The results of the physical properties of pan bread are shown in Table 4 and Figure 1(a). The results indicated that supplementation of wheat flour with buttermilk significantly ($P \leq 0.05$) enhanced the weight and volume of pan bread compared to nonsupplemented control. Strikingly, loaf weight and volume enhancement during baking of pan bread containing buttermilk is a desirable quality attribute as consumers are often attracted to bread with high weight and volume believing that it has more substance for the same price. The loaf weight concomitantly ($P \leq 0.05$) increased with increase in the concentration of buttermilk reaching the maximum at 100% buttermilk supplementation. Increase in bread weight may be due to increased water absorption capacity of buttermilk containing dough (Table 2). The highest ($P \leq 0.05$) loaf volume (2883.33± 28.86 cm) and loaf specific volume (5.92 ± 0.07 cm^2/g) were observed at 30% buttermilk supplementation followed by control (0% buttermilk), 60% and 100% buttermilk. Although

TABLE 4: Physical characteristics of pan breads made using wheat flour fortified with different concentrations (0, 30, 60, and 100% of added water) of buttermilk.

Physical properties	Fortification rate			
	0%	30%	60%	100%
Loaf weight (g)	483.67 ± 1.08^c	487.67 ± 1.51^b	496.00 ± 2.00^a	498.33 ± 2.30^a
Loaf volume (cm)	2650.00 ± 12.24^b	2883.33 ± 28.86^a	2633.33 ± 17.22^b	2566.67 ± 10.01^c
Loaf specific volume (cm^2/g)	5.48 ± 0.01^b	5.91 ± 0.07^a	5.31 ± 0.06^c	5.15 ± 0.07^d

[a–d] Mean values of triplicate samples ± SD. Means not sharing a common superscript (s) in a row are significantly different at $P \leq 0.05$ as assessed by Duncan's Multiple Range Test.

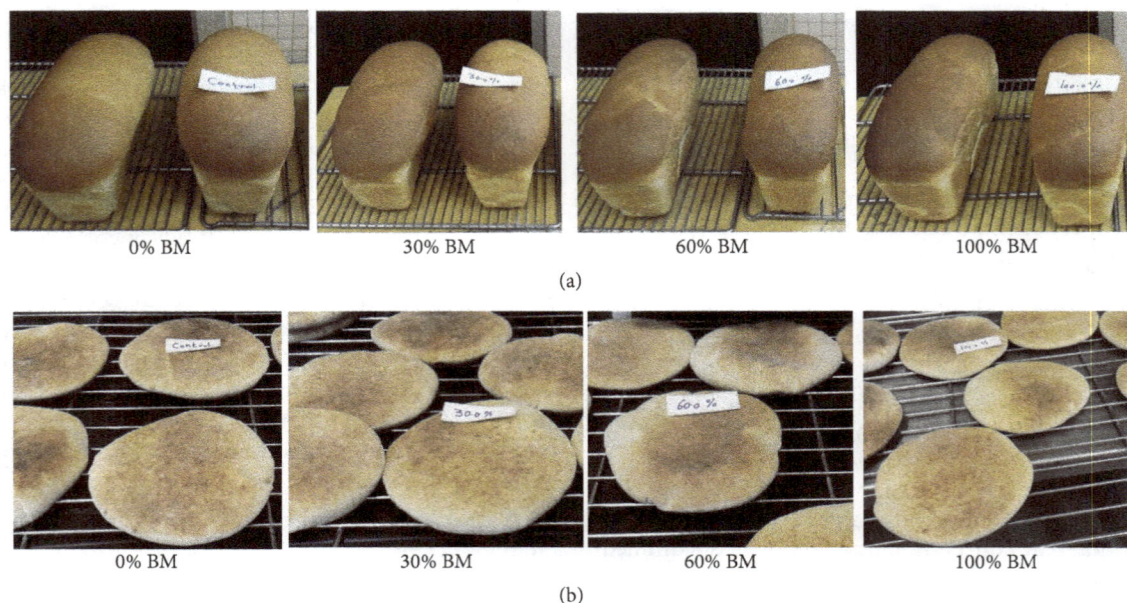

FIGURE 1: Physical appearance of pan (a) and pita (b) breads containing different levels of buttermilk (BM). Control; 0% BM, 30% butter milk; 30% BM, 60% buttermilk; 60% BM, and 100% buttermilk; 100% BM.

they reduced by buttermilk supplementation at high percentages (60 and 100% buttermilk), the values of the loaf bread specific volume are still within the range 3.5–6.0 cm^2/g that characterize the regular bread as identified by grain products research institutes [36]. The decrease ($P \leq 0.05$) in loaf volume and specific volume at 60 and 100% buttermilk of the bread may be attributed to the reduction in the wheat structure forming proteins and a low ability of the dough to entrap air. In addition, the higher resistance of dough observed during dough handling and preparation might affect the gas retention in the dough and bread during baking and hence reduce the volume of the loaves. Moktan and Ojha [12] reported that increase in the percentage of germinated horse gram decreased the loaf volume and specific volume of bread but increased the loaf weight. Gani et al. [34] stated that addition of whey proteins, whey protein hydrolysates, casein, and casein hydrolysates significantly reduced the loaf specific volume.

On the other hand, the physical properties of pita bread also improved following the supplementation of wheat flour with different percentages of buttermilk (Figure 1(b)).

Madenci and Bilgiçli [19] studied the effects of supplementation of wheat flour with whey protein concentrate powder and buttermilk powder on the quality of flat bread. They found that incorporation of buttermilk at 8% decreased the thickness and increased diameter and spread ratio of flat bread compared to controls. In addition, Madenci et al. [37] reported a decrement in thickness of lavash and an increment in diameter/spread of lavash with the usage of whey protein concentrate and buttermilk in the formulation. Overall, our results demonstrated that incorporation of buttermilk at 30% and above in bread mixture could improve the physical properties of the bread and thus it is well recommended to incorporate buttermilk in pan and pita bread formulations.

3.3. Chemical Composition of Pan and Pita Bread. The chemical composition of pan and pita bread made from wheat flour supplemented with different concentrations of buttermilk (0, 30, 60, and 100% of added water) is shown in Table 5. Generally, incorporation of buttermilk in bread formulation increased protein, oil, and ash contents and reduced the carbohydrate contents of both types of bread. The increases

TABLE 5: Chemical composition of pan and pita breads made using wheat flour fortified with different concentrations (0, 30, 60, and 100% of added water) of buttermilk.

Parameters	Fortification rate			
	0%	30%	60%	100%
Pan bread				
Protein	13.65 ± 0.30^b	13.79 ± 0.21^b	14.21 ± 0.01^a	14.38 ± 0.23^a
Oil	4.17 ± 0.13^c	4.45 ± 0.09^b	4.55 ± 0.11^b	4.79 ± 0.05^a
Ash	2.69 ± 0.10^a	2.78 ± 0.00^a	2.75 ± 0.00^a	2.92 ± 0.01^a
Carbohydrate	79.53 ± 0.35^a	79.12 ± 0.10^a	78.44 ± 0.13^b	77.89 ± 0.16^c
Pita bread				
Protein	13.41 ± 0.12^b	13.47 ± 0.39^b	15.04 ± 0.69^a	15.17 ± 0.58^a
Oil	1.54 ± 0.10^c	2.12 ± 0.06^b	2.19 ± 0.04^b	2.98 ± 0.19^a
Ash	1.71 ± 0.00^b	1.75 ± 0.05^b	2.06 ± 0.11^a	1.98 ± 0.04^a
Carbohydrate	83.26 ± 0.45^a	82.29 ± 0.49^a	80.69 ± 0.34^b	79.85 ± 0.25^b

[a-c]Mean values of triplicate samples ± SD. Means not sharing a common superscript (s) in a row are significantly different at $P \leq 0.05$ as assessed by Duncan's Multiple Range Test.

in protein, fat, and ash contents might result from the added buttermilk that contains appreciable amounts of these constituents, whereas the proportional reduction in carbohydrate is expected as the carbohydrate is calculated by subtracting protein, oil, and ash contents from 100%. The highest protein, fat, and ash contents of both types of bread were observed at 100% buttermilk supplementation while the lowest values were found in control samples suggesting concentration dependent effects of buttermilk on these constituents. In agreement with our findings, numerous reports indicated that supplementation of wheat flour with various legume flours or whey protein concentrate powder and buttermilk powder has concomitantly increased protein, oil, and ash contents and reduced carbohydrate content of fortified pan and flat breads [12, 18, 19, 38]. Overall, improvement of protein content in pan and pita bread following buttermilk supplementation could be of nutritional importance as the protein of animal sources in known by its good nutritional quality compared to cereal proteins. Thus, supplementation of bread with buttermilk could potentially improve the nutritional quality of pan and pita breads, those regularly consumed by people in Saudi Arabia, and hence could improve the nutritional and health status of humans.

3.4. *Sensory Attributes of Pan and Pita Bread.* Sensory assessment is a vital measure for quality assessment in a newly developed food product to attract consumers and to meet their requirements [29]. The choice of a food product depends on various factors such as character, mood and experience, and attitudes like sensory properties, health and nutrition, and price and value [8]. The sensory attributes of pan and pita bread prepared from wheat flour supplemented with or without different concentrations of buttermilk are presented in Table 6. For pan bread, the surface and crumb color were not affected by incorporation of buttermilk in the bread formulation. This is opposite to the findings of Gani et al. [34] where the color difference breads supplemented with milk protein concentrates and hydrolysates

increased significantly with increasing levels of concentrates and hydrolysates; this was attributed to higher degree of Maillard browning which is influenced by the distribution of water and the reaction of reducing sugars and amino acids with increasing levels of concentrates and hydrolysates. *This study may* be attributed to the low concentrates of protein in breads supplemented of buttermilk which were less than the Gani et al. [34] study. However, inclusion of 60 and 100% of buttermilk significantly ($P \leq 0.05$) reduced the degree of symmetry, the degree of cracking, and crumb texture compared to untreated control and that supplemented with 30% buttermilk. The overall evaluation also indicated that 30% buttermilk and control pan bread are superior to 60% and 100% buttermilk containing samples. Although it is not significant, 30% buttermilk containing pan bread outscores the control and those formulated with 60 and 100% buttermilk in all sensory attributes. These findings demonstrated that incorporation of buttermilk in pan bread formulations at 30% replacement of added water could enhance the sensory characteristics of pan bread.

For pita bread, the addition of buttermilk affected most of the sensory attributes with the exception of degree of cracking that was not significantly affected by incorporation of buttermilk. Both control and 30% buttermilk containing pita bread showed the highest scores of all sensory attributes compared to that containing 60 and 100% buttermilk. Similar to that of pan bread, the overall acceptability of pita bread was observed in control and 30% buttermilk containing pita bread. These findings suggested that incorporation of buttermilk in pita bread at a concentration of 30% enhanced the sensory properties of pita bread. Similarly, inclusion of various types of legumes, nuts, and mushroom flour and protein isolates in bread was reported to affect the sensory quality of the products as the supplementation rate elevated [6, 9, 12, 29, 32, 38]. In addition, supplementation of wheat flour with various concentrations of whey protein isolate and buttermilk powder concurrently affected the sensory attributes of the bread [18, 19]. Overall our findings demonstrated that

TABLE 6: Sensory attributes of pan and pita breads made using wheat flour fortified with different concentrations (0, 30, 60, and 100% of added water) of buttermilk.

Parameters	Fortification rate			
	0%	30%	60%	100%
Pan bread				
Surface color	9.40 ± 0.29^a	9.70 ± 0.27^a	9.30 ± 0.25^a	9.42 ± 0.37^a
Degree of symmetry	9.50 ± 0.10^a	9.60 ± 0.09^a	9.00 ± 0.17^b	8.60 ± 0.29^b
Degree of cracking	9.70 ± 0.38^a	9.80 ± 0.42^a	9.10 ± 0.16^b	8.80 ± 0.28^b
Crumb color	9.60 ± 0.19^a	9.50 ± 0.70^a	9.50 ± 0.30^a	9.40 ± 0.14^a
Crumb texture	9.50 ± 0.10^a	9.60 ± 0.19^a	9.30 ± 0.12^b	9.10 ± 0.17^b
Overall evaluation	47.70 ± 0.54^a	48.10 ± 1.28^a	46.20 ± 0.69^b	45.30 ± 0.70^b
Pita bread				
Surface color	4.80 ± 0.22^a	4.70 ± 0.18^a	4.10 ± 0.06^b	4.60 ± 0.14^a
Degree of symmetry	4.90 ± 0.31^a	4.80 ± 0.12^a	4.30 ± 0.18^b	4.90 ± 0.01^a
Degree of cracking	4.70 ± 0.18^a	4.70 ± 0.22^a	4.50 ± 0.52^a	4.80 ± 0.42^a
Crumb color	5.00 ± 0.05^a	4.90 ± 0.31^a	4.30 ± 0.12^b	4.50 ± 0.12^b
Crumb texture	4.90 ± 0.31^a	4.30 ± 0.17^b	3.80 ± 0.13^c	4.10 ± 0.11^b
Taste and flavor	4.80 ± 0.22^a	4.80 ± 0.12^a	4.40 ± 0.11^b	4.10 ± 0.19^b
Symmetry of top and bottom layers	4.80 ± 0.22^a	4.60 ± 0.10^a	4.20 ± 0.13^b	3.80 ± 0.28^b
Overall acceptance	4.84 ± 0.18^a	4.61 ± 0.09^a	4.22 ± 0.03^b	4.39 ± 0.08^b

[a-c]Mean values of triplicate samples $\pm$ SD. Means not sharing a common superscript (s) in a row are significantly different at $P \leq 0.05$ as assessed by Duncan's Multiple Range Test.

supplementation of bread formula with buttermilk at 30% substitution of added water is recommended for improving the nutritional and sensorial attributes of pan and pita bread.

4. Conclusion

The present study focused on the utilization of buttermilk in pan and pita bread making to improve the rheological and nutritional qualities of bread without major effects on the consumer acceptability of the products. The results revealed that incorporation of 30% buttermilk in pan and pita bread formulations significantly improved the rheological properties (water absorption capacity, dough development time, gelatinization temperature, and peak viscosity), physical properties (bread weight, volume, and specific volume), and sensory quality of pan and pita bread. Therefore, supplementation of bread with 30% buttermilk is recommended and could potentially improve the nutritional and sensory qualities of pan and pita breads, those regularly consumed by people in Saudi Arabia, and thus could improve the nutritional and health status of those communities.

Conflicts of Interest

No potential conflicts of interest exist.

References

[1] M. A. Saccotelli, A. Conte, K. R. Burrafato, S. Calligaris, L. Manzocco, and M. A. Del Nobile, "Optimization of durum wheat bread enriched with bran," *Food Science & Nutrition*, vol. 5, no. 3, pp. 689–695, 2017.

[2] W. Siebel, *Future of Flour – A compendium of Flour Improvement*, L. Popper, W. Schafer, and W. Freund, Eds., Verlag Agri Media, Hamburg, Germany, 2006.

[3] M. O. Aljobair, "Assessment of the Bread Consumption Habits Among the People of Riyadh, Saudi Arabia," *Pakistan Journal of Nutrition*, vol. 16, no. 5, pp. 293–298, 2017.

[4] V. A. Jideani and F. C. Onwubali, "Optimisation of wheat-sprouted soybean flour bread using response surface methodology," *African Journal of Biotechnology*, vol. 8, no. 22, pp. 6364–6373, 2009.

[5] B. Iuliana, S. Georgeta, S. Violeta, and A. Iuliana, "Effect of the addition of wheat bran stream on dough rheology and bread quality," *The Annals of the University Dunareade Josof Galati Fascicle VI Food Technology*, vol. 36, no. 1, pp. 39–52, 2012.

[6] L. Campbell, S. R. Euston, and M. A. Ahmed, "Effect of addition of thermally modified cowpea protein on sensory acceptability and textural properties of wheat bread and sponge cake," *Food Chemistry*, vol. 194, pp. 1230–1237, 2016.

[7] H. A. Eissa, A. S. Hussein, and B. E. Mostafa, "Rheological properties and quality evaluation of Egyptian Balady bread and biscuits supplemented with flours of ungerminated and germinated legume seeds or mushroom," *Polish Journal of Food Nutrition Sciences*, vol. 57, no. 4, pp. 487–496, 2007.

[8] O. L. Erukainure, J. N. C. Okafor, A. Ogunji, H. Ukazu, E. N. Okafor, and I. L. Eboagwu, "Bambara–wheat composite flour: rheological behavior of dough and functionality in bread," *Food Science & Nutrition*, vol. 4, no. 6, pp. 852–857, 2016.

[9] I. Mohammed, A. R. Ahmed, and B. Senge, "Dough rheology and bread quality of wheat–chickpea flour blends," *Industrial Crops and Products*, vol. 36, no. 1, pp. 196–202, 2012.

[10] P. D. Ribotta, S. A. Arnulphi, A. E. León, and M. C. Añón, "Effect of soybean addition on the rheological properties and breadmaking quality of wheat flour," *Journal of the Science of Food and Agriculture*, vol. 85, no. 11, pp. 1889–1896, 2005.

[11] D. Sabanis and C. Tzia, "Effect of rice, corn and soy flour addition on characteristics of bread produced from different wheat cultivars," *Food and Bioprocess Technology*, vol. 2, no. 1, pp. 68–79, 2009.

[12] K. Moktan and P. Ojha, "Quality evaluation of physical properties, antinutritional factors, and antioxidant activity of bread fortified with germinated horse gram (Dolichus uniflorus) flour," *Food Science & Nutrition*, vol. 4, no. 5, pp. 766–771, 2016.

[13] V. Conway, S. F. Gauthier, and Y. Pouliot, "Buttermilk: Much more than a source of milk phospholipids," *Animal Frontiers*, vol. 4, no. 2, pp. 44–51, 2014.

[14] R. Kumar, M. Kaur, A. K. Garsa, B. Shrivastava, V. P. Reddy, and A. Tyagi, "Natural and Cultured Buttermilk," in *Fermented milk and dairy products*, A. K. Puniya, Ed., pp. 203–225, CRC Press/Taylor and Francis, 2015.

[15] V. Conway, P. Couture, C. Richard, S. F. Gauthier, Y. Pouliot, and B. Lamarche, "Impact of buttermilk consumption on plasma lipids and surrogate markers of cholesterol homeostasis in men and women," *Nutrition, Metabolism & Cardiovascular Diseases*, vol. 23, no. 12, pp. 1255–1262, 2013.

[16] K. L. Fuller, T. B. Kuhlenschmidt, M. S. Kuhlenschmidt, R. Jiménez-Flores, and S. M. Donovan, "Milk fat globule membrane isolated from buttermilk or whey cream and their lipid components inhibit infectivity of rotavirus in vitro," *Journal of Dairy Science*, vol. 96, no. 6, pp. 3488–3497, 2013.

[17] S. C. Larsson, S.-O. Andersson, J.-E. Johansson, and A. Wolk, "Cultured milk, yogurt, and dairy intake in relation to bladder cancer risk in a prospective study of Swedish women and men," *American Journal of Clinical Nutrition*, vol. 88, no. 4, pp. 1083–1087, 2008.

[18] A. A. Hassan, H. A. M. El-Shazly, A. M. Sakr, and W. A. Ragab, "Influence of Substituting Water with Fermented Skim Milk, Acid Cheese Whey or Buttermilk on Dough Properties and Baking Quality of Pan Bread," *World Journal of Dairy Food Science*, vol. 8, no. 1, pp. 100–117, 2013.

[19] A. B. Madenci and N. Bilgiçli, "Effect of Whey Protein Concentrate and Buttermilk Powders on Rheological Properties of Dough and Bread Quality," *Journal of Food Quality*, vol. 37, no. 2, pp. 117–124, 2014.

[20] AACC, *Approved Methods of the AACC*, American Association of Cereal Chemists, St. Paul, MN, USA, 10th edition, 2000.

[21] E. I. Mousa and I. S. Al-Mohizea, "Bread baking in Saudi Arabia," *CerealFoodWorld*, vol. 32, no. 9, pp. 614–617, 1987.

[22] AOAC, *Official Methods of Analysis*, Association of Official Analytical Chemists – International, Gaithersburg, MD, USA, 18th edition, 2005.

[23] E. Hallen, Ş. Ibanoğlu, and P. Ainsworth, "Effect of fermented/germinated cowpea flour addition on the rheological and baking properties of wheat flour," *Journal of Food Engineering*, vol. 63, no. 2, pp. 177–184, 2004.

[24] M. Mashayekh, M. R. Mahmoodi, and M. H. Entezari, "Effect of fortification of defatted soy flour on sensory and rheological properties of wheat bread," *International Journal of Food Science & Technology*, vol. 43, no. 9, pp. 1693–1698, 2008.

[25] B. Bilgin, O. Daðlioðlu, and M. Konyali, "Functionality of bread made with pasteurized whey and /or buttermilk," *Italian Journal of Food Science*, vol. 18, no. 3, pp. 277–286, 2006.

[26] A. A. Anton, K. A. Ross, O. M. Lukow, R. G. Fulcher, and S. D. Arntfield, "Influence of added bean flour (Phaseolus vulgaris L.) on some physical and nutritional properties of wheat flour tortillas," *Food Chemistry*, vol. 109, no. 1, pp. 33–41, 2008.

[27] M. G. E. Gadallah, I. R. S. Rizk, H. E. Elsheshetawy, S. H. Bedeir, and A. M. Abouelazm, "Impact of Partial Replacement of Wheat Flour with Sorghum or Chickpea Flours on Rheological Properties of Composite Blends," *Journal of Agricultural and Veterinary Sciences*, vol. 10, no. 1, pp. 83–98, 2017.

[28] I. Pasha, S. Rashid, F. M. Anjum, M. T. Sultan, M. M. Nasir Qayyum, and F. Saeed, "Quality evaluation of wheat-mungbean flour blends and their utilization in baked products," *Pakistan Journal of Nutrition*, vol. 10, no. 4, pp. 388–392, 2011.

[29] M. Majeed, M. U. Khan, M. N. Owaid et al., "Development of oyster mushroom powder and its effects on physicochemical and rheological properties of bakery products," *Journal of Microbiology, Biotechnology and Food Sciences*, vol. 6, no. 5, pp. 1221–1227, 2017.

[30] A. A. Adebowale, M. T. Adegoke, S. A. Sanni, M. O. Adegunwa, and G. O. Fetuga, "Functional properties and biscuit making potentials of sorghum-wheat flour composite," *American Journal of Food Technology*, vol. 7, no. 6, pp. 372–379, 2012.

[31] M. Hrušková, I. Švec, and I. Jurinová, "Chemometrics of wheat composites with hemp, teff, and chia flour: Comparison of rheological features," *International Journal of Food Science*, vol. 2013, Article ID 968020, 2013.

[32] E. Pejcz, A. Mularczyk, and Z. Gil, "Technological characteristics of wheat and non-cereal flour blends and their applicability in bread making," *Journal of Food and Nutrition Research*, vol. 54, no. 1, pp. 69–78, 2015.

[33] L. S. Zaidul, A. Abd Karim, D. M. A. Manan, N. A. Nik Norulaini, and A. K. M. Omar, "Gelatinization Properties of Sago and Wheat Flour Mixtures," *ASFAN Food Journal*, vol. 12, no. 1, pp. 199–209, 2003.

[34] A. Gani, A. A. Broadway, F. A. Masoodi et al., "Enzymatic hydrolysis of whey and casein protein- effect on functional, rheological, textural and sensory properties of breads," *Journal of Food Science and Technology*, vol. 52, no. 12, pp. 7697–7709, 2015.

[35] A. A. Wani, D. S. Sogi, P. Singh, P. Sharma, and A. Pangal, "Dough-handling and cookie-making properties of wheat flour-watermelon protein isolate blends," *Food and Bioprocess Technology*, vol. 5, no. 5, pp. 1612–1621, 2012.

[36] L.-Y. Lin, H.-M. Liu, Y.-W. Yu, S.-D. Lin, and J.-L. Mau, "Quality and antioxidant property of buckwheat enhanced wheat bread," *Food Chemistry*, vol. 112, no. 4, pp. 987–991, 2009.

[37] B. Madenci, S. Turker, and N. Bilgicli, "Effect of some dairy by-product on physical, chemical and sensory properties of lavash bread," in *Proceedings of the III*, pp. 309–312, 2012.

[38] J. Ndife, L. O. Abdulraheem, and U. M. Zakari, "Evaluation of the nutritional and sensory quality of functional breads produced from whole wheat and soya bean flour blends," *African Journal of Food Science*, vol. 5, no. 8, pp. 466–472, 2011.

7

A Note on Fatty Acids Profile of Meat from Broiler Chickens Supplemented with Inorganic or Organic Selenium

Marta del Puerto,[1] M. Cristina Cabrera,[1,2] and Ali Saadoun[2]

[1]*Department of Animal Production & Pastures, Nutrition and Food Quality Laboratory,
Faculty of Agronomy, University of the Republic (UDELAR), E. Garzón 809, Montevideo, Uruguay*
[2]*Physiology & Nutrition, Faculty of Sciences, University of the Republic (UDELAR),
Iguá 4225, Montevideo, Uruguay*

Correspondence should be addressed to Ali Saadoun; asaadoun@fcien.edu.uy

Academic Editor: Rosana G. Moreira

This investigation evaluated, in broiler chickens *Pectoralis* and *Gastrocnemius* muscles, the effect of the dietary supplementation with sodium selenite (0.3 ppm) versus selenomethionine (0.3 ppm), on the fatty acids composition, lipids indices, and enzymes indexes for desaturase, elongase, and thioesterase. The selenium reduced, in both muscles, the content of atherogenic fatty acids, C14:0 and C16:0, while it increased the C18:1 level. On the other hand, selenium increased, in both muscles, the content of C18:3n3 and EPA, but not DPA and DHA. No selenium effect was detected for PUFA/SFA, n-6, n-3, n-6/n-3, and atherogenic and thrombogenic indices. As for the enzyme indexes, a selenium effect is only detected for thioesterase. Taken together, the results highlight the potential effect of dietary selenium, mainly selenomethionine, in the modulation of the composition of fatty acids in chicken meat, in particular, reducing the content of atherogenic fatty acids and increasing the health promoting n-3 PUFA.

1. Introduction

Being a nonruminant animal, chickens tend to use, without significant changes, the lipids and the fatty acids present in a diet in order to fulfil its physiological needs for growth and muscle development [1]. When the chicken diet includes corn, soya meal, sunflower meal, and other green foods, it contributes to a relatively high content of fatty acids from the polyunsaturated fatty acids (PUFA) in the meat.

However, the presence of high levels of PUFA in chicken meat makes it more susceptible to the oxidation process and to undergoing alterations in smell, taste, and nutritional value. To counteract this concern, it is generally advised to add dietary antioxidants in chicken feed, to maintain the lipid stability in meat [2]. One of them, selenium, is a constituent of glutathione peroxidase, an essential enzyme in nutrients metabolism, and a first line of defense against the oxidation process [3]. In the feeding of farm animals, selenium is added

in diet in inorganic as well as organic form. Until recently the form of added selenium in farm animals diet had been inorganic, either selenite or selenate [4]. Nowadays, however organic selenium attracts more interest as a supplement. This change could be linked to the fact that, apparently, organic selenium has a higher absorption rate than inorganic selenium in chickens [5].

It seems that selenium also has an interesting interaction with lipids metabolism. Indeed, Schäfer et al. [6] found that a selenium deficiency can interfere with the normal conversion of α-linolenic (ALA) into eicosapentaenoic (EPA) and docosahexaenoic (DHA) acids, which can result in an increased omega-6 : omega-3 ratio in the liver of rats. Furthermore, some studies showed that the supplementation with selenium in the diet of beef, pig, and chicken modified the fatty acid profiles of their meat [6–10]. Similar results were obtained for milk and colostrum [11]. This effect concerning selenium could be an interesting way of modifying the fatty

TABLE 1: Chemical composition of experimental diets.

Items (%)	Experimental diets		
	Control	Se-Met (0.30 ppm Se)	Se-Na (0.30 ppm Se)
Ground corn	60.0	60.0	60.0
Soybean meal	32.5	32.5	32.5
Meat and bone meal	4.20	4.20	4.20
Calcium carbonate	1.00	1.00	1.00
Dicalcium phosphate	0.80	0.80	0.80
Sunflower oil	0.50	0.50	0.50
Salt	0.30	0.30	0.30
Lysine	0.25	0.25	0.25
Dl-Methionine	0.10	0.10	0.10
Premix*	0.25	0.25	0.25
Chemical composition			
Crude protein, (%)**	20.0	20.0	20.0
ME, (MJ/kg)	2931	2931	2931
Crude fiber, (%)**	3.70	3.70	3.70
Ca, (%)**	1.20	1.20	1.20
P, (%)**	0.42	0.42	0.42
Se, (ppm)**	0.01	0.29	0.30

*Mineral and vitamin premix. Provided (per 1 kilogram): 8.000.000 IU of vitamin A, 1.300.000 IU of vitamin D3, 16.000 IU of vitamin E, 5 g of vitamin K, 3.5 g of vitamin B2, 6.5 g of calcium D-pantothenate, 20 g of niacin, 0.3 g of folic acid, 8.5 mg of vitamin B12, 350 g of choline chloride, 0.3 g of vitamin B1, 0.6 g of vitamin B6, 60 g of Mn, 25 g of Zn, 16 g of Fe, 1 g of Cu, 1 g of I, and 60 mg of Co. ME: metabolizable energy. **Analyzed values.

acid profile of chicken meat, in addition to its protective effect against the oxidation process.

Therefore, the aim of this study is to determine the effect of the diet supplementation with selenium in inorganic form (sodium selenite, Se-Na), in comparison to the organic form (selenium methionine, Se-Met), on the fatty acids composition and some lipids indices (in terms of the health needs of consumers) of chicken meat from *Pectoralis* and *Gastrocnemius* muscles. Likewise, some enzyme indexes for desaturase, elongase, and thioesterase will be calculated in an attempt to detect any effect of the two kinds of selenium, on the lipids metabolism in the two muscles.

2. Materials and Methods

2.1. Animals and Diets. The animal care and handling were approved by the Committee on Experimental Animals of the Universidad de la República, Montevideo, Uruguay (CHEA), before the beginning of the experiment. Two hundred one-day-old male Ross birds were obtained from a commercial hatchery and reared until thirty-five days on a floor pen with wood shavings, in a climate-controlled room with a photoperiod of 23 hours of light/day. They were fed ad libitum with a commercial corn-soya diet (219 g·kg^{-1} crude protein and 13.35 MJ·kg^{-1} of metabolizable energy). Tap water was given ad libitum. At thirty-five days, ninety birds were selected on homogenous weight basis and assigned randomly into three groups of thirty birds each. The birds were located in ten experimental pens (90 cm × 90 cm) with wood shavings as litter. Each pen located three birds, fed ad libitum, with one of the experimental diets, until slaughtering. A corn-soya based diet (Table 1) was formulated to meet the nutrient requirements for finishing male broilers [12] and was considered as the basal diet, not supplemented with selenium (Table 1). The other two diets were supplemented with selenium (Table 1) from an inorganic source (0.3 ppm Se, as sodium selenite, Se-Na in text and tables) or an organic source (0.3 ppm Se, as selenomethionine, Se-Met in text and tables). Dry matter, crude protein, fat, and fibre analysis of feed were carried out according to AOAC [13]. At fifty-six-days-old, all the birds were slaughtered at our own experimental abattoir. Preharvest handling, transportation (transportation time was 3 minutes), and slaughtering procedures were in accordance with the good animal welfare practices approved by the CHEA rules. The birds were slaughtered, after a fasting time of 16 hours (overnight), by cutting the jugular vein until total bleeding (3 min) to cause the least possible stress to the animal. The carcasses were cooled and maintained at 4°C for 24 hours postmortem. After that, the Pectoralis and Gastrocnemius muscles were withdrawn and stored at −80°C until analysis.

2.2. Analytical Determinations. The intramuscular lipids were extracted according to Folch et al. [16]. Briefly, a sample of 4 grams of Pectoralis and Gastrocnemius muscles (free of dissectible visible fat) was homogenized at 35000 rpm with a Virtis 45 during 1 min with 80 ml of chloroform : methanol (2 : 1). Afterward, the homogenate was filtered, moved in a separating funnel, mixed by inversion for one minute, and decanted overnight. The lower phase (chloroform containing lipids) was recuperated in a glass balloon, evaporated at

45°C with a light vacuum in a Rotavapor (IKA basic). The balloon was dried in an oven at 35°C for 60 min and cooled at ambient temperature overnight in a vacuum desiccator. The balloon was weighted to determine the percentage of lipids of each sample. The methylation of fatty acids followed the procedure described by Ichihara et al. [17], using methanolic KOH. The determination of fatty acids by gas chromatography followed the procedure described by Eder [18], using fused-silica capillary column CPSIL-88 of 100 m installed in a split/splitless chromatograph Clarus 500 (Perkin Elmer Instruments, USA). A FID detector and automatic injection of $1\,\mu l$ of sample with a split of 50% were used. Hydrogen was used as carrier gas. The thermal conditions were injector/detector temperatures 250°C/250°C and oven held at 90°C for one minute after the injection of the sample; after that, the oven temperature was increased to 225°C at 15°C/min. Fatty acids methylated esters (FAMEs) were determined comparing the retention time to authentic standards (Sigma Corp, USA). Individual FAME was quantified as a percentage of total analysed FAMEs.

2.3. Calculus of Lipids Indices. The calculus of lipids indices was performed from the data of the fatty acid composition of intramuscular lipids, obtained here. The following indices were calculated.

2.3.1. Index of Atherogenicity (IA). Compute the relationship between the sum of the main saturated (proatherogenic) and the unsaturated (antiatherogenic) fatty acids. It was calculated according to Ulbricht and Southgate [19] as follows:

$$IA = \frac{(4 \times C14{:}0 + C16{:}0)}{[\sum MUFA + \sum (n\text{-}6) + \sum (n\text{-}3)]}. \qquad (1)$$

2.3.2. Index of Thrombogenicity (IT). Estimate the potential to form clots in the blood vessels, determined by the relationship between the prothrombogenic (saturated) and the antithrombogenic fatty acids (sum of MUFA and PUFA). It was calculated according to Ulbricht and Southgate [19] as follows:

$$IT = \frac{(C14{:}0 + C16{:}0 + C18{:}0)}{[0.5 \times \sum MUFA + 0.5 \times \sum (n\text{-}6) + 3 \times \sum (n\text{-}3) + \sum (n\text{-}3) / \sum (n\text{-}6)]}. \qquad (2)$$

2.3.3. Hypocholesterolemic/Hypercholesterolemic Ratio (h/H). Compute the relation between unsaturated fatty acids (MUFA and PUFA) and the saturated fatty acids 14:0 and 16:0.

The h/H ratio was calculated according to Fernández et al. [20], as follows:

$$\frac{h}{H} = \frac{(C14{:}1 + C16{:}1 + C18{:}1 + C20{:}1 + C22{:}1 + C18{:}2 + C18{:}3 + C20{:}3 + C20{:}4 + C20{:}5 + C22{:}4 + C22{:}5 + C22{:}6)}{(C14{:}0 + C16{:}0)}. \qquad (3)$$

2.4. Enzyme Activity Index. The enzyme activity of desaturase, elongase, and thioreductase was estimated by relating the amount of the specific substrate to the corresponding product of the respective enzyme. The calculated ratios were 16:1n-7 to 16:0 and 18:1n9 to 18:0. The activity of stearoyl-CoA desaturase (delta-9-desaturase) was estimated by calculating the ratio 16:1n-7 + 18:1n-9 to 16:0 + 18:0. The delta-5 desaturase + delta-6 desaturase sum was used as an index for the estimation of the formation of long chain n-6 and n-3 starting from the corresponding precursors C18:2n6 and C18:3n3 [14]. Also, the ratio 18:0 to 16:0 was calculated to estimate the elongase activity. The thioesterase was estimated as the ratio of C16:0 to C14:0 [15].

2.5. Statistical Analysis. Data are presented as mean ± SEM. Fatty acid content, lipids indices, and enzyme indexes were analysed by ANOVA with a GLM procedure using diet and muscle type as fixed factors and the interaction diet × muscle ($P < 0.05$). Also, one-way ANOVA was used to determine the effect of the diet for each muscle separately, as well as a Tukey-Kramer post hoc test ($P < 0.05$).

3. Results and Discussion

Selenium is an essential micronutrient, which is implicated in various physiological animal functions like growth, fertility, and immunity responses. It plays a crucial role in the defense against the accumulation of hydroperoxides from cellular metabolism [21, 22]. However, it is necessary to consider the toxicity risk of selenium when added in high doses in animal feed [22]. The FDA [23] has approved and advised that the supplementation of selenium in complete feed for chicken, swine, turkey, sheep, cattle, and duck should not exceed a level of 0.3 ppm. The European Food Safety Authority recommended similar levels of selenium supplementation in animal feed [24]. In the present investigation, the organic and the inorganic selenium were added in doses of 0.3 ppm (Table 1). The use of such level ensures the applicability of the results in chicken nutrition, in order to produce meat which is enriched in selenium but contains no risk for consumers.

Nevertheless, another role of selenium has been observed in relation to the modification of the fatty acids composition of chicken meat. Indeed, Haug et al. [21] showed a different repartition of fatty acids, in particular, the n-3 PUFA ones. In the present investigation, the lipids and fatty acids composition of the *Pectoralis* muscle showed a lower content in total lipids than *Gastrocnemius* muscle (Table 2). This is a usual and expected result for chicken [25]. However, the supplementation with selenium did not show any main effect, although there is a significant interaction between muscle and diet. This means that selenium could affect the lipids content depending on the muscle type (Table 2).

The comparison between the fatty acids compositions of the two muscles showed that the *Pectoralis* muscle contained more SFA than the *Gastrocnemius* muscle, principally the C16:0 and the C18:0. When the main effect of diet is considered, it seems that the C14:0 and the C16:0, but not C18:0, showed a reduced level in the two muscles in comparison to the control (Table 2). In the case of C16:0, only Se-Met was effective. This is an interesting finding because it seems that this decreasing effect is specific for those fatty acids, C14:0 and C16:0, which promote the occurrence of cardiovascular diseases in human [26]. This point needs to be considered in future investigations. Concerning MUFA, it seems that the *Gastrocnemius* muscle showed more MUFA than the *Pectoralis* muscle (Table 2) and, as expected, the C18:1 is the most represented fatty acid within the MUFA. Furthermore, the selenium as Se-Met promoted more content of C18:1 when compared to Se-Na and control (Table 2). For PUFA, no significant main effects were observed for muscle or diet (Table 2). However, when the fatty acids were considered individually, it appears that the *Pectoral* showed more PUFA than the *Gastrocnemius* muscle for the most valuable fatty acids for human health such as DPA and DHA. Meanwhile, the two essential PUFA C18:2n6 (linoleic acid) and C18:3n3 (α-linolenic acid) showed a higher level in *Gastrocnemius* muscle. It is known that these two fatty acids come exclusively from diet and cannot be synthetized by the tissues. When the main effect of diet is considered, it seems that the C18:3n3 showed a higher level when the selenium, independently of its chemical source, is added in feed in comparison to the control (Table 2). These results are in accord with the findings of Haug et al. [21], Ševčíková et al. [27], and Kralik et al. [28]. Furthermore, a significant interaction has been obtained, suggesting that the selenium could modify the distribution of this fatty acid in chicken meat, depending on the muscles. A similar, but limited, main effect was obtained for C18:2n6. Indeed, only Se-Met increased the level of this fatty acid, in comparison to Se-Na and control (Table 2). The differences in the level of C18:3n3, and on a smaller scale, of C18: 2n6, in comparison to the control, could be explained by the reduced degradation of these fatty acids by the oxidation processes. This protection could be done by the action of glutathione and the enzyme glutathione peroxidase (GPx), through the elimination of free radicals capable of initiating and propagating the fatty acids oxidation, particularly the PUFA ones [29].

Like discussed before, selenium in the chicken diet has been associated in previous investigations with an increase in n-3 fatty acids in *Gastrocnemius* muscle [21, 27] and in *Pectoralis* muscle [28]. Taken together, it was observed, in those investigations, that the increase of the fatty acids level,

after supplementation by selenium, happened mainly for the α-linolenic acid (α-C18:3n3), EPA, DPA, and DHA in *Gastrocnemius* muscle. Unfortunately, in the present investigation, DPA and DHA did not show any significant effect of selenium on their level in comparison to the control. For EPA, the results showed a significant diet main effect when Se-Na is used in comparison to the Se-Met and the control (Table 2). However, the absolute values were very low and incite us to be cautious on the interpretation about the effectiveness of selenium for this fatty acid. Contrarily to our finding and those of Ševčíková et al. [27], Haug et al. [21], and Kralik et al. [28], Zduńczyk et al. did not find any effects of selenium on the fatty acids composition of *Pectoralis* muscle [30]. As it was found in the present study with chicken, the supplementation with selenium affected the fatty composition in other species and products, independently of its chemical form. In ruminant species, the effect of the supplementation with selenium, independently of its form, showed controversial results concerning the modification of lipids content and the fatty acids composition of meat [31]. More investigations should be done to clarify this interesting point linking the supplementation with selenium to the modification of the composition of meat in some specific fatty acids, such as C18:3n3, EPA, and DHA, which have been generally associated with health concerns for human [32].

In the present investigation, the focus on human nutrition and health has been considered through the calculus of some lipids indices associated with the fatty acids composition. These indices are generally used to rank foods in regard to their potential effect on the promotion of cardiovascular diseases. In the present investigation, the PUFA/SFA, n-6, n-3, and the n-6/n-3 ratio yield no different main effects when the two muscles were considered (Table 3). On the contrary, the atherogenic (IA) and thrombogenic (IT) indices showed that the *Pectoralis* muscle had undesirable IA and IT indices when compared to the *Gastrocnemius* muscle. For these two indices, the desirable value has to be as low as possible. The hypocholesterolaemic versus hypercholesteraemic indices (h/H), as opposed to the IA and IT indices, must be as high as possible in order to protect consumer from the hypercholesterolaemia, a factor which promotes the atherosclerosis syndrome in humans [33]. The h/H obtained in the present investigation had a better value in the *Pectoralis* muscle than in the *Gastrocnemius* muscle (Table 3). However, there is no effect of the supplementation of selenium, in its two forms, on the health scope. For all these considered indices, no main effects have been observed concerning selenium, independently of its form (Table 3). Nonetheless, there is an interaction (muscle × diet) for the n-3 fatty acids, the n-6/n-3 ratio, and the h/H indices. These last results encourage us to carry on, in order to have a better understanding of the action of selenium in the repartition of fatty acids in chicken meat.

Finally, one of the goals of the present investigation was to compare the effect of feed supplementation with organic and inorganic selenium on the activity of desaturase, elongase, and thioesterase enzymes in the two muscles, through the

TABLE 2: Fatty acids composition (g/100 g of total fatty acids) of *Pectoralis* and *Gastrocnemius* muscles of chickens supplemented in diet with inorganic (Se-Na) or organic (Se-Met) selenium.

| | Pectoralis (Pe) | | | Gastrocnemius (G) | | | Main effects | | |
	Control	Se-Na	Se-Met	Control	Se-Na	Se-Met	Muscle (M)	Diet (D)	M × D
Lipids %	1.77 ± 0.11	1.41 ± 0.17	1.13 ± 0.05	2.44 ± 0.31	3.01 ± 0.27	3.09 ± 0.24	$P < 0.01$ G > Pe	NS	$P < 0.04$
Fatty acids %									
C14:0	0.69 ± 0.10	0.60 ± 0.17	0.62 ± 0.08	0.73 ± 0.04	0.61 ± 0.18	0.61 ± 0.13	NS	$P < 0.03$ Se < C	NS
C14:1	0.12 ± 0.03	0.10 ± 0.01	0.09 ± 0.01	0.17 ± 0.03	0.12 ± 0.01	0.11 ± 0.02	$P < 0.01$ G > Pe	$P < 0.01$ Se < C	NS
C16:0	26.6 ± 3.20	27.0 ± 2.54	26.6 ± 1.61	26.8 ± 0.62	25.6 ± 2.91	24.0 ± 1.60	$P < 0.01$ G > Pe	$P < 0.05$ Se-Met < C	NS
C16:1	4.21 ± 1.92	4.22 ± 0.85	3.78 ± 0.21	5.58 ± 1.51	5.54 ± 1.08	4.52 ± 0.62	$P < 0.04$ Pe > G	$P < 0.05$ Se < C	NS
C18:0	10.2 ± 2.29	9.99 ± 0.72	9.82 ± 0.47	8.65 ± 1.93	8.80 ± 1.34	9.18 ± 1.04	$P < 0.01$ Pe > G	NS	NS
C18:1	33.6 ± 2.10	34.5 ± 1.73	35.1 ± 2.69	35.2 ± 3.21	37.5 ± 2.19	37.1 ± 3.27	$P < 0.01$ G > Pe	$P < 0.03$ Se-Met > C	NS
C18:2n6	15.7 ± 1.92	14.4 ± 1.10	16.0 ± 2.43	16.7 ± 1.26	15.8 ± 1.15	17.5 ± 2.37	$P < 0.01$ G > Pe	$P < 0.03$ Se-Met > Se-Na	NS
C20:1	0.18 ± 0.09	0.20 ± 0.08	0.16 ± 0.02	0.18 ± 0.02	0.20 ± 0.10	0.33 ± 0.13	$P < 0.02$ G > Pe	$P < 0.05$ Se-Met > C	$P < 0.01$
C18:3n3	0.35 ± 0.06	0.46 ± 0.09	0.43 ± 0.12	0.44 ± 0.02	0.42 ± 0.02	0.81 ± 0.11	$P < 0.01$ G > Pe	$P < 0.01$ Se > C	$P < 0.01$
C20:3n6	0.27 ± 0.14	0.22 ± 0.08	0.20 ± 0.10	0.15 ± 0.05	0.15 ± 0.06	0.21 ± 0.06	$P < 0.01$ Pe > G	NS	NS
C20:3n3	0.23 ± 0.12	0.30 ± 0.09	0.20 ± 0.08	0.19 ± 0.10	0.22 ± 0.02	0.16 ± 0.05	$P < 0.01$ Pe > G	NS	NS
C22:1	0.64 ± 0.31	0.65 ± 0.17	0.60 ± 0.13	0.46 ± 0.28	0.40 ± 0.11	0.35 ± 0.03	$P < 0.01$ Pe > G	NS	NS
C20:4n6 ARA	3.47 ± 1.65	3.36 ± 0.82	3.23 ± 0.83	2.17 ± 0.97	2.19 ± 0.63	2.25 ± 0.35	$P < 0.01$ Pe > G	NS	NS
C22:4n6	1.03 ± 0.61	0.96 ± 0.32	0.81 ± 0.35	0.62 ± 0.30	0.56 ± 0.20	0.56 ± 0.07	$P < 0.01$ Pe > G	NS	NS
C20:5n3 EPA	0.05 ± 0.02	0.08 ± 0.04	0.04 ± 0.01	0.07 ± 0.07	0.05 ± 0.02	0.03 ± 0.01	NS	$P < 0.03$ Se-Met < Se-Na	$P < 0.03$
C22:5n3 DPA	0.35 ± 0.17	0.26 ± 0.08	0.26 ± 0.12	0.18 ± 0.11	0.17 ± 0.06	0.19 ± 0.04	$P < 0.01$ Pe > G	NS	NS
C22:6n3 DHA	0.26 ± 0.13	0.23 ± 0.09	0.20 ± 0.08	0.13 ± 0.02	0.14 ± 0.06	0.13 ± 0.03	$P < 0.01$ Pe > G	NS	NS
Unidentified	2.10 ± 0.20	2.45 ± 1.18	1.87 ± 0.82	1.52 ± 0.48	1.50 ± 0.25	2.01 ± 0.75	—	—	—
SFA	37.5 ± 1.07	37.6 ± 0.99	37.0 ± 0.74	36.2 ± 0.54	35.0 ± 2.10	33.8 ± 0.80	$P < 0.01$ Pe > G	NS	NS
MUFA	38.7 ± 3.15	39.6 ± 1.96	39.8 ± 2.71	41.6 ± 4.51	43.8 ± 3.20	42.4 ± 3.79	$P < 0.01$ G > Pe	NS	NS
PUFA	21.1 ± 1.87	19.7 ± 1.23	20.8 ± 1.45	20.3 ± 1.16	19.4 ± 0.82	21.5 ± 1.34	NS	NS	NS

Data are mean ± SEM. NS: not significant, C: control, Se: effect of selenium from Se-Na and Se-Met, SFA: saturated fatty acids, MUFA: monounsaturated fatty acids, and PUFA: polyunsaturated fatty acids.

TABLE 3: Calculated lipids indices in terms of the health needs of consumers for *Pectoralis* and *Gastrocnemius* muscles of chickens supplemented in diet with inorganic (Se-Na) or organic (Se-Met) selenium.

| | *Pectoralis* (Pe) | | | *Gastrocnemius* (G) | | | Main effects | | |
	Control	Se-Na	Se-Met	Control	Se-Na	Se-Met	Muscle (M)	Diet (D)	M × D
P/S	0.58 ± 0.03	0.54 ± 0.02	0.58 ± 0.05	0.57 ± 0.02	0.58 ± 0.05	0.65 ± 0.04	NS	NS	NS
n-6	20.5 ± 3.31	19.0 ± 2.17	20.2 ± 2.68	19.6 ± 2.57	18.7 ± 1.67	20.5 ± 2.61	NS	NS	NS
n-3	1.23 ± 0.22	1.33 ± 0.19	1.13 ± 0.11	1.02 ± 0.12	0.98 ± 0.07	1.32 ± 0.07	NS	NS	$P < 0.02$
n-6/n-3	18.2 ± 2.82	14.9 ± 1.60	18.0 ± 0.79	19.7 ± 1.24	19.7 ± 2.33	15.6 ± 0.96	NS	NS	$P < 0.01$
IA	0.49 ± 0.01	0.49 ± 0.01	0.48 ± 0.02	0.48 ± 0.02	0.45 ± 0.01	0.41 ± 0.01	$P < 0.01$ Pe > G	NS	NS
IT	1.14 ± 0.05	1.15 ± 0.05	1.13 ± 0.04	1.09 ± 0.02	1.04 ± 0.09	0.96 ± 0.03	$P < 0.03$ Pe > G	NS	NS
h/H	2.23 ± 0.014	2.25 ± 0.02	2.17 ± 0.12	2.45 ± 0.21	2.23 ± 0.09	2.61 ± 0.12	$P < 0.01$ G > Pe	NS	$P < 0.04$

Data are mean ± SEM. NS: not significant, P/S: PUFA/SFA, IA: atherogenicity, IT: thrombogenicity, and h/H: hypocholesterolemic/hypercholesterolemic ratio.

TABLE 4: Enzymes indexes of fatty acid metabolism estimated on the basis of fatty acid composition of *Pectoralis* and *Gastrocnemius* muscles of chickens supplemented in diet with inorganic (Se-Na) or organic (Se-Met) selenium.

| Enzyme indexes | *Pectoralis* muscle (Pe) | | | *Gastrocnemius* muscle (G) | | | Main effects | | |
	Control	Se-Na	Se-Met	Control	Se-Na	Se-Met	Muscle (M)	Diet (D)	M × D
Δ-9 desaturase									
16:1/16:0	0.15 ± 0.01	0.16 ± 0.01	0.14 ± 0.01	0.21 ± 0.01	0.22 ± 0.02	0.19 ± 0.01	$P < 0.001$ G > Pe	NS	NS
18:1/18:0	3.46 ± 0.34	3.47 ± 0.11	3.59 ± 0.13	4.22 ± 0.37	4.38 ± 0.33	4.11 ± 0.27	$P < 0.002$ G > Pe	NS	NS
16:1 + 18:1/16:0 + 18:0	1.03 ± 0.03	1.04 ± 0.02	1.07 ± 0.03	1.15 ± 0.05	1.27 ± 0.07	1.26 ± 0.05	$P < 0.001$ G > Pe	NS	NS
Δ-5 + Δ-6 desaturases	19.8 ± 2.62	20.6 ± 1.05	18.7 ± 1.60	12.7 ± 1.16	13.5 ± 0.94	12.5 ± 0.55	$P < 0.001$ G < Pe	NS	NS
Elongase 18:0/16:0	0.39 ± 0.04	0.37 ± 0.02	0.37 ± 0.01	0.32 ± 0.02	0.34 ± 0.01	0.39 ± 0.02	NS	NS	NS
Thioesterase 16:0/14:0	38.8 ± 0.68	46.5 ± 2.55	43.3 ± 1.37	36.6 ± 0.42	43.4 ± 2.59	40.0 ± 2.01	NS	$P < 0.001$ Se-Na > Control	NS

Results are means ± SEM. Δ-9 desaturase indexes, Δ-5 + Δ-6 desaturases, elongase, and thioesterase indexes were calculated according to [14, 15].

calculus of a few indexes related to the activity of those enzymes. The enzyme activities were estimated by relating the amount of the specific substrate to the corresponding product of the respective enzyme. These indexes can be used as surrogates of the measure of the true enzyme activities [34]. The delta-9-desaturase can be estimated by the ratio C16:1/C16:0 specifically for the C16:1 and by the ratio C18:1/C18:0. This last one is specific for the desaturation of C18:1, the main MUFA present in chicken meat [27]. Meanwhile, the total delta-9 desaturase index (for both C16:1 and C18:1) can be estimated by the sum of the previous two indexes [14, 35]. Furthermore, there is a convenient way to estimate the total activities of delta-5-desaturase and delta-6-desaturase, by using a similar approach. These two desaturases are the key enzymes catalysing the formation of n-6 and n-3 PUFA starting from their precursors C18:2n6 and C18:3n3 [14]. The elongase and the thioesterase enzymes can be estimated by the same procedure based on the calculus of the ratios C18:0/C16:0 and C16:0/C14:0, respectively. The thioesterase is responsible for terminating the fatty acids synthesis and release of the neosynthetized fatty acids, mainly C16:0 and C14:0. In the current investigation, a main muscle effect has been obtained only for the desaturase enzymes, showing that *Gastrocnemius* muscle has more desaturases activities than the *Pectoralis* muscle (Table 4). This observation could explain why the *Gastrocnemius* muscle contains more MUFA, such as C14:1, C16:1, and C18:1 (Table 2), which is probably due to the action of delta-9-desaturase. Likewise, more C18:2n6 and C18:3n3 could be due to the action of the delta-5 and delta-6 desaturase (Table 2). When the effect of selenium is analysed, only the index for thioesterase showed a significant main effect. The results showed that the Se-Na induced a higher index in comparison to the control (Table 4). Note that a higher index for thioesterase means a lower release of de novo synthetized C14:0 and C16:0. Perhaps this result explains why there is a significant main effect due to the selenium, since meat has a lower level of those two atherogenics fatty acids. The two forms of selenium and Se-Met reduced significantly the level of C14:0 and C16:0, respectively (Table 2).

4. Conclusions

It seems that selenium, independently of its chemical form, reduced the level of the two most atherogenic fatty acids, C14:0 and C16:0, present in chicken meat. At the same time, the Se-Met raised the beneficial polyunsaturated fatty acids, such as the C18:2n6 and particularly the C18:3n3. This interesting and favourable unbalanced repartition between the atherogenic fatty acids and the beneficial PUFA, caused by the selenium in chicken meat, seems to be promising and should be carefully considered in future investigations. Furthermore, even though the effect of selenium on the enzymes implicated in the fatty acid synthesis, desaturation, and elongation remains inconclusive in the present investigation, it seems it would be of great interest to include it in future studies. Selenium, and particularly Se-Met, could be used, within approved doses, as tools to modify the profile of fatty acids in chicken meat, in order to move towards a more convenient composition of fatty acids regarding the human health.

Competing Interests

The authors declare that there is no conflict of interests.

Acknowledgments

The author del Puerto has been grant holder from ANII (National Agency for Research and Innovation, Uruguay) during the investigation. The authors gratefully acknowledge the technical assistance of Carmen Figarola and Pablo Flaskbard at Faculty of Agronomy for chicken management. The authors would like to thank Zulma Saadoun for correcting the English language of this manuscript.

References

[1] A. Saadoun and B. Leclerq, "In vivo lipogenesis of genetically lean and fat chickens: effects of nutritional state and dietary fat," *Journal of Nutrition*, vol. 117, no. 3, pp. 428–435, 1987.

[2] P. F. Surai, "Natural antioxidants in poultry nutrition: new developments," in *Proceedings of the 16th European Symposium on Poultry Nutrition*, pp. 669–676, Strasbourg, France, August 2007.

[3] T. I. Perez, M. J. Zuidhof, R. A. Renema, J. M. Curtis, Y. Ren, and M. Betti, "Effects of vitamin E and organic selenium on oxidative stability of ω-3 enriched dark chicken meat during cooking," *Journal of Food Science*, vol. 75, no. 2, pp. T25–T34, 2010.

[4] M. P. Lyons, T. T. Papazyan, and P. F. Surai, "Selenium in food chain and animal nutrition: lessons from nature—review," *Asian-Australasian Journal of Animal Sciences*, vol. 20, no. 7, pp. 1135–1155, 2007.

[5] Y. Wang, X. Zhan, X. Zhang, R. Wu, and D. Yuan, "Comparison of different forms of dietary selenium supplementation on growth performance, meat quality, selenium deposition, and antioxidant property in broilers," *Biological Trace Element Research*, vol. 143, no. 1, pp. 261–273, 2011.

[6] K. Schäfer, A. Kyriakopoulos, H. Gessner, T. Grune, and D. Behne, "Effects of selenium deficiency on fatty acid metabolism in rats fed fish oil-enriched diets," *Journal of Trace Elements in Medicine and Biology*, vol. 18, no. 1, pp. 89–97, 2004.

[7] A. S. Netto, M. A. Zanetti, G. R. D. Claro, M. P. De Melo, F. G. Vilela, and L. B. Correa, "Effects of copper and selenium supplementation on performance and lipid metabolism in confined brangus bulls," *Asian-Australasian Journal of Animal Sciences*, vol. 27, no. 4, pp. 488–494, 2014.

[8] A. S. C. Pereira, M. V. D. Santos, G. Aferri et al., "Lipid and selenium sources on fatty acid composition of intramuscular fat and muscle selenium concentration of Nellore steers," *Revista Brasileira de Zootecnia*, vol. 41, no. 11, pp. 2357–2363, 2012.

[9] E. González and J. F. Tejeda, "Effects of dietary incorporation of different antioxidant extracts and free-range rearing on fatty acid composition and lipid oxidation of Iberian pig meat," *Animal*, vol. 1, no. 7, pp. 1060–1067, 2007.

[10] S. F. Zanini, C. A. A. Torres, N. Bragagnolo, J. M. Turatti, M. G. Silva, and M. S. Zanini, "Effect of oil sources and vitamin E levels in the diet on the composition of fatty acids in rooster thigh and chest meat," *Journal of the Science of Food and Agriculture*, vol. 84, no. 7, pp. 672–682, 2004.

[11] S. Salman, D. Dinse, A. Khol-Parisini et al., "Colostrum and milk selenium, antioxidative capacity and immune status of dairy cows fed sodium selenite or selenium yeast," *Archives of Animal Nutrition*, vol. 67, no. 1, pp. 48–61, 2013.

[12] M. Larbier and B. Leclercq, "Nutrition et alimentation des volailles," *Editions Quae*, vol. 355, 1992.

[13] AOAC International, *Official Methods of Analysis*, Association of Official Analitical Chemists, Washington, DC, USA, 18th edition, 2005.

[14] A. Dal Bosco, C. Mugnai, S. Ruggeri, S. Mattioli, and C. Castellini, "Fatty acid composition of meat and estimated indices of lipid metabolism in different poultry genotypes reared under organic system," *Poultry Science*, vol. 91, no. 8, pp. 2039–2045, 2012.

[15] A. Haug, N. F. Nyquist, M. Thomassen, A. T. Høstmark, and T-K. K. Østbye, "N-3 fatty acid intake altered fat content and fatty acid distribution in chicken breast muscle, but did not influence mRNA expression of lipid-related enzymes," *Lipids in Health and Disease*, vol. 13, pp. 92–102, 2014.

[16] J. Folch, M. Lees, and G. H. Sloane-Stanley, "A simple method for isolation and purification of total lipides from animal tissues," *The Journal of Biological Chemistry*, vol. 226, no. 1, pp. 497–509, 1957.

[17] K. Ichihara, C. Yamaguchi, Y. Araya, A. Sakamoto, and K. Yoneda, "Preparation of fatty acid methyl esters by selective methanolysis of polar glycerolipids," *Lipids*, vol. 45, no. 4, pp. 367–374, 2010.

[18] K. Eder, "Gas chromatographic analysis of fatty acid methyl esters," *Journal of Chromatography B: Biomedical Sciences and Applications*, vol. 671, no. 1-2, pp. 113–131, 1995.

[19] T. L. V. Ulbricht and D. A. T. Southgate, "Coronary heart disease: seven dietary factors," *The Lancet*, vol. 338, no. 8773, pp. 985–992, 1991.

[20] M. Fernández, J. A. Ordóñez, I. Cambero, C. Santos, C. Pin, and L. D. L. Hoz, "Fatty acid compositions of selected varieties of Spanish dry ham related to their nutritional implications," *Food Chemistry*, vol. 101, no. 1, pp. 107–112, 2006.

[21] A. Haug, S. Eich-Greatorex, A. Bernhoft et al., "Effect of dietary selenium and omega-3 fatty acids on muscle composition and quality in broilers," *Lipids in Health and Disease*, vol. 6, article no. 29, 2007.

[22] Y. Wang, L. Jiang, Y. Li, X. Luo, and J. He, "Effect of different selenium supplementation levels on oxidative stress, cytokines, and immunotoxicity in chicken thymus," *Biological Trace Element Research*, vol. 172, no. 2, pp. 488–495, 2016.

[23] D. E. Ullrey, "Basis for regulation of selenium supplements in animal diets," *Journal of Animal Science*, vol. 70, no. 12, pp. 3922–3927, 1992.

[24] EFSA, "Safety and efficacy of selenium compouns (E8) as feed additives for all species: sodi-um selenite, based on a dossier submitted by Todini and Co Spa," *EFSA Journal*, vol. 14, p. 4442, 2016.

[25] G. Castromán, M. del Puerto, A. Ramos, M. C. Cabrera, and A. Saadoun, "Organic and con-ventional chicken meat produced

in Uruguay: colour, Ph, fatty acids composition and oxidative status," *American Journal of Food and Nutrition*, vol. 1, pp. 12–21, 2013.

[26] J. Praagman, J. W. J. Beulens, M. Alssema et al., "The association between dietary saturated fatty acids and ischemic heart disease depends on the type and source of fatty acid in the European Prospective Investigation into Cancer and Nutrition-Netherlands cohort," *The American Journal of Clinical Nutrition*, vol. 103, no. 2, pp. 356–365, 2016.

[27] S. Ševčíková, M. Skřivan, G. Dlouhá, and M. Koucký, "The effect of selenium source on the performance and meat quality of broiler chickens," *Czech Journal of Animal Science*, vol. 51, no. 10, pp. 449–457, 2006.

[28] Z. Kralik, G. Kralik, M. Grčević, P. Suchý, and E. Straková, "Effects of increased content of organic selenium in feed on the selenium content and fatty acid profile in broiler breast muscle," *Acta Veterinaria Brno*, vol. 81, no. 1, pp. 31–35, 2012.

[29] M. Betti, B. L. Schneider, W. V. Wismer, V. L. Carney, M. J. Zuidhof, and R. A. Renema, "Omega-3-enriched broiler meat: 2. Functional properties, oxidative stability, and consumer acceptance," *Poultry Science*, vol. 88, no. 5, pp. 1085–1095, 2009.

[30] Z. Zduńczyk, R. Gruzauskas, A. Semaskaite, J. Juskiewicz, A. Raceviciute-Stupeliene, and M. Wroblewska, "Fatty acid profile of breast muscle of broiler chickens fed diets with different levels of selenium and vitamin E," *Archiv fur Geflugelkunde*, vol. 75, no. 4, pp. 264–267, 2011.

[31] Y. Mehdi and I. Dufrasne, "Selenium in cattle: a review," *Molecules*, vol. 21, no. 5, p. 545, 2016.

[32] A. P. Simopoulos, "The importance of the omega-6/omega-3 fatty acid ratio in cardiovascular disease and other chronic diseases," *Experimental Biology and Medicine*, vol. 233, no. 6, pp. 674–688, 2008.

[33] M. Rafieian-Kopaei, M. Setorki, M. Doudi, A. Baradaran, and H. Nasri, "Atherosclerosis: process, indicators, risk factors and new hopes," *International Journal of Preventive Medicine*, vol. 5, no. 8, pp. 927–946, 2014.

[34] B. Vessby, I.-B. Gustafsson, S. Tengblad, M. Boberg, and A. Andersson, "Desaturation and elongation of fatty acids and insulin action," *Annals of the New York Academy of Sciences*, vol. 967, pp. 183–195, 2002.

[35] S. B. Smith, T. S. Hively, G. M. Cortese et al., "Conjugated linoleic acid depresses the δ9 desaturase index and stearoyl coenzyme A desaturase enzyme activity in porcine subcutaneous adipose tissue," *Journal of Animal Science*, vol. 80, no. 8, pp. 2110–2115, 2002.

8

Seasonal Microbial Conditions of Locally Made Yoghurt (Shalom) Marketed in Some Regions of Cameroon

Lamye Glory Moh,[1] Lunga Paul Keilah,[2] Pamo Tedonkeng Etienne,[3] and Kuiate Jules-Roger[1]

[1]Department of Biochemistry, University of Dschang, Dschang, Cameroon
[2]Department of Biochemistry, University of Yaounde 1, Yaounde, Cameroon
[3]Department of Animal Production, University of Dschang, Dschang, Cameroon

Correspondence should be addressed to Kuiate Jules-Roger; jrkuiate@yahoo.com

Academic Editor: Salam A. Ibrahim

The microbial conditions of locally made yoghurt (shalom) marketed in three areas of Cameroon were evaluated during the dry and rainy seasons alongside three commercial brands. A total of ninety-six samples were collected and the microbial conditions were based on total aerobic bacteria (TEB), coliforms, yeasts, and moulds counts as well as the identification of coliforms and yeasts using identification kits. Generally, there was a significant increase ($p \leq 0.05$) in total aerobic and coliform counts (especially samples from Bamenda), but a decrease in yeast and mould counts of the same samples during the rainy season when compared to those obtained during the dry season. These counts were mostly greater than the recommended standards. Twenty-one Enterobacteriaceae species belonging to 15 genera were identified from 72 bacterial isolates previously considered as all coliforms. *Pantoea* sp. (27.77%) was highly represented, found in 41% (dry season) and 50% (rainy season) of samples. In addition, sixteen yeast species belonging to 8 genera were equally identified from 55 yeast isolates and *Candida* sp. (76.36%) was the most represented. This result suggests that unhygienic practices during production, ignorance, warmer weather, duration of selling, and inadequate refrigeration are the principal causes of higher levels of contamination and unsafe yoghurts.

1. Introduction

Yoghurt consumption has become very popular in Cameroon ever since the production of locally made yoghurt started. Yoghurt in itself is a very nutritious diet [1] for people across all age groups. Yoghurt quality varies from one producer to another as there is no well-described standard for its production. In Cameroon, it is generally produced with leftover "shalom yaourt" or any commercial brand of yoghurt (Camlait or Dolait) for fermentation [2]. Consumers are becoming more inquisitive about the quality of these fermented products due to episodes of diarrhoea they experienced at times. Its high and easily assimilable nutritive value provides a suitable environment for microbial contamination, proliferation, and spoilage. Microbial contamination can lead to food poisoning outbreaks and unsatisfactory products [3] and this is an enormous economic problem worldwide.

Unsafe food is still an important threat in most developing countries, especially in Africa [4, 5]. Microbial contamination and foodborne microbial diseases constitute a large and growing public health concern. In fact, most countries with case-reporting systems have documented significant increases over the past few decades in foodborne microbial diseases incidence [6]. Milk is a highly nutritious food that serves as an excellent growth medium for a wide range of microorganisms [7]. Through microbial activity alone, approximately one-fourth of the world's food supply is lost [8]. Undesirable microbes that can cause spoilage of dairy products include Gram-negative psychrotrophs, coliforms, lactic acid bacteria, yeasts, and moulds. For this reason, increased emphasis should be placed on the microbiological examination of dairy products.

Food safety challenges in Africa include unsafe water and poor environmental hygiene, weak foodborne disease surveillance, inability of small and medium scale producers to provide safe food, outdated food regulations, and inadequate law enforcement, as well as insufficient cooperation among stakeholders. The World Health Organization requires that small scale dairy processing plants in the developing countries urgently comply with the Codex Alimentarius principles [5]. Few African countries have enacted foodborne disease surveillance systems; in Cameroon regulations concerning hygienic control of dairy products have been issued, but they are rarely enforced, and the hygienic condition of the milk chain is not sufficiently controlled (Njomaha, personal communication). Enterobacteriaceae and coliform bacteria within this family represent two of the most common groups of indicator organism used by the food industry [9]. Enterobacteriaceae, a large and heterogeneous family of Gram-negative bacteria, may constitute health hazards to the consumers [10] if present in yoghurt. They are useful indicators of overall GMP, but not necessarily faecal contamination [9]. The presence of indicators organisms usually indicates that a potential problem or failure in the process has occurred, whereas their absence in food provides a degree of assurance that the hygiene and food manufacturing process has been carried out appropriately. The most important index of microbiological quality is total bacterial counts, coliforms, yeasts, and moulds and detection of specific pathogens and their toxins as recorded by Kwee et al. [10]. E. coli, a coliform, is considered as normal flora of intestinal tracts of humans and animals. They have been used as indicator organisms for bacteriological quality of milk and its products [11]. Generally the presence of higher number of coliforms indicates heavy contamination caused by unsanitary conditions and poor production [12]. Although bacteria can be food spoilage organisms, yeasts and filamentous fungi are often involved in the deterioration of yoghurts [13, 14]. They are a major cause of yoghurt spoilage [15, 16] and their growth is favoured by the low pH of yoghurt [13, 17]. The presence of yeasts and moulds in milk and dairy products is undesirable even in small amounts due to the resulting objectionable changes that lower the products quality [18]. They are responsible for off-flavours, loss of texture quality due to gas production, and package swelling and shrinkage [19]. More so, moulds and yeasts growing in yoghurt utilize some of the acid and produce a corresponding decrease in the acidity, making the food environment more susceptible for proteolysis and putrefaction by bacteria [15, 20].

A lot of work has been done on the hygienic quality of yoghurt or locally made yoghurt in most part of the world [21–26]. Most of them concluded that the locally made types were of lower hygienic quality when compared to the commercial brands. In Cameroon, little or nothing has been done on the identification of these microbes, and no published studies exist on microbial quality of yoghurts produced and sold in Northwest and West Regions (Dschang and Bafoussam) of Cameroon. Even when an attempt is done in some of the above places or other regions of the country on dairy products, it often ends at the level of colony count which can be misleading at times without identification. Cameroon has two seasons which are the rainy and dry seasons. The rainy season begins in March and ends

around October depending on the part of the country, while the dry season begins in October/November and ends in March. Most streets, especially in the rural areas, are dirt roads, very dry, and dusty during dry season and become very muddy during the rainy season; thus, this study was carried out during the two seasons. Research in the field of safety/quality evaluation of market yoghurt/shalom is essential to create awareness among common people about the existing situation and protect the consumer's health and rights. Therefore, this study was conducted to evaluate the microbial quality of locally made yoghurts (shalom) available in some regions of Cameroon.

2. Materials and Methods

2.1. Collection of Samples. Samples of locally made yoghurts were collected from 15 (dry seasons) and 14 (rainy seasons) producers at three different occasions from some Regions (Bamenda, Bafoussam, and Dschang) of Cameroon from 2012 to 2013. This gave a total of eighty-seven samples: forty-five and forty-two during the dry (November–January) and rainy (May–early August) seasons, respectively. Concurrently, three commercial brands of yoghurt, (BB), (AA), and (CC), available in Cameroon were equally collected on the same day from a well-known sale point in Dschang, giving an overall sample size of ninety-six. Samples were collected in sterile and labeled containers and transported under aseptic conditions in an ice packed container, at 4–7°C to the Laboratory of Microbiology and Antimicrobial Substances, Faculty of Science, University of Dschang.

2.2. Microbiological Analysis

2.2.1. Preparation of Materials. All media were obtained in dehydrated forms and prepared according to the manufacturer's instructions. Glassware such as Petri-dishes, test tubes, pipettes, flasks, and bottles was sterilized in a hot oven at 170°C for two hours, whereas distilled water was sterilized by autoclaving for 15 min at 121°C [27].

2.2.2. Preparation of Serial Dilutions. This was done according to APHA [28] in which 1 ml of yoghurt from a homogenous sample was serially diluted into 9 mL of sterile distilled water to prepare eightfold dilutions from 10^{-1} to 10^{-8}. 50 μl of diluted samples were spread over prepared dried plates with different media.

2.2.3. Enumeration of Total Aerobic Bacteria (TEB). Nutrient agar (Oxoid) was used to determine the total aerobic bacterial count [29] and appropriate dilutions were pour-plated. The cultured plates were incubated aerobically at 37°C for 24 h. TEB was counted after the colonies were evaluated [30].

2.2.4. Enumeration of Coliform Bacteria. MacConkey agar (Oxoid) was used to determine the coliform count [29, 31]. The cultured plates were incubated aerobically at 37°C for 24 h after pour plating of the appropriate dilutions. The colonies were evaluated and counted at the end of the incubation [30].

TABLE 1: The microbial quality of yoghurt samples as a function of production area and producer during the dry season (I).

| Samples | Location | Microbial colony counts (log 10 CFU/ml) | | | | |
		Total bacterial count	Coliform counts	Total fungal counts	Yeast counts	Mould counts
V	Dschang	11.63 ± 0.10^{ij}	4.33 ± 0.16^{defg}	4.27 ± 0.13^{c}	4.02 ± 0.21^{bc}	3.56 ± 0.24^{cdef}
P	Dschang	9.28 ± 0.00^{cd}	5.12 ± 0.09^{g}	6.12 ± 0.13^{i}	4.97 ± 0.11^{c}	4.26 ± 0.11^{f}
G	Dschang	11.04 ± 0.11^{f}	4.29 ± 0.20^{defg}	5.56 ± 0.24^{h}	3.73 ± 3.23^{b}	1.10 ± 1.90^{ab}
NR	Bamenda	11.56 ± 0.02^{ij}	4.40 ± 0.03^{efg}	4.63 ± 0.07^{ef}	4.47 ± 0.19^{bc}	4.05 ± 0.18^{def}
S	Bamenda	11.78 ± 0.02^{k}	4.00 ± 0.19^{cdef}	4.25 ± 0.07^{bc}	4.17 ± 0.11^{bc}	2.40 ± 0.17^{bcde}
D	Bamenda	9.62 ± 0.02^{e}	3.88 ± 0.50^{cde}	4.34 ± 0.13^{cd}	4.27 ± 0.09^{bc}	2.35 ± 2.05^{bcd}
MR	Bamenda	8.80 ± 0.06^{a}	4.73 ± 0.09^{fgh}	4.91 ± 0.09^{g}	4.87 ± 0.08^{bc}	3.60 ± 0.30^{cdef}
MB	Bamenda	9.40 ± 0.17^{d}	3.53 ± 0.40^{cd}	5.04 ± 0.12^{g}	4.94 ± 0.15^{bc}	4.11 ± 0.13^{ef}
PA	Bamenda	9.16 ± 0.02^{bc}	3.30 ± 0.00^{c}	0.00 ± 0.00^{a}	0.00 ± 0.00^{a}	0.00 ± 0.00^{a}
SY	Bamenda	9.17 ± 0.07^{bc}	4.82 ± 0.16^{fgh}	4.93 ± 0.06^{g}	4.91 ± 0.07^{bc}	2.40 ± 2.07^{bcde}
PV	Bamenda	9.15 ± 0.05^{b}	3.53 ± 0.40^{cd}	4.43 ± 0.03^{cde}	4.38 ± 0.15^{bc}	2.20 ± 1.90^{bc}
T	Bafoussam	11.36 ± 0.14^{h}	6.90 ± 0.08^{i}	4.54 ± 0.13^{de}	4.54 ± 0.13^{bc}	0.00 ± 0.00^{a}
Ce	Bafoussam	11.18 ± 0.04^{g}	1.25 ± 2.18^{b}	4.02 ± 0.10^{b}	4.02 ± 0.10^{bc}	0.00 ± 0.00^{a}
C	Bafoussam	11.30 ± 0.08^{h}	5.37 ± 0.04^{h}	4.83 ± 0.07^{fg}	4.63 ± 0.13^{bc}	3.50 ± 0.17^{cdef}
K	Bafoussam	11.30 ± 0.05^{h}	4.31 ± 0.19^{defg}	4.61 ± 0.04^{ef}	4.11 ± 0.35^{bc}	2.53 ± 2.20^{bcdef}
AA	Commercial 1	9.54 ± 0.05^{e}	4.57 ± 0.31^{efgh}	4.84 ± 0.49^{fg}	4.84 ± 0.49^{bc}	0.00 ± 0.00^{a}
BB	Commercial 2	11.35 ± 0.03^{h}	0.00 ± 0.00^{a}	0.00 ± 0.00^{a}	0.00 ± 0.00^{a}	0.00 ± 0.00^{a}
CC	Commercial 3	11.48 ± 0.11^{i}	4.81 ± 0.05^{fgh}	4.65 ± 0.10^{ef}	4.65 ± 0.10^{bc}	0.00 ± 0.00^{a}

Values are mean ± SD of 3 determinants. Along the columns, values with the same letter (a, b, c, d, e, f, and g) are not significantly different ($p > 0.05$).

2.2.5. Enumeration of Yeast and Moulds. Sabouraud Dextrose agar (Oxoid) (supplemented with 0.5 g/l chloramphenicol) was used to determine yeast and mould counts [32]. After the pour-plated plates were incubated aerobically at 25°C for 3–5 days, the developed colonies were evaluated and counted.

2.3. Isolation and Identification of Microorganisms. The distinguished colonies on the incubated plates were picked and purified by repeated subculturing done by streaking on the appropriate media with a sterile loop (the strategy consisted of picking 1 colony to represent every visibly different morphology on each plate) using the streak method. Purified colonies were prepared in their respective broth: Mueller Hinton broth for coliforms and Sabouraud Dextrose broth for yeast and moulds. From these preparations, 0.5 ml of each was pipitted into 0.5 ml of glycerol and stored in a freezer at −5°C awaiting identification. All the bacterial cultures were subcultured prior to their use in further experiments and the obtained fresh cultures were used for biochemical tests.

By microscopic observation of each culture following incubation, the purity of isolates was confirmed and preliminary identifications were done according to Bergey's Manual [33, 34]. Proper identification to species level was carried out on the basis of biochemical tests with API 20E (for identification of Enterobacteriaceae and other nonfastidious Gram-negative rods) and API 20 C AUX (for the identification of yeast) (bioMerieux, Marcy l'Etoile, France) according to the instructions of the manufacturer.

2.4. Statistical Analysis. Data were subjected to the one-way analysis of variance (ANOVA), and differences between samples at $p \leq 0.05$ were determined by Waller Duncan test using the Statistical Package for the Social Sciences (SPSS) version 11.0. The results were presented as an average of the logarithm ($\log_{10}$) of colony forming unit (cfu)/ml (log 10 cfu/ml) in the samples as mean ± SD of the replicates.

3. Results and Discussion

3.1. Total Aerobic Bacterial Counts. The total aerobic bacterial counts of all the samples (Table 1) during the dry season in Cameroon (November–January) were very high when compared to other results. Samples from Dschang had bacterial counts of 9.28 ± 0.00 to 11.63 ± 0.10; Bamenda, 8.80 ± 0.06 to 11.78 ± 0.02; Bafoussam 11.18 ± 0.04 to 11.36 ± 0.14; and commercial brands 9.54 ± 0.05 to 11.48 ± 0.11 (log 10 cfu/ml). Samples from Bafoussam had the highest total bacterial count (with no variation within 75% of samples), followed by those from Bamenda, commercial brands, and Dschang. As shown in Table 2, total aerobic bacterial counts varied significantly ($p \leq 0.05$) with each other as well as the commercial brands during the rainy season. These counts were still very high when compared to 6.77 obtained by Al-Tahiri [35] with yoghurts produced by modern dairies in Jordan and 7.86 from Younus et al. [21] obtained from dahi (locally made yoghurt in Pakistan and India).

During the rainy season, samples from Dschang had bacterial counts of 8.70 ± 0.09 to 11.55 ± 0.06; Bamenda, 9.17 ± 0.13 to 11.73 ± 0.01; Bafoussam, 11.02 ± 0.08 to 11.75 ± 0.03; and commercial brands, 10.34 ± 0.12 to 11.34 ± 0.06 (log 10 cfu/ml). On average, bacterial count of all samples was

TABLE 2: The microbial quality of yoghurt samples as a function of production area and producer during the rainy season (II).

Samples	Location	Microbial colony counts (log 10 CFU/ml)				
		Total bacterial count	Coliform counts	Total fungal counts	Yeast counts	Mould counts
P	Dschang	8.70 ± 0.09^a	5.55 ± 0.07^{hi}	5.02 ± 0.05^h	4.87 ± 0.10^{fg}	4.30 ± 0.21^e
G	Dschang	11.55 ± 0.06^h	4.49 ± 0.35^{cdef}	4.06 ± 0.20^c	4.05 ± 0.18^c	0.00 ± 0.00^a
SY	Bamenda	11.64 ± 0.02^{hi}	4.43 ± 0.08^{cdef}	4.57 ± 0.06^{def}	4.39 ± 0.02^d	3.96 ± 0.21^{de}
NR	Bamenda	9.17 ± 0.13^b	3.87 ± 0.33^c	4.54 ± 0.05^{de}	4.06 ± 0.15^c	4.43 ± 0.01^e
PV	Bamenda	11.18 ± 0.05^f	4.10 ± 0.30^{cd}	3.94 ± 0.18^{bc}	3.66 ± 0.39^b	3.40 ± 0.17^d
PA	Bamenda	11.23 ± 0.04^{fg}	5.46 ± 0.27^{ghi}	0.00 ± 0.00^a	0.00 ± 0.00^a	0.00 ± 0.00^a
D	Bamenda	11.03 ± 0.11^e	3.76 ± 0.40^c	4.46 ± 0.10^d	4.34 ± 0.12^d	3.77 ± 0.00^{de}
MB	Bamenda	9.55 ± 0.04^c	3.84 ± 0.27^c	3.73 ± 0.37^b	3.70 ± 0.34^b	2.35 ± 2.05^c
S	Bamenda	11.06 ± 0.04^e	4.27 ± 0.18^{cde}	4.08 ± 0.15^c	4.08 ± 0.15^c	0.00 ± 0.00^a
MR	Bamenda	11.73 ± 0.01^i	5.51 ± 0.03^{hi}	4.41 ± 0.06^d	4.41 ± 0.06^d	0.00 ± 0.00^a
T	Bafoussam	11.19 ± 0.04^f	4.76 ± 0.15^{defg}	5.05 ± 0.02^h	5.05 ± 0.02^g	0.00 ± 0.00^a
Ce	Bafoussam	11.75 ± 0.03^i	5.10 ± 0.02^{fghi}	4.92 ± 0.08^{gh}	4.92 ± 0.08^{fg}	0.00 ± 0.00^a
C	Bafoussam	11.02 ± 0.08^e	0.00 ± 0.00^a	4.61 ± 0.09^{def}	4.49 ± 0.31^{de}	1.30 ± 2.25^b
K	Bafoussam	11.27 ± 0.02^{fg}	1.20 ± 2.07^b	4.52 ± 0.11^d	4.52 ± 0.11^{de}	0.00 ± 0.00^a
AA	Commercial 1	11.02 ± 0.17^e	4.85 ± 0.04^{efgh}	4.50 ± 0.18^d	4.50 ± 0.18^{de}	0.00 ± 0.00^a
BB	Commercial 2	11.34 ± 0.06^g	0.00 ± 0.00^a	0.00 ± 0.00^a	0.00 ± 0.00^a	0.00 ± 0.00^a
CC	Commercial 3	10.34 ± 0.12^d	4.34 ± 0.09^{cde}	4.77 ± 0.17^{efg}	4.77 ± 0.17^{ef}	0.00 ± 0.00^a

Values are mean ± SD of 3 determinants. Along the columns, values with the same letter (a, b, c, d, e, f, g, and h) are not significantly different ($p > 0.05$).

significantly higher ($p \leq 0.05$) than the commercial brands. Samples from Bafoussam had the highest aerobic bacterial count, followed by those from Dschang and then Bamenda (Table 2). However, 25% of samples from Bamenda and 50% of samples from Dschang had total bacterial counts less than that of the least commercial brand. Generally, there was a significant increase ($p \leq 0.05$) in total count during the rainy season when compared to those obtained during the dry season (especially samples from Bamenda). This could be due to the low turnover of yoghurt during this season since a considerable number of people ignorantly consume it just to quench taste, regardless of the nutritional value. Consequently, temperature fluctuation resulting from longer selling/storage time may offer a favourable environment for the multiplication of these bacteria.

High bacterial count is also expected because of the presence of starter cultures, which are mainly lactic acid bacteria. The standard aerobic bacterial count is 10^6–10^7 cfu/ml [36, 37], corresponding to 6-7 in log 10 cfu/ml. Thus, the results of this study showed that total aerobic bacterial count in all samples was very high relative to the standard values. Very high count however is used as an indication of postpasteurisation contamination [38], due to inadequate hygienic measures during production. In most foods, the total bacterial count is often an indication for the sanitary quality, safety, and utility of foods. It may reflect the conditions under which the product is manufactured such as contamination of raw materials and ingredients, the effectiveness of processing, and the sanitary conditions of equipment and utensils at the processing plants [11]. Storage in unhygienic conditions and prolonged storage time can also contribute to this [39].

This signals the paramount need for a sensitization of yoghurt producers involved in this study as per the hygienic conditions of their production processes.

3.2. Coliform Counts. A majority of the samples had coliform counts higher than 10^2 cfu/ml (2 in log 10 cfu/ml) (Table 1) which is the maximum determined in most of the international standards (Kucukoner and Tarakci, 2003). The coliform count of samples from Dschang during the dry season ranged from 4.29 ± 0.20 to 5.12 ± 0.09; Bamenda, 3.30 ± 0.00 to 4.82 ± 0.16; Bafoussam, 1.25 ± 2.18 to 6.90 ± 0.08; and commercial brand, 0.00 ± 0.00 to 4.81 ± 0.05 (log 10 cfu/ml). Coliform count of samples varied significantly ($p \leq 0.05$) from each other. Samples from Dschang (66.66%), Bamenda (62.50%), and Bafoussam (50%) were significantly lower ($p \leq 0.05$) than the commercial brands. Generally, during the rainy season counts were still greater than 10^2 cfu/ml (Table 2). Samples from Dschang had coliform counts from 4.49 ± 0.35 to 5.55 ± 0.07; Bamenda, 3.76 ± 0.40 to 5.51 ± 0.03; Bafoussam, 0.00 ± 0.00 to 5.10 ± 0.02; and commercial brands, 0.00 ± 0.00 to 4.85 ± 0.04 (log 10 cfu/ml). Coliform counts generally increased in 47.05% and decreased in 41.17% during the rainy season. Like in total bacterial count, there was a significant increase ($p \leq 0.05$) in coliform count during the rainy season when compared to those obtained during the dry season in samples from Bamenda, while those from Bafoussam decreased. Interestingly, sample BB (commercial) was void of coliforms during both seasons as well as C from Bafoussam during the rainy season (Tables 1 and 2).

These results are in line with the work of Moreira et al. [40] who reported that warmer weather and inadequate refrigeration are the principal causes of higher levels of contamination. Coliforms detection or enumerations are often used as parameters for evaluating yoghurt quality in different countries [12, 41, 42]. It is an indicator of poor hygiene, inadequate processing, or postprocessing contamination in yoghurt as established and recommended by public health authorities worldwide, used as indicator organisms for bacteriological quality of milk and its products [11]. The high levels of coliform counts (3.30 ± 0.00 to 6.90 ± 0.08 log 10 cfu/ml) in both the locally made varieties and even the commercial brands (except BB) might indicate a low level of hygiene and improper sanitation during/after the manufacturing process [43] or insufficient preheating during production. It also shows negligence in sanitary measures especially the commercial brands which are regarded as quality-controlled products. Meanwhile, the absence of coliforms in some samples as mentioned above is an indication of Good Manufacturing Practices (GMP) employed by the producers and retailers [25, 44]. Coliforms are not supposed to be present in yoghurt in such high levels because of pasteurisation and controlled hygienic procedures [32]. The presence of coliforms in these yoghurts could pose an adverse effect in consumers' health and suggests negligence on the part of the producers or the yoghurt vendors as well as quality controllers. According to the standard stipulated by the National Agency of Food and Drug Administration Control (NAFDAC), E. coli and coliforms generally must not be detectable in any 100 ml of yoghurt sample [45]. Other probabilities of contamination can be from contaminated water source and equipment used or due to contamination at storage and display/sale outlet [46].

3.3. Coliform Species and Other Enterobacteriaceae. A total of seventy-two (72) bacterial isolates (40 during the dry season (I) and 32 during the rainy season (II)) were identified to specie level from ninety-six (96) yoghurt samples (Table 3). They were distributed as follows: 27 (14 I and 13 II), 14 (10 I and 4 II), 25 (13 I and 12 II), and 6 (3 I and 3 II) from Bamenda, Dschang, Bafoussam, and commercial brands, respectively. Twenty-one (21) bacteria species belonging to 15 genera were identified, with the number of occurrences indicated in parentheses: *Pantoea* sp. *1* (15), *Pantoea* sp. *3* (3), *Pantoea* sp. *4* (2), *Klebsiella pneumonia* ssp. *pneumonia* (11), *Klebsiella oxytoca* (1), *Providencia stuartii* (5), *Providencia rettgeri* (2), *Providencia alcalifaciens/rustigianii* (1), *Shigella* sp. (7), *Enterobacter aerogenes* (3), *Enterobacter cloacae* (3), *Serratia plymuthica* (5), *Escherichia coli 1* (3), *Burkholderia cepacia* (3), *Citrobacter freundii* (2), *Pseudomonas oryzihabitans* (2), *Moellerella wisconsensis* (2), *Raoultella ornithinolytica* (1), *Pasteurella pneumotropica/Mannheimia haemolytica* (1), *Ochrobactrum anthropi* (1), *and Proteus penneri* (1) (Table 3).

Generally, the percentage of isolates was higher during the dry season (55.55%) when compared to those of the rainy season (44.44%). *Pantoea* sp. (27.77%) was the highest represented species with 21.42%, 20.00%, 46.13%, and 0.00% (dry season) as well as 38.46%, 0.00%, 25.00%, and 33.33% (rainy season) from Bamenda, Dschang, Bafoussam, and commercial brands, respectively. It was present in all the batches (except commercial samples during the dry season

and Dschang during the rainy season) and found in 7 out of 17 (dry season) and 8 out of 16 (rainy season) samples. This was followed by *Klebsiella* sp. (16.66%), absent in Bamenda throughout but occupying 20%, 23.07%, and 33.33% (dry season) when compared to 75.00%, 8.33%, and 33.33% (rainy season) isolates from Dschang, Bafoussam, and commercial brands, respectively. The third candidate was *Providencia* sp. (11.11%), absent in the commercial samples throughout the season with 21.48%, 10.00%, and 0.00% (dry season) as well as 15.38%, 0.00%, and 8.33% (rainy season) from Bamenda, Dschang, and Bafoussam, respectively. The fourth was *Shigella* sp. (9.72%), absent in Dschang during both seasons, Bafoussam, and commercial brands (rainy season) but present in Bamenda during the two seasons. Notwithstanding, the lowest frequency of occurrence (1.38%) was recorded by *Raoultella ornithinolytica, Pasteurella pneumotropica/Mannheimia haemolytica, Ochrobactrum anthropi, and Proteus penneri* as shown in Table 3.

It can be observed in Table 4 that some of these bacteria were present only in one of the places or yoghurt brand, for example, species like *Pantoea* sp. *3, Pantoea* sp. *4, Raoultella ornithinolytica and Providencia alcalifaciens/rustigianii* (Bamenda), *Ochrobactrum anthropi, Providencia rettgeri, and Pasteurella pneumotropica/Mannheimia haemolytica* (Dschang in dry season) as well as *Klebsiella oxytoca, Citrobacter freundii,* and *Escherichia coli 1* in samples from Bafoussam during the rainy season (II). Also, the specie *Proteus penneri* was only observed in the commercial sample (CC). The rest of the species were present in 2 or 3 places and in one season or the other (Table 4).

There were differences in the bacteriological load of batches of yoghurt samples assessed. This together with the isolation of indicator organisms shows failure of GMP in industries and local producers that manufactured these yoghurts from which they were isolated [47]. The genus *Pantoea* includes several species of which others can cause disease in humans such as tumors [48, 49]. However, *E. coli,* an index organism, indicates the presence of other pathogenic microorganisms and has been linked to diarrhoeal diseases, urethrocystitis, prostatitis, and pyelonephritis [23]. More so, Enteropathogenic *E. coli* have also been incriminated as a potential food poisoning agent and are associated with infantile diarrhoea and gastroenteritis in adults. *E. coli* might had entered into some of these yoghurt samples through water used in production, unhygienic hawking habits, and storage environment and not necessarily failure of GMP. Meanwhile, *Klebsiella* sp., another coliform, may be an indicator of product contamination through faecal contaminated water or raw materials [50].

Coliforms have been related to bacterial pneumonia cases more severe than those produced by *Streptococcus pneumonia* and urinary tract infection. This is the case of *K. pneumoniae* and *K. oxytoca* which are opportunistic pathogens and have been linked over the years as the main cause of septicaemia, pneumonia, urinary tract infections, and soft tissue infections [51, 52]. *Burkholderia cepacia* is also known for pneumonia or bacterial infections that occur in patients with impaired immune systems or chronic lung disease, particularly cystic fibrosis (CF). Infection with *Shigella* sp. is normally limited

TABLE 3: Enterobacteriaceae found in yoghurts samples from two different batches as a function of sample group and season.

Place	Producers	Enterobacteriaceae species/season				Total number of each Enterobacteriaceae from each group/season	
		I	% of identification	II	% of identification	I	II
Bamenda	NR	Pantoea sp. 1, Enterobacter cloacae Raoultella ornithinolytica,	79.30 95.27 91.05	Pantoea sp. 3 Pantoea sp. 3,	75.89 94.90	Pantoea sp. 1 (2), Pantoea sp. 4 (1), Raoultella ornithinolytica (1), Shigella sp. (2), Serratia plymuthica (1), Burkholderia cepacia (1), Providencia stuartii (2), Providencia alcalifaciens/rustigianii (1), Pseudomonas oryzihabitans (1)	Pantoea sp. 1 (1), Pantoea sp. 3 (3), Pantoea sp. 4 (1), Enterobacter cloacae (2), Shigella sp. (3), Serratia plymuthica (1), Burkholderia cepacia (1), Providencia stuartii (2), Moellerella wisconsensis (1)
	MR	Enterobacter cloacae, Shigella sp.	95.27 70.17	Serratia plymuthica, Shigella sp., Burkholderia cepacia	91.39 66.01 81.30		
	D	Serratia plymuthica	91.39	Shigella sp., Pantoea sp. 4	69.04 75.58		
	PV	Shigella sp., Providencia stuartii, Providencia alcalifaciens/rustigianii	69.04 86.08 61.94	Shigella sp.	69.04		
	SY	Pantoea sp. 4 Pantoea sp. 1	86.48 80.43	Providencia stuartii Pantoea sp. 3	91.13 75.89		
	S	Pseudomonas oryzihabitans	82.77	Moellerella wisconsensis	89.38		
	MB	Burkholderia cepacia Providencia stuartii	81.30 86.08	Providencia stuartii Pantoea sp. 1	71.56 73.72		
Dschang	P	Moellerella wisconsensis Pasteurella pneumotropica/Mannheimia haemolytica Pantoea sp. 1	88.87 70.36 73.72	Klebsiella pneumonia ssp. pneumonia Klebsiella pneumonia ssp. pneumonia Providencia rettgeri	97.69 98.88 100	Moellerella wisconsensis (1), Klebsiella pneumonia ssp. pneumonia (2), Pasteurella pneumotropica/Mannheimia haemolytica (1), Pantoea sp. 1 (2), Providencia rettgeri (1), Ochrobactrum anthropi (1), Enterobacter cloacae (1), Pseudomonas oryzihabitans (1)	Klebsiella pneumonia ssp. pneumonia (3), Providencia rettgeri (1)
	V	Enterobacter cloacae Pseudomonas oryzihabitans Klebsiella pneumonia ssp. pneumonia	81.23 85.72 97.69	/	/		
	G	Klebsiella pneumonia ssp. pneumonia Providencia rettgeri Ochrobactrum anthropi Pantoea sp. 1	97.69 100 92.49 79.30	Klebsiella pneumonia ssp. pneumonia	97.69		

Table 3: Continued.

Place	Producers	Enterobacteriaceae species/season				Total number of each Enterobacteriaceae from each group/season	
		I	% of identification	II	% of identification	I	II
		Klebsiella pneumonia ssp. pneumonia	98.85	Serratia plymuthica	91.39		
	T	Klebsiella pneumonia ssp. pneumonia	97.69	Citrobacter freundii	99.99		
		Klebsiella pneumonia ssp. pneumonia	98.01	Citrobacter freundii	99.99		
Bafoussam		Enterobacter aerogenes	78.96	Klebsiella oxytoca	98.46	Klebsiella pneumonia ssp. pneumonia (4), Enterobacter aerogenes (2), Serratia plymuthica (1), Pantoea sp. 1 (6), Burkholderia cepacia (1), Shigella sp. (1)	Serratia plymuthica (2), Citrobacter freundii (2), Klebsiella oxytoca (1), Escherichia coli 1 (3), Pantoea sp. 1 (3), Providencia stuartii (1)
				Serratia plymuthica	86.82		
	Ce	Pantoea sp. 1	90.19	Escherichia coli 1	99.84		
				Escherichia coli 1	92.99		
		Pantoea sp. 1	90.49	Escherichia coli 1	99.02		
				Providencia stuartii	91.13		
		Pantoea sp. 1	80.43				
	C	Pantoea sp. 1	90.49	Pantoea sp. 1	90.19		
		Serratia plymuthica	91.39				
		Burkholderia cepacia	71.24				
		Shigella sp.	69.04				
	K	Pantoea sp. 1	90.75	Pantoea sp. 1	90.75		
		Pantoea sp. 1	90.75	Pantoea sp. 1	81.83		
Commercial	AA	Enterobacter aerogenes	83.44			Enterobacter aerogenes (1), Klebsiella pneumonia ssp. pneumonia (1), Shigella sp. 1 (1)	Klebsiella pneumonia ssp. pneumonia (1), Pantoea sp. 1 (1), Proteus penneri (1)
		Klebsiella pneumonia ssp. pneumonia	98.67	Klebsiella pneumonia ssp. pneumonia	98.67		
	CC	Shigella sp.	83.19	Pantoea sp. 1	83.19		
				Proteus penneri	77.49		

[1] The letters I and II refer to batches of the dry and rainy seasons, respectively. [2] The sixteen manufacturers/producers are designated as NR, MR, D, PV, SY, S, MB, P, V, G, T, Ce, C, K, AA, and CC. [3] Numbers in parentheses are the number of isolates identified as that species.

TABLE 4: The distribution of Enterobacteriaceae as a function of sample group and season.

| Gram negative bacteria isolated from yoghurt | Yoghurt samples/season | | | | | | | | Frequency of occurrence (I) | % of occurrence (I) | Frequency of occurrence (II) | % of occurrence (II) |
| | Bamenda | | Dschang | | Bafoussam | | Commercial | | | | | |
	(I)	(II)	(I)	(II)	(I)	(II)	(I)	(II)				
Pantoea sp. 1	+	+	+	−	+	+	−	+	3	75	3	75
Pantoea sp. 3	−	+	−	−	−	−	−	−	0	00	1	25
Pantoea sp. 4	+	+	−	−	−	−	−	−	1	25	1	25
Raoultella ornithinolytica	+	−	−	−	−	−	−	−	1	25	0	00
Shigella sp.	+	+	−	−	+	−	+	−	3	75	1	25
Serratia plymuthica	+	+	−	−	+	+	−	−	2	50	2	50
Burkholderia cepacia	+	+	−	−	+	−	−	−	2	50	1	25
Providencia stuartii	+	+	−	−	−	+	−	−	1	25	2	50
Providencia rettgeri	−	−	+	+	−	−	−	−	1	25	1	25
Providencia alcalifaciens/rustigianii	+	−	−	−	−	−	−	−	1	25	0	00
Pseudomonas oryzihabitans	+	−	+	−	−	−	−	−	2	50	0	00
Moellerella wisconsensis	−	+	+	−	−	−	−	−	1	25	1	25
Klebsiella pneumonia ssp. *pneumonia*	−	−	+	+	+	−	+	+	3	75	2	50
Klebsiella oxytoca	−	−	−	−	−	+	−	−	0	00	1	25
Pasteurella pneumotropica/Mannheimia haemolytica	−	−	+	−	−	−	−	−	1	25	0	00
Ochrobactrum anthropi	−	−	+	−	−	−	−	−	1	25	0	00
Enterobacter aerogenes	−	−	−	−	+	−	+	−	2	50	0	00
Enterobacter cloacae	−	+	+	−	−	−	−	−	1	25	1	25
Citrobacter freundii	−	−	−	−	−	+	−	−	0	00	1	25
Escherichia coli 1	−	−	−	−	−	+	−	−	0	00	1	25
Proteus penneri	−	−	−	−	−	−	−	+	0	00	1	25

The letters I and II refer to batches of the dry and rainy seasons, respectively; + = positive, − = negative.

to the distal ileum and colon, and common symptoms include diarrhoea, fever, nausea, vomiting, stomach cramps, and flatulence. In cases of *Shigella*-associated dysentery, the epithelial cells of the intestinal mucosa in the caecum and rectum are destroyed. *Shigella* has also been implicated as one of the causes of reactive arthritis. Apart from that, *E. cloacae* and *C. freundii* are associated with illnesses such as necrotizing enterocolitis, diarrhoea, meningitis, urinary tract infections, intra-abdominal and ophthalmic infections, and septicaemia [53–55]. Generally, the presence of *Enterobacter* sp. in yoghurt or other foods may be caused by poor environmental conditions due to dust and contaminated water used in production. These species have been known to be inhabitants of dairy products [50]. Meanwhile, *Citrobacter* sp. has been shown to carry various virulence determinants found in other pathogens. *Providencia alcalifaciens*, a food poisoning agent, causes diarrhoea [56] especially in children and *Providencia stuartii* is also linked to infective endocarditis [57]. In addition, *Providencia rettgeri* is the cause of ocular infections, including keratitis, conjunctivitis, and endophthalmitis [58] as well as travelers' diarrhoea. *P. alcalifaciens*, *P. rettgeri*, and *P. stuartii* have generally been implicated in gastroenteritis. Applebaum and Campbell [59] reported that *Ochrobactrum anthropi* was responsible for an infection in humans known as pancreatic abscess. In addition, *Raoultella ornithinolytica* has been known to cause enteric fever-like syndrome [60], giant renal cyst leading to colic obstruction [61], and *R. ornithinolytica* bacteremia [62]. *Pseudomonas* is found in soil, water, plants, and animal and is present in small percentage in the normal intestinal flora and on the human skin [63]. Lastly *P. penneri* has the ability to cause major infectious diseases and nosocomial outbreaks [64] and carries similar pathogenic determinants like *P. mirabilis* and *P. vulgaris* [65]. It usually infects urinary tract, blood, neck, and ankle [65, 66]. Thus, these yoghurts predispose their consumers to a vast array of diseases whose causative agents are supposed to be susceptible to pasteurisation. However, their presence in postpasteurised yoghurt may be as a result of inadequate heating process, the use of contaminated water, postproduction contamination, and the presence of poor sanitary behaviours during packaging and storage conditions at the production areas as well as unhygienic hawking habits.

3.4. Yeast Counts. Generally, there was no significant difference ($p > 0.05$) in yeast counts amongst the samples during the dry season (Table 1). The present data show that 87.5% and 66.66% of samples from Bamenda and commercial brands had yeast, while it was present in 100% of samples from Dschang and Bafoussam during the dry season. Samples from Dschang had counts from 3.73 ± 3.23 to 4.97 ± 0.11; Bamenda 0.00 ± 0.00 to 4.94 ± 0.15; Bafoussam, 4.02 ± 0.10 to 4.63 ± 0.13; and commercial brands, 0.00 ± 0.00 to 4.84 ± 0.49 (log 10 cfu/ml). Samples collected from all the localities had yeast count (4.02 ± 0.21 to 4.97 ± 0.11 (log 10 cfu/ml)) higher than 3 log 10 cfu/ml which is the international standards [35, 37, 67]. There was a decrease in yeast counts during the rainy season (Table 2) in samples from Bamenda and an increase in those from Bafoussam when compared to most of the samples collected during the dry season (Table 1). Still, 87.5% and

66.66% of samples from Bamenda and commercial brands had yeast, as well as 100% of samples from Dschang and Bafoussam. The yeast count of samples from Dschang was from 4.05 ± 0.18 to 4.87 ± 0.10; Bamenda, 0.00 ± 0.00 to 4.41 ± 0.06; Bafoussam, 4.49 ± 0.31 to 5.05 ± 0.02; and commercial brands, 0.00 ± 0.00 to 4.77 ± 0.17 (log 10 cfu/ml). Throughout the seasons, samples from Bafoussam had the highest yeast counts, followed by those from Dschang, Bamenda, and commercial brands the least. The higher yeast count during the dry season might reflect the ability of more yeast to grow during warmer weather [39], increased species diversity, and alteration in microbial flora leading to higher levels of contamination. This explains why there was an increase in yeast count during the dry season when compared to yoghurt sold in the rainy season.

High counts of yeast and mould have also been reported in yoghurts [23, 68–70]. Though higher than the international standard in most cases, yeast counts in this study corroborated those reported in Australia [71, 72], Nigeria [73], and Egypt [74]. However, examples of yeast occurrences in yoghurts with more than 6 log 10 cfu/ml [75, 76] and 3 log 10 cfu/ml or lesser have also been recorded from various countries such as UK, Canada, USA, and the Netherlands [77–79]. The high levels of moulds and yeast obtained in this study are attributed to poor handling and production [20, 66, 80]. Certain yeasts play an important role in the spoilage of fermented products. Since milk is pasteurised before yoghurt production, the presence of yeasts in yoghurt is caused by inappropriate pasteurisation and/or recontamination processes during manufacture [75].

3.5. Mould Counts. The control samples were void of moulds, while 100%, 87.5%, and 50% of samples from Dschang, Bamenda, and Bafoussam had moulds, respectively, during the dry season (Table 1). Yoghurt samples from Dschang and Bamenda had mould counts of 0.00 ± 0.00 to 4.26 ± 0.11 and Bafoussam 2.53 ± 2.20 to 3.50 ± 0.17 (log 10 cfu/ml). The control samples were still void of moulds during the rainy season with a reduction in the spread and count of moulds in all the regions. In this season, 50%, 57.14%, and 25% of samples from Dschang, Bamenda, and Bafoussam had moulds, respectively. A maximum of 2 log 10 cfu/ml of mould is allowed in yoghurt [81]. Yoghurts having initial mould counts > 2 log 10 cfu/ml tend to spoil quickly and may even spoil before refrigeration [40]. This standard was not met in 27.78% (Table 1) and 29.41% (Table 2) of all samples. There was a clear association between levels of yeast and moulds contamination and season, as counts were higher during the dry season and lower during the rainy season. The presence of high yeast and mould counts in examined yoghurt samples may also indicate inefficient preheating process during manufacturing, using unsatisfactory sterilized plastic cups in packing or inefficient chilling on storage [82]. However, it could as well be attributed to contamination from air and the old yoghurt or commercial yoghurt used as starter culture during production. Mould and yeast contamination causes deterioration and influences the biochemical characters and flavour of the product and its appearance is commercially undesirable and often results in downgrading of the product.

Spoilage becomes evident when yeast populations reach 5 to 6 (log 10 cfu/ml) and is first recognized as a swelling of the yoghurt package due to gas production by yeast fermentation. The yoghurt acquires a yeasty, fermented odour and flavour and a gassy appearance which eventually ruptures; colonies of yeasts on the undersurface of the package lid can be seen at times [13].

3.6. *Yeast Species.* Fifty-five yeast isolates (25 during the dry season (I) and 30 during the rainy season (II)) were identified to specie level from ninety-six yoghurt samples (Table 5). They were distributed as follows: 28 (11 I and 17 II), 9 (7 I and 2 II), 12 (5 I and 7 II), and 6 (2 I and 4 II) from Bamenda, Dschang, Bafoussam, and commercial brands, respectively. Sixteen yeast species belonging to 8 genera were identified, with the number of occurrences indicated in parentheses: *Candida zeylanoides* (15), *Candida kruzei/inconspicua* (14), *Candida dubliniensis* (6), *Candida lusitaniae* (3), *Candida boidinii* (2), *Candida albicans 1* (1), *Candida albicans 2* (1), *Trichosporon asahii* (3), *Stephanoascus ciferrii* (2), *Kodamaea ohmeri* (2), *Rhodotorula mucilaginosa 1* (1), *Rhodotorula mucilaginosa 2* (1), *Pichia angusta* (1), *Cryptococcus laurentii* (1), *Cryptococcus humicola* (1), and *Kloeckera* sp. (1) (Table 5).

Generally, the percentage of isolates was higher during the rainy season (54.54%) when compared to that of the dry season (45.45%). *Candida* sp. was the highest among the isolates (76.36%), with the highest percentage contributed by *Candida zeylanoides* (27.27%), *Candida kruzei/inconspicua* (25.45%), and *Candida dubliniensis* (10.90%). Other species that were detected in lower percentages were *Candida lusitaniae* (5.45%), *Candida boidinii* (3.63%), *Candida albicans 1* (1.81%), and *Candida albicans 2* (1.81%). *Candida zeylanoides* occupying the highest percentage was absent in the commercial samples (dry and rainy seasons) and those from Dschang and Bafoussam (dry season). It was mostly represented during the dry season with 27.27% (present in all samples) and 42.85% and 20.00% in samples from Bamenda, Dschang, and Bafoussam, respectively, though it was higher in Bamenda (45.45%) during the rainy season. *Candida kruzei/inconspicua* was absent in Dschang but represented 45.45%, 40.00%, and 50.00% (dry season) as compared to 23.52%, 14.28%, and 25.00% (rainy season) in samples from Bamenda, Bafoussam, and commercial brands, respectively. On the other hand, *Candida dubliniensis* either present during the dry (Bamenda) or rainy season (Bafoussam) or even during both seasons (commercial brands) having the highest percentage in Bafoussam (28.57%). With the exception of *Candida lusitaniae* which was found in 3 of the 4 groups of samples (Dschang, Bafoussam, and commercial brand), the rest were present either in just one or two of these groups or in just one of the seasons (Table 6). There was no association between total yeast count and number of species found but the diversity was greater during the rainy season when compared to those of the dry season. This could lead to increased mutualism in the breakdown/utilization of the food substrate and thus enhancing spoilage [40] during the rainy season. Different species of yeasts were found in the same manufacturer's yoghurt on different occasions (dry and rainy seasons) suggesting that the contamination

was not systematic. Only *Candida kruzei/inconspicua* was found in at least one sample from all the different groups (except those from Dschang). It was found in 7 out of 14 (dry season) and in 6 out of 13 samples containing yeasts (Table 6). Several species of *Candida* have been reported as contaminant in yoghurt [83, 84]. They mainly involve deterioration [13, 14] and also responsible for off-flavours and loss of texture quality due to gas production during lactose assimilation [19]. *C. albicans* is a member of the normal flora of the skin and oral cavity; its presence in the samples may be due to high sugar content of yoghurt [72]. With the presence of these yeasts species, these yoghurts may expose their consumers to possible risk of fungal infections, of which candidiasis is the most deleterious and life threatening [85]. Some reports suggest that the public health significance of yeast contaminants in foods is negligible as few known pathogenic yeasts, such as *Candida albicans* and *Cryptococcus neoformans,* are not transmitted through foods [14, 86]. However, the public health safety of yeasts in foods may need some rethinking as there have been occasional reports of gastroenteritis from foods, wherein yeasts were suspected to be the causative agent [87]. Some representatives of the genus *Rhodotorula* cause staining and give a bitter taste to the products. Warmer weather, inadequate refrigeration, and improper storage are the principal causes of higher levels of contamination, increased diversity, and change in yeast mycoflora [40]. Also yeast species mainly representatives of the genera *Candida* and *Rhodotorula* have been known to decrease the quality of dairy products by lactose assimilation [84, 88, 89]. Spoilage of yoghurts by yeasts has emerged as a major problem in the dairy industry [72, 78]. Interestingly, 7 (12.72%) of the yeast species (*Kloeckera* sp., *Trichosporon asahii, Cryptococcus humicola,* and *Kodamaea ohmeri*) could utilize lactose which is an important technological property in milk fermentation. The yeast species identified in the present study might have originated from the ingredients used as well as processing equipment that might not had been properly cleaned and sanitized. Starter cultures of lactic acid bacteria used to ferment the yoghurt are another potential source of yeast contamination. This suggests that overall improved and high quality of hygienic precautions should be adopted to avoid contamination especially during the production of yoghurt.

4. Conclusion

In view of the above results, locally made yoghurt samples obtained from Bamenda, Dschang, Bafoussam, and even some of the commercial samples constitute a high risk of health hazard to the consumers especially during the rainy season. The findings of this study warrant the need to undertake safety measures to avoid potential threats and apply educational programs for dairy products producers about the risk of contamination, prevention, and reduction of these pathogens from the yoghurt. Nevertheless, strict hygienic measures need to be applied during production, storage, and distribution of the yoghurts. License given to small dairy producers must be issued after the assurance of a minimum level of GMP. Periodical inspection must be

TABLE 5: Yeast species in yoghurts samples from two different batches as a function of sample group and season.

Place	Producers	Yeast species/season — I	% of identification	Yeast species/season — II	% of identification	Total number — I	Total number — II
Bamenda	NR	Candida kruzei/inconspicua	95.33	Rhodotorula mucilaginosa 1	99.95	Candida kruzei/inconspicua (5), Candida dubliniensis (2), Candida zeylanoides (3), Trichosporon asahii (1)	Candida kruzei/inconspicua (4), Candida zeylanoides (8), Candida boidinii (2), Rhodotorula mucilaginosa 1 (1), Rhodotorula mucilaginosa 2 (1), Stephanoascus ciferrii (1)
		Candida dubliniensis	100	Candida kruzei/inconspicua	95.33		
				Candida kruzei/inconspicua	64.12		
				Candida boidinii	93.38		
		Trichosporon asahii	100	Candida zeylanoides	97.96		
	MR	Candida zeylanoides	99.94	Candida kruzei/inconspicua	95.33		
		Candida kruzei/inconspicua	95.33				
	PV	Candida kruzei/inconspicua	95.33	Rhodotorula mucilaginosa 2	99.89		
				Candida zeylanoides	99.94		
				Candida zeylanoides	97.96		
	SY	Candida kruzei/inconspicua	95.14	Candida zeylanoides	99.94		
		Candida zeylanoides	99.98	Stephanoascus ciferrii	84.98		
				Candida zeylanoides	95.58		
				Candida zeylanoides	95.33		
	MB	Candida zeylanoides	99.98	Candida kruzei/inconspicua	99.94		
		Candida kruzei/inconspicua	61.36	Candida zeylanoides	99.98		
		Candida dubliniensis	100	Candida zeylanoides	99.54		
				Candida boidinii	93.38		
Dschang	P	Candida zeylanoides	99.94	Kodamaea ohmeri	81.82	Candida zeylanoides (3), Pichia angusta (1), Candida lusitaniae (1), Kodamaea ohmeri (1), Trichosporon asahii (1)	Kodamaea ohmeri (1), Trichosporon asahii (1)
		Pichia angusta	94.91				
		Candida zeylanoides	99.98				
		Candida lusitaniae	82.17				
	V	Kodamaea ohmeri	74.44	—	/		
	G	Trichosporon asahii	99.98	Trichosporon asahii	99.98		
		Candida zeylanoides	99.94				

TABLE 5: Continued.

Place	Producers	Yeast species/season I	% of identification	II	% of identification	Total number of each yeast from each group/season I	II
Bafoussam	T	Cryptococcus humicola	76.21	Candida dubliniensis	100	Candida kruzei/inconspicua (2), Candida zeylanoides (1), Candida albicans 2 (1), Cryptococcus humicola (1)	Candida kruzei/inconspicua (1), Candida dubliniensis (2), Candida albicans 1 (1), Candida lusitaniae (1), Cryptococcus laurentii (1), Kloeckera sp. (1)
	Ce	Candida kruzei/inconspicua	89.93	Candida kruzei/inconspicua Candida albicans 1	89.93 96.06		
	C	Candida kruzei/inconspicua Candida zeylanoides	95.33 99.94	Cryptococcus laurentii Kloeckera sp	100 65.10		
	K	Candida albicans 2	66.76	Candida dubliniensis Candida lusitaniae	100 62.97		
Commercial	AA	Candida kruzei/inconspicua	95.14	Candida lusitaniae Candida kruzei/inconspicua Stephanoascus ciferrii	97.88 95.14 62.16	Candida kruzei/inconspicua (1), Candida dubliniensis (1),	Candida lusitaniae (1), Candida kruzei/inconspicua (1), Candida dubliniensis (1), Stephanoascus ciferrii (1)
	CC	Candida dubliniensis	100	Candida dubliniensis	100		

[1] The letters I and II refer to batches of the dry and rainy seasons, respectively. [2] Numbers in parentheses are the number of isolates identified as that species.

TABLE 6: The distribution of yeast species as a function of sample group and season.

Yeast species isolated from yoghurt	Yoghurt samples/season								Frequency of occurrence (I)	% of occurrence (I)	Frequency of occurrence (II)	% of occurrence (II)
	Bamenda		Dschang		Bafoussam		Commercial					
	(I)	(II)	(I)	(II)	(I)	(II)	(I)	(II)				
Candida kruzei/inconspicua	+	+	–	–	+	+	+	+	3	75	3	75
Candida boidinii	–	+	–	–	–	–	–	–	0	00	1	25
Candida dubliniensis	+	–	–	–	–	+	+	+	2	50	2	50
Candida zeylanoides	+	+	+	–	+	–	–	–	3	75	1	25
Candida lusitaniae	–	–	+	–	–	+	–	+	1	25	3	50
Candida albicans 1	–	–	–	–	–	+	–	–	0	00	1	25
Candida albicans 2	–	–	–	–	+	–	–	–	1	25	0	00
Rhodotorula mucilaginosa 1	–	+	–	–	–	–	–	–	0	00	1	25
Rhodotorula mucilaginosa 2	–	+	–	–	–	–	–	–	0	00	1	25
Stephanoascus ciferrii	–	+	–	–	–	–	–	+	0	00	2	50
Trichosporon asahii	+	–	+	+	–	–	–	–	2	50	1	25
Kodamaea ohmeri	–	–	+	+	–	–	–	–	1	25	1	25
Pichia angusta	–	–	+	–	–	–	–	–	1	25	0	00
Cryptococcus laurentii	–	–	–	–	–	+	–	–	0	00	1	25
Cryptococcus humicola	–	–	–	–	+	–	–	–	1	25	0	00
Kloeckera sp.	–	–	–	–	–	+	–	–	0	00	1	25

The letters I and II refer to batches of the dry and rainy seasons, respectively; + = positive, – = negative.

done by specialists on the production sites to ensure that this minimum level of GMP is respected and sanctions should be applied where necessary. Moreover, regulation of small scale (locally made) yoghurt production in Cameroon should be a part of a strategy to enhance production of safe and high quality yoghurts. Finally, branded yoghurts are supposed to be products of high standard but in this case these products are not safe for consumption (except BB). There is equal need for a HACCP (Hazard Analysis and Critical Control Points) program for the production of yoghurt in Cameroon.

Conflicts of Interest

The authors declare that they have no conflicts of interest.

References

[1] L. C. Durga, D. Sharda, and M. P. Sastry, "Effect of storage conditions on keeping quality, riboflavin and niacin of plain and fruit yoghurt," *Indian Journal of Dairy Science*, vol. 39, no. 4, pp. 404–409, 1986.

[2] G. M. Lamye, K. A. L. Suffo, T. E. Pamo, and J.-R. Kuiate, "Physical and chemical quality appraisal of locally made yoghurt marketed in some regions of cameroon," *World Journal of Food Science and Technology*, vol. 1, no. 2, pp. 84–92, 2017.

[3] A. Sofu and F. Y. Ekinci, "Estimation of storage time of yogurt with artificial neural network modeling," *Journal of Dairy Science*, vol. 90, no. 7, pp. 3118–3125, 2007.

[4] D. K. Sandrou and I. S. Arvanitoyannis, "Application of Hazard Analysis Critical Control Point (HACCP) system to the cheese-making industry: A review," *Food Reviews International*, vol. 16, no. 3, pp. 327–368, 2000.

[5] WHO, "World Health Organization, Food safety and health: A strategy for the WHO African region," Resolution AFR/RC57/4, Regional Committee for Africa, Brazzaville, Republic of Congo, 2007, Fifty-seventh session.

[6] WHO., *World Health Organization, General Information Related to Microbiological Risks in Food*, 2012, http://www.who.int/food-safety/micro/general/en/index.html.

[7] M. Rajagopal, B. G. Werner, and J. H. Hotchkiss, "Low pressure CO2 storage of raw milk: Microbiological effects," *Journal of Dairy Science*, vol. 88, no. 9, pp. 3130–3138, 2005.

[8] L. Jespersen and M. Jakobsen, "Specific spoilage organisms in breweries and laboratory media for their detection," *International Journal of Food Microbiology*, vol. 33, no. 1, pp. 139–155, 1996.

[9] C. Baylis, M. Uyttendaele, H. Joosten, and A. Davies, "The Enterobacteriaceae and their significance to the food industry," Tech. Rep., The ILSI Europe Emerging Microbiological Issues Task Force, 2011.

[10] W. S. Kwee, T. W. Dommett, J. E. Giles, R. Roberts, and R. A. D. Smith, "Microbiological paramters during powdered milk manufacture," *Australian Journal of Dairy Technology*, vol. 41-43, 1986.

[11] ICMSF, *International Commission on Microbiological Specification for Food, Microbial Ecology of Foods*, vol. 1-2, Univerisity of Toronto Press, Toronto, Canada, 1986.

[12] Brasil, Ministério da Agricultura, Pecu aria e Abastecimento Resolução n°5, de13 de novembro de 2000. Oficializar os padrões de identidade e qualidade de leites fermentados. Diário Oficialda União, 2000.

[13] G. H. Fleet, "Yeasts in dairy products," *Journal of Applied Bacteriology*, vol. 68, no. 3, pp. 199–211, 1990.

[14] G. Fleet, "Spoilage yeasts," *Critical Reviews in Biotechnology*, vol. 12, no. 1-2, pp. 1–44, 1992.

[15] F. Canganella, M. Ovidi, S. Paganini, A. M. Vettraino, L. Bevilacqua, and L. D. Trovatelli, "Survival of undesirable microorganisms in fruit yoghurts during storage at different temperatures," *Food Microbiology*, vol. 15, no. 1, pp. 71–77, 1998.

[16] F. Cappa and P. S. Cocconcelli, "Identification of fungi from dairy products by means of 18S rRNA analysis," *International Journal of Food Microbiology*, vol. 69, no. 1-2, pp. 157–160, 2001.

[17] H. Rohm, F. Lechner, and M. Lehner, "Microflora of Austrian natural-set yogurt," *Journal of Food Protection*, vol. 53, no. 6, pp. 478–480, 1990.

[18] K. G. Abdel hameed, "Evaluation of chemical and microbiological quality of raw goat milk in Qena province," *Assiut Veterinary Medical Journal*, vol. 57, no. 129, pp. 131–144, 2011.

[19] R. Foschino, C. Garzaroli, and G. Ottogalli, "Microbial contaminants cause swelling and inward collapse of yoghurt packs," *Lait*, vol. 73, no. 4, pp. 395–400, 1993.

[20] S. B. Oyeleke, "Microbial assessment of some commercially prepared yoghurt retailed in Minna, Niger State," *African Journal of Microbiology Research*, vol. 3, no. 5, pp. 245–248, 2009.

[21] S. Younus, T. Masud, and T. Aziz, "Quality evaluation of market yoghurt/dahi," *Pakistan Journal of Nutrition*, vol. 1, no. 5, pp. 226–230, 2002.

[22] V. László, "Microbiological quality of commercial dairy products," in *Communicating Current Research and Educational Topics and Trends in Applied Microbiology*, A. Méndez-Vilas, Ed., 2007.

[23] J. Okpalugo, K. Ibrahim, K. Izebe, and U. Inyang, "Aspects of microbial quality of some milk products in Abuja Nigeria," *Tropical Journal of Pharmaceutical Research*, vol. 7, no. 4, p. 1169, 2008.

[24] A. M. A. Isam, A. E. Eshraga, E. A. Y. Abu, and E. B. Efadil, "Physicochemical, microbiological and sensory characteristics of yoghurt produced from camel milk," *Journal of Environmental, Agricultural and Food Chemistry*, vol. 10, no. 6, pp. 2305–2313, 2011.

[25] B. Igbabul, J. Shember, and J. Amove, "Physicochemical, microbiological and sensory evaluation of yoghurt sold in Makurdi metropolis," *African Journal of Food Science and Technology*, vol. 5, no. 6, pp. 129–135, 2014.

[26] M. A. Njoya, P. Y. Mahbou, C. W. Nain, and H. Imele, "Physicochemical, microbiological, and sensory properties of pinapple (*Ananascomosus*(L.) Merr.) Flavoured Yoghurt," *International Journal of Agriculture Innovations and Research*, vol. 4, no. 6, 2016.

[27] R. T. Marshall, *Standard methods for examination of dairy products*, Public Health Assoc., Inc, Washington, DC, USA, 16th edition, 1992.

[28] A.P.H.A., *Standards Methods for Examination of Dairy Products*, American Public Health Association, Washington, DC, USA, 17th edition, 2004.

[29] A. H. Kawo, E. M. Omole, and J. Naaliya, "Quality assessment of some processed yoghurt products sold in Kano Metropolis, Kano, Nigeria," *BEST Journal*, vol. 3, no. 1, pp. 96–99, 2006.

[30] W. F. Harrigan and M. E. McCance, *Laboratory methods in foods and dairy microbiology*, Academic press, London, UK, 1976.

[31] G. L. Christen, P. M. Davidson, J. S. Mc Alliser, and L. A. Roth, "Coliform and other indicator bacteria," in *Standard Methods*

for the Examination of Dairy Products, R. T. Marshall, Ed., p. 450, Port City Press, Baltimore, MD, USA, 16th edition, 1992.

[32] W. F. Harrigan, MacCance., and E. Margaret, *Laboratory Methods in Microbiology*, Academic Press, New York, NY, USA, 1976.

[33] O. Kandler and N. Weiss, "Regular, nonsporing Gram- positive rods," in *Bergey's Manual of Systematic Bacteriology*, P. Sneath, N. Mair, M. Sharpe, and J. Holt, Eds., vol. 2 of *section 14*, pp. 1208–1234, Williams & Wilkins, Baltimore, MD, USA, 1986.

[34] K. Schleifer, "Gram- positive cocci," in *Bergey's Manual of Systematic Bacteriology*, P. Sneath, N. Mair, M. Sharpe, and J. Holt, Eds., vol. 2 of *section 12*, pp. 1043–1071, Williams & Wilkins, Baltimore, MD, USA, 1986.

[35] R. Al-Tahiri, "A comparison on microbial conditions between traditional dairy products sold in karak and same products produced by modern dairies," *Pakistan Journal of Nutrition*, vol. 4, no. 5, pp. 345–348, 2005.

[36] Codex Alimentarius, *CODEX Standard for Fermented Milks 242-2003*, 2nd edition, 2003.

[37] L. A. Rodrigues, M. B. T. Ortolani, and L. A. Nero, "Microbiological quality of yoghurt commercialized in viçosa, Minas Gerais, Brazil," *African Journal of Microbiology Research*, vol. 4, no. 3, pp. 210–213, 2010.

[38] A. Y. Tamine and K. Robinson, *Yoghurt. Science and Technology*, Institute of Applied Science, 2004.

[39] M. M. Sarkar, T. N. Nahar, M. K. Alam, M. M. Rahman, M. H. Rashid, and M. A. Islam, "Chemical and bacteriological quality of popular Dahi available in some selected areas of Bangladesh," *Bangladesh Journal of Animal Science*, vol. 41, no. 1, pp. 47–51, 2012.

[40] S. R. Moreira, R. F. Schwan, E. P. De Carvalho, and A. E. Wheals, "Isolation and identification of yeasts and filamentous fungi from yoghurts in Brazil," *Brazilian Journal of Microbiology*, vol. 32, no. 2, pp. 117–122, 2001.

[41] Brasil, "Ministério da Saúde, Agência Nacional de Viilância Sanitária," Resolução RDC nº12, de 02 de janeiro de 2001. Aprova o regulamento técnico sobre padrões microbiológicos para alimentos. Diário Oficial da União, 2001.

[42] V. M. Marshall, "Yoghurt: Science & Technology Tamine, A.Y., Robinson, R.K.; Woodhead Publishing Ltd., 1999, xv+606 pages, ISBN 85573 3994, £136," *Food Chemistry*, vol. 70, no. 2, p. 273, 2000.

[43] G. A. Birollo, J. A. Reinheimer, and C. G. V. Vinderola, "Viability of lactic acid microflora in different types of yoghurt," *Food Research International*, vol. 33, no. 9, pp. 799–805, 2000.

[44] *Dairy substitutes, Encyclopedia of science and technology*, Mac Graw Hill publishers, New York, Mac Graw, 5th edition, 1997.

[45] I. E. Mbaeyi-Nwaoha and N. I. Egbuche, "Microbiological evaluation of sachet water and street-vended yoghurt and Zobo drinks sold in Nsukka Metropolis," *International Journal of Biological and Chemical Science*, vol. 6, no. 4, pp. 1703–1717, 2012.

[46] Y. Karagul, C. Wilson, and H. White, "Formulation and processing of yoghurt," *Journal of Dairy Science*, vol. 87, pp. 543–550, 2004.

[47] FAO/WHO, "Risk management and food safety," Report of a joint FAO/WHO Expert Consultation paper no 65, FAO Food and Nutrition, Rome, Italy, 1997.

[48] P. A. D. Grimont and F. Grimont, "Genus pantoea," in *Bergey's Manual of Systematic Bacteriology*, D. J. Brenner, N. R. Krieg, and J. T. Staley, Eds., vol. 2 of *The Proteobacteria*, pp. 713–720, Springer-Verlag, New York, NY, USA, 2nd edition, 2005.

[49] D. M. Weinthal, I. Barash, M. Panijel, L. Valinsky, V. Gaba, and S. Manulis-Sasson, "Distribution and replication of the pathogenicity plasmid pPATH in diverse populations of the gall-forming bacterium Pantoea agglomerans," *Applied and Environmental Microbiology*, vol. 73, no. 23, pp. 7552–7561, 2007.

[50] K. Talaro and A. Talaro, *Foundation in Microbiology*, W.M.C Brown publisher, Dubuque, Iowa, USA, 2006.

[51] J. Cai, Z. Wang, C. Cai, and Y. Zhou, "Characterization and identification of virulent Klebsiella oxytoca isolated from abalone (Haliotis diversicolor supertexta) postlarvae with mass mortality in Fujian, China," *Journal of Invertebrate Pathology*, vol. 97, no. 1, pp. 70–75, 2008.

[52] G. Kovtunovych, T. Lytvynenko, V. Negrutska, O. Lar, S. Brisse, and N. Kozyrovska, "Identification of Klebsiella oxytoca using a specific PCR assay targeting the polygalacturonase pehX gene," *Research in Microbiology*, vol. 154, no. 8, pp. 587–592, 2003.

[53] K. K. Lai, "Enterobacter sakazakii infections among neonates, infants, children, and adults: Case reports and a review of the literature," *Medicine*, vol. 80, no. 2, pp. 113–122, 2001.

[54] J. Van Acker, F. De Smet, G. Muyldermans, A. Bougatef, A. Naessens, and S. Lauwers, "Outbreak of necrotizing enterocolitis associated with Enterobacter sakazakii in powdered milk formula," *Journal of Clinical Microbiology*, vol. 39, no. 1, pp. 293–297, 2001.

[55] W.-L. Yu, H.-S. Cheng, H.-C. Lin, C.-T. Peng, and C.-H. Tsai, "Outbreak investigation of nosocomial Enterobacter cloacae bacteraemia in a neonatal intensive care unit," *Infectious Diseases*, vol. 32, no. 3, pp. 293–298, 2000.

[56] P. M. Hawkey, "Proteus, providencia and Morganella spp," in *Principles and Practice of Clinical Bacteriology*, S. H. Gillespie and P. M. Hawkey, Eds., pp. 391–396, John Wiley & Sons Ltd., London, UK, 2nd edition, 2006.

[57] P. R. Krake and N. Tandon, "Infective endocarditis due to Providenda stuartii," *Southern Medical Journal*, vol. 97, no. 10, pp. 1022–1023, 2004.

[58] A. F. Koreishi, B. A. Schechter, and C. L. Karp, "Ocular Infections Caused by Providencia rettgeri," *Ophthalmology*, vol. 113, no. 8, pp. 1463–1466, 2006.

[59] P. C. Applebaum and D. B. Campbell, "Pancreatic abscess associated with Achromobacter group Vd biovar 1," *Journal of Clinical Microbiology*, vol. 12, no. 2, pp. 282-283, 1980.

[60] V. P. Morais, M. T. Daporta, A. F. Bao, M. G. Campello, and G. Quindós-Andrés, "Enteric fever-like syndrome caused by Raoultella ornithinolytica (Klebsiella ornithinolytica)," *Journal of Clinical Microbiology*, vol. 47, no. 3, pp. 868-869, 2009.

[61] B. Vos and M. Laureys, "Giant renal cyst as cause of colic obstruction," *Revue Médicale de Bruxelles*, vol. 30, no. 2, pp. 107–109, 2009.

[62] N. Mau and L. A. Ross, "Raoultella ornithinolytica bacteremia in an infant with visceral heterotaxy," *The Pediatric Infectious Disease Journal*, vol. 29, no. 5, pp. 477-478, 2010.

[63] E. Jawetz, J. L. Melnick, E. A. Adelberg, G. F. Brooks, J. S. Butel, and L. N. Ornston, *Medical Microbiology*, Prentice-Hall International Inc, Englewood Cliffs, NJ, USA, 18th edition, 1989.

[64] C. M. O'Hara, F. W. Brenner, and J. M. Miller, "Classification, identification, and clinical significance of Proteus, Providencia, and Morganella," *Clinical Microbiology Reviews*, vol. 13, no. 4, pp. 534–546, 2000.

[65] S. T. Castro, C. R. Rodríguez, B. E. Perazzi, M. Radice, M. Paz Sticott, and H. Muzio, "Comparison of different methods in order to identify Proteus spp," *Revista Argentina de Microbiología*, vol. 38, pp. 119–124, 2006.

[56] R. Cantón, M. P. Sánchez-Moreno, and M. I. M. Reilly, "Proteus penneri," *Enfermedades Infecciosas y Microbiología Clínica*, vol. 24, no. 1, pp. 8–13, 2006.

[57] E. Kü üköner and Z. Tarak i, "Influence of different fruit additives on some properties of stirred yoghurt during storage," *Journal of Agricultural Science*, vol. 13, no. 2, pp. 97–101, 2003.

[58] V. O. Ifeanyi, E. O. Ihesiaba, O. M. Muomaife, and C. Ikenga, "Assessment of microbiological quality of yogurt sold by street vendors in onitsha metropolis, anambra state, nigeria," *British Microbiology Research Journal*, vol. 3, no. 2, pp. 198–205, 2013.

[59] N. De, T. M. Goodluck, and M. Bobai, "Microbiological quality assessment of bottled yogurt of different brands sold in Central Market, Kaduna Metropolis, Kaduna, Nigeria," *International Journal of Current Microbiology and Applied Science*, vol. 3, no. 2, pp. 20–27, 2014.

[70] B. Digbabul, J. Shember, and J. Amove, "Physicochemical, microbiological and sensory evaluation of yoghurt sold in Makurdi metropolis," *African Journal of Food Science and Technology*, vol. 5, no. 6, pp. 129–135, 2014.

[71] G. H. Fleet and M. A. Mian, "The occurrence and growth of yeasts in dairy products," *International Journal of Food Microbiology*, vol. 4, no. 2, pp. 145–155, 1987.

[72] V. R. Suriyarachchi and G. H. Fleet, "Occurrence and growth of yeasts in yogurts," *Applied and Environmental Microbiology*, vol. 42, no. 4, pp. 574–579, 1981.

[73] M. D. GREEN and S. N. IBE, "Yeasts as Primary Contaminants in Yogurts Produced Commercially in Lagos, Nigeria," *Journal of Food Protection*, vol. 50, no. 3, pp. 193–198, 1987.

[74] M. S. A. Haridy, "Occurrence of yeasts in yogurt, cheese and whey," *Cryptogamie Mycologie*, vol. 14, pp. 255–262, 1993.

[75] H. Rohm, F. Eliskases-Lechner, and M. Bräuer, "Diversity of yeasts in selected dairy products," *Journal of Applied Bacteriology*, vol. 72, no. 5, pp. 370–376, 1992.

[76] N. Van Uden and I. D. Carmo-Sousa, "Presumptive tests with liquid media for coliform organisms in yogurt in the presence of lactose fermenting yeasts," *Dairy Industry*, vol. 22, p. 1029, 1957.

[77] D. R. Arnott, C. L. Duitschaever, and D. H. Bullock, "Microbiological evaluation of yogurt produced commercially in Ontario," *Journal of Milk Food Technology*, vol. 37, no. 1, pp. 11–13, 1974.

[78] J. G. Davis, "The microbiology of yogurt," in *Lactic acid bacteria in beverages and food*, J. G. Carr, C. V. Cutting, and G. C. Whiting, Eds., pp. 245–266, Academic press, London, UK, 1975.

[79] N. M. Saad, M. K. Moustafa, and A. A. H. Ahmed, "Microbiological qualityof yoghurt produced in Assiut City," *Assiut Veterinary Journal*, vol. 19, pp. 87–91, 1987.

[80] E. R. Amakoromo, H. C. Innocent-Adiele, and H. O. Njoku, "Microbiological quality of a yoghurt-like product from african yam bean," *Journal of Natural Sciences*, vol. 10, no. 6, pp. 6–9, 2012.

[81] Anonymous, *Yogurt standard (TS1330), Turkish Standards Institute*, Necatibey Cad. 112., Bakanlıklar, Ankara, Turkey, 1989.

[82] A. M. Saudi, H. A. El- Essawy, and N. M. Hafez, "Aspects on the microbiological quality of yoghurt plastic containers," *VeterinaryMedical Journal Giza*, vol. 37, no. 3, pp. 397–406, 1989.

[83] A. H. Rose, *Economic Microbiology, Fermented Foods*, vol. 7, Academic Press, London, UK, 1982.

[84] B. J. Wood, *Microbiology of fermented foods*, vol. 1, 2, Elsevier, London, UK, 1985.

[85] M. S. G. Chekem, P. K. Lunga, J. D. D. Tamokou et al., "Antifungal properties of Chenopodium ambrosioides essential oil against candida species," *Pharmaceuticals*, vol. 3, no. 9, pp. 2900–2909, 2010.

[86] R. Hurley, J. de Louvois, and A. Mulhall, "Yeasts as human and animal pathogens," in *The Yeasts*, A. H. Rose and J. S. Harrison, Eds., vol. 1, p. 207, Academic Press, London, UK, 2nd edition, 1987.

[87] E. C. TODD, "Foodborne Disease in Canada - a 5-year Summary," *Journal of Food Protection*, vol. 46, no. 7, pp. 650–675, 1983.

[88] C. P. Kurtzman, "Classification and general properties of yeasts," in *The Yeasts, Biotechnology and Bioanalysis*, H. Verachtert and R. De Mot, Eds., p. 184, Academic Press, London, UK, 1990.

[89] D. A. Mossel, "Experience with some methods for the enumeration and identification of yeasts occurring in foods," in *Biology and activities of yeasts*, F. D. Skinner, S. M. Passinfre, and R. K. Divenpart, Eds., p. 279, Academic Press, London, UK, 1980.

9

The Content and Bioavailability of Mineral Nutrients of Selected Wild and Traditional Edible Plants as Affected by Household Preparation Methods Practiced by Local Community in Benishangul Gumuz Regional State, Ethiopia

Andinet Abera Hailu and Getachew Addis

Ethiopian Public Health Institute, P.O. Box 5654, Addis Ababa, Ethiopia

Correspondence should be addressed to Andinet Abera Hailu; andinet_abera@yahoo.com

Academic Editor: Elad Tako

Edible parts of some wild and traditional vegetables used by the Gumuz community, namely, *Portulaca quadrifida, Dioscorea abyssinica, Abelmoschus esculentus,* and *Oxytenanthera abyssinica,* were evaluated for their minerals composition and bioavailability. Mineral elements, namely, Ca, Fe, Zn, and Cu, were analyzed using Shimadzu atomic absorption spectrophotometer. Effects of household processing practices on the levels of mineral elements were evaluated and the bioavailability was predicted using antinutrient-mineral molar ratios. Fe, Zn, Ca, Cu, P, Na, and K level in raw edible portions ranged in (0.64 ± 0.02–27.0 ± 6.2), (0.46 ± 0.02–0.85 ± 0.02), (24.49 ± 1.2–131.7 ± 8.3), (0.11 ± 0.01–0.46 ± 0.04), (39.13 ± 0.34–57.27 ± 0.94), (7.34 ± 0.42–20.42 ± 1.31), and (184.4 ± 1.31–816.3 ± 11.731) mg/100 g FW, respectively. Although statistically significant losses in minerals as a result of household preparation practices were observed, the amount of nutrients retained could be valuable especially in communities that have limited alternative sources of these micronutrients. The predicted minerals' bioavailability shows adequacy in terms of calcium and zinc but not iron.

1. Introduction

Numerous wild edible plant species have been used by different communities in Ethiopia, mainly as supplement to conventional foods [1–8]. However, the biodiversity is threatened through replacement of forests with agricultural expansion and deforestation without cultivation and domestication of potential species [2]. These situations could exacerbate local food shortages and aggravate widespread malnutrition in the country. Diversification of production and consumption habits to include a broader range of plant species, particularly those currently identified as underutilized, could significantly contribute to improve health and nutrition, livelihoods, and ecological sustainability. Edible wild and traditional vegetables have played an important role in supplementing staple foods by supplying trace elements, vitamins, and minerals [9]. As wild food plants grow in natural conditions, they are easily accessed and freely harvested for their human food and nutrition values. They are relevant in household food security and nutrition in some rural areas, particularly during seasonal food shortage periods, and provide good nutritional supplies, notably micronutrients [10].

Micronutrient deficiencies affect billions of people globally. Although less prevalent in higher-income populations, these deficiencies do occur in such groups, especially among premature infants, children, and the elderly. Several sources revealed that edible wild plant and traditional vegetable species increase the nutritional quality by providing minerals, fiber, vitamins, and essential fatty acids and enhance taste and color in rural diets [11]. Underutilized green leafy vegetables are a good source of many nutrients like iron, calcium, ascorbic acid, and β-carotene that could help in overcoming micronutrient malnutrition and easily accessed by the community at a low cost. Because micronutrient deficiencies (such as vitamin A) are associated with low intake of foods such as vegetables, as opposed to starchy (energy rich) staples which provide the majority of energy intake in typical African diets, increment in energy production and consumption will likely do little to ameliorate the problem of micronutrient deficiency unless identification, proper evaluation, and domestication of nutritionally potential lesser known vegetables are integrated into the diets of the population. Moreover, the roles of edible wild plants and lesser known crops in human nutrition are potentially valuable to maintain a balance between population growth and agricultural productivity, particularly in the tropical and subtropical areas of the world [12]. Hence, continuous search for new source of nutrient especially from plant foods is a basis for selecting promising species for further studies on green leafy vegetables to meet the nutritional requirements. Evaluation of the nutrient and antinutrient compositions of wild edible plants helps to identify foods rich in minerals and acquiring knowledge on the methods of appropriate preparation to enhance bioavailability of nutrients. As the consumption of plant products low in bioavailable minerals is high in rural communities of Ethiopia [13], the presence of antinutritional factors that limits the optimal utilization of wild and traditional vegetables and the extent to which the household food preparation methods could reduce them need investigations. Therefore, the objective of this work was to assess mineral contents and their bioavailability as a function of local household preparation methods of some wild and traditional vegetables of the Gumuz ethnic community located in Benishangul Gumuz state western Ethiopia.

2. Materials and Methods

The study plants were collected from Agalo Meti district in Kemash zone of Benishangul Gumuz Regional state located at 530 km west of the capital Addis Ababa.

2.1. Selection of Species. Selection of species for our investigation was based on ethnobotanical study findings in the study area [4]. *Dioscorea abyssinica* (tuber), *Oxytenanthera abyssinica* Munro (young shoot), *Abelmoschus esculentus* (L.) Moench (immature pod), and *Portulaca quadrifida* (aerial part) were selected based on their wider utilization by the community in the study area. The first two species are wild and the last two exist both under cultivation and wild stand. The selected vegetables were recommended for domestication according to an association in Benishangul Gumuz called *"Tikuret le Gumuz Limat Mahiber"* for the reason that they are widely consumed by the community and because they are most suited for domestication. Each species was collected at the time when edible parts are acceptable for consumption by the local community. Voucher specimens of the selected edible plants were collected and processed for verification. The specimens were identified by a botanist from Bioversity International and Ethiopian Public Health Institute (EPHI) using standard procedures and deposited in the National Herbarium of Addis Ababa University (AAU).

2.2. Sample Collection and Preparation. Edible parts of the selected vegetables were harvested manually from their natural habitat in different areas of Agalo Meti woreda. Five-kilogram edible parts of the respective species were collected from 20 different plants of the same species to ensure the representativeness. Sample preparation for laboratory analyses is illustrated in Figure 1.

2.3. Mineral Analysis. The methods of [14] were used to determine minerals. The sample extract solution was transferred to polyethylene bottle and stored until use for determinations of minerals. Blank was prepared without sample by taking the same amount of reagents under the same condition. The minerals, namely, calcium, iron, zinc, and Cu, were analyzed using Shimadzu atomic absorption spectrophotometer (AA-6800/"AA Wizard" software). Sodium and potassium contents were determined using flame photometer (Jenway, PF 7, Essex, UK) according to the method described by AOAC 2005, 966.16 and 965.30, respectively. Phosphorus was determined using UV-visible spectrophotometer (CECIL Instruments, Cambridge England, deuterium F 500 mA, power T3. 15 A) based on method 970.39 [14]. Absorbance of standard, blank, and samples was read at 660 nm using UV-visible spectrophotometer. Absorbance versus concentration calibration curve was constructed and the equation obtained was used to calculate the unknown phosphorus concentration in the samples:

Phosphorus in mg/100 gm

$$= \frac{(A_s - A_B) * \text{dilution factor} * \text{extracted volume} * 100}{\text{Slope} * \text{weight of sample} * 1000}, \quad (1)$$

where A_s is the absorbance of sample, A_B is the absorbance of blank, and Slope is calculated from the calibration curve.

2.4. Determinations of Antinutritional Factors

2.4.1. Phytic Acid. The phytate content was determined as described by [15]. Briefly, 0.5 gm of dried sample was extracted with 10 mL of 0.2 N HCl for 1 hour at an ambient temperature and centrifuged (3000 rpm for 30 m). The clear supernatant was used for the phytic acid determination. Two mL of wade reagent (0.03% $FeCl_3 \cdot 6H_2O$ and 0.3% sulfosalicylic acid) was added to 3 mL of the supernatant sample solution. The solution mixture was homogenized and centrifuged (3000 rpm/10 minutes). The absorbance was measured at 500 nm using UV-visible spectrophotometer. The amount of phytic acid was calculated using phytic acid standard curve prepared in the same condition and the result was expressed as phytic acid mg/100 g dry weight.

2.4.2. Oxalate. The oxalate content of the samples was determined using titration method [16, 17]. Two g of finely ground samples was placed in a 250 mL conical flask containing 190 mL of distilled water. Ten mL 6 M HCl solution was added to each of the samples and the suspension was digested at 100°C for 1 h. The samples were then cooled and made up to 250 mL mark of the flask. The samples were filtered

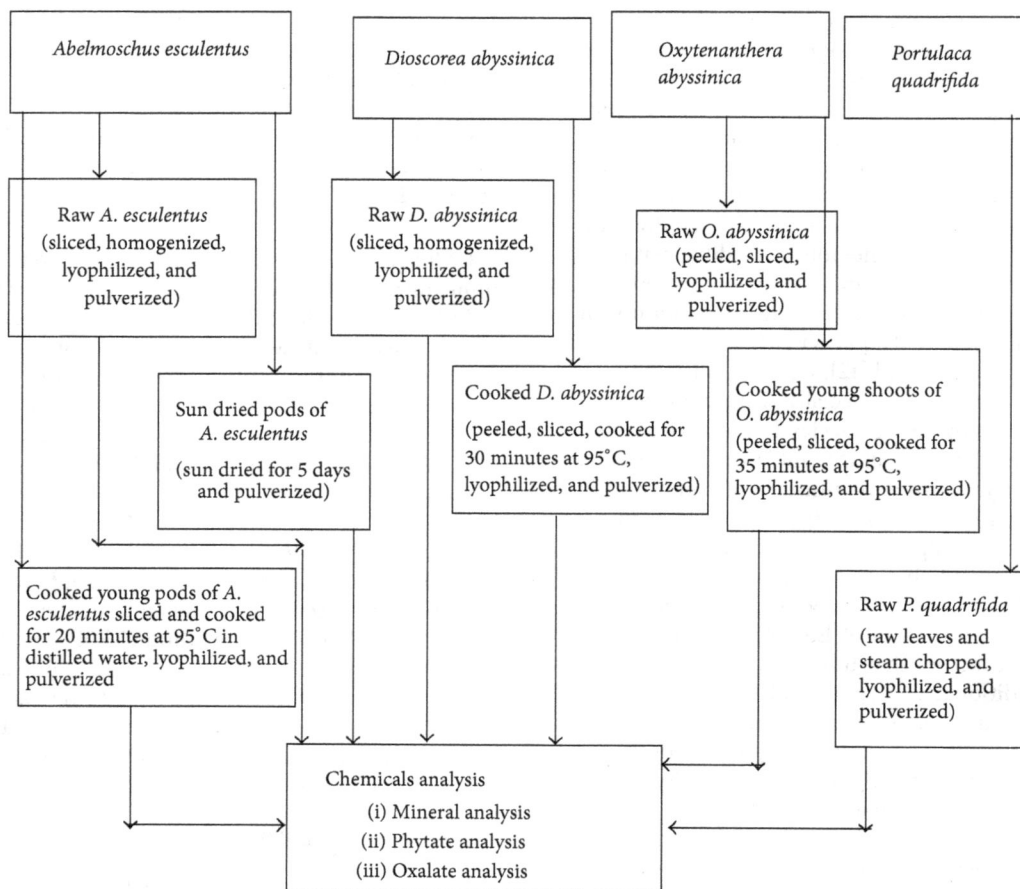

FIGURE 1: Scheme of sample preparation for chemical analysis.

and duplicate portion of 125 mL of the filtrate was measured into beaker and four drops of methyl red indicator was added, followed by the addition of concentrated NH_4OH solution (drop wise) until the solution was changed from pink to yellow color. Each portion was then heated to 90°C, cooled, and filtered to remove the precipitate containing ferrous ion. Each of the filtrates was again heated to 90°C and 10 mL of 5% $CaCl_2$ solution was added to each of the samples with stirring consistently. After cooling, the samples were left overnight. The solutions were then centrifuged at 2500 rpm for 5 min. The supernatants were decanted and the precipitates were completely dissolved in 10 mL 20% H_2SO_4. The total filtrates resulting from digestion of 2 g of each of the samples were made up to 200 mL. Aliquots of 125 mL of the filtrates were heated until near boiling and then titrated against 0.02 M standardized $KMnO_4$ solutions to a pink color which persisted for 30 sec. The oxalate contents of each sample were calculated [18]. The analysis was carried out in duplicate, and the results were expressed in dry basis.

2.5. Minerals Bioavailability. Molar ratios of antinutrient/minerals were used to predict the minerals bioavailability [19]. The suggested critical values used to predict the

bioavailability were calcium : phytate < 6, phytate : iron > 1, phytate : zinc > 15, and phytate : calcium/zinc > 0.5 [19, 20].

2.6. Statistical Analysis. Descriptive statistics such as means and standard deviation were calculated using SPSS version 16 software. One-way analysis of variance (ANOVA) was used to see the effect of household processing methods on mineral contents and the bioavailability of minerals in selected vegetables. Multiple comparison tests using least significant difference technique (LSD, $P < 0.05$) were applied to compare the means of each parameter between different household preparation practices using SPSS version 16 software. Paired comparison t-test was used to determine if there was a significant mean difference between raw and processed vegetables for each parameter.

3. Results and Discussion

3.1. Results. From the results presented in Table 1, it is noticeable that the concentration of different macroelements (Ca, P, Na, and K) and trace elements (Fe, Zn, and Cu) of the wild and traditional vegetables was high and the vegetables studied could be regarded as appreciably important

TABLE 1: Mineral content (mg/100 g, fresh basis) of raw and processed vegetables.

Species		Fe	Zn	Ca	Cu	P	Na	K
	RAS	2.33 ± 0.37^b	$0.68 \pm .04^b$	131.7 ± 8.3^b	0.11 ± 0.01^b	39.14 ± 0.95^b	7.82 ± 0.28^b	184.4 ± 1.3^b
A. esculentus	CAS	0.68 ± 0.08^a	0.55 ± 0.03^a	104.23 ± 4.9^a	0.08 ± 0.01^a	$36.28 \pm .067^a$	4.4 ± 0.29^a	155.4 ± 1.7^a
	SAS	2.58 ± 0.16^b	0.54 ± 0.02^a	140.4 ± 8.13^b	0.08 ± 0.01^a	51.85 ± 0.34^c	7.51 ± 0.92^b	183.3 ± 53.4^b
D. abyssinica	RDP	27.0 ± 6.24^b	0.46 ± 0.02^b	43.19 ± 2.0^a	0.46 ± 0.04^b	40.68 ± 2.7^b	8.94 ± 0.54^b	341.16 ± 3.6^b
	CDP	1.79 ± 0.06^a	0.29 ± 0.06^a	40.74 ± 6.8^a	0.19 ± 0.02^a	23.69 ± 2.4^a	5.88 ± 0.75^a	128.49 ± 3.9^a
P. quadrifida	RPO	8.06 ± 0.11	0.84 ± 0.06	117.99 ± 10.8	0.14 ± 0.01	39.13 ± 0.34	20.42 ± 1.31	816.3 ± 11.7
O. abyssinica	ROA	0.64 ± 0.02^a	0.85 ± 0.02^b	24.49 ± 1.2^a	0.11 ± 0.01^b	57.27 ± 0.94^b	7.34 ± 0.42^b	456.2 ± 12.3^b
	COA	0.6 ± 0.06^a	0.54 ± 0.02^a	22.94 ± 4.21^a	0.07 ± 0.01^a	39.7 ± 1.89^a	4.20 ± 0.55^a	273.2 ± 1.6^a

RAS = raw, CAS = cooked, and SAS = sun dried; RDP = raw, CDP = cooked, RPO = raw, ROA = raw, and COA = cooked; values are expressed as mean ± SD of three determinations; mean values in the same column corresponding to each species followed by different superscript letters were considered significant at $P < 0.05$.

TABLE 2: Antinutrient/mineral molar ratios of raw and processed vegetables.

Species		[Phy]/[Zn][1]	[Ca]/[Phy][2]	[Phy]/[Fe][3]	[Phy × Ca]/[Zn][4]	[Ox]/[Ca][5]
	RAS	6.077 ± 0.45^a	51.11 ± 0.77^a	1.50 ± 0.001^a	1.73 ± 0.12^a	0.17 ± 0.001^b
A. esculentus	CAS	5.50 ± 0.23^a	57.20 ± 0.35^b	4.07 ± 0.35^b	1.57 ± 0.06^a	0.15 ± 0.001^a
	SAS	7.48 ± 0.03^b	53.04 ± 0.77^a	1.68 ± 0.06^a	2.11 ± 0.1^b	0.15 ± 0.001^a
D. abyssinica	RDP	3.79 ± 0.19^a	35.0 ± 0.50^a	0.06 ± 0.001^a	0.10 ± 0.01^a	1.04 ± 0.001^b
	CDP	3.66 ± 0.60^a	45.63 ± 0.84^a	0.51 ± 0.01^b	0.09 ± 0.01^a	0.28 ± 0.001^a
P. quadrifida	RPO	14.22 ± 0.39	15.06 ± 0.12	1.28 ± 0.02	3.21 ± 0.07	0.72 ± 0.002
O. abyssinica	ROA	2.03 ± 0.10^a	22.64 ± 0.51^a	2.22 ± 0.11^a	0.12 ± 0.01^a	0.75 ± 0.002^b
	COA	3.74 ± 0.10^b	24.55 ± 0.09^a	1.81 ± 0.13^a	0.22 ± 0.0^b	0.50 ± 0.01^a
Critical values		15.0	6.0	1.0	0.5	2.5

RAS = raw, CAS = cooked, and SAS = sun dried; RDP = raw, CDP = cooked, RPO = raw, ROA = raw, and COA = cooked. Values in the same column followed by the same superscript corresponding to the same species are not significantly different ($P < 0.05$).
[1] (mg phytate/MW of phytate : mg Zn/MW of Zn), [2] (mg Ca/MW of Ca : mg phytate/ MW of phytate), [3] (mg phytate/MW of phytate : mg Fe/MW of Fe), [4] (mol/Kg of phytate × mol/Kg of Ca : mol/Kg of Zn), and [5] (mg oxalate/MW of oxalate : mg Ca/MW of Ca).

sources of these essential elements. Potassium is abundant in all vegetables analyzed followed by calcium, phosphorus, and sodium from the macroelements. From the trace elements, iron level was relatively higher followed by zinc and copper.

3.2. Effect of Processing on Minerals Content of Young Pods of Abelmoschus esculentus.

As presented in Table 2, the iron and zinc contents of raw A. esculentus (okra) were 2.3 mg/100 g and 0.68 mg/100 g, respectively. Iron content in immature pods of raw A. esculentus obtained in the present study was higher than findings of [21] for different A. esculents varieties reported within the range of 0.87–0.97 mg/100 g FW but the zinc content (0.68 ± 0.04 mg/100 g FW) obtained in this study was slightly lower compared to the value reported (1.29–1.37 mg/100 g FW) in their study. The variations in the contents of the minerals in A. esculentus may be due to the varietal difference or genetic factor, environmental factor, and their interactions [21]. Significant loss of iron and zinc was observed in the cooked A. esculentus but not statistically significant between raw and sun dried young pods of A. esculentus. This could be due to leaching of the minerals into the cooking water. Sun drying is the gradual loss of water through evaporation and cannot support leaching. It has to also be noted that minerals are not volatile.

3.3. Effect of Cooking on Minerals Content of Dioscorea abyssinica.

Iron level in raw (whole) Dioscorea abyssinica was 27 mg/100 g FW. However, cooking and peeling the bark of the tuber significantly reduced the iron content to 1.79 mg/100 g FW. The iron content of D. abyssinica obtained in this study in the raw sample was higher when compared with the value reported for Dioscorea pentaphylla (8 mg/100 g FM) [22]. Zinc concentration of D. abyssinica was also significantly reduced during cooking and debarking from 0.5 mg/100 g to 0.3 mg/100 g FW. The statistically significant ($P < 0.05$) reduction of iron and zinc concentration upon cooking and peeling of the tuber may be because of the presence of minerals in the outer nonedible part of the tuber which was removed by peeling after boiling the tuber. The same reason may apply to the minerals which showed enormous reduction in the boiled and peeled D. abyssinica. Reduction in zinc content was also reported in peeled and boiled tubers of Dioscorea cayenensis [23]. The potassium level in raw tubers of D. abyssinica was 341.16 mg/100 g. Statistically significant losses were observed when the tubers were cooked compared to the raw samples. This loss may be attributed to the leaching out of minerals including potassium into the cooking water [24]. The high potassium content in this tuber could help to maintain normal blood pressure and can be labeled as heart protective vegetable.

3.4. Effect of Cooking on Minerals Content of Young Shoots of Oxytenanthera abyssinica. The minerals content of raw and cooked juvenile shoots of *Oxytenanthera abyssinica* is presented in Table 1. The iron and zinc contents of juvenile shoots of *O. abyssinica* were 0.6 mg/100 g and 0.9 mg/100 g, respectively. The zinc content was higher in raw bamboo shoots compared to other study vegetables. All minerals tested were decreased in the cooked samples with the reduction being statistically significant in Zn, Cu, P, Na, and K but Fe and Ca were not significantly ($P < 0.05$) lost in the cooked samples. The variation in the degree of loss might be related to the chemical forms of minerals in the food matrix. The respective potassium contents determined in raw and cooked samples of *O. abyssinica* were 456.2 mg/100 g and 273.2 mg/100 g on fresh weight basis. The potassium content of raw *O. abyssinica* obtained in the current study is well agreed with the value reported for other bamboo species such as *Bambusa tulda* (408 mg/100 g, FW) and *Dendrocalamus hamiltonii* (416 mg/100 g, FW) [25]. Raw shoots of *O. abyssinica* had the highest phosphorus content compared to other study vegetables. However, cooking has significantly reduced the phosphorus level. Phosphorus is a major component of bones and teeth [26].

Studies confirmed that there is no known benefit of high sodium consumption. Sodium intakes more than 1 g per day tend to aggravate a genetically determined susceptibility to hypertension, and intakes above 7 g/day may induce hypertension even in individuals who have no specific genetic susceptibility [27]. In this context, all the vegetables contained safe sodium levels.

3.5. Potential of the Study Vegetables in Meeting the RDA Requirements of Some Minerals. The recommended dietary allowance (RDA) for iron is 10 mg/day for adults [28]. The raw immature pods of *A. esculentus* could provide 116.5% of the RDA requirement for iron if 500 g fresh vegetable is consumed per day, considering it would be fully bioavailable. The cooked and sun dried *A. esculentus* could provide 34% and 129% RDA requirement for iron from 500 g FW meal. Cooked *D. abyssinica* and bamboo shoots can also provide 89.5% and 30% RDA requirement of iron from fresh 500 g meal, respectively, and only 125 g fresh leaves and steams of *Portulaca quadrifida* can provide 100% RDA requirement for iron. This RDA calculation did not consider the inhibitory effect of different antinutritional factors that affect the bioavailability of iron. The RDA for zinc is 12 mg/day for lactating woman [28]. Five hundred grams of meal (fresh basis) from cooked *A. esculentus* could provide 23% of the zinc RDA requirement for lactating mothers. Similarly, cooked *D. abyssinica* and juvenile shoots of *O. abyssinica* in the cooked form could provide 12.1 and 22.5% RDA requirement for zinc, respectively. *Portulaca quadrifida* could provide 35% of RDA for zinc considering the requirement for lactating woman. Calcium content ranged from 22.9 mg/100 g FW in cooked bamboo shoot to 140.4 mg/100 g FW in sun dried *A. esculentus*. Raw immature pods of *A. esculentus* contained 131.7 mg/100 g fresh weight basis. The calcium level obtained in this study for *A. esculentus* grown in Ethiopia was higher than the values reported for different *A. esculentus* varieties in Nigeria [21]. Raw aerial part of *P. quadrifida* contained higher (118 mg/100 g FW) calcium content next to sun dried and raw *A. esculentus*. The calcium content in raw and cooked *O. abyssinica* was 24.5 mg/100 g and 22.9 mg/100 g FW, respectively. The calcium content reported in India for the different bamboo species was within the range of 21.17–180.69 mg/100 g (29); the present value obtained for *O. abyssinica* was similar to *Bambusa balcooa* (24.01 mg/100 g). The calcium content shows no significant ($P > 0.05$) variation between raw and boiled *D. abyssinica* and *O. abyssinica* (Table 1). However, significant reduction was observed when *A. esculentus* was cooked but sun drying did not affect the calcium content. The RDA for calcium is 1000 mg/day for adults [28]. Consumption of 500 g cooked *A. esculentus*, cooked *D. abyssinica*, and *O. abyssinica* and raw *P. quadrifida* can contribute 52.12, 20.37, 11.5, and 59.0% RDA requirement for calcium, respectively. *Abelmoschus esculentus* and *P. quadrifida* are good source of calcium. Copper was the element found in a trace amount in all wild and traditional vegetables analyzed which ranged from 0.07 to 0.46 mg/100 g FW. This is important due to the fact that copper is required by body at a trace level for many metabolic activities (25). Phosphorus content in this study ranged from 23.7 mg/100 g to 57.3 mg/100 g FW. Raw *A. esculentus*, *D. prehensilis*, and *O. abyssinica* contained 39.1, 40.7, and 57.3 mg/100 g phosphorus, respectively. Cooking resulted in significant losses (7.3–41.76%) of phosphorus with variable degree of proportion among the vegetables (Table 1). Similar findings were reported in the earlier studies [23]. RDA for phosphorus is 700 mg/day for adults [28]. Consumption of 500 g cooked *A. esculentus*, *D. abyssinica*, and *O. abyssinica* and raw *P. quadrifida* could contribute by meeting 25.9, 16.9, 27.9, and 28.4% RDA requirement for phosphorus, respectively.

3.6. Bioavailability of Minerals. Antinutritional components such as oxalates, tannins, polyphenols, and phytic acid (myoinositol hexaphosphate) present in plant foods are known to have adverse effects on human nutrition by inhibiting iron [29] and zinc [30] absorption. The molar ratios along with the suggested critical values for predicting the bioavailability calcium, iron, and zinc are presented in Table 2.

3.7. Molar Ratio of Phytate to Zinc. The calculated phytate/zinc molar ratios for raw and processed WEPs were within the range of 2.03–14.22, which were in the range of the suggested critical level (<15 regarded as favorable for zinc absorption) [19]. Ratios ≥15 are associated with low zinc bioavailability [19]. According to WHO cut-offs phytic acid to zinc mole ratio ≥15, 5–15, and <5 is equal to zinc bioavailability as low (10–15%), moderate (30–35%), and high (50–55%), respectively [31]. In this context, *A. esculentus* and *P. quadrifida* had moderate zinc bioavailability, whereas the molar ratios suggested that high zinc bioavailability could be achieved from *D. abyssinica* and juvenile shoots of *O. abyssinica*. However, the critical phytate : zinc molar ratio may also depend on dietary calcium levels because of the kinetic synergism between calcium and zinc ions resulting in Ca : Zn : Phy complex which is less soluble than phytate complexes formed by either of the ions alone [32], suggesting that Ca : phy/Zn molar ratio is better predictor of zinc bioavailability than Phy : Zn molar ratio alone.

3.8. Molar Ratio of Calcium × Phytate/Zinc. Calcium × phytate/zinc molar ratios of cooked and sun dried *A. esculents* and *P. quadrifida* were above the critical level (0.5 mol/Kg) as indicated in Table 2, suggesting that calcium interference was more likely to affect zinc bioavailability. The Ca × Phy/Zn molar ratio reported in the earlier study for different *A. esculentus* varieties in Nigeria was within the range of 0.293–0.436 [21]. Higher calcium content in *A. esculentus* that grows in Ethiopia compared to Nigeria could be attributed to genetic differences, environmental variability, and interaction of the genetic factor with the environment. Considering both Phy/Zn and Ca × Phy/Zn molar ratios, zinc could adequately be absorbed in the body from *D. abyssinica* and shoots of *O. abyssinica.*

3.9. Molar Ratio of Phytate to Iron. As indicated in Table 2, phytate/iron molar ratios were >1 (indicative of poor iron bioavailability) for all raw and processed study plants except *D. abyssinica.* This might be due to the reported higher phytate content and insufficient phytic acid degradation (4.5–27.5%) by boiling alone. Phytate reduction (13–33%) after boiling *Dioscorea* sp. was reported by [23]. *Dioscorea abyssinica* could be a better source of bioavailable iron.

3.10. Molar Ratio of Calcium to Phytate. The calcium/phytate molar ratios in all the raw and processed study plants were >6, which is regarded as favorable for calcium absorption [33], predicting that a good calcium bioavailability could be achieved from all the selected vegetables. In addition to phytic acid, oxalic acid in insoluble form is responsible for interference of divalent metals absorption particularly calcium by forming insoluble salts [34]. The oxalate/calcium molar ratios of all the selected raw and processed vegetables were below the critical level of 2.5 known to significantly impair calcium bioavailability suggesting that they are good calcium bioresources for the local populace.

4. Conclusions

Immature pods of *Abelmoschus esculentus*, aerial parts of *Portulaca quadrifida*, and juvenile shoots of *Oxytenanthera abyssinica* consumed by the Gumuz community indicated that they are appreciably important sources of essential minerals. Except *D. abyssinica*, the wild and traditional plant parts have significant micronutrient compositions. However, cooking significantly reduced some of the minerals. Reducing duration of cooking time and using other processing methods such as fermentation (*D. abyssinica* and *O. abyssinica*) might alleviate the deterioration of the nutrients. The predicted mineral bioavailability shows adequacy in terms of calcium and zinc (moderately bioavailable) but not in iron. Hence, there is a need for some enhancers to increase iron absorption in all species. The study results further revealed that

A. esculentus and *P. quadrifida* are rich sources of bioavailable calcium. With the exception of *D. abyssinica,* the vegetables are rich in potassium and can contribute in maintaining normal blood pressure and its heart protective role.

To be able to justify the overall nutritional value of the wild and semiwild edible vegetables, proper assessment of the type and concentration of their antinutrients is necessary. The rate in reduction of antinutritional factors depended upon the type of processing (cooking and sun drying) and vegetable. However, the reduction of phytic acid, oxalate, and tannins by traditional cooking methods alone was not adequate to the level that could improve iron and zinc bioavailability. The vegetables have higher moisture content and are mostly used after cooking as side dishes by the community. The combination effects of these factors might reduce the actual impact of antinutritional factors in impairing bioavailability of nutrients and/or causing ill health. However, appropriate processing methods that are known to reduce the antinutritional factors (such as fermentation) could be encouraged in the community.

Conflict of Interests

The authors declare that there is no conflict of interests regarding the publication of this paper.

Acknowledgments

The authors thank Ethiopian Public Health Institute, Center for International Forestry Research (CIFOR), and Addis Ababa University for their financial and technical supports.

References

[1] G. Addis, Z. Asfaw, and Z. Woldu, "Ethnobotany of wild and semi-wild edible plants of Konso ethnic community, South Ethiopia," *Ethnobotany Research & Applications*, vol. 11, pp. 121–142, 2013.

[2] G. Addis, K. Urga, and D. Dikasso, "Ethnobotanical study of edible wild plants in some selected districts of Ethiopia," *Human Ecology*, vol. 33, no. 1, pp. 83–118, 2005.

[3] A. Assefa and T. Abebe, "Wild edible trees and shrubs in the semi-arid lowlands of Southern Ethiopia," *Jounal of Science and Development*, vol. 1, no. 1, pp. 5–19, 2011.

[4] T. Awas, *Plant diversity in Western Ethiopia: ecology, ethnobotany and conservation [Ph.D. thesis]*, University of Oslo, 2007, https://www.duo.uio.no/handle/10852/11779.

[5] D. H. Feyssa, J. T. Njoka, Z. Asfaw, and M. M. Nyangito, "Seasonal availability and consumption of wild edible plants in semiarid Ethiopia: implications to food security and climate change adaptation," *Journal of Horticulture and Forestry*, vol. 3, no. 5, pp. 138–149, 2011.

[6] G. A. Getachew, Z. Asfaw, V. Singh, Z. Woldu, J. J. Baidu-Forson, and S. Bhattacharya, "Dietary values of wild and semi-wild edible plants in Southern Ethiopia," *African Journal of Food, Agriculture, Nutrition and Development*, vol. 13, no. 2, pp. 121–141, 2013.

[7] M. Giday, T. Teklehaymanot, A. Animut, and Y. Mekonnen, "Medicinal plants of the Shinasha, Agew-awi and Amhara

peoples in northwest Ethiopia," *Journal of Ethnopharmacology*, vol. 110, no. 3, pp. 516–525, 2007.

[8] E. L. Molla, Z. Asfaw, E. Kelbessa, and P. Van Damme, "Wild edible plants in Ethiopia: a review on their potential to combat food insecurity," *Afrika Focus*, vol. 24, no. 2, pp. 71–121, 2011.

[9] A. Romojaro, M. Á. Botella, C. Obón, and M. T. Pretel, "Nutritional and antioxidant properties of wild edible plants and their use as potential ingredients in the modern diet," *International Journal of Food Sciences and Nutrition*, vol. 64, no. 8, pp. 944–952, 2013.

[10] J. A. Garí, *Agro Biodiversity Strategies to Combat Food Insecurity and Hiv/Aids Impact in Rural Africa: Advancing Grassroots Responses for Nutrition, Health and Sustainable Livelihoods*, Population and Development Service, FAO, Rome, Italy, 2003.

[11] E. Yildirim, A. Dursun, and M. Turan, "Determination of the nutrition contents of the wild plants used as vegetables in Upper Coruh Valley," *Turkish Journal of Botany*, vol. 25, no. 6, pp. 367–371, 2001.

[12] S. Gupta, A. Jyothi Lakshmi, M. N. Manjunath, and J. Prakash, "Analysis of nutrient and antinutrient content of underutilized green leafy vegetables," *LWT—Food Science and Technology*, vol. 38, no. 4, pp. 339–345, 2005.

[13] M. Umeta, C. E. West, and H. Fufa, "Content of zinc, iron, calcium and their absorption inhibitors in foods commonly consumed in Ethiopia," *Journal of Food Composition and Analysis*, vol. 18, no. 8, pp. 803–817, 2005.

[14] Association of Official Analytical Chemists international (AOAC), *Determinatination of Lead, Cadmium, and Minerals in Foods by Atomic Absorption Spectrophotometery Method (999.11/985.35)*, Association of Official Analytical Chemists, Gaithersburg, Md, USA, 2005.

[15] I. A. Vaintraub and N. A. Lapteva, "Colorimetric determination of phytate in unpurified extracts of seeds and the products of their processing," *Analytical Biochemistry*, vol. 175, no. 1, pp. 227–230, 1988.

[16] S. A. Adeniyi, C. L. Orjiekwe, and J. E. Ehiagbonare, "Determination of alkaloids and oxalates in some selected food samples in Nigeria," *African Journal of Biotechnology*, vol. 8, no. 1, pp. 110–112, 2009.

[17] H. M. Inuwa, V. O. Aina, B. Gabi, I. Aimola, and A. Toyin, "Comparative determination of antinutritional factors in groundnut oil and palm oil," *Advance Journal of Food Science and Technology*, vol. 3, no. 4, pp. 275–279, 2011.

[18] P. M. Zarembski and A. Hodgkinson, "The determination of oxalic acid in food," *Analyst*, vol. 87, no. 1038, pp. 698–702, 1962.

[19] E. R. Morris and R. Ellis, "Usefulness of the dietary phytic acid/zinc molar ratio as an index of zinc bioavailability to rats and humans," *Biological Trace Element Research*, vol. 19, no. 1-2, pp. 107–117, 1989.

[20] L. Hallberg, M. Brune, and L. Rossander, "Iron absorption in man: ascorbic acid and dose-dependent inhibition by phytate," *The American Journal of Clinical Nutrition*, vol. 49, no. 1, pp. 140–144, 1989.

[21] F. Adetuyi and A. Osagie, "Nutrient, antinutrient, mineral and zinc bioavailability of okra *Abelmoschus esculentus* (L) Moench Variety," *American Journal of Food and Nutrition*, vol. 1, no. 2, pp. 49–54, 2011.

[22] S. Shanthakumari, V. R. Mohan, and J. de Britto, "Nutritional evaluation and elimination of toxic principles in wild yam (*Dioscorea* spp.)," *Tropical and Subtropical Agroecosystems*, vol. 8, no. 3, pp. 319–325, 2008.

[23] P. E. Akin-Idowu, R. Asiedu, B. Maziya-Dixon, A. Odunola, and A. Uwaifo, "Effects of two processing methods on some nutrients and anti-nutritional factors in yellow yam (*Dioscorea cayenensis*)," *African Journal of Food Science*, vol. 3, no. 1, pp. 022–025, 2009.

[24] P. Kakati, S. C. Deka, D. Kotoki, and S. Saikia, "Effect of traditional methods of processing on the nutrient contents and some antinutritional factors in newly developed cultivars of green gram [*Vigna radiata* (L.) Wilezek] and black gram [*Vigna mungo* (L.) Hepper] of Assam, India," *International Food Research Journal*, vol. 17, no. 2, pp. 377–384, 2010.

[25] N. Chongtham, M. S. Bisht, and S. Haorongbam, "Nutritional properties of bamboo shoots: potential and prospects for utilization as a health food," *Comprehensive Reviews in Food Science and Food Safety*, vol. 10, no. 3, pp. 153–168, 2011.

[26] J. J. Otten, J. P. Hellwig, and L. D. Meyers, *Institute of Medicine (É-U). Dietary DRI Reference Intakes: The Essential Guide to Nutrient Requirements*, National Academies Press, Washington, DC, USA, 2006.

[27] O. I. Io, "Evaluation of iron, zinc, sodium and phytate contents of commonly consumed indigenous foods in Southwest Nigeria," *Journal of Nutrition & Food Sciences*, vol. 2, no. 10, 2012.

[28] E. Whitney and S. R. Rolfes, *Understanding Nutrition*, Thomson Learning, Belmont, Calif, USA, 11th edition, 2008.

[29] R. F. Hurrell, M. Reddy, and J. D. Cook, "Inhibition of non-haem iron absorption in man by polyphenolic-containing beverages," *British Journal of Nutrition*, vol. 81, no. 4, pp. 289–295, 1999.

[30] R. S. Gibson, L. Perlas, and C. Hotz, "Improving the bioavailability of nutrients in plant foods at the household level," *Proceedings of the Nutrition Society*, vol. 65, no. 2, pp. 160–168, 2006.

[31] J. Goicoechea, World Health Organization, and Regional Office for Europe, *Primary Health Care Reforms*, World Health Organization, Geneva, Switzerland, 1996.

[32] H. W. Lopez, F. Leenhardt, C. Coudray, and C. Remesy, "Minerals and phytic acid interactions: is it a real problem for human nutrition?" *International Journal of Food Science and Technology*, vol. 37, no. 7, pp. 727–739, 2002.

[33] P. E. Akin-Idowu, O. A. Odunola, R. Asiedu, B. Maziya-Dixon, and A. O. Uwaifo, "Variation in nutrient and antinutrient contents of tubers from yellow yam (*Dioscorea cayenensis*) genotypes grown at two locations," *Journal of Food, Agriculture and Environment*, vol. 6, no. 3-4, pp. 95–100, 2008.

[34] L. G. Hassan, B. U. Bagudo, A. A. Aliero, K. J. Umar, and N. A. Sani, "Evaluation of nutrient and anti-nutrient contents of *Parkia biglobosa* (L.) flower," *Nigerian Journal of Basic and Applied Sciences*, vol. 19, no. 1, 2011.

10

Preparation of Modified Films with Protein from Grouper Fish

M. A. Valdivia-López, A. Tecante, S. Granados-Navarrete, and C. Martínez-García

Departamento de Alimentos y Biotecnología, Facultad de Química, Universidad Nacional Autónoma de México, 04510 Ciudad de México, Mexico

Correspondence should be addressed to M. A. Valdivia-López; mavald@unam.mx

Academic Editor: Melvin Pascall

A protein concentrate (PC) was obtained from Grouper fish skin and it was used to prepare films with different amounts of sorbitol and glycerol as plasticizers. The best performing films regarding resistance were then modified with various concentrations of $CaCl_2$, $CaSO_4$ (calcium salts), and glucono-δ-lactone (GDL) with the purpose of improving their mechanical and barrier properties. These films were characterized by determining their mechanical properties and permeability to water vapor and oxygen. Formulations with 5% (w/v) protein and 75% sorbitol and 4% (w/v) protein with a mixture of 15% glycerol and 15% sorbitol produced adequate films. Calcium salts and GDL increased the tensile fracture stress but reduced the fracture strain and decreased water vapor permeability compared with control films. The films prepared represent an attractive alternative for being used as food packaging materials.

1. Introduction

Biodegradable films are an attractive development for the food industry. Their use is associated with the broad range of properties they possess. Such properties are helpful to keep food in optimum conditions during transport and storage and constitute an interesting answer to the demand of consumers for higher quality and long-shelf life products, while reducing disposable packaging material and increasing recyclability [1]. The extensive list of biodegradable film ingredients available allows targeting a broad range of potential functional properties [2].

A biodegradable film is defined as a thin continuous layer made from biodegradable materials [3], that is, materials that can be degraded by enzyme action of living organisms, such as bacteria, yeast, and fungi [4]. Some critical conditions such as abundance and availability of raw materials are needed to make the production of biodegradable polymers feasible. Protein is one resource that meets these characteristics [5]. Proteins, polysaccharides, and lipids have been used as film-forming materials. However, proteins have been widely chosen because they are abundant, are available in plant and animal sources, and form stable networks [6]. Also, protein films are better than those prepared with polysaccharides because proteins are composed of 20 different amino acids, and they have a particular structure which offers a broad range of functional properties [7].

Fish skin is a good source of inexpensive collagen, which is the main support protein that constitutes the structures of the body of animals, vertebrates, and invertebrates and is concentrated in specialized connective tissues: skin, tendon, and bone [8]. Castañeda [9] studied the properties of fish skin proteins demonstrating that it is useful to form biodegradable films. The results obtained revealed the films with 5% protein concentrate (PC) and 75% sorbitol (plasticizer) to present the best structural characteristics, and although they exhibited acceptable mechanical properties, their barrier properties to moisture migration were not good. Water vapor loss is one of the more severe problems in food preservation, and it can cause adverse effects on texture, nutritive value, scalability, and integrity of food products.

Several properties of fish skin films, such as mechanical properties, permeability, light absorption, transparency, antimicrobial activity, and antioxidant ability, are influenced by the addition of active substances [10]. For example, Park and collaborators [11] modified biodegradable soy films, adding $CaCl_2$ and $CaSO_4$ (calcium salts) and glucono-δ-lactone (GDL). They concluded that calcium salts and GDL reduced the water vapor permeability and improved their mechanical properties. Also, Zactiti and Kieckbusch [12] added calcium salts to alginate films and observed lower elongation, higher tensile strength, and a considerable reduction in water solubility and water vapor permeability.

In Mexico around 11300 tons of Grouper fish (*Epinephelus marginatus*) are captured per year [13], signifying one of the major domestic fisheries. Grouper is consumed as fresh and frozen fillet, and the skin, which is about 10% of the weight of the fish, is not used as a commercial product; this means that about 1300 tons of skin of this fishery are discarded. Fish skins are a major by-product of fishing and aquaculture. Thus, the fish skin could provide a valuable source of protein [14]. Many fish skin films produced from fish-processing coproducts have been studied, such as the skins of cuttlefish [15], blue shark [14], bigeye snapper, and brownstripe red snapper [6], showing, in general, poor mechanical properties and high water vapor permeability which are the main drawbacks for applications. Therefore, the aim of the present work was to study film formation from collagen fractions extracted from the skin of Grouper fish (*Epinephelus marginatus*) using plasticizers and different concentrations of calcium salts and glucono-δ-lactone and to evaluate the effect on its mechanical and barrier properties.

2. Materials and Methods

2.1. Compositional Analysis of Grouper Fish Skin. The skin of Grouper fish (*Epinephelus marginatus*) was obtained from a local market in Mexico City. The skin of three different batches, each one with three replicas, was analyzed for moisture, fat, and total crude protein content. Assays were done using AOAC methods [16]: moisture in a vacuum oven (931.04); fat, goldfish (920.85); and total crude protein, Kjeldahl (981.10).

2.2. Protein Concentrate. Protein was extracted from fresh fish skin according to the procedure of Batista [17], with some modifications. The skin was cut into small pieces, soaked in 0.1 M NaOH (pH 12 ± 0.5) with skin to water ratio of 1:10 (w/v), respectively, and the mixture was stirred 120 min at 45°C. After this time, the suspension was centrifuged (Beckman J2 centrifuge J2-mark M2) 15 min at 4°C and 5000 rpm, and the supernatant was recovered. At this point, the protein content expressed as total crude protein, and soluble protein was determined in the supernatant, while the residue, composed of milled skin and scales, was examined for total crude protein (Kjeldahl AOAC 981.10). One milliliter of 10% sodium hexametaphosphate was added to 50 mL of supernatant, pH was adjusted to 2.5 with 2 M HCl, and the liquid was kept two hours in refrigeration for complete precipitation of proteins. After this, proteins were separated by centrifugation at 5000 rpm for 15 minutes. The residue constitutes the protein concentrate (PC); its amount of total crude and soluble proteins was determined and the PC pellet was frozen until further needed. This PC was used for film formation. The yield of protein extraction was determined in every one of the extracted fractions from measurements of total crude protein and soluble protein [18]. Fish skin

and protein concentrate were weighed and the volume of supernatants was measured.

2.3. Molecular Weight of Extracted Proteins. The protein fractions in the concentrate were separated by sodium dodecylsulfate-polyacrylamide gel electrophoresis (SDS-PAGE) according to Laemmli [19]. A 10% polyacrylamide gel was used for high molecular weights (36–200 kDa Sigma Marker Sigma) and a 12% gel for low molecular weights (20–66 kDa Sigma Marker). The solution, with 1 and 2 mg/mL PC (pH 12 ± 0.1), was mixed with buffer in a 1:1 (v/v) ratio and heated for 3 min in boiling water. Gels were loaded with the treated sample (20 mL) and with molecular weight markers (4 mL) and were run in an electrophoresis chamber (Bio-Rad) at 100 volts for 90–120 min. After this, gels were stained overnight with Coomassie blue solution and washed with a 10:10:80% (v/v/v) methanol/acetic acid/water solution. The washed gels were scanned in a densitometer (Bio-Rad, Model GS700) and the molecular weights of the separate bands were determined with the Quantity-One software (Bio-Rad).

2.4. Solubility of the Protein Concentrates at Different pH. The variation of solubility with pH was determined following the method of Saeed and Cheryan [20]. The pH of twelve aqueous solutions (10 mL) with 3% (w/v) PC was adjusted (Thermo Electron Corporation, USA) to values in the range of 1.0 to 12.0 with 1.0 N HCl and NaOH. Protein solutions were stirred for 30 min. Solutions were then centrifuged (Labtronic Scientific, Model H-1650) at 2500 rpm for 15 min and the PC of the supernatant was determined using Lowry's method [18].

2.5. Protein Concentrate Films. Films were prepared following the method of Sobral and collaborators [21]. Five grams of protein concentrate (PC) was dispersed in 100 mL distilled water under continuous stirring and pH was adjusted to 11.5 ± 0.2 with 1.0 N NaOH. Sorbitol and glycerol were added to different films, in concentrations of 50 and 75% (w/w) referred to the total amount of protein in the PC. An additional formulation included 4% (w/v) PC, with a mixture of 15% (w/w) glycerol and 15% (w/w) sorbitol [22]. The formulations were identified as 5PC-50G, 5PC-75G, 5PC-50S, 5PC-75S, and 4PC-15G/15S, where the first part indicates the concentration of PC and the second one the concentration of plasticizer. Different solutions were stirred for 10 min and then heated to 70°C for 20 min in a water bath. After this time, the solutions were filtered and sonicated (Branson 3510, Bransonic® ultrasonic) for 15 min. Finally, 50 mL of solution was poured into Teflon-covered pans 12 cm in diameter and dried at room temperature for approximately 48 hours (Table 1).

Films were considered adequate when they were easily detached manually from the container surface, nonsticky and flexible enough to handling. Then, these selected films were modified to improve their mechanical properties, permeability to water vapor and oxygen. The modification was done by adding $CaCl_2$, $CaSO_4$ (calcium salts), and glucono-δ-lactone (GDL) in concentrations of 0.1, 0.2, and 0.3% (w/w), of the amount of total protein in the PC [11]. The salts were added

TABLE 1: Films formulations.

Identification	Protein concentrate (%)	Plasticizer
5PC-50G	5	50% glycerol
5PC-75G	5	75% glycerol
5PC-50S	5	50% sorbitol
5PC-75S	5	75% sorbitol
4PC-15G/15S	4	15% glycerol + 15% sorbitol

TABLE 2: Modified films.

Identification	Protein concentrate (%)	Plasticizer	Modified salts
5PC-75S	5	75% sorbitol	0.1% $CaCl_2$
5PC-75S	5	75% sorbitol	0.2% $CaCl_2$
5PC-75S	5	75% sorbitol	0.3% $CaCl_2$
5PC-75S	5	75% sorbitol	0.05% $CaSO_4$
5PC-75S	5	75% sorbitol	0.1% $CaSO_4$
5PC-75S	5	75% sorbitol	0.2% $CaSO_4$
5PC-75S	5	75% sorbitol	0.3% $CaSO_4$
5PC-75S	5	75% sorbitol	0.1% GDL
5PC-75S	5	75% sorbitol	0.2% GDL
5PC-75S	5	75% sorbitol	0.3% GDL
4PC-15G/15S	4	15% glycerol + 15% sorbitol	0.05% $CaSO_4$
4PC-15G/15S	4	15% glycerol + 15% sorbitol	0.1% GDL

after the plasticizer; a salt solution of known concentration was added depending on the amount of protein present (Table 2). Modified and unmodified films were analyzed. The latter are designated as the control.

2.6. Water Vapor Permeability. Water vapor permeability (WVP) was determined according to the ASTM E96-95 method [23]. This is a gravimetric procedure in which the amount of water adsorbed by anhydrous calcium chloride is determined. Acrylic cells, previously taken to constant weight, were used. A known amount of desiccant (*ca.* 35 g) was placed in the cell leaving a head space of about 1 cm. The films were fixed to the rim of the cell with a pressing ring leaving a known area for water vapor transmission. Cells were placed in a chamber with a relative air humidity of 62±2% and maintained at room temperature (≈23 ± 2°C). The increment of mass and temperature of each cell was recorded every 24 h for five days. Determinations were done in triplicate.

2.7. Oxygen Permeability. Oxygen permeability was determined in a stainless steel cell (CSI-135 Permeability Tester) according to the ASTM D1434-82 method [24]. The oxygen transmission coefficient was obtained by monitoring the change in volume generated by the transfer of oxygen through the film as a result of an applied differential gas pressure.

The cell was operated at a manometric pressure of 4 psi (27571 Pa), 293.25 K, and a barometric pressure of 77994 Pa.

2.8. Mechanical Resistance. Before testing the films for mechanical resistance they were first conditioned at 62 ± 2% relative humidity and 23±2°C for 48 h. Relative humidity was generated with a saturated solution of $Mg(NO_3)_2 \cdot 6H_2O$ and measured with a hygrometer (Oakton, Japón). Measurements were made with a testing machine (Sintech 1/S, MTS, USA) using a load cell of 100 N according to the ASTM D 882 method [25]. Film strips 8 cm long and 1 cm wide were examined. Their average thickness was measured with a micrometer (Mitutoyo, Japan) on both ends and in the middle of the strip. Strips were stretched at 250 mm/min until they broke up. Force-time data were transformed into true stress, (1) σ_T, and Hencky strain, (2) ε_H. Young's modulus was determined from the slope of the linear portion of σ_T versus ε_H plot:

(1) True stress (σ_T): $\sigma_T = \sigma(1 + e)$,

(2) Hencky (ε_H): $\varepsilon_H = \ln(1 + e)$.

In these equations σ is the nominal stress and e is the Cauchy strain.

2.9. Statistical Analysis. All experiments were carried out in triplicate. Statistics on an entirely randomized design were determined with the SPSS 10.0 for Windows procedure. Differences were considered to be statistically significant at $p < 0.05$.

3. Results and Discussion

3.1. Yield of Protein Concentrate. The extraction process did not include a purification step to obtain an entirely pure protein concentrate. Hence, yields of total protein, 63.81%, and soluble protein, 45.33%, in the concentrate were small. Impurities in the concentrate can include salts normally present in the skin and sodium hexametaphosphate used to make precipitation easier; such impurities were possibly solubilized and precipitated together with proteins. The average composition of the three different batches was on a dry basis: 65.21 ± 2.85% moisture, 6.3 ± 0.18% fat, and 76.51 ± 4.54% total crude protein.

3.2. Molecular Weight of Protein Fractions. Protein patterns of the protein concentrate separated with 10 and 12% polyacrylamide (Figure 1) shows low and high molecular weight (MW) proteins. Bands between 31 and 66 kDa are visible; five of them show high intensity, with MW in the range of 33–48 kDa (Figure 1(a)); bands below 30 kDa separated with 12% polyacrylamide are shown in Figure 1(b). Proteins with MW below 23 kDa are not clearly observed and those present in high concentration are in the range of 31–48 kDa. Several bands >48 kDa are also visible, being the brightest for 58–60, 71, 95, and 194 kDa. The well-defined high molecular weight bands in Figure 1(b) correspond to 95 and 194 kDa, although their concentration is lower than those of the low

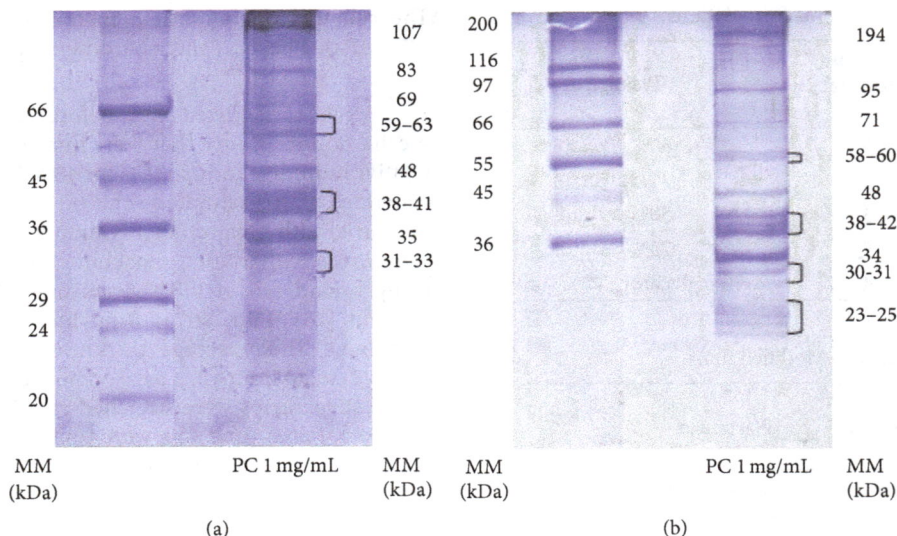

FIGURE 1: Protein fractions in the concentrate from SDS-PAGE with (a) a low molecular marker and 10% polyacrylamide and (b) a high molecular weight marker and 12% polyacrylamide, stained with Coomassie blue. PC: protein concentrate in a 1 mg/mL solution.

molecular weight proteins. Several authors have character-ized fish gelatin extracted from various fish species, showing similar protein patterns. Norziah et al. [26] characterized fish gelatin extracted from residues of surimi production and obtained two bands of similar molecular weights, identified as α-collagen (100 kDa) and β-collagen (200 kDa). The β component is formed when two simple collagen strands (α units) are cross-linked to each other by covalent bonds. Limpisophon et al. [14] also identified these two bands as α and β collagen in characterizing gelatin extracted from blue shark (*Prionace glauca*).

3.3. Solubility of the Protein Concentrates for Different pH.
Protein solubility depends on pH; above or below the iso-electric point (pI) the net charge is negative or positive, respectively, and water molecules can interact with these charges thereby contributing to solubility. Figure 2 shows the change in solubility of the fish skin protein content of two batches as a function of pH in the range of 1 to 12. Both batches show a similar trend with no statistical difference, within the pI in the range of 2 to 4 pH units. As pH increases, the solubility of the protein concentrates increases because there are more negative charges enabling electrostatic repulsion with the solvent. The maximum solubility, 43.17%, occurred at pH 11 and 12. The purpose of knowing the solubility of the PC was to set the pH for film preparation. The pH 11.5 was chosen for film development considering the obtained data.

In other studies [27] collagen extracted from bigeye snapper skin with acid and pepsin exhibited the maximum solubility between pH values of 4 and 5, respectively. Also, Kittiphattanabawon et al. [28] observed that the maximum solubility of acid extracted collagen from big eye snapper skin was between pH 2 and pH 5. Likewise, the results in collagen obtained from chicken by-products show that the maximum collagen solubility was at pH 2.

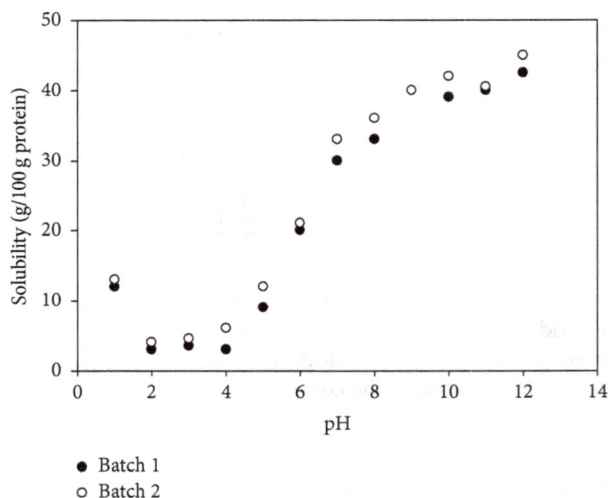

FIGURE 2: Solubility of the protein concentrates of Grouper skin batches for different pH.

3.4. Films Characteristics.
Different formulations were tested to evaluate the effect of plasticizer type and concentration on film formation. Low molecular weight plasticizers incorpo-rate easier into the protein matrix and in consequence they have a good performance on film formation [29]. The films of formulation 5PC-50G firmly adhered to the pan surface and broke up quickly. Formulation 5PC-75G produced films that could not be adequately formed because the plasticizer concentration was excessive; films remained moist for several days and could not be detached from the pan surface. Formu-lation 4PC-15G/15S produced films that were easily separated from the container surface but were less flexible than those with 5PC-75S. This behavior is attributed to the lower

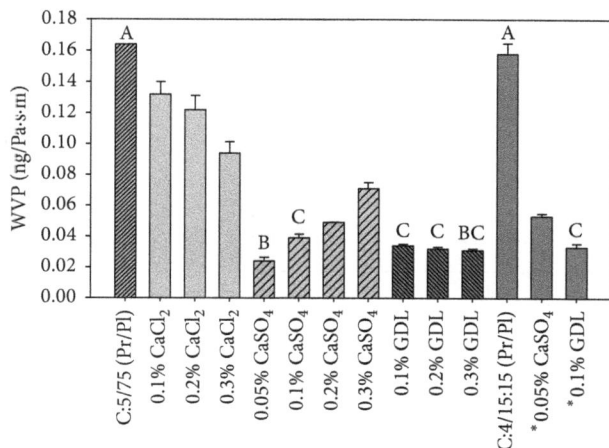

FIGURE 3: Water vapor permeability of unmodified and modified films prepared with formulations 5PC-75S and 4PC-15G/15S. Note: values in a column with different letters are significantly different at $p < 0.05$.

concentration of plasticizer. Films prepared 5PC-50S were brittle, stiff, and therefore unsuitable for subsequent analyses. Films with the best forming characteristics were 5PC-75S and 4PC-15G/15S. Film flexibility is mainly determined by protein-protein and protein-water interactions and may be controlled by the concentration and type of plasticizer, which reduces the intermolecular interactions between adjacent protein chains. As a consequence, chain mobility increases and films become flexible preventing rupture during handling and storage [30]. Our results show that it is possible to obtain films from Grouper fish skin proteins and that the incorporation of different kind of plasticizers into fish skin films resulted in more or less film flexibility and moisture.

Film formation has been proven with proteins extracted from different species of fish including Atlantic sardine (*Sardina pilchardus*) [31], red snapper (*Lutjanus vitta*) [6], and Nile tilapia (*Oreochromis niloticus*) [30].

3.5. Water Vapor Permeability. Figure 3 shows the effect of calcium salts and GDL on WVP of 5PC-75S and 4PC-15G/15S films. Each modifier produced a different effect. The addition of calcium chloride ($CaCl_2$) decreased WVP of films; the higher the salt concentration, the lower the WVP. For 5PC-75S the WVP of the control film was reduced from 0.164 ng/Pa·s·m to 0.094 ng/Pa·s·m, that is, almost 42% when 0.3% $CaCl_2$ was added. Calcium sulfate ($CaSO_4$) had a greater impact on WVP than calcium chloride; the WVP of the control film was reduced to 0.070 ng/Pa·s·m, that is, around 57%, for 0.3% $CaSO_4$. However, the effect of calcium sulfate was opposite to that of calcium chloride as lower concentrations of the former resulted in lower WVP. The addition of $CaSO_4$ concentrations as low as 0.05% resulted in a WVP of 0.024 ng/Pa·s·m, which represents a reduction of almost 85% regarding that of the control. This suggests that the concentration of calcium chloride should be increased if a WVP similar to that with calcium sulfate wants to be obtained. Park et al. [11] reported statistically significant

reductions in WVP of soy protein films modified with $CaSO_4$. However, films modified with $CaCl_2$ did not show significant decreases relative to control films. The authors explained that negative charges given by carboxyl groups ($-COO^-$) predominated on the protein chain and their interaction with divalent cations Ca^{2+} resulted in a more stable network. These ionic interactions not only reduce the mobility of protein segments but also increase their hydrophobicity as the interaction between Ca^{2+} and the negatively charged carboxyl groups prevents cations to interact with the water decreasing the solubility of proteins and thus WVP through the polymer.

The Hofmeister series can explain the difference between the effect caused by $CaCl_2$ and $CaSO_4$ on the WVP in our films as protein-protein interactions and protein crystallization are some of the physical behaviors that obey this series [32]. The series was originally developed as a measure of the efficiency of various anions to precipitate globular proteins. The effect of ions is usually related to their position in the series; $SO_4^{2-} > HPO_4^{2-} > CH_3COO^- > Cl^-$, which shows that the sulfate ion results in increased protein stability and lower solubility, as a result of greater protein-protein attraction, than the chloride ion [33]. The more significant effect caused by $CaSO_4$ in comparison with $CaCl_2$ on WVP of the protein films can be attributed to the fact that the SO_4^{2-} ion is a better sequestrant of solvent water molecules and hence prevents the formation of hydrogen bonds on proteins surface. Therefore, effective protein-protein interactions occur that result in lower solubility and less water diffusion through the protein network.

The addition of glucono-δ-lactone (GDL) also reduced the WVP on 5PC-75S control films. WVP values were 0.034, 0.032, and 0.031 ng/Pa·s·m for 0.1, 0.2, and 0.3%, respectively, without being significantly different (Figure 3). GDL is a cyclic ester, gradually hydrolyzed in water to gluconic acid forms widely used in the food industry as acidulants [34]. The observed decrease in WVP of films modified with GDL may be due to an increased hydrophobicity and hence a reduction in solubility attributed to the charged carboxyl groups that adversely reduce the action of protons produced by GDL [11]. Therefore, the neutralized protein molecules can aggregate due to a decreased electrostatic repulsion and prevent carboxyl groups to interact with the water. The 5PC-75S films that exhibited the lower WVP were those modified with 0.05% $CaSO_4$ and 0.1, 0.2, and 0.3% GDL. However, the WVP values of the latter were not significantly different. Figure 3 also shows the WVP of 4PC-15G/15S control films and films modified with 0.05% $CaSO_4$ and 0.1% GDL, which were the modifiers and concentrations with the greater effect on the WVP of 5PC-75S films. The WVP of 4PC-15G/15S control films was 0.158 ng/Pa·s·m. Although it is lower than for 5PC-75S control films, there is not a significant difference between the two formulations because, even if the formulation 4PC-15G/15S had a lower concentration of protein, it also contains a smaller amount of plasticizers. Therefore, it is possible to assume that similar interactions occurred in films of both formulations. The WVP of the 4PC-15G/15S modified films significantly in comparison with their control. Calcium

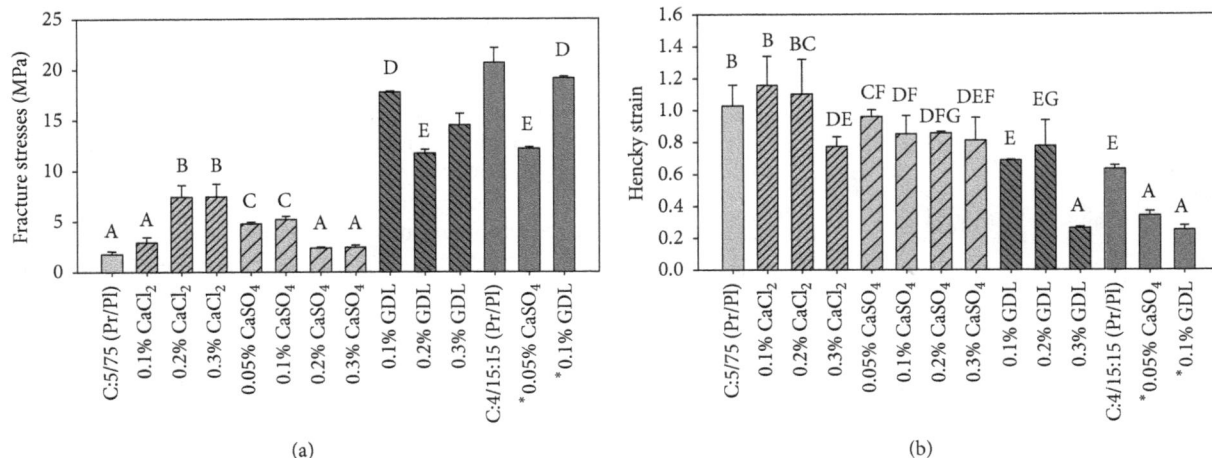

FIGURE 4: Fracture stresses (a) and Hencky strain (b) of films prepared with formulations 5PC-75S and 4PC-15G/15S.Note: values in a column with different letters are significantly different at $p < 0.05$.

sulfate reduced WVP to 0.053 ng/Pa·s·m and GDL reduced it to 0.033 ng/Pa·s·m, which represent reductions of around 66 and 79%, respectively, regarding the control.

Comparing the results of all the films analyzed the formulation 5PC-75S with 0.05% $CaSO_4$ and with different concentrations of GDL produced films with the best barrier against water vapor transmission, regardless of the concentration of protein and plasticizer. On the other side, the WVP of the unmodified films were of the same order of magnitude as those of protein films from other fish species. For example, the WVP of films made of skin proteins of Alaskan pink salmon was 0.169 ng/Pa·s·m [35], which is very similar to 0.164 ng/Pa·s·m for our 5PC-75S control film. However, our values for films modified with $CaSO_4$ and GDL are lower by one or two orders of magnitude compared with films based on other protein concentrates, or synthetic polymers except polyester, which is still lower by one or two orders of magnitude, compared to the modified films in this study. These modifications are promising because modified films are expected to provide greater resistance to water transmission to the matrix they are covering, than films made up only of protein and plasticizer.

3.6. Oxygen Permeability.

The stability of foods is affected by the presence or absence of oxygen. This gas affects the shelf life of foods because it participates in oxidation reactions, microorganism growth, changes in color, and respiration of fruits and vegetables [36]. Therefore, the oxygen permeability of protein films is essential for establishing their functionality as food protectants. Polymers containing groups that can self-associate by hydrogen or ionic bonds, such as proteins, produce films with excellent properties against oxygen permeability [37]. The average oxygen permeability of PC films of Grouper fish was 1.09×10^{-17} mol·mm/mm²·s·Pa. This corresponds to 3.27×10^{-9} cm³·cm/cm²·s·cmHg or 32.7 ± 0.79 barrers (1 barrer = 10^{-10} cm³·cm·cm^{-2}·s^{-1}·cmHg^{-1}). The oxygen permeability of low density polyethylene at 298 K is 2.20 barrers [38]. This oxygen permeability is 15-fold lower

than that of PC films of Grouper fish. Polymers with oxygen permeability below 38.9 cm³·μm/m²·d·kPa (0.060 barrers) at 23°C are considered good barriers to oxygen [39]. In general, protein-based films are considered good oxygen barriers. Oxygen permeabilities reported for films of various proteins are lower than 38.9 cm³·μm/m²·d·kPa [35].

3.7. Mechanical Properties.

The mechanical properties of the films provide an indication of their integrity under stresses associated with processing, handling, and storage. Figure 4(a) shows the fracture stresses of unmodified (control) and modified 5PC-75S and 4PC-15G/15S films. The addition of 0.2 and 0.3% calcium chloride ($CaCl_2$) to 5PC-75S films increased their fracture stress from 1.6 MPa, for the control, to around 7.3 MPa. The addition of 0.1% of this salt was not sufficient for obtaining a tensile fracture stress significantly different from that of the control and this also occurred for 0.2 and 0.3% $CaSO_4$. On the contrary, the addition of 0.05 and 0.1% of this salt increased the fracture stress to about 5.0 MPa, which is lower than the increase with 0.2 and 0.3% calcium chloride. Therefore, low concentrations of calcium sulfate were sufficient to make films resistant to fracture upon stretching. Films modified with 0.1, 0.2, and 0.3% GDL showed fracture stresses between 11.7 and 17.8 MPa, which represents a significant increase over the 1.6 MPa for the control. These high values suggest the existence of greater protein interaction in modified films as compared with the control and those modified with calcium salts. According to Park et al. [11] GDL promotes protein aggregation because hydrophobicity is increased and solubility decreases, so films become more resistant but less flexible. The fracture stress of unmodified 4PC-15G/15S films was 20.7 MPa, which suggests a greater chain-chain interaction between proteins as the plasticizer concentration was lower than for unmodified 5PC-75S films. The addition of 0.05% $CaSO_4$ and 0.1% GDL to 4PC-15G/15S films also improved the resistance of these materials to stretching. The fracture stresses of 5PC-75S films with 0.2% GDL and 4PC-15G/15S films with 0.05% $CaSO_4$

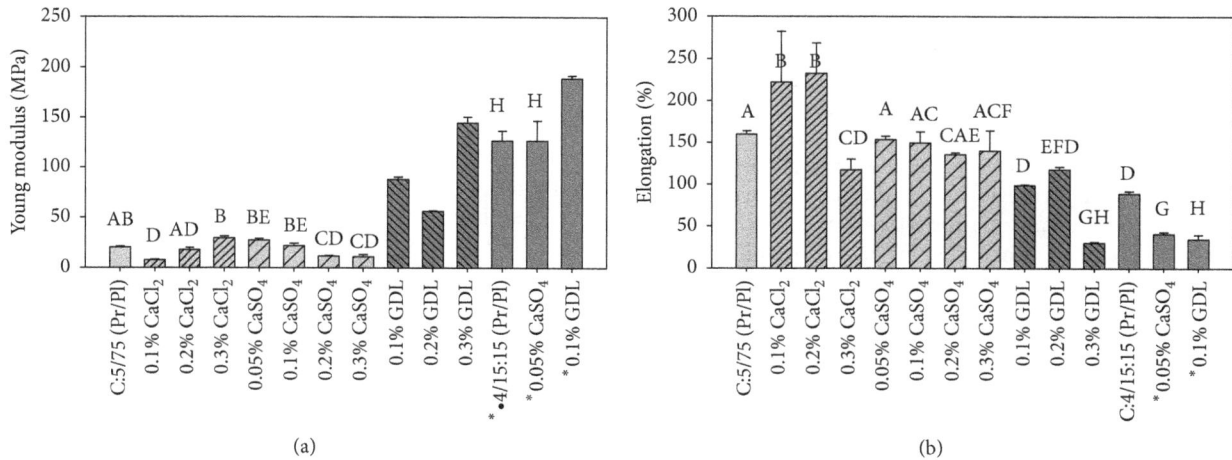

FIGURE 5: Young's modulus (a) and elongation percentage (b) of films prepared with formulations 5PC-75S and 4PC-15G/15S. Note: values in a column with different letters are significantly different at $p < 0.05$.

were not significantly different. The same happened between 5PC-75S and 4PC-15G/15S films with 0.1 GDL.

The strain is given by a pure number, because it compares the shape of the material before and after deforming it. It is an important feature, because if a material can be stretched considerably before breaking up, this is an indication that it can withstand the applied load. Figure 4(b) shows the trend of fracture strain, expressed as Hencky strain, for the two formulations of modified and unmodified films. In general, the addition of increasing concentrations of calcium salts and GDL resulted in significant reduction in fracture strain, with 0.3% GDL being the most noteworthy. The same trend was observed for 4PC-15G/15S films, but with a significant decrease in the control compared to the control of 5PC-75S films.

Young's modulus is the slope of the linear part of the stress-strain curve. It indicates the resistance of the material to deformation which is related to its stiffness. Figure 5(a) shows Young's modulus for modified and unmodified 5PC-75S and 4PC-15G/15S films. Films modified with GDL were the most rigid of all the series according to their Young's modulus (Figure 5(a)) and fracture stress, regardless of the formulation. However, they did not withstand large deformations before breaking up because being more rigid due to increased interactions between polypeptides chains, they are more susceptible to deformation because of the reduced mobility of protein chains. This was the case, for example, for 4PC-15G/15S films modified with GDL and $CaSO_4$. The fracture strains for 5PC-75S films modified with 0.1 and 0.2% $CaCl_2$ were 1.16 and 1.10, respectively, because protein chains show more mobility.

The latter exhibited more resistance to deformation; 189 069 MPa for 0.1% GDL and 126 930 and 126 871 MPa for control and 0.05% $CaSO_4$, respectively, without significant difference between both of them. These films also showed high fracture stresses, and this is again attributed to the lower concentration of plasticizer, which made them stiffer. In the case of 5PC-75S films modified with calcium salts, Young's modulus did not change considerably in comparison with

the control. However, the film modified with 0.05% $CaSO_4$, which had a lower WVP, exhibited the greatest Young's modulus, 27.4 MPa, in comparison with the control and films modified with calcium salts. This behavior confirms the presence of an increased cross-linking between proteins in the formulation. Therefore, these films were less elastic than those modified with calcium sulfate and calcium chloride.

Elongation is another way of expressing the flexibility of films to traction. Figure 5(b) shows the percentage of elongation for the unmodified and modified formulations. 5PC-75S films can stretch over 100% of their original length, while a maximum elongation of about 232% was observed for films modified with 0.1 and 0.2% calcium chloride. The percentage of elongation for 4PC-15G/15S modified with 0.1% GDL and 0.05% $CaSO_4$ was about 34.4%, while that for the control film was 88.5%. The 4PC-15G/15S and 5PC-75S films modified with GDL showed high fracture stress and Young's modulus, and therefore less percentage of elongation.

4. Conclusions

It was possible to produce films with proteins obtained from the skin of Grouper fish using an adequate proportion of PC and plasticizers. This means that proteins can form ordered three-dimensional networks capable of interacting with the plasticizer and water. The extraction yield of the protein concentrate from different batches is small. 5PC-75S films showed the best physical properties. Film formation was also possible with less protein, that is, 4% and a mixture of equal amounts of sorbitol and glycerol as plasticizers. The addition of different concentrations of calcium chloride, calcium sulfate, and GDL modifies mechanical and barrier properties differently and to different extents. Permeability to oxygen was not detected over a period of 24 hours. This result could be convenient to retard chemical, physical, and microbiological degradation in foods and offers an alternative to the use of a biodegradable packaging. Protein films of Grouper fish skin were better barriers to water vapor and oxygen compared with protein films from other natural sources. Traction tests evidenced the greater resistance of films modified with GDL, which were less deformable and resistant to physical changes during handling. In general, the 5PC-75S formulation exhibits a high stretching capacity than 4PC-15G/15S. Films modified with GDL and CaSO$_4$ represent a significant advance in film technology based on proteins. The lack of oxygen permeability significantly reduced WVP and acceptable mechanical properties compared to other proteins are attractive properties for the materials studied here.

Competing Interests

The authors declare that they have no competing interests.

Acknowledgments

Thanks are due to Mariana Ramírez-Glly for her technical support with mechanical tests.

References

[1] K. Petersen, P. V. Nielsen, G. Bertelsen et al., "Potential of biobased materials for food packaging," *Trends in Food Science and Technology*, vol. 10, no. 2, pp. 52–68, 1999.

[2] C. Bourlieu, V. Guillard, B. Vallès-Pamiès, S. Guilbert, and N. Gontard, "Edible moisture barriers: how to assess of their potential and limits in food products shelf-life extension?" *Critical Reviews in Food Science and Nutrition*, vol. 49, no. 5, pp. 474–499, 2009.

[3] J. Krochta, E. Balwin, and M. Nísperos-Carriedo, *Edible Coatings and Films to Improve Food Quality*, CRS Press, Boca Raton, Fla, USA, 1994.

[4] M. Avella, J. J. De Vlieger, M. E. Errico, S Fischer, P. Vacca, and M. G. Volpe, "Biodegradable starch/clay nanocomposite films for food packaging applications," *Food Chemistry*, vol. 93, no. 3, pp. 467–474, 2005.

[5] R. N. Tharanathan, "Biodegradable films and composite coatings: past, present and future," *Trends in Food Science & Technology*, vol. 14, no. 3, pp. 71–78, 2003.

[6] A. Jongjareonrak, S. Benjakul, W. Visessanguan, T. Prodpran, and M. Tanaka, "Characterization of edible films from skin gelatin of brownstripe red snapper and bigeye snapper," *Food Hydrocolloids*, vol. 20, no. 4, pp. 492–501, 2006.

[7] S. Guilbert, B. Cuq, and N. Gontard, "Recent innovations in edible and/or biodegradable packaging materials," *Food Additives & Contaminants*, vol. 14, no. 6-7, pp. 741–751, 1997.

[8] E. Foegeding, T. C. Lanier, and H. O. Hultin, "Characteristics of edible muscle tissue," in *Food Chemistry*, O. R. Fennema, Ed., pp. 879–942, Marcel Dekker, New York, NY, USA, 1996.

[9] P. K. Castañeda, *Desarrollo y evaluación de propiedades de películas proteínicas de pesquerías de la subclase Elasmobranchii [Ph.D. thesis]*, Partial Fulfillment of the Requirements for the Diploma of Bachelor of Science (B.Sc.), Facultad de Química, UNAM, Mexico City, Mexico, 2011.

[10] C. Pires, C. Ramos, G. Teixeira et al., "Characterization of biodegradable films prepared with hake proteins and thyme oil," *Journal of Food Engineering*, vol. 105, no. 3, pp. 422–428, 2011.

[11] S. K. Park, C. O. Rhee, D. H. Bae, and N. S. Hettiarachchy, "Mechanical properties and water-vapor permeability of soy-protein films affected by calcium salts and glucono-δ-lactone,"

[12] E. M. Zactiti and T. G. Kieckbusch, "Potassium sorbate permeability in biodegradable alginate films: effect of the antimicrobial agent concentration and crosslinking degree," *Journal of Food Engineering*, vol. 77, no. 3, pp. 462–467, 2006.

[13] CONAPESCA, "Anuario estadístico de acuacultura y pesca," 2008, http://www.conapesca.sagarpa.gob.mx/wb/cona/anuario_2008.

[14] K. Limpisophon, M. Tanaka, W. Weng, S. Abe, and K. Osako, "Characterization of gelatin films prepared from under-utilized blue shark (*Prionace glauca*) skin," *Food Hydrocolloids*, vol. 23, no. 7, pp. 1993–2000, 2009.

[15] M. S. Hoque, S. Benjakul, and T. Prodpran, "Properties of film from cuttlefish (*Sepia pharaonis*) skin gelatin incorporated with cinnamon, clove and star anise extracts," *Food Hydrocolloids*, vol. 25, no. 5, pp. 1085–1097, 2011.

[16] AOAC, *Official Methods of Analysis*, Association of Official Analytical Chemists, Washington, DC, USA, 16th edition, 1995.

[17] I. Batista, "Recovery of proteins from fish waste products by alkaline extraction," *European Food Research and Technology*, vol. 210, no. 2, pp. 84–89, 1999.

[18] O. H. Lowry, N. J. Rosebrough, A. L. Farr, and R. J. Randall, "Protein measurement with the Folin phenol reagent," *The Journal of Biological Chemistry*, vol. 193, no. 1, pp. 265–275, 1951.

[19] U. K. Laemmli, "Cleavage of structural proteins during the assembly of the head of bacteriophage T4," *Nature*, vol. 227, no 5259, pp. 680–685, 1970.

[20] M. Saeed and M. Cheryan, "Sunflower protein concentrates and isolates low in polyphenols and phytate," *Journal of Food Science*, vol. 53, no. 4, pp. 1127–1143, 1988.

[21] P. J. A. Sobral, F. C. Menegalli, M. D. Hubinger, and M. A. Roques, "Mechanical, water vapor barrier and thermal properties of gelatin based edible films," *Food Hydrocolloids*, vol. 15, no. 4-6, pp. 423–432, 2001.

[22] F. M. Vanin, P. J. A. Sobral, F. C. Menegalli, R. A. Carvalho, and A. M. Q. B. Habitante, "Effects of plasticizers and their concentrations on thermal and functional properties of gelatin-based films," *Food Hydrocolloids*, vol. 19, no. 5, pp. 899–907, 2005.

[23] ASTM E 96-95, "Standard test method for water vapor transmission of materials," in *Proceedings of the 1995 Annual Book of ASTM Standards*, vol. 04.06, pp. 868–876, ASTM International (EUA), 2002.

[24] ASTM, "Standard test method for determining gas permeability characteristics of plastic film and sheeting," ASTM D1434-82, Annual Book of ASTM Standards, ASTM International (EUA), 2003.

[25] ASTM D 882, "Standard test method for tensile properties of thin plastic sheeting," in *1997 Annual Book of ASTM Standards*, vol. 14.02, pp. 161–170, ASTM International (EUA), 2002.

[26] M. H. Norziah, A. Al-Hassan, A. B. Khairulnizam, M. N. Mordi, and M. Norita, "Characterization of fish gelatin from surimi processing wastes: thermal analysis and effect of transglutaminase on gel properties," *Food Hydrocolloids*, vol. 23, no. 6, pp. 1610–1616, 2009.

[27] A. Jongjareonrak, S. Benjakul, W. Visessanguan, and M. Tanaka, "Isolation and characterization of collagen from bigeye snapper (*Priacanthus macracanthus*) skin," *Journal of the Science of Food and Agriculture*, vol. 85, no. 7, pp. 1203–1210, 2005.

[28] P. Kittiphattanabawon, S. Benjakul, W. Visessanguan, T. Nagai, and M. Tanaka, "Characterisation of acid-soluble collagen from

skin and bone of bigeye snapper (*Priacanthus tayenus*)," *Food Chemistry*, vol. 89, no. 3, pp. 363–372, 2005.

[29] T. Bourtoom, M. S. Chinnan, P. Jantawat, and R. Sanguan-deekul, "Effect of plasticizer type and concentration on the properties of edible film from water-soluble fish proteins in surimi wash-water," *Food Science and Technology International*, vol. 12, no. 2, pp. 119–126, 2006.

[30] T. M. Paschoalick, F. T. Garcia, P. J. A. Sobral, and A. M. Q. B. Habitante, "Characterization of some functional properties of edible films based on muscle proteins of Nile Tilapia," *Food Hydrocolloids*, vol. 17, no. 4, pp. 419–427, 2003.

[31] B. Cuq, N. Gontard, J.-L. Cuq, and S. Guilbert, "Selected functional properties of fish myofibrillar protein-based films as affected by hydrophilic plasticizers," *Journal of Agricultural and Food Chemistry*, vol. 45, no. 3, pp. 622–626, 1997.

[32] Y. Zhang and P. S. Cremer, "Interactions between macro-molecules and ions: the Hofmeister series," *Current Opinion in Chemical Biology*, vol. 10, no. 6, pp. 658–663, 2006.

[33] R. A. Curtis and L. Lue, "A molecular approach to biosepara-tions: protein-protein and protein-salt interactions," *Chemical Engineering Science*, vol. 61, no. 3, pp. 907–923, 2006.

[34] M. Trop and A. Kushelevsky, "The reaction of glucono delta lactone with proteins," *Journal of Dairy Science*, vol. 68, pp. 2534–2535, 1985.

[35] B.-S. Chiou, R. J. Avena-Bustillos, P. J. Bechtel, S. H. Imam, G. M. Glenn, and W. J. Orts, "Effects of drying temperature on barrier and mechanical properties of cold-water fish gelatin films," *Journal of Food Engineering*, vol. 95, no. 2, pp. 327–331, 2009.

[36] G. L. Robertson, *Food Packaging, Principles and Practice*, CRC Press, New York, NY, USA, 2nd edition, 2006.

[37] S. Damodaran and A. D. Paraf, "Food proteins. An overview," in *Food Proteins and Their Applications*, pp. 529–545, Marcel Dekker, New York, NY, USA, 1997.

[38] Permeability Coefficient of Common Polymers (Plastics), 2016, http://www.faybutler.com/pdf_files/HowHoseMaterialsAffect-Gas3.pdf.

[39] S.-I. Hong and J. M. Krochta, "Oxygen barrier performance of whey-protein-coated plastic films as affected by temperature, relative humidity, base film and protein type," *Journal of Food Engineering*, vol. 77, no. 3, pp. 739–745, 2006.

11

Kinetics and Quality of Microwave-Assisted Drying of Mango (*Mangifera indica*)

Ernest Ekow Abano

Department of Agricultural Engineering, School of Agriculture, College of Agricultural and Natural Sciences, University of Cape Coast, Cape Coast, Ghana

Correspondence should be addressed to Ernest Ekow Abano; ekowabano@yahoo.com

Academic Editor: Qingrong Huang

The effect of microwave-assisted convective air-drying on the drying kinetics and quality of mango was evaluated. Both microwave power and pretreatment time were significant factors but the effect of power was more profound. Increase in microwave power and pretreatment time had a positive effect on drying time. The nonenzymatic browning index of the fresh samples increased from 0.29 to 0.60 while the ascorbic acid content decreased with increase in microwave power and time from 3.84 mg/100g to 1.67 mg/100g. The effective moisture diffusivity varied from 1.45×10^{-9} to 2.13×10^{-9} m^2/s for microwave power range of 300-600 W for 2 to 4 minutes of pretreatment. The Arrhenius type power-dependent activation energy was found to be in the range of 8.58–17.48 W/mm. The fitting of commonly used drying models to the drying data showed the Midilli et al. model as the best. Microwave power of 300 W and pretreatment time of 4 minutes emerged as the optimum conditions prior to air-drying at 7°C. At this ideal condition, the energy savings as a result of microwave application was approximately 30%. Therefore, microwave-assisted drying should be considered for improved heat and mass transfer processes during drying to produce dried mangoes with better quality.

1. Introduction

Mango (*Mangifera indica* L.) is one of the most popular fruits among millions of people in many countries worldwide. In Ghana, mango is grown in savannah and transitional areas by smallholder farmers and continues to remain a seasonal crop. Statistics point to annual production of 95,460 tonnes in 2013 [1]. In the diet of humans, mango plays important role; it provides the diet with colour, phytochemicals, and nutrients. The average mango composition is water (83 g/100 g), carbohydrate (15.2 g/100 g), sugar (13.7 g/100 g), fibre (1.8 g/100 g), fats (0.38 g/100 g), proteins (0.510 g/100 g), vitamins (mainly vitamin A, 389 mg/100 g, and vitamin C, 36 mg/100 g), and minerals (mainly potassium 168 mg/100 g and phosphorous 14 mg/100 g) [2]. The fruit is an excellent source of antioxidants including ascorbic acid. It provides about 50% of the recommended daily intake of vitamin C [3] and contains high amounts of beta-carotene, which is responsible for the typical yellow colour of the mangoes. Beta-carotene is very beneficial for humans as it is a provitamin A and antioxidant [4]. The pulp is found to contain pigment carotenoids, polyphenols,

and omega-3 and omega-6 polyunsaturated fatty acids [5] Fresh ripe mango contains more than 80 g/100 g water within a soft-pulpy cell wall structure, which is responsible for the fast decay after harvest. Therefore, the right postharvest processing intervention is required to prolong the shelf-life of mangoes.

Drying is among the methods for the purpose to produce high quality dried products, which can be consumed directly or used as ingredient for the preparation of chutneys, cakes, muesli, and oat granola. Conventional air-drying has been widely used in industrial drying of food products but this method is energy-intensive and time-consuming and often produces poor quality products. Many authors have reported that this method leads to degradation of products flavour, colour, nutrients, and case hardening, due to their long drying times and high temperatures employed in practice [6, 7]. Hence, combination of advanced drying methods with the conventional hot air is often recommended to reduce long drying times and poor product quality associated with conventional hot air-drying. Combination of osmotic dehydration preceded by microwave-assisted hot air-drying of mango has been studied [8]. The effect of gluten coating on osmotic dehydration of mango cubes has similarly been investigated [9]. Application of microwave prior to conventional drying of mango (*Mangifera indica* L.) is uncommon in scientific literature.

Microwave offers advantages that have been employed prior to or with conventional drying in food processing technologies. Several researchers have provided strong evidence that microwave-assisted drying is ideal for fruits and vegetables [10–14], which speed up drying process, increase mass transfer, and produce good quality products. Therefore, in this present work, the effect of microwave power and time as pretreatment to convective air-drying of mango slices was investigated.

2. Materials and Method

2.1. Sample Preparation. Fresh Kent mango fruits were obtained from the Abura Market, Cape Coast, Ghana. Selection of mango samples was based on visual assessment of uniform colour and geometry. The mangoes were washed under running tap water, peeled, and sliced into sizes of 50 mm × 25 mm × 10 mm using a stainless steel knife and immediately kept at −24°C to slow down physiological and chemical changes. Prior to the test, the samples were allowed to warm up to room temperature conditions. The initial moisture content of the mangoes was obtained from drying 5 g samples in an oven at 105°C for 24 hours according to the AOAC (1990) method. The initial total soluble solids were determined to be 16.7% with a refractometer (ABBE 98 490, Holland).

2.2. Experimental Design. A two-factor, 3-level factorial design was used for the experiment. The effect of two independent variables, microwave power X_1 (300–600 W) and pretreatment time X_2 (2–4 min), on three response variables, drying time, ascorbic acid, and nonenzymatic browning, was evaluated (Table 2).

2.3. Microwave Pretreatment. A domestic Samsung microwave machine with varied powers was used to predry 100 g of mango slices at power intensities of 300 W, 450 W, and 600 W for 2, 3, and 4 minutes prior to convective hot air-drying. After the microwave pretreatment, the slices were removed, weighed, and immediately subjected to convective air-drying. 100 g of mango slices without microwave pretreatment was used to serve as the control.

2.4. Drying Equipment and Drying Procedure. Microwave pretreated samples were transferred to a hot air cabinet dryer (GENLAB Oven, Model SDO225, 240 AC 1PH, 540 × 920 × 440 mm, 2 kW) set at temperature of 70°C and air circulation of 0.5 m/s. This temperature was chosen based on previous optimization studies [14]. The dryer was run idle for 1 hr earlier to the drying experiment. During drying, the masses of the samples were monitored every 30 min at the initial stages and later changed to 1 hr at the later stages of drying until constant mass was reached by a digital balance with

an accuracy of ±0.001 g. For measuring the weight of the sample during experimentation, the tray with sample was taken out of the drying chamber, weighed on the digital top pan balance, and placed back into the chamber within 10 seconds.

2.5. Determination of Ascorbic Acid (Vitamin C). Ascorbic acid content of the samples before and after drying was assayed colorimetrically following Roe and Kuether [19]. Two grams of dried mango slices was ground finely using a mortar and a pestle and placed in a 25 mL volumetric flask with 4% oxalic acid solution. The mixture was centrifuged and 10 mL of the supernatant was transferred into a conical flask after which bromine water was added in drops with constant mixing until the extract turns orange yellow. The solution was made up to 25 mL with 4% oxalic acid solution. Similarly, 10 mL of the stock ascorbic acid solution was converted into dehydroform by bromination. Again, 10 mL of standard dehydroascorbic acid solution was pipetted into a series of tubes. Aliquot (2 mL) of brominated sample extract was similarly pipetted out differently. The volume in each tube was made up to 3 mL by adding distilled water. One millilitre of DNPH (2,4-dinitrophenylhydrazine) reagent was added, followed by 1-2 drops of thiourea to each tube. A blank was set with water instead of ascorbic acid solution. The content of the tubes was placed on a shaker to mix and incubated at 37°C for 3 hours in a water bath. After incubation, the orange red osazone crystals formed were dissolved by adding 7 mL of 80% sulphuric acid. A graph of ascorbic acid concentration versus absorbance at 540 nm was plotted ($R^2 = 0.9973$) and used to calculate the ascorbic acid content in the sample.

2.6. Nonenzymatic Browning Determination. A method previously reported by [14] was used to evaluate the nonenzymatic browning of the dried mango slices. The extent of browning was measured as absorbance at 440 nm. Brown pigment formed was extracted from the test portions of the dried mango slices. Two-gram sample was ground into fine powder, after which 50 mL of ethanol (60%, v/v) was added and allowed to stand for 12 hours. The mixture was stirred slowly and filtered through 0.45 μm nylon filter membrane. Browning index of filtrates was estimated by a spectrophotometer against 60% ethanol as blank. All samples were extracted in triplicate.

2.7. Drying Kinetics. The drying kinetics of mango slices were expressed in terms of empirical models, where the experimental data obtained were plotted in the form of a dimensionless moisture ratio (MR) against drying time in minutes. The MR of the mango slices was determined using

$$\mathrm{MR} = \frac{M - M_e}{M_o - M_e}, \qquad (1)$$

where MR is the moisture ratio, M_o is the initial moisture content (g water/g dry matter), M is the moisture content at any time (g water/g dry matter), and M_e is the equilibrium moisture content (g water/g dry matter) [20].

TABLE 1: Mathematical models that were applied to the experimental data.

Model name	Model expression	Reference
Page	$MR = \exp(-kt^n)$	[15]
Henderson and Pabis	$MR = a\exp(-kt)$	[16]
Logarithmic	$MR = a\exp(-kt) + c$	[17]
Midilli et al.	$MR = a\exp(-kt^n) + bt$	[18]

TABLE 2: Three-level factorial design for two factors and results of DT, BI, and AA.

MP (W)	MT (min)	DT (min)	BI (Abs)	AA (mg/g)
450	3	690	0.45265	1.98316
300	4	690	0.4375	2.04337
300	3	810	0.42425	2.05638
600	3	570	0.5293	1.793878
300	2	870	0.3875	2.13061
450	2	750	0.44065	2.028827
600	2	630	0.52375	1.904337
600	4	570	0.6033	1.660969
450	4	630	0.48565	1.91352
450	3	730	0.4518	1.933673
450	3	730	0.4304	2.032653
Control		990	0.28645	2.203570

MP is microwave power; MT is microwave pretreatment time.

Three empirical drying models widely used in scientific literature, Page, Henderson and Pabis, and Logarithmic, were fitted to the experimental data set (MR, t) shown in Table 1 to describe the drying kinetics of mango slices. A nonlinear regression procedure of SPSS 20.0 [21] was used to determine the drying rate constant, k, and coefficients (a, c, n) in the empirical models. The modelling was characterized by the reduced chi-square (χ^2), root mean square error (RMSE), and the determination coefficient (R^2) [22] displayed in (2), (3), and (4), respectively. Consider

$$\chi^2 = \frac{\sum_{i=1}^{N}\left(MR_{exp,i} - MR_{pred,i}\right)^2}{N - z}, \quad (2)$$

$$RMSE = \sqrt{\frac{1}{N}\sum_{i=1}^{N}\left(MR_{exp,i} - MR_{pred,i}\right)^2}, \quad (3)$$

$$R^2 = 1 - \left[\frac{\sum_{i=1}^{N}\left(MR_{pred,i} - MR_{exp,i}\right)^2}{\sum_{i=1}^{N}\left(MR_{pred,i} - MR_{pred,i}\right)^2}\right], \quad (4)$$

where $MR_{exp,i}$ and $MR_{pred,i}$ are the experimental and predicted moisture ratio, respectively, N is the number of observations, and z is the number of constants in the drying model.

2.8. Determination of Moisture Diffusivity. Fick's second law of diffusion, which characterizes moisture migration during thin layer drying of food materials, was used to calculate the

effective moisture diffusivity, considering a constant moisture diffusivity, infinite slab geometry, and uniform initial moisture distribution [23]:

$$MR = \frac{8}{\pi^2}\sum_{n=0}^{\infty}\frac{1}{(2n+1)}\exp\left(-\frac{(2n+1)\pi^2}{4L^2}D_{eff}t\right), \quad (5)$$

where D_{eff} is the effective moisture diffusivity (m²/s) and L is half the thickness of slice of the sample (m). Equation (5) can be simplified to the following for long drying times:

$$MR = \frac{8}{\pi^2}\exp\left(-\frac{\pi^2 D_{eff}t}{4L^2}\right). \quad (6)$$

D_{eff} of the mango slices was obtained from the slope (K) of the graph of ln MR against the drying time. ln MR versus drying time (t) results in a straight line with a negative slope and K is related to D_{eff} by

$$K = \frac{\pi^2 D_{eff}t}{4L^2}. \quad (7)$$

2.9. Calculation of Activation Energy. According to Pillai et al. [24], for the standard microwave oven drying procedure, the internal temperature of sample is not an assessable variable. Therefore, the use of Arrhenius-type equation is considered for illustrating the relationship between the diffusivity coefficient and the ratio of the microwave power output to sample thickness instead of temperature for the calculation of the activation energy. The activation energy is found as modified from the revised Arrhenius. The equation as suggested by [25] is represented as

$$D_{eff} = D_o\exp\left[\frac{E_a q}{P}\right], \quad (8)$$

where D_o is the constant in the Arrhenius equation (m²/s), E_a is the activation energy (W/mm), P is the microwave power (W), and q is the sample thickness (mm). Equation (8) can be rearranged as

$$\ln(D_{eff}) = \ln(D_o) - \frac{E_a q}{P}. \quad (9)$$

The activation energy for moisture diffusion was obtained from the graph of $\ln(D_{eff})$ against q/P.

2.10. Energy Consumption during Drying. The energy consumption in kWh of the microwave and the hot air dryer was calculated using

$$E_c = \frac{P \times t}{1000}, \quad (10)$$

where E_c is the energy consumption in kWh, P is the power rating of either the microwave equipment or the convective air dryer in W, and t is the drying time in hours.

2.11. Optimization of the Drying Process. The optimization of the drying process was performed using a multivariate response method [26] with

$$\text{DI} = \left[\prod_{i=1}^{3} di\,(Y_i) \right]^{1/3}. \tag{11}$$

di represents the desirability for the various responses: drying time, ascorbic acid content, and nonenzymatic browning (Y_i). The DI ranges between 0 and 1. Zero is the least preferred value while 1 is the most desired. Maximizing DI is the goal of optimization analysis. The optimization process incorporates goals and priorities for the factors and the responses. For this present study, the goal for the factors was at any level within the range of the design values, but, in the case of the responses, minimum values of drying time (DT), nonenzymatic browning index (BI), and maximum values of ascorbic acid (AA) were desired.

2.12. Statistical Analysis. A quadratic model was fitted to the average values of the responses to get the regression equations with design expert software [27]. The statistical significance of the model term was evaluated at 95% probability. The 3D plots for the factors were generated for the various responses. The accuracy of the model to describe the response variables was diagnosed by the determination coefficients (R^2) values and the nonsignificance of the lack of fit test.

3. Results and Discussion

3.1. Effect of Microwave Power and Pretreatment Time on Drying Kinetics of Mango Slices. Microwave power and pretreatment time increment had a positive effect on drying time for the 70°C dried samples as shown in Figures 1, 2, and 3. The initial average moisture content of the mangoes (*Mangifera indica*) was found to be 4.65 g moisture/g dry matter, which decreased to 0.03 g moisture/g dry matter (d.b.) after drying. The drying process generally followed a falling rate regime and the increase in microwave power and pretreatment time significantly ($P \leq 0.05$) accelerated the drying process and increased energy efficiency. The estimated effect for each factor and the interaction between the variables were estimated (Table 3). Variation in the estimated coefficients shows that there were different contributions of the factors to drying time. Microwave power contributed 1.67 times higher than pretreatment time. As microwave power and pretreatment time increased, more moisture was removed and in the end resulted in the reduction in drying time (Figure 4). Drying time reduced from 870 min to 570 min as the microwave power and pretreatment time increased from 300 to 600 W and 2 to 5 min, respectively. This means that there was significant savings in time as microwave power pretreatment time increased.

The results corroborate what was reported by Rayaguru and Routray [28], Bai-Ngew et al. [29], and Karaaslan and Tunçer [30] for microwave-drying of *Pandanus amaryllifolius* leaves, durian chips, and spinach, respectively. Formation of porous structure in the tissues of mango as a result of

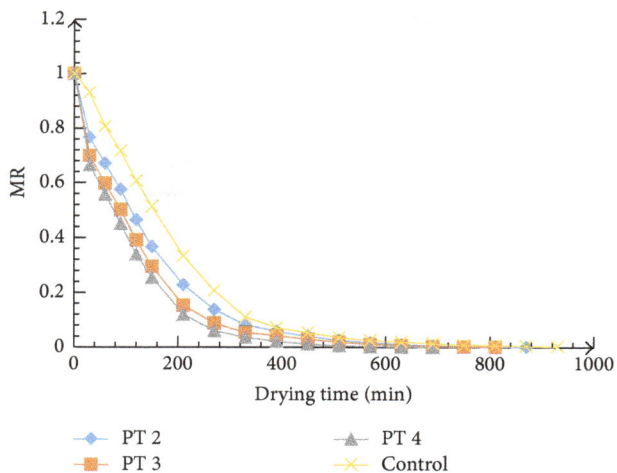

FIGURE 1: Variation of MR versus drying time at microwave power 300 W.

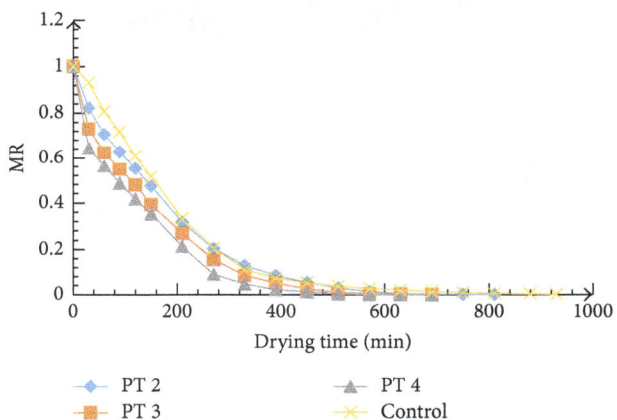

FIGURE 2: Variation of MR versus drying time at microwave power 450 W.

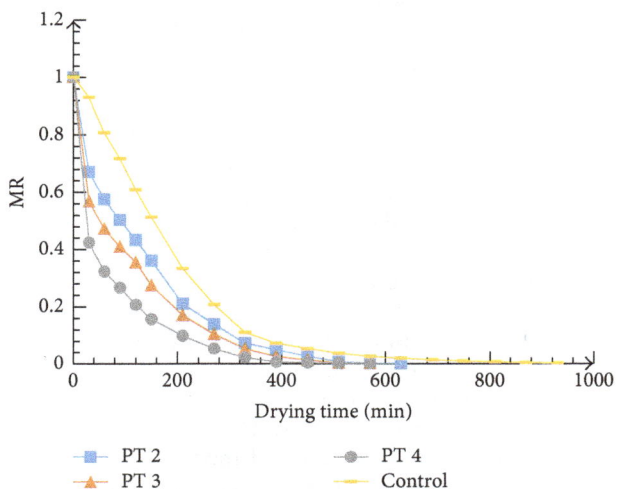

FIGURE 3: Variation of MR versus drying time at microwave power 600 W.

TABLE 3: Analysis of variance (ANOVA) for the effects of microwave power and time on drying time.

Source	Coefficient estimates	Sum of squares	Degree of freedom	Mean square	F value	P value, Pro > F
Intercept	711.05	—	1	—	—	—
Model	—	86302.39	5	17260.48	31.78	0.0009[*]
X_1	−100.00	60000.00	1	60000.00	110.47	0.0001[*]
X_2	−60.00	21600.00	1	21600.00	39.77	0.0015[*]
$X_1 X_2$	30.00	3600.00	1	3600.00	6.63	0.0498[*]
X_1^2	−12.63	404.21	1	404.21	0.74	0.4278[**]
X_2^2	−12.63	404.21	1	404.21	0.74	0.4278[**]
Lack of fit	—	1649.12	3	549.71	1.03	0.5268[**]
R^2	0.9695	—	—	—	—	—

[*]Significant (<0.0500); [**]not significant. Lack of fit is not significant at P value >0.0500.

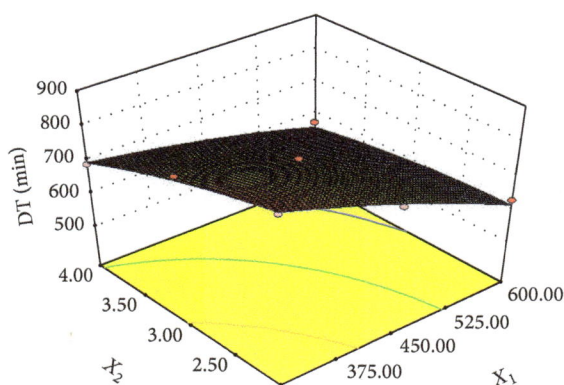

FIGURE 4: Effect of microwave power (X_1) and microwave time (X_2) on drying time.

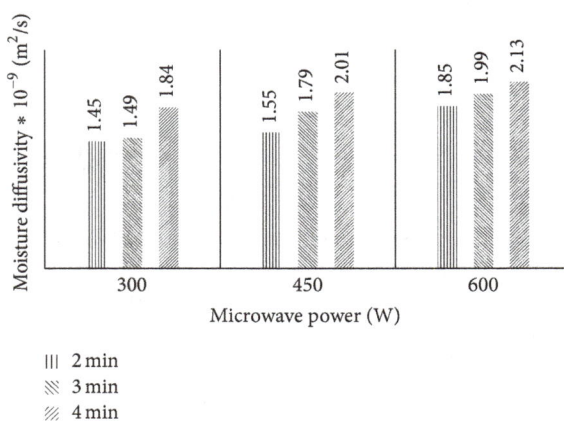

||| 2 min
\\\ 3 min
/// 4 min

FIGURE 5: Effect of microwave power and time on moisture diffusivity of mango slices.

electromagnetic waves application has been noted to be the plausible reason for the accelerated drying with microwave. In comparison with the control, microwave pretreatment enhanced heat and mass transfer within the mango tissues resulting in increased drying rates and energy utilization. The decrease in drying time with an increase in the microwave power density has been reported for other food materials, including tomato pomace [31], onions [32], apple pomace [33], and potatoes slices [34].

3.2. Effect of Microwave Power and Pretreatment Time on Moisture Diffusivity.

The variation of ln(MR) against drying time plot for the various microwave power and pretreatment time used to calculate the various effective moisture diffusivity, D_{eff}, had determination coefficient greater than 0.98. The effect on microwave power and pretreatment time on D_{eff} is evident (Figure 5). Effective moisture diffusivity coefficient increased with microwave power and pretreatment time. At a microwave power of 300 W, the D_{eff} values increased from 1.45×10^{-9} m^{2}s^{-1} to 1.84×10^{-9} m^{2}s^{-1} for samples pretreated for 2 to 4 min and dried at 70°C. A similar increase was observed for the D_{eff} values of the various microwave pretreatment conditions (Figure 5). The D_{eff} values obtained for microwave-assisted drying of mango slices lie within the general range of 10^{-12}–10^{-8} m^{2}s^{-1} for drying of food

materials [17]. Increased heat energy as a result of increase in microwave power is reported to enhance the activity of the water molecules leading to higher moisture diffusivity [35]. Microstructure observation with scanning electron microscope (SEM) for microwave-assisted drying revealed creation of microchannel on the surface of the test samples. This shows that the higher the microwave power is and the longer the samples were treated, the higher the formation of porous structure is within the tissue of the mango slices to enhance heat and mass transfer, leading to higher moisture diffusivity rates.

The values of the moisture diffusivity were used to estimate the activation energy for moisture diffusion, E_a. Figure 6 shows the variation of ln(D_{eff}) against q/P for various pretreated mango slices dried at 70°C. The activation energy, which is the energy needed to initiate internal moisture diffusion, is an indication of the temperature sensitivity of D_{eff}. The activation energy obtained for drying process was 13.52 W/mm for 2 min, 17.48 W/mm for 3 min, and 8.58 W/mm for 4 min pretreated samples. The activation energy values obtained in this study were lower than the 46.91 W/mm reported for microwave-vacuum-drying of tomato slices [36] and 21.6 W/g for microwave-drying of mango ginger [37] but generally higher than the 5.54 W/mm

TABLE 4: Parameters and statistical results for the various drying models for microwave pretreated samples at 300 W.

Model name	PT	Model constants	R^2	RMSE	χ^2
Page	2	$k = 0.006, n = 1.031$	0.9971	0.0167	0.0003
	3	$k = 0.011, n = 0.947$	0.9946	0.0217	0.0005
	4	$k = 0.014, n = 0.928$	0.9940	0.0231	0.0006
Henderson and Pabis	2	$k = 0.007, a = 0.994$	0.997	0.0183	0.0004
	3	$k = 0.008, a = 0.973$	0.995	0.0217	0.0005
	4	$k = 0.009, a = 0.970$	0.994	0.0231	0.0006
Logarithmic	2	$k = 0.007, a = 1.000$, and $c = -0.010$	0.9971	0.0167	0.0003
	3	$k = 0.008, a = 0.974$, and $c = -0.002$	0.9946	0.0217	0.0005
	4	$k = 0.009, a = 0.973$, and $c = -0.005$	0.9940	0.0231	0.0006
Midilli et al.	2	$k = 0.005, n = 1.056, a = 0.979$, and $b = -4.2 \times 10^{-6}$	0.9971	0.0167	0.0003
	3	$k = 0.010, n = 0.963, a = 0.981$, and $b = -5.76 \times 10^{-6}$	0.9953	0.0203	0.0005
	4	$k = 0.013, n = 0.926, a = 0.984$, and $b = -17.96 \times 10^{-6}$	0.9947	0.0216	0.0006

TABLE 5: Parameters and statistical results for the various drying models for microwave pretreated samples at 450 W.

Model name	PT	Model constants	R^2	RMSE	χ^2
Page	2	$k = 0.003, n = 1.095$	0.996	0.0209	0.0005
	3	$k = 0.008, n = 0.960$	0.990	0.0306	0.0011
	4	$k = 0.016, n = 0.868$	0.981	0.0405	0.0019
Henderson and Pabis	2	$k = 0.006, a = 1.009$	0.994	0.0250	0.0007
	3	$k = 0.007, a = 0.967$	0.991	0.0294	0.0010
	4	$k = 0.008, a = 0.938$	0.981	0.0414	0.0020
Logarithmic	2	$k = 0.005, a = 1.036$, and $c = -0.042$	0.997	0.0177	0.0004
	3	$k = 0.006, a = 0.985$, and $c = -0.029$	0.993	0.0258	0.0008
	4	$k = 0.007, a = 0.951$, and $c = -0.021$	0.982	0.0396	0.0018
Midilli et al.	2	$k = 0.003, n = 1.106, a = 0.975$, and $b = -2.237 \times 10^{-5}$	0.9972	0.01715	0.0003
	3	$k = 0.010, n = 0.917, a = 0.977$, and $b = -6.329 \times 10^{-5}$	0.9938	0.02449	0.0008
	4	$k = 0.019, n = 0.820, a = 0.977$, and $b = -6.734 \times 10^{-5}$	0.9865	0.03464	0.0016

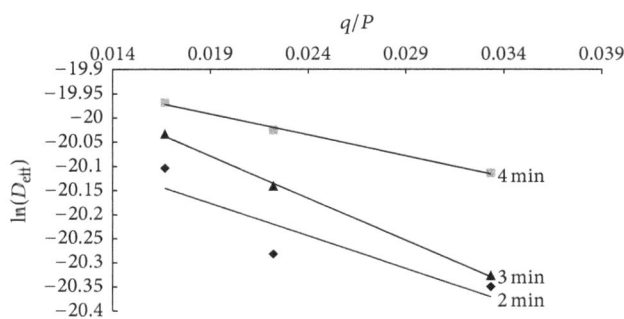

FIGURE 6: Variation of $\ln(D_{\text{eff}})$ against q/P for the various microwave pretreatment time.

for okra [25]. The values were however within activation energy values of 13.6 W/mm and 14.945 W/mm reported for *Pandanus* leaves [28] and potatoes [34], respectively.

3.3. Modelling of the Drying Curves. The dimensionless moisture ratio against drying time for the experimental data at various pretreatment times and air temperatures was fitted to the Page, Henderson and Pabis, Logarithmic, and Midilli et al. models available in the literature. The results of such a fitting of the experimental data for microwave-assisted drying at 70°C are displayed in Tables 4, 5, and 6, which show the values of the estimated constants with their corresponding statistical R^2, RMSE, and χ^2 values characterizing each fitting. From the results obtained, it is evident that the experimental data fitted adequately to the models used in this study. The correlation coefficients obtained are in the range of 0.995–0.9972. This means that the three models could satisfactorily describe the microwave-assisted convective air-drying of mango slices. The relatively high values of correlation coefficients, low root mean square errors, and low reduced chi-square indicate a good predicting capacity for the temperature tested over the entire duration of the drying process. Among the models examined, the Midilli et al. model was observed to be the most appropriate one for all the experimental data with the highest value for the coefficient of determination (R^2) and lowest reduced chi-square (χ^2) and RMSE. The estimated parameters and statistical analysis of the models examined for the microwave powers at different times are illustrated in Tables 4, 5, and 6. It was observed that the value of drying rate constant (k) for all the models tested increased with

TABLE 6: Parameters and statistical results for the various drying models for microwave pretreated samples at 600 W.

Model name	PT	Model constants	R^2	RMSE	χ^2
Page	2	$k = 0.015, n = 0.865$	0.988	0.0327	0.0013
	3	$k = 0.039, n = 0.717$	0.985	0.0340	0.0014
	4	$k = 0.109, n = 0.576$	0.993	0.0215	0.0005
Henderson and Pabis	2	$k = 0.007, a = 0.940$	0.987	0.0348	0.0014
	3	$k = 0.009, a = 0.904$	0.969	0.0504	0.0030
	4	$k = 0.016, a = 0.919$	0.949	0.0601	0.0042
Logarithmic	2	$k = 0.007, a = 0.949,$ and $c = -0.015$	0.987	0.0338	0.0013
	3	$k = 0.009, a = 0.898,$ and $c = -0.011$	0.969	0.0496	0.0008
	4	$k = 0.018, a = 0.904,$ and $c = -0.035$	0.956	0.0562	0.0018
Midilli et al.	2	$k = 0.021, n = 0.780, a = 0.984,$ and $b = 7.6 \times 10^{-6}$	0.99202	0.02672	0.0010
	3	$k = 0.066, n = 0.585, a = 0.994,$ and $b = -8.19 \times 10^{-6}$	0.9933	0.023204	0.00078
	4	$k = 0.15, n = 0.493, a = 0.999,$ and $b = -9.11 \times 10^{-5}$	0.9968	0.01519	0.00003

TABLE 7: Analysis of variance (ANOVA) for the effects of microwave power and microwave time on AA.

Source	Coefficient estimates	Sum of squares	Degree of freedom	Mean square	F value	P value, Pro > F
Intercept	1.98	—	1	—	—	—
Model	—	0.17	5	0.034	25.97	0.0014[*]
X_1	−0.15	0.13	1	0.13	95.65	0.0002[*]
X_2	−0.074	0.033	1	0.033	25.06	0.0041[*]
$X_1 X_2$	−0.039	0.006094	1	0.006094	4.61	0.0846[**]
X_1^2	−0.047	0.005506	1	0.005506	4.16	0.0968[**]
X_2^2	−0.0005774	0.0000008	1	0.0000008	0.000638	0.9808[**]
Lack of fit	—	0.001714	3	0.00005712	0.23	0.6224[**]
R^2	0.9287	—	—	—	—	—

[*]Significant (<0.0500); [**]not significant. Lack of fit is not significant at P value >0.0500.

microwave pretreatment time. This implies that drying rate potential increased with increase in microwave pretreatment time. Murthy and Manohar [37] found the Midilli et al. model to best explain the thin layer drying behaviour of stone apple slices pretreated with microwave prior to hot air-drying at different temperatures (40–70°C) in a forced convection dryer. Our findings are consistent with microwave-assisted drying of yam cubes [14].

3.4. Effect of Microwave Power and Pretreatment Time on Ascorbic Acid. The effect of microwave power and pretreatment time on ascorbic acid (AA) of mangoes dried at 70°C is shown in Figure 7. Both power and time were significant model terms on the ascorbic acid (AA) content (Table 7). Increment in microwave power and longer pretreatment time resulted in more reduction of the AA content. The AA content of the fresh mango decreased from 3.84 mg/100 g dry matter to 1.794 mg/100 g dry matter after drying the various pretreated samples, representing 53% loss in vitamin C. In comparison with the control (2.204 mg/100 g), the AA content of the respective microwave pretreated sample was lower (2.056, 1.983, and 1.794) for 300 W, 450 W, and 600 W powers, respectively. The power of the microwave had about twice negative effect on AA degradation compared to the treatment time (Table 7). The reduction of ascorbic acid content observed for the microwave pretreated samples may

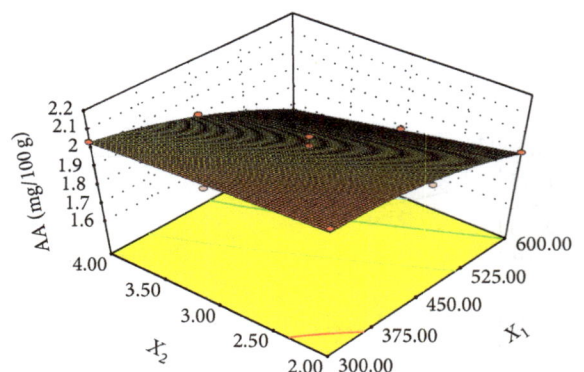

FIGURE 7: Effect of microwave power (X_1) and pretreatment time (X_2) on AA.

be due to the destruction of vitamin C by the electromagnetic waves. The thermal effect of the microwave power coupled with the irreversible oxidative reactions due to longer drying times during hot air-drying may have contributed to the damage of about half of the AA content.

Losses of ascorbic acid during microwave-drying have been reported [38]. The degradation of AA in this present study is in agreement with results obtained by Zheng and Lu for microwave pretreatment of different parts of green

TABLE 8: Analysis of variance (ANOVA) for the effects of microwave power and microwave time on BI.

Source	Coefficient estimates	Sum of squares	Degree of freedom	Mean square	F value	P value, Pro > F
Intercept	0.45	—	1			
Model	—	0.035	3	0.012	42.76	0.0001*
X_1	0.068	0.028	1	0.028	99.82	0.0001*
X_2	0.029	0.005078	1	0.005078	18.35	0.0036*
$X_1 X_2$	0.007387	—	1	0.0002336	—	—
X_1^2	0.032	0.002799	1	0.002799	10.12	0.0155*
X_2^2	0.015	—	1	0.0002767	—	—
Lack of fit	—	0.001619	5	0.0003238	2.04	0.3612**
R^2	0.9483					

*Significant (<0.0500); **not significant. Lack of fit is not significant at P value >0.0500.

asparagus [39]. Ascorbic acid losses between 10% and 50% are reported for drying food stuff [40]. In a related study involving microwave-drying of okra fruit, AA reduction between 43% and 63% was stated [41].

3.5. Effect of Microwave Power and Pretreatment Time on Nonenzymatic Browning. Nonenzymatic browning is another quality indicator in drying. Whereas browning is desirable in some processing food, excessive browning is undesirable in dried mangoes. The effect of microwave power and pretreatment time on nonenzymatic browning is clear (Figure 8). Brown pigment formation significantly increased with both microwave power and pretreatment time from 0.39 to 0.60 at the various condition studied. As expected, microwave power had a negative profound effect compared to pretreatment time on BI (Table 8). The control however had less browning (0.29) than all the microwave-assisted dried mangoes. This agrees with a study by [42] for dried tomato quarters at 50°C to 90°C, where the author observed that brown pigment formation increased with temperature from 0.58 to 0.68. Browning index equal to 0.60 was considered critical based on visual assessment by consumers. This increasing trend of nonenzymatic browning as a result of microwave power and treatment time is attributed to the occurrence and reaction between nitrogenous constituents and reducing sugars, nitrogenous constituents and organic acids, and sugars and organic acids [42, 43].

3.6. Optimization of the Drying Parameters. The optimal microwave-assisted drying condition for mango slices was established using overall desirability index explained earlier. The maximum predicted DT, BI, and AA were 695.79 min, 0.45 Abs units, and 2.04 mg/100 g, respectively. These predicted values are closer to their corresponding experimental values of 870 min, 0.60 Abs units, and 2.13 mg/100 g. The overall desirability of 0.701 shown in Figure 9 was obtained for the microwave effect, the kinetics, and the quality of dried mango slices. In the range of the factors used, 95% confidence prediction gave optimal microwave power of 300 W and pretreatment time of 4 min. At this best condition, the DT was 695.80 min, the BI was 0.45 Abs units, and the AA was 2.04 mg/100 g dry matter.

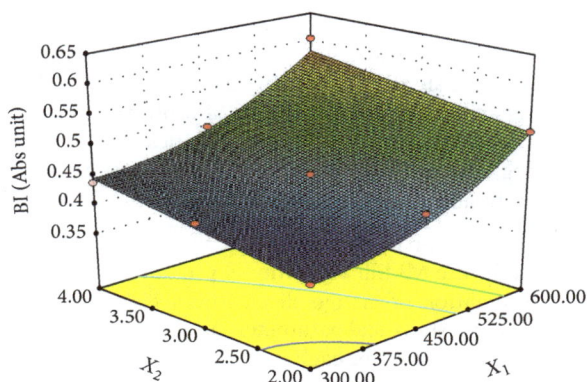

FIGURE 8: Effect of microwave power (X_1) and pretreatment time (X_2) on BI.

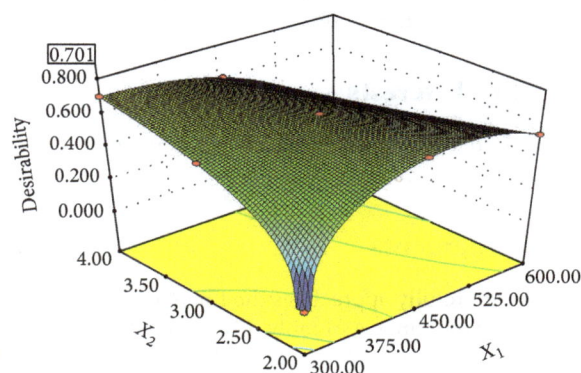

FIGURE 9: Effect of drying temperature and frying time on 3D plot of the desirability index for the optimal frying time.

3.7. Energy Consumption during Microwave-Assisted Drying. The control samples took about 990 min to dry from 4.64 g water/g dry matter to 0.04 g water/g dry matter and consumed 33 kWh of energy in the process. This value was compared with the energy consumption from the optimized microwave pretreatment conditions of 300 W/4 min to make the energy savings known as a result of microwave application. A reduction of 29.70% energy requirement was achieved because of microwave application at 300 W for

4 min. Reduction in energy consumption due to microwave application may be due to the energy efficiency advantage of microwave over the conventional method, which causes moisture migration from within rather than from the surface of the samples. Similar trends were also observed by [44] and [34] for microwave-drying of parsley and potato slices, respectively.

4. Conclusion

The following conclusions can be drawn from the present work where microwave-assisted air-drying of mango slices has been studied:

(i) Moisture migration during drying of mango slices occurs in the falling rate regime.

(ii) Both microwave power and pretreatment time played a significant role in characterizing the drying behaviour of mango slices. Microwave power had a higher effect on drying kinetics, ascorbic acid degradation, and formation of brown pigment than pretreatment time during drying.

(iii) Of the four empirical drying models tested in this study, the Midilli et al. model provided the best representation of mango slices. The decision was based on three statistical parameters adopted to evaluate the goodness of fit for each model.

(iv) Microwave power of 300 W and pretreatment time of 4 min were found to be the ideal conditioning prior to convective air-drying at 70°C. This condition reduced the energy requirement for air-drying of mango slices by approximately 30%.

Conflict of Interests

The author declares that there is no conflict of interests regarding the publication of this paper.

Acknowledgments

The author gratefully appreciates the financial support of the Cape Coast Municipal Assembly, Mr. Amos Kofi Ziempuo, and Lawrence Arthur at the Department of Agricultural Engineering for their technical support towards the study.

References

[1] FAOSTAT, "Food and Agriculture Organization FAOSTAT database collections, agricultural data, and food and agriculture organization of the United Nations" 2014, http://faostat.fao.org/.

[2] G. Pamplona and M. D. Roger, *Encyclopedia of Foods and Their Healing Power*, vol. 2, S. L Safeliz, Madrid, Spain, 2007.

[3] T. Djioua, F. Charles, F. Lopez-Lauri et al., "Improving the storage of minimally processed mangoes (Mangifera indica L.) by hot water treatments," *Postharvest Biology and Technology*, vol. 52, no. 2, pp. 221–226, 2009.

[4] I. Pott, M. Marx, S. Neidhart, W. Mühlbauer, and R. Carle, "Quantitative determination of β-carotene stereoisomers in fresh, dried, and solar-dried mangoes (Mangifera indica L.)," *Journal of Agricultural and Food Chemistry*, vol. 51, no. 16, pp. 4527–4531, 2003.

[5] USDA, National Nutrient Database for Standard Reference, SR-23, Fruits Reports, 2009, Raw Mango, p. 449, 2010.

[6] C. Contreras, M. E. Martín-Esparza, A. Chiralt, and N. Martínez-Navarrete, "Influence of microwave application on convective drying: effects on drying kinetics, and optical and mechanical properties of apple and strawberry," *Journal of Food Engineering*, vol. 88, no. 1, pp. 55–64, 2008.

[7] R. Vadivambal and D. S. Jayas, "Changes in quality of microwave-treated agricultural products—a review," *Biosystems Engineering*, vol. 98, no. 1, pp. 1–16, 2007.

[8] A. Andrés, P. Fito, A. Heredia, and E. M. Rosa, "Combined drying technologies for development of high-quality shelf-stable mango products," *Drying Technology*, vol. 25, no. 11, pp. 1857–1866, 2007.

[9] M. Azam, M. A. Haq, and A. Hasnain, "Osmotic dehydration of mango cubes: effect of novel gluten-based coating," *Drying Technology*, vol. 31, no. 1, pp. 120–127, 2013.

[10] F. Prothon, L. M. Ahrné, T. Funebo, S. Kidman, M. Langton, and I. Sjöholm, "Effects of combined osmotic and microwave dehydration of apple on texture, microstructure and rehydration characteristics," *LWT—Food Science and Technology*, vol. 34, no. 2, pp. 95–101, 2001.

[11] R. Schiffmann, "Microwave processes for the food industry," in *Handbook of Microwave Technology for Food Application*, A. Datta and R. Anantheswaran, Eds., pp. 299–352, Marcel Dekker, New York, NY, USA, 2001.

[12] M. Zhang, J. Tang, A. S. Mujumdar, and S. Wang, "Trends in microwave-related drying of fruits and vegetables," *Trends in Food Science & Technology*, vol. 17, no. 10, pp. 524–534, 2006.

[13] A. Figiel, "Drying kinetics and quality of vacuum-microwave dehydrated garlic cloves and slices," *Journal of Food Engineering*, vol. 94, no. 1, pp. 98–104, 2009.

[14] E. E. Abano and R. S. Amoah, "Microwave and blanch-assisted drying of white yam (Dioscorea rotundata)," *Food Science & Nutrition*, vol. 3, no. 6, pp. 586–596, 2015.

[15] G. E. Page, *Factors influencing the maximum rate of air drying shelled corn in thin layers [M.S. thesis]*, Purdue University, West Lafayette, Ind, USA, 1949.

[16] A. Yagcioglu, A. Degirmencioglu, and F. Cagatay, "Drying characteristics of the laurel leaves under different drying conditions," in *Proceedings of the 7th International Congress on Agricultural Mechanization and Energy*, A. Bascetincelik, Ed., pp. 565–569, Adana, Turkey, 1999.

[17] I. Doymaz, "Drying of thyme (Thymus vulgaris L.) and selection of a suitable thin-layer drying model," *Journal of Food Processing and Preservation*, vol. 35, no. 4, pp. 458–465, 2010.

[18] A. Midilli, H. Kucuk, and Z. Yapar, "A new model for single-layer drying," *Drying Technology*, vol. 20, no. 7, pp. 1503–1513, 2002.

[19] J. H. Roe and C. A. Kuether, "The determination of ascorbic acid in food materials through the 2, 4-dinitrophenylhydrazine derivative of ascorbic acid," *Journal of Biology Geomathematics*, vol. 147, pp. 399–407, 1953.

[20] B. Özbek and G. Dadali, "Thin-layer drying characteristics and modelling of mint leaves undergoing microwave treatment," *Journal of Food Engineering*, vol. 83, no. 4, pp. 541–549, 2007.

[21] SPSS, *SPSS 20.0 for Windows Inc*, SPSS, Chicago, Ill, USA, 2013.

[22] A. Vega-Galvez, E. Uribe, R. Lemus, and M. Miranda, "Hot air drying characteristics of aloe vera (*Aloe barbadensis* Miller) and influence of temperature on kinetic parameters," *Lebensmittel-Wissenschaft & Technologie*, vol. 40, pp. 1698–1707, 2007.

[23] J. Crank, *The Mathematics of Diffusion*, Clarendon Press, Oxford, UK, 2nd edition, 1975.

[24] M. G. Pillai, R. Iyyaswami, R. Lima, and T. Miranda, "Moisture diffusivity and energy consumption during microwave druing of plaster of Paris," *Chemical Product and Process. Modeling*, vol. 5, no. 1, pp. 1934–2659, 2010.

[25] G. Dadali, D. K. Apar, and B. Özbek, "Estimation of effective moisture diffusivity of okra for microwave drying," *Journal of Drying Technology*, vol. 25, no. 9, pp. 1445–1450, 2007.

[26] R. Meyers and D. Montgomery, *Response Surface Methodology: Process and Product Optimization Using Designed Experiments*, John Wiley & Sons, New York, NY, USA, 2002.

[27] Stat-Ease, *Design, Expert. Design Expert 8.0.7.1*, Stat-Ease, Minneapolis, Minn, USA, 2008.

[28] K. Rayaguru and W. Routray, "Effect of drying conditions on drying kinetics and quality of aromatic *Pandanus amaryllifolius* leaves," *Journal of Food Science and Technology*, vol. 47, no. 6, pp. 668–673, 2010.

[29] S. Bai-Ngew, N. Therdthai, and P. Dhamvithee, "Characterization of microwave vacuum-dried durian chips," *Journal of Food Engineering*, vol. 104, no. 1, pp. 114–122, 2011.

[30] S. N. Karaaslan and I. K. Tunçer, "Development of a drying model for combined microwave-fan-assisted convection drying of spinach," *Biosystems Engineering*, vol. 100, no. 1, pp. 44–52, 2008.

[31] M. Al-Harahsheh, A. H. Al-Muhtaseb, and T. R. A. Magee, "Microwave drying kinetics of tomato pomace: effect of osmotic dehydration," *Chemical Engineering and Processing: Process Intensification*, vol. 48, no. 1, pp. 524–531, 2009.

[32] D. Arslan and M. M. Özcan, "Study the effect of sun, oven and microwave drying on quality of onion slices," *LWT—Food Science and Technology*, vol. 43, no. 7, pp. 1121–1127, 2010.

[33] A. Motevali, S. Minaei, M. H. Khoshtaghaza, M. Kazemi, and A. Mohamad Nikbakht, "Drying of pomegranate arils: comparison of predictions from mathematical models and neural networks," *International Journal of Food Engineering*, vol. 6, no. 3, 2010.

[34] H. Darvishi, "Energy consumption and mathematical modeling of microwave drying of potato slices," *Agricultural Engineering International: CIGR Journal*, vol. 14, no. 1, 2012.

[35] M. A. Ghazavi, A. Maleki, and M. Moradi-Jalal, "Mathematical modeling, moisture diffusion and energy efficiency of thin layer drying of potato," *International Journal of Agricultural and Crop Science*, vol. 5, no. 15, pp. 1663–1669, 2013.

[36] E. E. Abano, H. Ma, and W. Qu, "Influence of combined microwave-vacuum drying on drying kinetics and quality of dried tomato slices," *Journal of Food Quality*, vol. 35, no. 3, pp. 159–168, 2012.

[37] T. P. K. Murthy and B. Manohar, "Microwave drying of mango ginger (*Curcuma amada* Roxb): prediction of drying kinetics by mathematical modelling and artificial neural network," *International Journal of Food Science and Technology*, vol. 47, no. 6, pp. 1229–1236, 2012.

[38] P. H. M. Marfil, E. M. Santos, and V. R. N. Telis, "Ascorbic acid degradation kinetics in tomatoes at different drying conditions," *LWT—Food Science and Technology*, vol. 41, no. 9, pp. 1642–1647, 2008.

[39] H. Zheng and H. Lu, "Effect of microwave pretreatment on the kinetics of ascorbic acid degradation and peroxidase inactivation in different parts of green asparagus (*Asparagus officinalis* L.) during water blanching," *Food Chemistry*, vol. 128, no. 4, pp. 1087–1093, 2011.

[40] S. Sokhansanj and D. Jayas, *Drying of Food Stuffs (Handbook of Industrial Drying)*, CRC Press, Boca Raton, Fla, USA, 2nd edition, 1995.

[41] L. Mana, T. Orikasab, Y. Muramatsuc, and A. Tagawaa, "Impact of microwave drying on the quality attributes of Okra fruit," *Journal of Food Processing and Technology*, vol. 3, article 186, 2012.

[42] S. Cernîşev, "Effects of conventional and multistage drying processing on non-enzymatic browning in tomato," *Journal of Food Engineering*, vol. 96, no. 1, pp. 114–118, 2010.

[43] B. Zanoni, C. Peri, R. Nani, and V. Lavelli, "Oxidative heat damage of tomato halves as affected by drying," *Food Research International*, vol. 31, no. 5, pp. 395–401, 1998.

[44] Y. Soysal, "Microwave drying characteristics of parsley," *Biosystems Engineering*, vol. 89, no. 2, pp. 167–173, 2004.

12

A Comparative Evaluation of Carcass Quality, Nutritional Value, and Consumer Preference of *Oreochromis niloticus* from Two Impoundments with Different Pollution Levels in Zimbabwe

Vimbai R. Hamandishe [iD],[1] Petronella T. Saidi [iD],[1] Venancio E. Imbayarwo-Chikosi,[1] and Tamuka Nhiwatiwa[2]

[1]*Department of Animal Science, Faculty of Agriculture, University of Zimbabwe, P. O. Box MP167, Mt Pleasant Harare, Zimbabwe*
[2]*Department of Biological Sciences, Faculty of Science, University of Zimbabwe, P. O. Box MP167, Mt Pleasant Harare, Zimbabwe*

Correspondence should be addressed to Vimbai R. Hamandishe; vhamandishe@gmail.com

Academic Editor: Thierry Thomas-Danguin

The objective of the study was to determine the quality and consumer preferences of Nile tilapia (*Oreochromis niloticus*) from two water bodies with different pollution levels and trophic states. Water quality assessment of the two impoundments was carried out. Fish were sampled from hypereutrophic Lake Chivero and oligomesotrophic Lake Kariba for proximate analysis, carcass quality, and sensory evaluation. Conductivity, dissolved oxygen, transparency, ammonia, total phosphates, reactive phosphates, and chlorophyll a were significantly different (P<0.05). Fish from Lake Kariba had significantly higher condition factors and lower fillet yields, while fish of length 10-20 cm, from Lake Chivero had significantly more fat. Lake Chivero fish were darker, greener, and less red while Lake Kariba fish were lighter, less green, and less red. Raw fish from Lake Kariba were significantly firmer, were less green and redder, had a stronger typical fish odour, and were more acceptable than Lake Chivero fish. Lake Chivero fish had a stronger foreign fish odour than their counterparts. No statistical differences were observed on fillet cooking losses, cooked fish sensory parameters, and acceptability. The fish could, however, not be safe due to possibility of toxins in water and feed (algae) which may bioaccumulate and ultimately affect other attributes of fish quality.

1. Introduction

Fish meat quality is defined based on the sensory characteristics, chemical composition, and physical properties [1]. These quality attributes influence how the fish are perceived by the consumer [2]. Fish quality also involves safety aspects such as being free from harmful bacteria, parasites, biotoxins, pesticidal chemicals, heavy metals, and many other substances. According to FAO [3], wild fish in Zimbabwe are generally perceived to be of poor quality compared to aquaculture fish due to reasons such as off-flavours, spoilage, poor presentation, and linkage with polluted water bodies. Flos et al. [4] reported that the quality of fish is affected by exogenous and endogenous factors. Exogenous factors include diet composition, feeding frequency, and the fish environmental parameters such as salinity, pH, and temperature [2]. Endogenous factors are genetic and linked to the life stage, age, size, sex, and anatomical position in the fish [5]. In previous studies, differences in quality among Nile tilapia (*Oreochromis niloticus*) populations have been attributed to environmental factors [6, 7]. These factors influence body composition, sensory quality, and preferences in several fish species [7, 8].

Consumption of fish is associated with a number of benefits. As an important part of the diet, fish provides an affordable source of protein essential for human nutrition. More so considering that about 33 percent of Zimbabweans are undernourished, with 11 percent of children under the age of five moderately or severely underweight [3]. Because of their high protein content, polyunsaturated fats, vitamins and minerals, fish can play a major role in alleviating malnutrition particularly in young children, pregnant women and the elderly [9]. Fish consumption prevents cardiovascular diseases risk factors such as blood pressure, some types of cancer, Alzheimer's disease, and brain damage and does not increase obesity [10–12]. Despite the benefits of fish consumption, the *per capita* fish consumption in Zimbabwe, recorded at 3 kg/year in 2016, is low and below the SADC average (6.0 kg/year) [3, 13]. The Dietary Guidelines for Americans recommend consumption of 12 kg per year of seafood [14].

In Zimbabwe, Nile tilapia (*Oreochromis niloticus*) is the most common fish species and constitutes the dominant bream/tilapia caught by artisanal and commercial fishermen and is also produced intensively under aquaculture. It is now widely distributed in most reservoirs and river systems of Zimbabwe [3, 15]. Wild fish stocks occur in both Lake Kariba and Lake Chivero producing significant quantities of fish into the market. Lake Kariba is an oligomesotrophic water body, while Lake Chivero is hypereutrophic due to high levels of pollution with major sources of pollutants being sewage and industrial waste. These water bodies have different physical, chemical, and biological characteristics which influence the plankton, diversity of prey in the water bodies, and the substances to which the fish are exposed, thus offering different environments and nutrient sources.

Poor fish quality can be very detrimental to consumption, marketing, and acceptance of fish, yet fish is one of the best sources of animal protein [16]. Information should be available on the effect of different water quality status on fish attributes but is however scarce. This study hypothesized that the chemical composition, carcass quality, and sensory properties of fish from Lake Kariba and Lake Chivero are different due to the differences in water quality caused by extensive water pollution in Lake Chivero. In this study, this hypothesis of the link between fish quality and source water quality (exogenous factors) is tested. With the growing demand for fish and fish products, it is essential to determine the influence of water quality on fish since environment is considered the main factor controlling wild fish quality, its growth, and production [17]. This study was conducted to investigate carcass quality, nutritional properties, and acceptability of raw and cooked fish, *O. niloticus,* reared in impoundments with different trophic states.

2. Materials and Methods

2.1. Study Area. The study was carried out at the University of Zimbabwe, Department of Animal Science, and University Lake Kariba Research Station. Fish samples (*Oreochromis niloticus*) from Lake Chivero and Lake Kariba were used in this study. Lake Chivero is situated on the Manyame River and is the fourth largest impoundment in Zimbabwe. It is located 37 km southwest of the capital city of Zimbabwe, Harare, at latitude 17° 54′ S and longitude 30° 48′ E [18]. All major rivers such as Manyame, Mukuvisi, and Marimba and tributaries that discharge into Lake Chivero pass through the city of Harare and industrial sites, causing heavy water pollution [19]. The heavy pollution has resulted in the lake being classified as hypereutrophic. The lake stratifies in

summer and overturns in the beginning of the winter season of each year [20]. The lake is loaded with nutrients especially in winter and there is high growth of phytoplankton due to nutrient availability [21].

Lake Kariba, located at 16.5°S and 28.8°E with an altitude of 518 m above sea level, is the largest inland dam in Southern Africa. It is a warm oligomesotrophic lake with three distinguishable seasons which are a hot rainy season (November-March), cool dry season (May-August), and a very hot dry season (September-November) [22]. The lake is divided into five geographical and limnologically distinct basins, namely, Mlibizi (basin 1), Binga (basin 2), Sengwa (basin 3), Bumi (basin 4), and Sanyati (basin 5) [23]. The levels of pollution differ in these basins with Sanyati basin receiving runoff water from farms, sewage, and mining drainage effluent from Kwekwe through Sebakwe River [24, 25].

2.2. General Water Quality Parameters. Water samples were collected from Lake Chivero and Lake Kariba from five sites for each lake at the same time as fish sampling. Sampling was carried out in June and October. Water samples were collected using a Ruttner sampler mounted to a speed boat. Vertically integrated samples were collected into 500 ml sterile bottles and immediately placed on ice in cooler boxes. Water quality parameters, namely, pH, conductivity, dissolved oxygen, and percent oxygen, were measured using handheld pH/oxygen/conductivity meters equipped with a cellOx 325 Oxygen Sensor (WTW), a SenTix 20 probe (WTW), and a conductivity sensor (WTW). Analysis was done on filtered samples to measure total nitrogen, total phosphorus, ammonia nitrogen, nitrate nitrogen (NO3-N), and orthophosphate phosphorus (PO4-P) according to Bartram and Ballance [26].

2.3. Fish Samples. Fish samples (*Oreochromis niloticus*) were obtained from the two lakes using seine nets. Sampling was done in June and October to determine the effect of season on fish nutritional composition, sensory characteristics, and carcass quality. Fish were also bought from local fishermen to get the required numbers and sizes. The fish were immediately transferred to the coolers and kept on ice until they were collected for further analyses. The fish were filleted and half of the fillets, approximately 50 grams, were used for individual measurement of cooking loss, protein, water, and fat content, and colour determination. The skin and all visible parts of fat were removed prior to analysis. All the fishes were individually analysed. The other fillets were used for sensory evaluation. These fillets were wrapped in plastic bags and frozen at −20°C.

2.4. Biometric Traits. Biometric parameters were measured before degutting, descaling, skinning, beheading, and harvesting of fillets using standard methods. These were total length (TL) in centimeters (cm), total weight (TW), dressed weight (DW), liver weight (LW), and fillet weight (FW) in grams (g). Dissection was done using a sharp knife and scissors.

The following parameters were calculated:

Condition factor (CF)

$$= \left[\text{fish} \frac{TW}{(TL \times TL \times TL)} \right] \times 100$$

Hepatosomatic index (HSI%) $= 100 \times \left[\dfrac{LW\,(g)}{TW\,(g)} \right]$ (1)

Dressing index (DW%) $= 100 \times \left| \dfrac{DW\,(g)}{TW)\,g)} \right|$

Fillet weight (FW%) $= 100 \times \left[\dfrac{FW\,(g)}{TW\,(g)} \right]$

2.5. Nutritional Quality of Fish. A total of forty fish fillet samples were subjected to moisture, ash, fat, and protein analysis using standard methods which are detailed in Association of the Official Analytical Chemists [27].

2.6. Percentage Cooking Loss. A total of forty frozen fish fillets were thawed at 3°C for 12 hours. Using a Faizco SF-820 laboratory digital scale, 100 g was taken from each fillet for determination of cooking loss. Each fillet was placed in tight Polyvinyl chloride (PVC) bags and cooked in a water bath heated at 100°C for 20 minutes and then cooled to room temperature. Water released after cooking and cooling was manually separated by pouring out the water from the PVC bag and the weight of the fillet was taken.

The cooking loss was calculated using the following formula:

Cooking loss %

$$= 100 \times \left[\frac{\text{weight of cooked fish muscle}}{\text{total muscle weight}} \right] \quad (2)$$

2.7. Colour Determination. The fish fillets were analysed for colour by first thawing at 4°C for 4 hours. A high resolution digital Cannon Powershot SX400IS, with 16 megapixels, was used to take pictures of the fish fillets. The camera was secured on a tripod and lens faced downward towards the fillet sample. Images of the fillets were taken using day light conditions and uniform light intensity. The images were saved on a memory card and transferred to the computer where Adobe Photoshop extended CS6 was used as described by Yam and Papadakis [28]. The histogram window was used to determine the colour distributions along the x-axis and y-axis on the fish fillets. Five points, measuring 1.2×1.4 inches, were cropped on each fillet using Photoshop in order to obtain the L, a, and b values. The window displayed the statistics (mean, standard deviation, median, percentage, etc.) of the colour value, lightness (L) with lightness=0 for black and lightness $= 100$ for white, redness (a), and yellowness (b), for a selected area on the fillet image. The figures obtained were converted to L∗, a∗, and b∗ since the histogram values are not standard colour values.

The following formulae were used:

$$L^* = \frac{\text{Lightness}}{225} \times 100$$

$$a^* = \frac{240a}{255} - 120 \quad (3)$$

$$b* = \frac{240b}{255} - 120$$

2.8. Evaluation of Raw Fish. Thirty whole raw fish from Lake Kariba and from Lake Chivero with mean weight ranging from 350 g to 450 g were evaluated by 15 trained panelists from the Faculty of Science and Faculty of Agriculture. The fish skin, belly flaps, peritoneal area, fillet, and general characteristics were evaluated in terms of colour (greenness, blackness, and redness), texture (firmness), visible fat deposition, odour (fresh fish odour intensity, foreign odour intensity), and acceptability. The parameters were ranked using a five-point hedonic scale with descriptors for each fish part and parameter. Prior to the assessment, frozen samples from each source were thawed overnight at 4°C. Whole fresh fish samples from each source were placed on trays and simultaneously presented to the panelists.

2.9. Consumer Preference. A consumer preference test was performed on fish samples from the two lakes, Kariba and Chivero, using the method of Stone and Sidel [29] and Resurrecion [30]. A panel of 154 untrained individuals drawn from college students who ate fish was used. Frozen fish fillets were defrosted for two hours at room temperature and the upper portion was cut into four individual portions weighing 50 grams each. Sixty portions from each source were then prepared by steam cooking in plastic cooking bags for twenty minutes and boiling temperature. No flavouring or spices were added to the fish fillets. The panelists indicated their preferences based on four sensory attributes, namely, taste, colour, flavour, and texture on a hedonic scale of one to five based on like/dislike. Sensory judgments were scored as follows: 1: dislike extremely, 2: dislike moderately, 3: neither like nor dislike, 4: like moderately, and 5: like extremely. The panelists rinsed their mouths with water before and after each sample.

2.10. Data Analyses. The Statistical Analysis System (SAS) version 9.3 (SAS, 2010) was used to analyse data on carcass quality, nutritional composition, and sensory evaluation. The Shapiro-Wilk test was used to determine if all sample data had been drawn from normally distributed populations using the PROC UNIVARIATE PLOT NORMAL procedure of SAS. All data on water quality, carcass, and nutritional composition parameters were analysed using the PROC GLM of SAS using the following model:

$$y_{ijk} = \mu + L_i + S_j + (LS)_{ij} + e_{ijk} \quad (4)$$

where y_{ijk} were the observed water quality, nutritional composition, and carcass parameters and L_i, S_j, and $(LS)_{ij}$ were the fixed effects of the source, season, and interaction between source and season, respectively. The μ and e_{ijk} were the overall mean and random residuals, respectively. Means were separated using the adjusted Tukey's method. Differences were considered statistically significant at $P<0.05$. The Kruskal-Wallis test was used to determine the effect of source of fish and month of sampling on colour, texture, visible fat deposition, typical fish odour, and foreign odour with the PROC NPAR1WAY analysis of data. Sensory evaluation data were obtained as scores collected on a 5-point hedonic scale. The data were therefore not drawn from a normally distributed population as was confirmed by the Shapiro-Wilk test. Mean scores from the sensory evaluation of raw and cooked fish were presented using spider web plots. Hedonic scores for each of the organoleptic variables on fish from Lake Kariba and Lake Chivero were compared with the t-test Approximation in a Wilcoxon two-sample test derived from a PROC NPAR1WAY analysis of SAS. A principal component analysis was carried out with PROC PRINCOIMP of SAS to identify the major components influencing consumer preference for fish from Lake Chivero and Lake Kariba. The most important principal components were determined from the magnitude of the eigenvalues for Lake Chivero and Lake Kariba as well as cumulative proportion.

3. Results

3.1. Water Quality of Lake Kariba and Lake Chivero. Water quality results of the two lakes are shown for the months of June and October (Table 1). Conductivity, oxygen percent, temperature, ammonium, reactive phosphorus, and chlorophyll a were statistically different between the two lakes and were influenced by month, lake, and their interaction ($P<0.05$). Total phosphates, nitrates, and transparency were influenced by source only while pH was influenced by month ($P<0.05$). The two water bodies showed significant differences in water quality.

3.2. Carcass Quality. Carcass quality results are shown in Table 2. Fish from Lake Kariba had significantly higher condition factors ($P<0.05$) in both the months of June and October than fish from Lake Chivero. Fish from both lakes had lower condition factors in the month of June than in October. Fillet yield was significantly higher for fish from Lake Chivero than fish from Lake Kariba in the month of October ($P<0.05$). There were no significant differences in fish hepatosomatic indices.

3.3. Nutritional Composition. Data in Table 3 shows nutritional composition of fillets from fish of medium size, 10-20cm. Source and month of fish sampling had no significant effect ($P<0.05$) on protein, ash, and DM while the interaction between source and size of the fish influenced fat content. Fat content was significantly higher ($P<0.05$) in fish sampled in the month of October from Lake Chivero for the medium sized fish. There were no significant differences observed in fish of length >20 cm between the two water bodies.

3.4. Percentage Cooking Loss. There were no differences in cooking loss between fish fillets from the two lakes across the experimental months ($P>0.05$). Least square means for cooking losses for fillets from Lake Kariba and Lake Chivero were 20.74% and 20.76%, respectively.

3.5. Fillet Colour. The measured colour of *Oreochromis niloticus* samples evaluated in this study reflected variability in terms of lightness, a∗ and b∗ ($P<0.05$) (Table 4). Fish fillets from Lake Chivero were significantly darker (lower lightness) ($P<0.05$) than those from Lake Kariba for both months, just as fillets were lighter in October than in June in both lakes ($P<0.05$). Fish fillets from Lake Chivero were greener and less red than those from Lake Kariba and the latter were redder and less green as shown by the chromatic component a∗ values (ranges from -120 to 120, from green to red). There were no differences in chromatic component b∗ (ranging from -120 to 120, blue to yellow) between fish fillets from the two lakes. However, fish fillets from Lake Chivero had more yellowness in October than in June ($P<0.05$).

3.6. Evaluation of Raw Fish. Figure 1 shows the mean scores for the raw fish parameters. The skins of fish from Lake Kariba were significantly more red than skins from Lake Chivero ($P<0.05$) while Lake Chivero fish fillets and belly flaps were significantly more green ($P<0.05$) than Lake Kariba fillets. The texture of fish from Lake Kariba was perceived to be significantly firmer ($P<0.05$) than that of fish from Lake Chivero as these were scored higher than Lake Chivero. In terms of the typical fish odour, Lake Kariba fish had a significantly stronger ($P<0.05$) typical fish odour than fish from Lake Chivero. Foreign odour was significantly stronger in fish from Lake Chivero than the counterparts. Acceptability of fish from Lake Kariba was significantly higher ($P<0.05$) than that of fish from Lake Chivero.

3.7. Consumer Preference and Effect of Source. The major components that influenced consumer preference are shown in Table 5. The eigenvalues show that PC1 was the most important component for fish from both Lake Chivero and Lake Kariba. Although PC2 had loadings below one, it was also relatively important since when combined with PC1, they contributed the majority of the variability that was observed among the six variables (smell, taste, acceptability, tenderness, flavour, and colour). As such, only PC1 and PC2 were retained. These contributed a cumulative total of 62.2% and 61.6% of observed variability among the six organoleptic variables in Lake Kariba and Lake Chivero, respectively. PC1 comprised smell, taste, and acceptability for both lakes (Figure 2). All other parameters had little individual significance, if any, as far as influencing consumer preference was concerned. Results from t-test approximation of the two-tailed Wilcox two-sample test indicated that there were no significant differences ($P>0.05$) in organoleptic scores for smell, taste, flavour, colour, tenderness, and acceptability between cooked fish sourced from Lake Kariba and Lake Chivero (Figure 3).

TABLE 1: Least square means for water physicochemical parameters for Lake Chivero and Lake Kariba in June and October.

Parameter	Units	Lake Chivero		Lake Kariba		s.e.
		June	October	June	October	
pH		7.67	8.28	8.02	8.44	0.230
Conductivity	μS/cm	513.00^a	683.60^a	93.36^b	71.26^b	42.540
Oxygen percent	%	30.68^a	112.00^b	99.12^b	90.54^b	5.992
Dissolved oxygen	mg/l	2.57^a	7.98^b	7.93^b	6.72^b	0.381
Temperature	°C	16.40^a	25.38^a	24.28^b	27.58^b	0.467
Transparency	m	0.86^a	0.64^a	2.40^b	1.94^b	0.230
Ammonia	mg/l	998.36^a	216.74^b	50.40^b	29.70^b	156.761
Total phosphorus	mg/l	804.56^a	833.44^a	32.20^b	24.24^b	77.60
Nitrates	mg/l	248.24	111.94	37.72	32.22	53.636
Total nitrogen	mg/l	1246.64	1548.62	85.32	2023.84	816.739
Reactive phosphorus	mg/l	739.22^a	543.58^b	13.84^c	11.66^c	40.167
Chlorophyll A	μg/l	3.70^a	12.92^b	1.14^{ac}	2.22^{ac}	1.420

Least squares means with similar superscripts within rows are nonsignificantly different (P>0.05).

TABLE 2: Carcass quality of *Oreochromis niloticus* from Lake Kariba and Lake Chivero.

Variable %	Lake Chivero		Lake Kariba		s.e.
	June	October	June	October	
Condition factor	1.86^a	2.02^{ab}	2.12^{bc}	2.24^{cd}	0.047
Hepatosomatic index	0.96	1.23	1.03	0.76	0.112
Dressing index	89.40^a	88.30^{ac}	86.72^{ad}	84.83^{cd}	0.971
Fillet yield	38.26^{ab}	38.92^a	35.58^b	32.00^c	0.858

Least square means with similar superscripts within rows are not significantly different (p>0.05).

TABLE 3: Least squares means for nutritional composition (%) of fish of size 10-20 cm of total length from Lake Chivero and Lake Kariba.

Source	Season/month	EE	CP	Ash	DM
Chivero	June	1.86^b	18.20	1.15	20.06^a
	October	3.14^a	19.69	1.62	19.39^{ab}
Kariba	June	1.60^b	17.93	0.90	19.55^{ab}
	October	0.83^{bc}	18.32	0.81	18.45^b
	s.e.	0.274	0.435	0.278	0.272

LS means with similar superscripts across columns are nonsignificantly different (p>0.05) for the principal components.

TABLE 4: Least square means for lightness and chromatic components (s.e.).

Colour parameter	Lake Chivero		Lake Kariba	
	June	October	June	October
Lightness	49.00(1.077)^d	54.68(1.077)^b	60.83(1.077)^a	51.45(1.115)^c
a*	-0.08(0.445)^c	4.17(0.443)^a	2.42(0.443)^b	2.25(0.458)^b
b*	10.09(0.406)^d	11.79(0.406)^{bc}	12.70(0.406)^{ac}	13.54(0.420)^c

TABLE 5: Eigenvalues for the principal components of fish from Lake Chivero and Lake Kariba.

Principal component	Lake Kariba (1)			Lake Chivero (2)		
	Eigenvalue	Proportion	Cumulative proportion	Eigenvalue	Proportion	Cumulative proportion
1	**2.810**	**0.468**	**0.468**	**2.802**	**0.467**	**0.467**
2	**0.926**	**0.154**	**0.622**	**0.895**	**0.149**	**0.616**
3	0.788	0.131	0.753	0.764	0.127	0.743
4	0.592	0.099	0.852	0.727	0.121	0.864
5	0.448	0.075	0.927	0.503	0.084	0.948
6	0.436	0.073	1.000	0.309	0.052	1.000

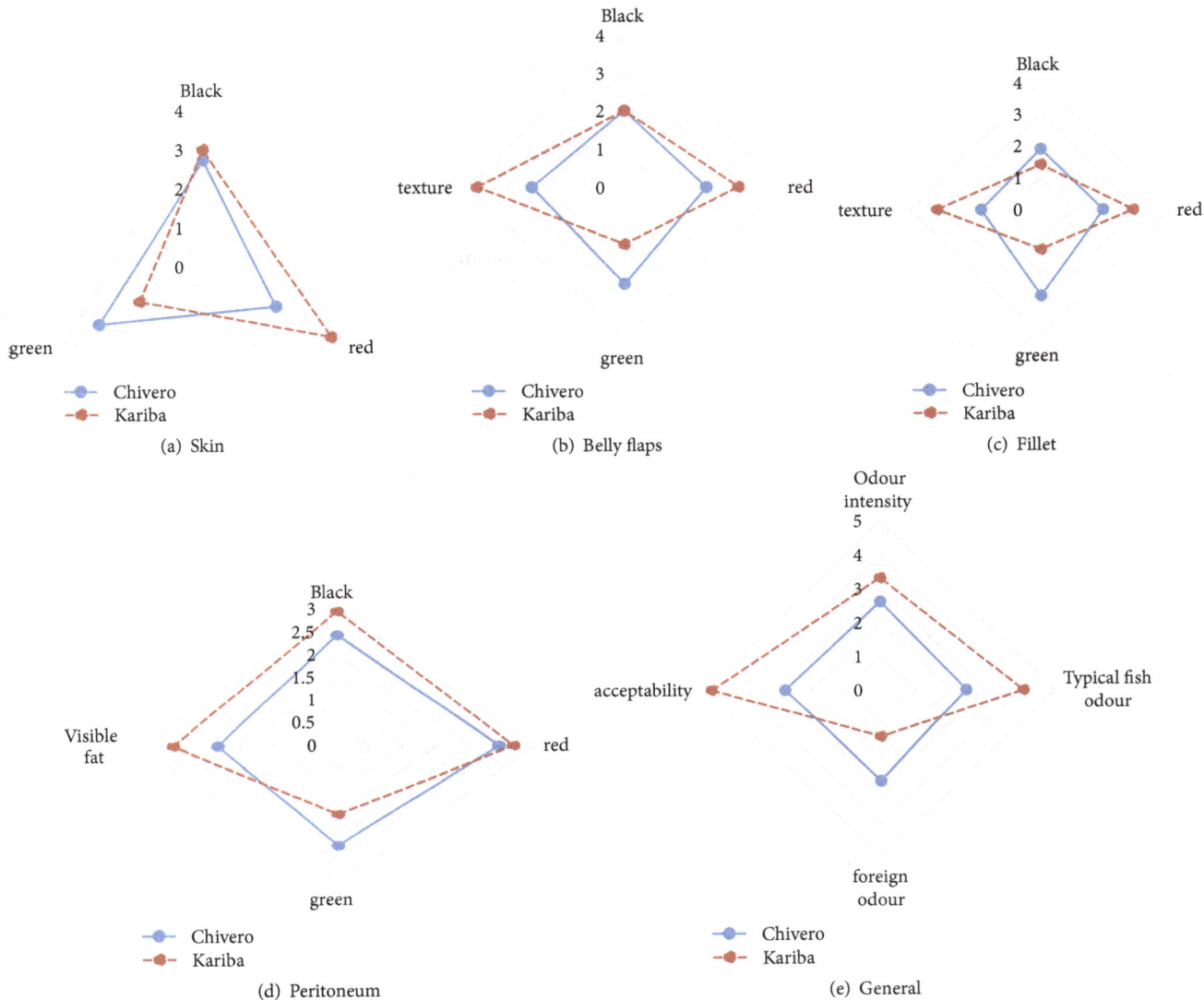

FIGURE 1: Comparison of raw Nile tilapia from Lake Kariba and Lake Chivero. (a) Colour of fish skin. (b) Colour and texture of belly flaps. (c) Appearance of peritoneum. (d) Colour and texture of fillets. (e) General odour and acceptability of the fish.

4. Discussion

Water quality results obtained in this study reflected the differences that exist between Lake Kariba and Lake Chivero. Differences in conductivity, dissolved oxygen, oxygen percent, temperature, ammonium, reactive phosphorus, and chlorophyll a were due to the extensive pollution in Lake Chivero. The pollution is caused by sewer and industrial effluents from neighboring city of Harare and industrial sites depositing raw effluent into the lake [31, 32]. High levels of degradable material are more oxygen demanding and this depletes dissolved oxygen in the water body [33]. The high level of nutrients, especially phosphorus, supports growth of different phytoplankton and macrophyte species including water hyacinth which deplete the water body of dissolved oxygen, thus the low oxygen levels in Lake Chivero compared to Lake Kariba [34, 35]. Dissolved oxygen is one of the major factors that affect fish growth [22].

Fish from Lake Kariba had a higher condition factor than from Lake Chivero probably due to the differences in water quality and type of food available in the two water bodies. Higher condition factors reflect a better nutritional status and better adaptation of the fish to its immediate environment [36]. However, the condition factor of fish from Lake Chivero was expected to be higher than that of fish from Lake Kariba to relate well with the high nutrient state which supports growth of phytoplankton, a source of food for the fish [21]. The observed poor condition of the fish might have been attributed to the stress resulting from pollution. Highly polluted water resources such as Lake Chivero are characterized by conditions that negatively affect fish development and proper functioning of internal organs such as

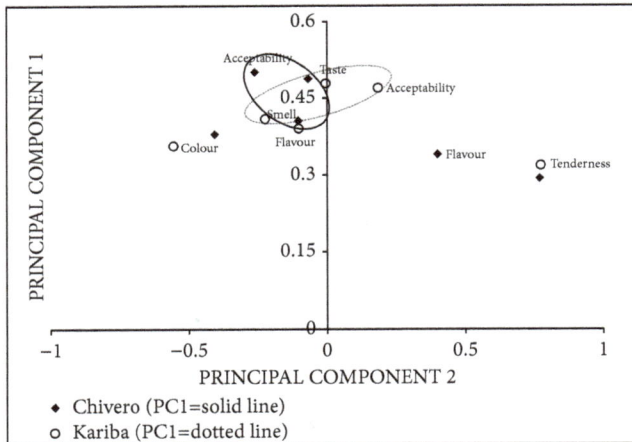

FIGURE 2: Component patterns for PC1 and PC2 for fish from Lake Kariba and Lake Chivero.

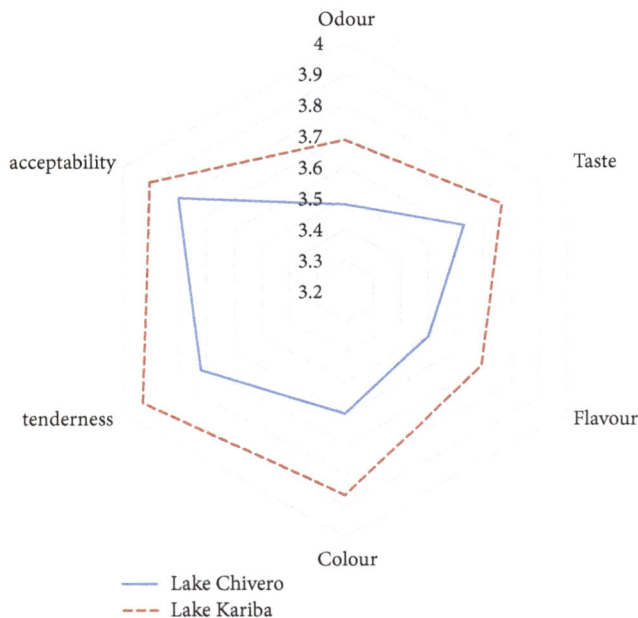

FIGURE 3: Comparison of sensory parameters for cooked fish from Lake Kariba and Lake Chivero.

the low dissolved oxygen [6]. In a study by Khallaf, Gala, and Authman [37], the level of pollution and season were the most important factors that influenced fish condition factor. Similar condition factors of 1.9-2.4 were reported by Asaminew, Tefera, and Tadesse [38] on *Oreochromis niloticus* from different water sources and Utete et al. [22] on the same species from Lake Chivero. Seasonal effects were also observed to influence fish condition factor. In the month of June, fish had lower condition factors than those sampled in October probably due to differences in water temperature. Water temperature ranging from 28°C to 32°C is optimum for survival, growth, and reproduction of *Oreochromis niloticus* [39]. In the month of June water temperatures were below the average optimal growth temperatures for fish as they averaged

16.5°C in Lake Chivero and 24°C in Lake Kariba. Condition factors were higher in the month of October possibly due to higher temperatures close to the optimal temperature for fish growth and survival.

Fish from Lake Chivero had a significantly higher fillet yield than fish from Lake Kariba in October. The variability of fillet weight while dressing-out percentage was not different could be attributed to differences in diet, fish age, sex, season of capture, and environmental conditions and to some extent techniques of filleting [40]. A higher fillet percentage from the fish is desirable since it leads to a higher yield of edible portions and subsequent reduction in the quantity of processing waste [41].

Muscle composition of fish flesh has a lot of influence on how the fish is generally perceived by the consumer in terms of its taste, flavour, and general acceptability [40]. The differences noted on ether extract and dry matter between fish from Lake Kariba and Lake Chivero are not surprising as nutritional value of most freshwater fish has been reported to differ between water bodies, fish size, and season [42]. Lake Chivero is highly polluted and so the feed resources and stress factors for fish in this lake are different when compared to Lake Kariba. Fish in the length class 10-20 cm from Lake Chivero had the highest fat content and the least was from Lake Kariba in the same month. Normal fat content for *Oreochromis niloticus* is 2.75±0.16% but however varies with the geographical region, diet, season, sexual maturity, and age [42]. Considering the fat content of all the fish samples, all the fish were found to be classified as lean. Lean fish have their fat content less than 5%.

The values of DM and ash obtained in this study compared well with the DM and ash of wild *Oreochromis niloticus* reported by Olopade, Taiwo, Lamidi, and Awonaike [43]. Protein content for all the fish from the two lakes in the two seasons was within the range that is classified as high which is greater than 15% [42]. When taking into account only the nutritional requirements for human health, fish from both sources can be reliable sources of nutrients if included in the diet but the fish may not be safe due to possibility of toxins and other poisonous substances that might accumulate in fish which include heavy metals and toxins.

Colour is an important fish quality parameter that has a significant influence on consumer acceptance and also its market value since consumers associate fish flesh clarity with product freshness [44]. The evaluation of fish fillets showed colour differentiation between fish from Lake Chivero and from Lake Kariba. The greenness and dark colour in fish fillets from Lake Chivero can probably be attributed to presence of pigments that are produced by phytoplankton and macrophytes, differences in fat content, water transparency, and higher deposition of melanin due to dietary effects [2, 45, 46]. These results are in agreement with the differences noted in the assessment of raw fish. Colour differences in cichlid species have been attributed to environmental variations which include water transparency [46]. Kop and Durmaz [45] indicated that there were pigments produced by phytoplankton and plants, namely, melanine, purines, pteridiums, and carotenoids which give colour to the skin and flesh of many fish species. The differences in water quality, therefore differences in phytoplankton species and biomass between Lake Chivero and Lake Kariba, could explain the differences in colour between the fish. Some authors reported a direct relationship between fat content and whiter fillets which was not consistent with the observations in this study [47].

Important differences were also noted on raw *Oreochromis niloticus* from Lake Chivero and Lake Kariba. Assessors perceived Lake Kariba fish to be significantly firmer than fish from Lake Chivero. Textural differences observed with fish from Lake Kariba being firmer can be due to differences in fat content, nutritional state of the fish, water content, and activity [48, 49]. Fish from Lake Chivero had more fat and more moisture than fish from Lake Kariba, which could have contributed to the differences in texture. The stronger foreign odour intensity of fish from Lake Chivero could have been due to compounds found in the fish environment. Some of the compounds include those that are produced by blue-green algae and actinomycetes called geosmin and 2-methylisoborneol [50]. These compounds have strong odour characteristics that influence the odour of fish. Differences in phytoplankton communities in the water bodies may explain the differences in fish odour [51].

In the consumer preference study, cooked fish was considered to have similar organoleptic characteristics, independent of origin. Although the raw fish had different properties such as texture, odour, and colour, these differences were not detected on cooked fish in the consumer preference study. Cooking could have altered some of the properties since cooking especially by boiling and steam cooking alters organoleptic properties of fish muscles [52]. The results indicate that some perceptions that consumers have on quality of fish can be a result of consumer beliefs and physical characteristics of the raw product.

5. Conclusions

The study has demonstrated the differences in fish meat quality that arise due to pollution of water bodies. The information generated can be used to determine the suitability of some ecosystems to produce fish that meet important fish meat quality specifications. Lake Kariba offered a better and adaptable environment for fish condition although fish from Lake Chivero produced higher fillet yields. Fillet colour was different as Lake Chivero fish were darker and greener. Despite the differences in fish environment, fish from both sources met human protein requirement. Typical fish odour was stronger in fish from Lake Kariba than fish from Lake Chivero while foreign fish odour was stronger in fish from Lake Chivero than fish from Lake Kariba. Acceptability of raw fish was higher for fish from Lake Kariba than from Lake Chivero while cooked fish from both sources were acceptable. However, the presence of toxins and microorganisms in water due to pollution will ultimately affect other attributes of quality. The impact of source on levels of contaminants such as heavy metals and biotoxins and microbiological status of fish needs to be investigated.

Conflicts of Interest

The authors declare that there are no conflicts of interest regarding the publication of this paper.

Acknowledgments

This work was supported by DAAD [Grant no. A1395763, 2016] and the University of Zimbabwe Research Board [Grant no. 91072, 2015]. The authors would like to acknowledge the support from staff in the Department of Biological Sciences, Department of Animal Science, and University Lake Kariba Research station who assisted in carrying out the various analyses in this study.

References

[1] P. J. Molnár, "A model for overall description of food quality," *Food Quality and Preference*, vol. 6, no. 3, pp. 185–190, 1995.

[2] M. G. Klanian and M. G. Alonso, "Sensory characteristics and nutritional value of red drum Sciaenops ocellatus reared in freshwater and seawater conditions," *Aquaculture Research*, vol. 46, no. 7, pp. 1550–1561, 2015.

[3] FAO, "Fishery and Aquaculture Country Profiles. Zimbabwe. Country Profile Fact Sheets," FAO Fisheries and Aquaculture Department, Rome, 2016.

[4] R. Flos, L. Reig, J. Oca, and M. Ginovart, "Influence of marketing and different land-based systems on gilthead sea bream (Sparus aurata) quality," *Aquaculture International*, vol. 10, pp. 189–206, 2002.

[5] S. Y. El-Zaeem, M. M. M. Ahmed, M. S. Salama, and W. N. A. El-Kader, "Flesh quality differentiation of wild and cultured Nile tilapia (Oreochromis niloticus) populations," *African Journal of Biotechnology*, vol. 11, no. 17, pp. 4085–4089, 2012.

[6] O. A. Oyelese, "Implications of carcass quality and condition factor to the processing of some selected fresh water fish families," *Journal of Fisheries International*, vol. 1, pp. 132–135, 2006.

[7] C. Nasopoulou, H. C. Karantonis, and I. Zabetakis, "Nutritional value of gilthead sea bream and sea bass," *Dynamic Biochemistry, Process Biotechnology and Molecular Biology*, vol. 5, pp. 32–40, 2011.

[8] Z. Sándor, Z. G. Papp, I. Csengeri, and Z. Jeney, "Fish meat quality and safety," *Tehnologija Mesa*, vol. 52, no. 1, pp. 97–105, 2011.

[9] S. T. Arannilewa, S. O. Salawu, A. A. Sorungbe, and B. B. Ola-Salawu, "Effect of frozen period on the chemical, microbiological and sensory quality of frozen tilapia fish (Sarotherodun galiaenus)," *African Journal of Biotechnology*, vol. 4, no. 8, pp. 852–855, 2005.

[10] F. J. A. Pérez-Cueto, Z. Pieniak, and W. Verbeke, "Attitudinal determinants of fish consumption in spain and poland," *Nutrición Hospitalaria*, vol. 26, no. 6, pp. 1412–1419, 2011.

[11] E. Weichselbaum, S. Coe, J. Buttriss, and S. Stanner, "Fish in the diet: A review," *Nutrition Bulletin*, vol. 38, pp. 128–177, 2013.

[12] D. Engeset, T. Braaten, B. Teucher et al., "Fish consumption and mortality in the European Prospective Investigation into Cancer and Nutrition cohort," *European Journal of Epidemiology*, vol. 30, no. 57, 2015.

[13] FAO, "Food Insecurity: when people must live with hunger and fear starvation," The State of Food Insecurity in The World, 2012.

[14] M. C. Nesheim, M. Oria, and P. T. Yih, *Committee on a Framework for Assessing the Health, Environmental, and Social Effects of the Food System; Food and Nutrition Board; Board on Agriculture and Natural Resources; Institute of Medicine; National Research Council; A Framework for Assessing Effects of the Food System*, National Academies Press (US), Washington DC, USA, 2017.

[15] T. A. Zengeya, A. J. Booth, and C. T. Chimimba, "Broad niche overlap between invasive nile tilapia Oreochromis niloticus and indigenous congenerics in Southern Africa: Should we be concerned?" *Entropy*, vol. 17, no. 7, pp. 4959–4973, 2015.

[16] H. L. Meiselman, "Criteria of food quality in different contexts," *Food Service Technology*, vol. 1, no. 2, pp. 67–77, 2001.

[17] F. A. El-Nemaki, A. A. El-nema, M. Z. Mohamed, and R. Olfat, "Impacts of different water resources on the ecological parameters and the quality of tilapia production at el-Abbassa fish farms in Egypt," in *Proceedings of the 8th International Symposium on Tilapia in Aquaculture*, pp. 491–512, 2008.

[18] L. Mhlanga, J. Day, M. Chimbari, N. Siziba, and G. Cronberg, "Observations on limnological conditions associated with a fish kill of Oreochromis niloticus in Lake Chivero following collapse of an algal bloom," *African Journal of Ecology*, vol. 44, no. 2, pp. 199–208, 2006.

[19] J. Nyamangara, N. Jeke, and J. Rurinda, "Long-term nitrate and phosphate loading of river water in the Upper Manyame Catchment," *Water SA*, vol. 39, no. 5, pp. 637–642, 2013.

[20] W. Rommens, J. Maes, N. Dekeza et al., "The impact of water hyacinth (Eichhornia crassipes) in a eutrophic subtropical impoundment (Lake Chivero, Zimbabwe). I. Water quality," *Archiv für Hydrobiologie*, vol. 158, no. 3, pp. 373–388, 2003.

[21] L. Mhlanga and W. Mhlanga, "Nitrate-induced changes and effect of varying total nitrogen to total phosphorus ratio on the phytoplankton community in Lake Chivero, Zimbabwe: microcosm experiments," *Water SA*, vol. 38, no. 4, pp. 607–614, 2012, http://www.scielo.org.za/scielo.php.

[22] B. Utete, L. Mutasa, N. Ndhlovu, and I. H. Tendaupenyu, "Impact of aquaculture on water quality in lake kariba, Zimbabwe," *International Journal of Aquaculture*, vol. 3, no. 4, pp. 11–16, 2013.

[23] G. Cronberg, "Phytoplankton in Lake Kariba," in *Advances in the Ecology of Lake Kariba*, J. Moreau, Ed., pp. 66–72, University of Zimbabwe Publications, Harare, Zimbabwe, 1997.

[24] C. H. Magadza, "The distribution of zooplankton in the Sanyati Bay, Lake Kariba; A multivariate analysis," *Hydrobiologia*, vol. 70, no. 1-2, pp. 57–67, 1980.

[25] R. A. Sanyanga and L. Mhlanga, "Limnology of Zimbabwe," *Limnology in Developing Countries*, pp. 117–170, 2004.

[26] J. Bartram and R. Balance, *Water Quality Monitoring. A Practical Guide to The Design and Implementation of Freshwater Quality Studies and Monitoring Programmes*, TJ Press, Padstow, Cornwall, Great Britain, 1996.

[27] AOAC, *Official Methods of Analysis*, Association of Official Analytical Chemists, Washington DC, USA, 16th edition, 1995.

[28] K. L. Yam and S. E. Papadakis, "A simple digital imaging method for measuring and analyzing color of food surfaces," *Journal of Food Engineering*, vol. 61, no. 1, pp. 137–142, 2004.

[29] H. Stone and J. L. Sidel, *Sensory Evaluation Practices*, Elsevier Academic Press, 3rd edition, 2004, http://www.fao.org/fishery/facp/ZWE/en.

[30] A. V. A. Resurreccion, "Consumer sensory testing for food product development," in *Developing New Food Products for a Changing Marketplace*, A. L. Brody and J. B. Lord, Eds., Taylor and Francis Group, Boca Raton, FL, USA, 2007.

[31] P. Tendaupenyu and C. H. D. Magadza, "Nutrient concentrations in the surface sediments of Lake Chivero, Zimbabwe: A shallow, hypereutrophic, subtropical artificial lake," *Lakes and Reservoirs*, vol. 22, no. 4, pp. 297–309, 2017.

[32] I. Chirisa, E. Bandauko, A. Matamanda, and G. Mandisvika, "Decentralised domestic wastewater systems in developing countries; the case study of Harare (Zimbabwe)," *Applied Water Science*, vol. 7, no. 3, pp. 1069–1078, 2017.

[33] R. Sharma and A. Capoor, "Seasonal variation in physical, chemical and biological parameters of Lake of water of Patna Bird sanctuary in relation to fish productivity," *World Applied Sciences Journal*, vol. 8, no. 1, pp. 129–132, 2010, http://citeseerx.ist.psu.edu/viewdoc/download.

[34] A. B. Fonge, B. G. Chuyong, A. S. Tening, A. C. Fobid, and N. F. Numbisi, "Seasonal occurrence, distribution and diversity of phytoplankton in the Douala Estuary, Cameroon," *African Journal of Aquatic Science*, vol. 38, no. 2, pp. 123–133, 2013.

[35] L. Mhlanga, W. Mhlanga, and P. Tendaupenyu, "Response of phytoplankton assemblages isolated for short periods of time in a hyper-eutrophic reservoir (Lake Chivero, Zimbabwe)," *Water SA*, vol. 40, no. 1, pp. 157–164, 2014.

[36] M. Ashraf, A. Zafar, A. Rauf, S. Mehboob, and N. A. Qureshi, "Nutritional values of wild and cultivated silver carp (Hypophthalmichthys molitrix) and grass carp (Ctenopharyngodon idella)," *International Journal of Agriculture and Biology*, vol. 13, no. 2, pp. 210–214, 2011.

[37] E. A. Khallaf, M. Galal, and M. Authman, "The biology of Oreochromis niloticus in a polluted canal," *Ecotoxicology*, vol. 12, no. 5, pp. 405–416, 2003, https://link.springer.com/content/pdf/10.1023%2FA%3A1026156222685.pdf.

[38] K. Asaminew, F. Tefera, and Z. Tadesse, "Adaptability, growth and reproductive success of the Nile tilapia, Oreochromis niloticus L. (Pisces: Cichlidae) stocked in Lake Small Abaya, South Ethiopia," *Ethiopian Journal of Biological Sciences*, vol. 10, no. 2, pp. 153–166, 2011, http://aau.edu.et/efasa/images/efasafolder/kassahun%20asaminew%20et%20al.pdf.

[39] K. B. Workagegn, "Evaluation of Growth Performance, Feed Utilization Efficiency and Survival Rate of Juvenile Nile tilapia, Oreochromis niloticus (Linnaeus, 1758) Reared at Different Water Temperature," *International Journal of Aquaculture*, vol. 2, no. 1, 2012.

[40] F. A. Mohamed, F. A. Khogali, A. H. Mohamed, O. O. Deng, and A. A. Mohammed, "Body weight characteristics and chemical composition of Nile tilapia (Oreochromis niloticus) collected from three different Sudanese dams," *International Journal of Fisheries and Aquatic Studies*, vol. 4, no. 5, pp. 507–510, 2016.

[41] M. Vandeputte, A. Puledda, A. S. Tyran et al., "Investigation of morphological predictors of fillet and carcass yield in European sea bass (Dicentrarchus labrax) for application in selective breeding," *Aquaculture*, vol. 470, pp. 40–49, 2017.

[42] F. Jim, P. Garamumhango, and C. Musara, "Comparative analysis of nutritional value of Oreochromis niloticus (Linnaeus), Nile tilapia, meat from three different ecosystems," *Journal of Food Quality*, vol. 2017, 2017.

[43] O. A. Olopade, I. O. Taiwo, A. A. Lamidi, and O. A. Awonaike, "Proximate Composition of Nile Tilapia (Oreochromis niloticus) (Linnaeus, 1758) and Tilapia Hybrid (Red Tilapia) from Oyan Lake, Nigeria," *Food Science and Technology*, vol. 73, no. 1, 2016.

[44] G. G. Fonseca, A. D. Cavenaghi-Altemio, M. de Fátima Silva, V. Arcanjo, and E. J. Sanjinez-Argandoña, "Influence of treatments in the quality of Nile tilapia (Oreochromis niloticus) fillets," *Food Science and Nutrition*, vol. 1, no. 3, pp. 246–253, 2013.

[45] A. Kop and Y. Durmaz, "The effect of synthetic and natural pigments on the colour of the cichlids (Cichlasoma severum sp., Heckel 1840)," *Aquaculture International*, vol. 16, no. 2, pp. 117–122, 2008.

[46] M. E. Maan and K. M. Sefc, "Colour variation in cichlid fish: Developmental mechanisms, selective pressures and evolutionary consequences," *Seminars in Cell Developmental Biology*, vol. 24, pp. 516–528, 2013.

[47] K. Grigorakis, K. D. A. Taylor, and M. N. Alexis, "Organoleptic and volatile aroma compounds comparison of wild and cultured gilthead sea bream (*Sparus aurata*): sensory differences and possible chemical basis," *Aquaculture*, vol. 225, pp. 109–119, 2003.

[48] U. B. Andersen, M. S. Thomassen, and A. M. B. Rora, "Texture Properties of farmed rainbow trout (Oncorhynchus mykiss): Effect of diet, muscle fat content and time of storage on ice," *Journal of the Science of Food and Agriculture*, vol. 74, pp. 347–353, 1997.

[49] J. Lu, T. Takeuchi, and H. Ogawa, "Flesh quality of tilapia Oreochromis niloticus fed solely on raw Spirulina," *Fisheries Science*, vol. 69, pp. 529–534, 2003.

[50] M. Yurkowski and J. L. Tabachek, "Geosmin and 2-methylisoborneol implicated as a cause of muddy odor and flavor in commercial fish from cedar lake, manitoba," *Canadian Journal of Fisheries and Aquatic Sciences*, vol. 37, no. 9, pp. 1449-1450, 1980.

[51] S. Percival, P. Drabsch, and B. Glencross, "Determining factors affecting muddy-flavour taint in farmed barramundi, Lates calcarifer," *Aquaculture*, vol. 284, no. 1-4, pp. 136–143, 2008.

[52] M. Y. Ali, F. I. Shams, M. Khanom, and M. G. Sarower, "Fillet yield, dress-out percentage and phenotypic relationship between different traits of grey Mullet (Liza parsia, Hamilton 1822)," *International Journal of Engineering and Applied Sciences*, vol. 4, no. 1, pp. 49–58, 2013.

13

The Bacteriological Quality, Safety, and Antibiogram of *Salmonella* Isolates from Fresh Meat in Retail Shops of Bahir Dar City, Ethiopia

Melkamnesh Azage and Mulugeta Kibret

Department of Biology, Science College, Bahir Dar University, P.O. Box 79, Bahir Dar, Ethiopia

Correspondence should be addressed to Mulugeta Kibret; mulugetanig@gmail.com

Academic Editor: Alejandro Castillo

The habit of raw meat consumption in addition to the poor hygienic standards and lack of knowledge contribute to food-borne diseases outbreaks. The objective of this research was to assess the bacterial quality and safety of fresh meat from retail Bahir Dar City Ethiopia. A total of 30 fresh meat samples were collected from butcher shops. Standard bacteriological methods were used to isolate and enumerate bacteria. Kirby-Bauer disk diffusion method was used for antimicrobial susceptibility testing of *Salmonella* isolates The mean counts of AMB, TC, and *S. aureus* were $\log_{10} 4.53$, 3.97, and $3.88 \log_{10}$cfu/g, respectively. *Salmonella* was isolated from 21 (70%) of the samples. *Salmonella* isolates in this study were highly susceptible to ciprofloxacin, gentamycin, and norfloxacin while they were resistant to erythromycin and tetracycline. High rate of multiple drug resistance was also noticed in *Salmonella* isolates. The microbial loads of meat were above the recommended microbial safety limits. Besides this, the isolation rate of *Salmonella* was high and high levels of drug resistance were documented for *Salmonella* isolates. Measures on handling and appropriate personal hygiene practices of workers in the retail shops are recommended to reduce the change of forborne disease outbreaks.

1. Introduction

Meat is consumed by many people worldwide because of its nutritive composition. The protein profile of meat consists of amino acids that have been described as excellent due to the presence of all essential ones required by the body [1]. It is considered to be spoiled when it is unfit for human consumption and subjected to changes by its own enzyme, microbial action, and any other factors [2]. Enteric bacteria species can cause infections in humans when undercooked meat products are consumed [3]. The microbiological quality of meat depends on the physical status of the animal at slaughter, the spread of contamination during slaughter and processing, the temperature, and other conditions of storage and distribution [4]. The need for microbial assessment of fresh meats consumption is emphasized and recommended to reduce possible contamination [5].

In Ethiopia, minced or raw beef consumption is usually used for the preparation of a popular traditional Ethiopian dish known as locally "KITFO" and mostly it is consumed raw or partially cooked. This habit is a potential cause for food-borne illnesses in addition to, the common factors such as overcrowding, poverty, inadequate sanitary conditions, and poor general hygiene [6]. The microbiological quality of meat and meat products is very important with regard to public health significance. There are several reports on outbreaks of food-borne illnesses because of consumption of meat [6, 7]. Moreover, antibiotic resistance levels are also elevated among food-borne pathogens such as *Salmonella*, *E. coli*, and *Shigella* [8–10].

The absence of organized slaughter house facility and the existence of small retail outlets have been the two biggest hurdles for hygienic production of meat [11]. It is essential to generate information about the quality of fresh beef sold in retail shops. Hence the present research work was undertaken to determine the bacteriological quality of the meat and determine the antimicrobial susceptibility profiles of *Salmonella* isolates from retail shops of Bahir Dar City, Ethiopia.

2. Materials and Methods

2.1. Description of the Study Area. This study was conducted in May 2015 in Bahir Dar town, which is the capital of Amhara National Regional State (ANRS) situated in the northern part of Ethiopia. Bahir Dar is located at $11°36'N$ latitude and $37°23'E$ longitude and has an elevation of 1840 m above sea level. The area of the town is about $160\,km^2$ and there are around 256,999 people living in there [12].

2.2. Sample Collection and Bacteriological Analysis. A cross-sectional study was conducted in retail meat shops to determine bacteriological quality and antibiogram of *Salmonella*. A total of 30 retail cut meat samples were collected from 30 purposively selected retail houses between 7:00 and 9:00 am. One kilogram of cut meat was aseptically collected with sterile glove and placed in a sterile glass beaker covered with aluminum foil. The samples were transported to the laboratory in ice box and bacteriological analysis was done within two hours of collection at Food Microbiology Laboratory of School of Chemical and Food Engineering, Bahir Dar University. The ambient temperature at the time of sample collection was 20°C.

Twenty-five grams of meat sample was mixed with 225 ml of 0.1% buffered peptone water (Merck, Darmstadt) and homogenized for 2 minutes by using stomacher (Seward Ltd., UK) [13]. Tenfold serial dilutions (10^{-2}–10^{-4}) were made from the homogenized sample and 1 ml from each sample of each dilution was taken and used for enumeration of aerobic mesophilic bacteria (AMB), total coliforms (TC), and *S. aureus* and the remaining homogenate was used for the isolation of *Salmonella*.

Enumeration of aerobic mesophilic bacteria was done using the pour plate techniques on plate count agar (Oxoid, England). One ml of homogenized sample was inoculated onto plate count agar, in triplicate and the plates were incubated aerobically at 32°C for a maximum of 48 hrs. After incubation, the plates having 30–300 colonies were counted using colony counter. Uninoculated media were incubated as negative control to check for sterility [14]. Violate Red Bile Agar (VRBA) (Oxoid, England) was used to count total coliforms after incubation at 30–37°C for 24–48 hrs, by using pour plate technique. All purplish red colonies were counted as coliforms [15]. For *Staphylococcus aureus* count, samples were spread-plated in triplicate plates of Mannitol Salt Agar (Oxoid, England) and incubated at 30–37°C for 48 hrs and yellow colonies were counted [14, 15].

2.3. Isolation of Salmonella. The homogenized sample was incubated at 37°C for 24 hrs and 1 ml of culture was transferred to 10 ml of selenite cysteine broth (SCB) (Himedia, India) and incubated further at 37°C for 24 hrs. A loop full of culture from selenite cysteine broth culture was subcultured onto Xylose lysine deoxycholate (XLD) agar (Oxoid, England) plate and incubated aerobically at 37°C for 24 hrs. After incubation, 2-3 characteristic colonies of *Salmonella* (red colonies with or without black center) were picked and stored on Tryptic Soya Agar (TSA) slants for further purification and used for biochemical characterization and antimicrobial susceptibility tests [16].

2.4. Antimicrobial Susceptibility Testing of Salmonella Isolates. In vitro antimicrobial susceptibility tests were performed on Mueller-Hinton agar (Oxoid, UK) using Kirby-Bauer disk disc diffusion technique [17]. The antimicrobials tested were ciprofloxacin (CIP, 5 μg), norfloxacin (NOR, 10 μg), amoxicillin (AMC, 30 μg), ampicillin (AMP, 10 μg), chloramphenicol (C, 30 μg), erythromycin (E, 15 μg), gentamicin (CN, 10 μg), nalidixic acid (NA, 30 μg), trimethoprim-sulfamethoxazole (SXT, 25 μg), cefoxitin (FOX, 30 μg), and tetracycline (TE, 30 μg) (Oxoid, UK). Morphologically identical 4–6 bacterial colonies from overnight culture were suspended in 5 ml nutrient broth and incubated for 4 hrs at 37°C. Turbidity of the broth culture was equilibrated to match 0.5 McFarland standards. The surface of Mueller-Hinton agar plate was evenly inoculated with the culture using a sterile cotton swab. The antibiotic discs were applied to the surface of the inoculated agar. After 18–24 hrs of incubation, the diameter of growth inhibition around the discs was measured and interpreted as sensitive, intermediate, or resistant according to Clinical and Laboratory Standards Institute [18]. Reference strain of *E. coli* ATCC 25922 was used as quality control for antimicrobial susceptibility tests.

2.5. Data Analysis. Data were analyzed using the statistical package for Social Science (SPSS) version 20 software by descriptive statistics. Results of bacterial counts were expressed in terms of mean log cfu/g and compared with Gulf Standards, 2002 [19] (Table 1). The isolation rate of *Salmonella* prevalence and antimicrobial susceptibility tests were expressed in terms of percentage.

3. Results and Discussion

In this study, the aerobic mesophilic bacteria counted in fresh meat ranged between 1.91 and 6.70 $\log_{10}$cfu/g having a mean value of 4.53 $\log_{10}$cfu/g (Table 2). All 30 samples of fresh meat had high counts of aerobic mesophilic bacteria. In the current study, the total coliform counts detected ranged between 1.40 and 6.50 $\log_{10}$cfu/g having a mean value of 3.97 $\log_{10}$cfu/g. The mean count of *S. aureus* in fresh meat in this study was 3.88 $\log_{10}$cfu/g and ranged between 1.42 and 8.47 $\log_{10}$cfu/g (Table 2).

Aerobic mesophilic count is one of the microbiological indicators for food quality and the presence of aerobic organisms reflects existence of favorable conditions for the multiplication of microorganisms [20]. Coliforms are indicators of water or food quality and their presence may be an indication of unhygienic condition [21]. The highest number of *S. aureus* on meat indicates the presence of cross-contamination, which usually related to human skin, hair, hand and discharge from nose, and clothing. High contamination of food with *S. aureus* has been related to improper personal hygiene of employees during handling and processing [12].

The results of this study are comparable to the findings of previous works [22–24]. Other researchers have reported higher and lower aerobic mesophilic, coliform, and *S. aureus*

TABLE 1: Guideline levels for determining microbial quality of ready-to-eat food (Gulf Standards and NSW Food Authority).

Microbial groups	Good	Acceptable	Unsatisfactory	Unacceptable and potentially dangerous
Aerobic mesophilic count	$<10^4$	$10^4-<10^6$	$\geq10^6$	N/A
Total coliform count	$<10^2$	10^2-10^4	$\geq10^4$	N/A
S. aureus count	$<10^2$	10^2-10^3	$10^3-<10^4$	$\geq10^4$
Pathogens	Not detected in 25 g of	—	—	Detected in 25 g of

TABLE 2: Bacterial counts of fresh meat in Bahir Dar town, May, 2015.

Bacterial counts	Minimum count ($\log_{10}$cfu/g)	Maximum count ($\log_{10}$cfu/g)	Mean ± SD ($\log_{10}$cfu/g)
AMC	1.91	6.70	4.53 ± 1.24
TC	1.40	6.50	3.97 ± 1.42
S. aureus	1.42	8.47	3.88 ± 1.81

TABLE 3: Antibiotic susceptibility pattern of Salmonella isolates in Bahir Dar town, May, 2015.

Antimicrobial Agents	Resistant No (%)	Intermediate No (%)	Sensitive No (%)
Ciprofloxacin	0 (0)	0 (0)	21 (100)
Nalidixic acid	0 (0)	2 (9.5)	19 (90.5)
Erythromycin	19 (90.5)	2 (9.5)	0 (0)
Ampicillin	5 (23.8)	4 (19)	12 (57.2)
Tetracycline	14 (66.7)	0 (0)	7 (33.3)
Trimethoprim-sulfamethoxazole	2 (9.5)	0 (0)	19 (90.5)
Gentamycin	0 (0)	0 (0)	21 (100)
Cefoxitin	2 (9.5)	7 (33.3)	12 (57.2)
Amoxicillin	2 (9.5)	2 (9.5)	17 (81)
Chloramphenicol	0 (0)	1 (4.8)	20 (95.2)
Norfloxacin	0 (0)	0 (0)	21 (100)

counts [25–27]. The differences might be as a result of differences in study areas, temperature, and personal hygiene practices of the vendors.

The total aerobic counts far exceed the prescribed microbiological safety limits of Gulf Standards [19]. The implication of the findings is that the product is not safe for human consumption, since the samples had counts of aerobic counts exceeding the acceptable limits [28]. In general most of the raw meats sold at butcher shops in this study were potentially hazardous for health.

Among 30 meat samples tested, 21 (70%) were positive for Salmonella. Salmonella isolates exhibited high level of resistance to erythromycin and tetracycline. The isolates were sensitive to ciprofloxacin, gentamycin, and norfloxacin. There were also intermediate levels of resistances to cefoxitin, ampicillin, and nalidixic acid (Table 3). Among 21 isolates of Salmonella, 15 (71.43%) were resistant to two or more antibiotics. Five of the isolates were resistant to three or more antibiotics (Table 4).

With regard to the antimicrobial susceptibility profiles of Salmonella isolates, all the Salmonella isolates showed high level of sensitivity (95–100%) to gentamycin, ciprofloxacin, norfloxacin, and chloramphenicol while high levels of resistance (66–90%) were documented against erythromycin and

TABLE 4: MDR pattern of salmonella isolates in Bahir Dar town, June, 2015.

Resistance pattern	Salmonella isolates No (%)
Resistant to two antibiotics	
E-TE	8 (38)
E- SXT	1 (4.8)
TE-AMP	1 (4.8)
Resistant to three antibiotics	
E-TE-AMP	2 (9.5)
Resistant to four antibiotics	
E-TE-AMP-AMC	1 (4.8)
E-TE-AMP- SXT	1 (4.8)
E-TE-FOX-AMC	1 (4.8)

tetracycline. From a study done in Ethiopia, Reda et al. [29] and Farzana et al. [30] reported comparable levels of sensitivity and resistance. This could be due to the fact that ciprofloxacin and norfloxacin are relatively expensive and newly introduced, compared to the other common antibiotics. The routine practice of giving antimicrobial agents to domestic livestock as a means of preventing and treating diseases, as well as promoting growth, is an important factor in the emergence of antibiotic-resistant bacteria that are subsequently transferred to humans through the food chain [31, 32]. Most infections with antimicrobial-resistant *Salmonella* are acquired by eating contaminated foods of animal origin [33, 34].

4. Conclusion

This study revealed high level of contamination in fresh meat as indicated by high aerobic mesophilic, *S. aureus,* and coliform counts which are above the recommended microbial safety limits. High bacterial loads and isolation of drug resistant *Salmonella* suggest a potential health risk to the consumers from the consumption of raw meat. These indicate poor handling and personal hygiene practices of workers in the retail shops and risk of food-borne disease. Investigation on antibiotic use in animal and animal feed is recommended.

Conflicts of Interest

The authors declare that there are no conflicts of interest regarding the publication of this paper.

References

[1] O. A. Olaoye, "Meat: an overview of its composition, biochemical changes and associated microbial agents," *International Food Research Journal*, vol. 18, no. 3, pp. 877–885, 2011.

[2] K. Bradeeba and P. K. Sivakumaar, "Assessment of microbiological quality of beef, mutton and pork and its environment in retail shops in Chidambaram, Tamil Nadu," *International Journal of Plant, Animal and Environmental Sciences*, vol. 3, pp. 91–97, 2013.

[3] C. N. Ateba and T. Setona, "Isolation of enteric bacterial pathogens from raw mince meat in mafikeng, north-west province," *South Africa Life Science Journal*, vol. 8, no. 2, pp. 1–7, 2011.

[4] A. J. Ilboudo, F. Tapsoba, A. Savadogo, M. Seydi, and A. S. Traore, "Improvement of the hygienic quality of farmhouse meat pies produced in Burkina Faso," *Advances in Environmental Biology*, vol. 6, no. 10, pp. 2627–2635, 2012.

[5] I.-O. E. Ukut, I. O. Okonko, I. S. Ikpoh et al., "Assessment of bacteriological quality of fresh meats sold in Calabar Metropolis, Nigeria," *Electronic Journal of Environmental, Agricultural and Food Chemistry*, vol. 9, no. 1, pp. 89–100, 2010.

[6] M. Birhaneselassie and D. Williams, "A Study of *Salmonella* carriage among asymptomatic food-handlers in Southern Ethiopia," *International Journal of Nutrition and Food Sciences*, vol. 2, no. 5, pp. 243–245, 2013.

[7] S. G. Bhandare, A. T. Sherikar, A. M. Paturkar, V. S. Waskar, and R. J. Zende, "A comparison of microbial contamination on sheep/goat carcasses in a modern Indian abattoir and traditional meat shops," *Food Control*, vol. 18, no. 7, pp. 854–858, 2007.

[8] M. Kibret and M. Tadesse, "The bacteriological safety and antimicrobial susceptibility of bacteria isolated from street-vended white lupin (Lupinus albus) in Bahir Dar, Ethiopia," *Ethiopian Journal of Health Sciences*, vol. 23, no. 1, pp. 19–26, 2013.

[9] L. Garedew, Z. Hagos, Z. Addis, R. Tesfaye, and B. Zegeye, "Prevalence and antimicrobial susceptibility patterns of *Salmonella* isolates in association with hygienic status from butcher shops in Gondar town, Ethiopia," *Antimicrobial Resistance and Infection Control*, vol. 4, no. 1, article 21, 2015.

[10] E. Khan, K. Jabeen, M. Ejaz, J. Siddiqui, M. F. Shezad, and A. Zafar, "Trends in antimicrobial resistance in Shigella species in Karachi, Pakistan," *Journal of Infection in Developing Countries*, vol. 3, no. 10, pp. 798–802, 2009.

[11] P. Kumar, J. Rao, Y. Haribabu, and . Manjunath, "Microbiological Quality of Meat Collected from Municipal Slaughter Houses and Retail Meat Shops from Hyderabad Karnataka Region, India," *APCBEE Procedia*, vol. 8, pp. 364–369, 2014.

[12] CSA (Central Statistical Authority of Ethiopia), "The 2010 Population and Housing Census of Ethiopia," 2011.

[13] R. Jacob Microbial quality of ready- to-eat foods available to populations of different demographics, A Master Thesis Submitted to the Faculty of Drexel University, Philadelphia, Pa, USA, 2010.

[14] M. U. D. Ahmad, A. Sarwar, M. I. Najeeb et al., "Assessment of microbial load of raw meat at abattoirs and retail outlets," *Journal of Animal and Plant Sciences*, vol. 23, pp. 745–748, 2013.

[15] A. Chaiba, F. F. Rhazi, A. Chahlaoui et al., "Microbiological quality of poultry meat on the Meknes Market (Morocco)," *Internet Journal of Food Safety*, vol. 9, pp. 67–71, 2007.

[16] WHO, "Global Salm-Surv: a global *Salmonella* surveillance and laboratory support project of the World Health Organization," 2003.

[17] A. Bauer, W. Kirby, J. Sheriss, and M. Turk, "Antibiotic susceptibility testing by standard single disk diffusion method," *American Journal of Clinical Pathology*, vol. 45, pp. 493–496, 1966.

[18] Clinical and Laboratory Standards Institute, *Performance Standards for Antimicrobial Susceptibility Testing; Seventeenth Information Supplement*, CLSI document M100-S17, Clinical and Laboratory Standards Institute, Wayne, Pa, USA, 2011.

[19] *Microbiological Criteria for Food Stuffs-part 1. GCC,* Gulf Region Standards, Riyadh, Saudi Arabia, 2000.

[20] M. E. Nyenje, C. E. Odjadjare, N. F. Tanih, E. Green, and R. N. Ndip, "Foodborne pathogens recovered from ready-to-eat foods from roadside cafeterias and retail outlets in alice, eastern cape province, South Africa: public health implications," *International Journal of Environmental Research and Public Health*, vol. 9, no. 8, pp. 2608–2619, 2012.

[21] N. Ejikeme and N. V. Ugochukwu, "Occurrence of *Staphylococcus aureus* in meat pie and egg roll sold in Umuahia Metropolis, Nigeria," *International Journal of Microbiology Immunology Research*, vol. 1, pp. 52–55, 2013.

[22] N. Badrie, A. Joseph, and A. Chen, "An observational study of food safety practices by street vendors and microbiological quality of street-purchased hamburger beef patties in Trinidad, West Indies," *Internet Journal of Food Safety*, vol. 3, pp. 25–31, 2001.

[23] A. Gebeyehu, M. Yousuf, and A. Sebsibe, "Evaluation of microbial load of beef of Arsi cattle in Adama Town, Oromia, Ethiopia," *Journal of Food Process Technology*, vol. 4, pp. 1–6, 2013.

[24] A. U. Nnachi and C. O. Ukaegbu, "Microbial quality of raw meat sold in Onitsha, Anambra State, Nigeria," *International Journal of Scientific Research*, vol. 3, pp. 214–218, 2014.

[25] R. Koffi-Nevr, M. Koussemon, and S. O. Coulibaly, "Bacteriological Quality of Beef Offered for Retail Sale in Cote d'Ivoire," *American Journal of Food Technology*, vol. 6, no. 9, pp. 835–842, 2011.

[26] N. Cohen, H. Ennaji, M. Hassa, and H. Karib, "The bacterial quality of red meat and offal in Casablanca (Morocco)," *Molecular Nutrition and Food Research*, vol. 50, no. 6, pp. 557–562, 2006.

[27] J. A. Khan and I. Shukla, "Re-emergence of chloramphenicol sensitive Salmonella Enteric serotype Typhi—a preliminary report," *BioNotes*, vol. 6, p. 50, 2004.

[28] M. D. Salihu, A. U. Junaidu, A. A. Magaji et al., "Bacteriological quality of traditionally prepared fried ground beef (Dambun nama) in Sokoto, Nigeria," *Advance Journal of Food Science and Technology*, vol. 2, no. 3, pp. 145–147, 2010.

[29] A. Reda, B. Seyoum, J. Yimam et al., "Antibiotic susceptibility patterns of Salmonella and Shigella isolates in Harer, Eastern Ethiopia," *Journal of Infectious Diseases and Immunity*, vol. 3, pp. 34–39, 2011.

[30] K. Farzana, M. R. Akram, and S. Mahmood, "Prevalence and antibiotics susceptibility patterns of some bacterial isolates from a street vended fruit product," *African Journal of Microbiology Research*, vol. 5, no. 11, pp. 1277–1284, 2011.

[31] L. Tollefson, S. F. Altekruse, and M. E. Potter, "Therapeutic antibiotics in animal feeds and antibiotic resistance," *Revue Scientifique et Technique*, vol. 16, no. 2, pp. 709–715, 1997.

[32] W. Witte, "Medical consequences of antibiotic use in agriculture," *Science*, vol. 279, no. 5353, pp. 996-997, 1998.

[33] F. J. Angulo, K. R. Johnson, R. V. Tauxe, and M. L. Cohen, "Origins and consequences of antimicrobial-resistant nontyphoidal *Salmonella*: implications for the use of fluoroquinolones in food animals," *Microbial Drug Resistance*, vol. 6, no. 1, pp. 77–83, 2000.

[34] P. D. Fey, T. J. Safranek, M. E. Rupp et al., "Ceftriaxone-resistant Salmonella infection acquired by a child from cattle," *New England Journal of Medicine*, vol. 342, no. 17, pp. 1242–1249, 2000.

14

Progress towards Sustainable Utilisation and Management of Food Wastes in the Global Economy

Purabi R. Ghosh,[1] **Derek Fawcett,**[1] **Shashi B. Sharma,**[2] **and Gerrard Eddy Jai Poinern**[1]

[1]*Murdoch Applied Nanotechnology Research Group, Department of Physics, Energy Studies and Nanotechnology, School of Engineering and Energy, Murdoch University, Murdoch, WA 6150, Australia*
[2]*Department of Agriculture and Food, 3 Baron-Hay Court, South Perth, WA 6151, Australia*

Correspondence should be addressed to Gerrard Eddy Jai Poinern; g.poinern@murdoch.edu.au

Academic Editor: Alfredo Cassano

In recent years, the problem of food waste has attracted considerable interest from food producers, processors, retailers, and consumers alike. Food waste is considered not only a sustainability problem related to food security, but also an economic problem since it directly impacts the profitability of the whole food supply chain. In developed countries, consumers are one of the main contributors to food waste and ultimately pay for all wastes produced throughout the food supply chain. To secure food and reduce food waste, it is essential to have a comprehensive understanding of the various sources of food wastes throughout the food supply chain. The present review examines various reports currently in the literature and quantifies waste levels and examines the trends in wastage for various food sectors such as fruit and vegetable, fisheries, meat and poultry, grain, milk, and dairy. Factors contributing to food waste, effective cost/benefit food waste utilisation methods, sustainability and environment considerations, and public acceptance are identified as hurdles in preventing large-scale food waste processing. Thus, we highlight the need for further research to identify and report food waste so that government regulators and food supply chain stakeholders can actively develop effective waste utilisation practices.

1. Introduction

Food is a basic human need, while food waste has been identified as a major challenge facing humanity today [1]. Currently, around 21,000 people die every day due to hunger related causes [2] and globally one in nine people go to bed each night hungry [3]. Nevertheless, approximately one-third of all food produced today goes to landfill [4]. The vast amount of food ending up as waste is not only a humanitarian problem, but also a serious economic and environmental problem [5–7]. The world has limited natural resources and environmental benign cost-effective solutions must be found to increase food production, improve distribution networks, and promote effective food supply chain management practices [8]. To alleviate the increasing demand for food production, it is necessary to significantly reduce food waste. Reducing food waste is an important factor that can significantly improve the overall efficiency of the food supply chain [6]. Researchers

in the field maintain that sustainable food production, intelligent management, and proper food distribution are the key factors that must be addressed if we expect to feed the predicted 12.3 billion people in 2100 [7, 9]. So, reducing food waste becomes a priority, since waste will continue to be generated throughout the food supply chain if no action is taken. Companies involved in the food supply chain and the population at large will continue to waste food as long as they can afford to waste. Importantly, food waste results in loss of time, effort, and the other resources that went into producing that food. Other resources lost include fertilizers, pesticides, and the soil and water. From an environmental perspective, food lost or discarded each year accounts for 3.3 billion tonnes of carbon dioxide emissions globally. The scale of food waste globally can be quite staggering and several significant examples are presented in Table 1 so that the reader can appreciate the magnitude of the problem. Thus, governments, industry, and communities must work

TABLE 1: Representative global examples of food loss (waste) [11].

Food loss (waste)	Reference
In the USA alone, annual food production consumes about 120 cubic kilometres of irrigation water. People throw away 30 percent of this food, which corresponds to 40 billion litres of water.	[184]
United Kingdom households waste an estimated 6.7 million MT of food every year, around one-third of the 21.7 million MT purchased. This means that approximately 32 percent of all food purchased per year is not eaten. Most of this (5.9 million tonnes or 88 percent) is currently collected by local authorities. Most of the food waste (4.1 million MT or 61 percent) is avoidable and could have been eaten if it had been better managed.	[185]
The amount of food lost or wasted every year is equivalent to more than half of the world's annual cereals crop (2.3 billion MT in 2009/2010). Only an estimated 43 percent of the cereal produced is available for human consumption, as a result of harvest and postharvest distribution losses and use of cereal for animal feed.	[186]
The water applied globally for irrigation to grow food that is wasted would meet the domestic needs of 9 billion people.	[187]
Annual food losses and waste are estimated at about 30 percent for cereals, 40 to 50 percent for root crops, 30 percent for fish, and 20 percent for oilseeds and meat.	[188]
On a global scale, just 43 percent of the fruits and vegetables produced are consumed and the remaining 57 percent are wasted.	[189]
Food waste accounts for roughly US$ 680 billion in industrialised countries and US$ 310 billion in developing countries.	[190]
Consumers in rich countries waste about 222 million MT of food every year, which is nearly equivalent to the entire net food production of 230 million MT of sub-Saharan Africa.	[185, 191]
Roughly one-third of food is lost or wasted. That translates into 1.3 billion MT each year, worth nearly one trillion US dollars, and is the equivalent of 6 to 10 percent of human-generated greenhouse gas emissions.	[192]
Food spoilage and waste account for annual losses of US$ 310 billion in developing countries, where nearly 65 percent of loss occurs at the production, processing, and postharvest stages.	[192]
In sub-Saharan Africa, up to 150 kg of the food produced per person is lost each year; depending on the crop, 15–35 percent of food harvested may be lost before it leaves the field.	[192]

collaboratively to achieve policy and cultural change towards prevention of food waste at all levels [10, 11]. Therefore, to keep pace with the ever increasing demand for food, it is essential to adopt a policy that says "no" to food waste.

Defining food waste is not always straightforward since distinguishing between edible and nonedible parts of food is subjective. In some parts of the world, a food judged edible may be considered nonedible in other parts. Naturally, not every part of an agricultural or livestock product is entirely edible and there will always be unavoidable nonedible parts such as citrus fruit zest, fruit stones, bones, and eggshells [12, 13]. In many cases, the difference between edible and nonedible is not clearly defined and depends on dietary habits (consumption of bread crusts, apple or potato peel, fat on meat, etc.), food culture, and geographic location. In the present study, food that is not consumed by the end user, which includes the nonedible parts of the food, is considered to be "food waste." All food products go through a life cycle, starting from the farm and progressing through processing, distribution, retail, and finally consumption and/or dumping as presented in Table 2. Inspecting Table 2 reveals that food waste occurs throughout the entire food supply chain. The degree of food waste depends on factors such as (1) developed and developing country [6, 14]; (2) prevailing weather conditions and pest management protocols [15]; (3) storage, transport facilities, and processing efficiency [16–18]; (4) market demand and visual appearance of produce [14]; (5) consumer acceptance of produce [19] and consumer affordability to waste [4].

Even a couple of decades ago, food waste was not considered to be a significant economic cost or a waste of natural resources [20]. However, growing public concerns about hunger, conserving the environment, and the effect of socioeconomic factors have accelerated research into food waste. Food waste research is aimed at finding better ways of using this natural and renewable resource [17]. Unfortunately, there will always be a certain amount of waste produced in the food supply chain. However, current levels of waste occurring in the food supply chain are much greater than other industries and arise from the lack of willingness or inability to coordinate the various activities along the chain [21–23]. Therefore, to make the food supply chain more sustainable and effectively manage food waste, a much deeper understanding of the current state of affairs is needed [24]. This not only means food waste itself, but also means taking into account associated factors like greenhouse gas emissions (GHG) and the use of other resources such as water, land, labour, money, and energy. After taking all these factors into consideration, it is also very important to make the various stages in the food supply chain such as production, distribution, and marketing more efficient and sustainable [25].

Generally speaking, the literature in the field often reports the importance of effective food waste management to reduce problems such as large waste volumes going to landfill, landfill gas emissions, landfill leakage contaminating waterways, and costs associated with transport and handling of wastes. Alternatively, many food wastes can be considered as a valuable source of nutrients with the potential to be processed into products to feed the world's increasing population [14, 26]. Recently, Mirabella et al. reported a range of nutrients available from fruit, vegetable, dairy, and meat and fish wastes

TABLE 2: Food wastes produced in the food supply chain as reported in the literature.

Food supply chain stage	Cause of food waste	Reference
Production and harvest	Crops left in ground; not meeting quality standard	[16, 17, 193]
	Overproduction to maintain supply	[17]
	No demand right at that time of harvest	[194]
	Wrong forecast/withdrawal of demand from retailers	[195]
	Fall of crops and livestock prices	[194]
	Failure to meet quality standards	[4]
	Lack of coordination within the supply chain	[6]
	Pests/diseases attacking/destroying crops	[173]
Storage	Lack of storage facilities	[6, 17]
	Livestock death and unsuitability for slaughter	[87]
	Lack of suitable refrigeration	[194]
	Shortened shelf-life promoting more food waste	[196]
Processing and handling	Trimming (shape, size) for attractive visual appearance	[6, 35]
	Crops nonedible or unsuitable for canning, livestock trimming during slaughtering or fish during canning/smoking, filleting	[6, 17]
	Dairy products during pasteurization and processing to milk based products	[87]
Transport and distribution	Excessive transportation	[197]
	Longer periods of inactivity and complex and expensive movements resulting in product damage	[197–199]
Retail	Products sorting to meet supermarket quality standard	[6]
	Products not donated due to safety standard	[194]
	Expiry of products such as meat and milk before being purchased	[200]
	Maintaining high standard and consumer attraction	[201]
	Packaging size not suitable for buyers	[87]
	Product/packaging damage and being not attractive to consumers	[202]
Consumer	Excessive awareness of "due date," "use by" date, "expiry date"	[194, 203]
	Buying behaviour and purchasing pattern	[15, 200]
	Family size, income, age, job pattern	[19, 204]
	Excessive buying without need	[58, 197, 201, 203]
	Misunderstanding/lack of knowledge about labelling	[173, 197, 205]
	Product purchased but not processed/cooked	[36]
	Surviving more on takeaway food while fridge is still full/no time to cook	[200]
	Cooked product not tasty enough to eat	[206, 207]
	Product expired and produce that is wilted/bruised/moulded and is thrown away	[36, 197]

that could be used in food products (gelling agent in confectionary, fat replacement in meat products, supplementary food products, and seafood flavours for soups) and beverage preservatives [1]. Food wastes have also been considered as a source of renewable energy with the potential to significantly reduce the current dependency on energy derived from fossil fuels [27, 28]. Using food waste as an alternative energy source has the advantage of reducing the amount of waste going to landfill and diminishing the associated problems of gas emissions and groundwater contamination [29, 30]. The use of food waste also alleviates the problem of land competition between food crops and crops for liquid biofuels [31].

The present review provides an overview of current research into terrestrial and aquatic food waste and progress towards utilising the waste. The review examines the various causes that result in food waste and also presents information regarding waste levels throughout the different stages in food supply chains operating in several regions around the world. Also discussed are the socioeconomic aspects of food

waste, the willingness to implement food waste initiatives that promote efficient and sustainable food chain management practices. In addition, probable future trends and initiatives for the implementation of effective ecofriendly and sustainable approaches for managing food wastes are outlined.

2. Terrestrial Food Waste in the Food Supply Chain and Current Waste Utilisation

2.1. Crop Waste in the Food Supply Chain and Current Waste Utilisation. Crop waste begins at the farm and continues throughout the food supply chain. Between farm and fork, food waste is produced by each of the six stages of the food chain as detailed in Table 2. In developed countries, food waste can be quite significant even at the agricultural or harvest stage. Food waste can result from factors such as produce sizing and aesthetic standards, produce quality regulations, production surpluses, and economic factors. For example, in 2009, Italian agricultural produce estimated to

TABLE 3: Amount/percent/value of fruit and vegetable waste in the world food supply chains.

World zone	Loss amount	Stage of waste	Calculation method	Reference
UK	36%	Household		[36]
	47% (veg. only)	Production, postharvest handling, processing		
Switzerland	11% (veg. only)	Retail	Share of losses calculated and estimated in percentage	[8]
	40% (veg. only)	Household		
Germany	43%	Household	Share of total footprint created	[208]
UK	8%	Food processing industries	Percentage	[36, 93]
14 European countries*	5–30%	Food processing industries	Percentage of total share	[94]
Sweden	4.3%	Retail	Percentage share of total delivered products in the retail stores	[209]
China	15%	Storage	Average loss in China calculated from data published by several researchers	[210]
	10%	Distribution		
China	25–35%	Storage	Percentage loss in 2011	[211]
Australia	US$ 810	Consumer waste	Average annual waste value per person	[19]
Africa	53% (incl. root and tuber)	Total supply chain	Percentage of total share	[62]
	10%	Production		
Sub-Saharan African	9%	Postharvest handling and storage	Percentage (by mass)	[6]
	25%	Processing and packaging		
	17%	Distribution		
	5%	Consumption		
South America	6.28%	Wholesale		[212]
Brazil	8.76%	Retail		
North America	48.7% (fresh and processed)	Supply chain	Total weight in lb. (pound) (data collected by USDA in 1995)	[59]
USA	18%	Retail	Estimated total value of food loss in 2008	[13]
	33%	Consumer		
Waterloo, Ontario, Canada	16%	Household	Average of reported food wastage percentages for online survey participants	[213]

*14 European countries: Portugal, Spain, France, Netherlands, Belgium, United Kingdom, Sweden, Finland, Denmark, Germany, Italy, Poland, Hungary, and Greece.

be 17.7 million tonnes was left in the ground and equated to around 3.25% of total produce production [32]. Surprisingly, some studies have indicated that the agrofood sector waste could be as large as 40% of the total production value [33], while studies in the Netherlands have revealed that annual food wastage costs are around € 4.4 billion (US$ 4.9 billion). End-consumers waste around €2.4 billion (US$ 2.7 billion) or about 10% of all food purchased and the remaining €2 billion (US$ 2.21 billion) was wasted through the various stages of the food supply chain [33, 34].

In a Swedish study, 16 different horticultural products including typical fruits and vegetables sold by retailers were responsible for wastes ranging from 0.4% to 6.3% of produce [35]. Similar studies have also found that fruits and vegetables are the main source of household food waste and equate to around one-third of purchased food products [36].

For instance, in the United Kingdom (UK), potatoes came first in a ranking of 100 fruits and vegetables and accounted for around 0.4 million tonnes (10%) of total waste produced annually [37]. Australians, for example, throw away around

AU$ 1.1 billion (US$ 0.84 billion) worth of fruits and vegetables each year making them the largest food waste category [19]. Studies have shown that fruits and vegetables are the most wasted food category among all terrestrial and aquatic food products in both developed and developing countries as seen in Table 3. Moisture content, temperature sensitivity, and delicate surface membranes make fruits and vegetables susceptible to spoilage during production, transportation, and storage. This susceptibility often leads to large amounts of waste throughout the food supply chain. For example, in Switzerland, around 47% of all vegetables produced are wasted in the food supply chain. And in Germany fruits and vegetables account for 43% of all household waste as seen in Table 3.

In many cases, the results of these studies are not comparable, since they did not assess the whole food supply chain (only looked at specific stages and waste types) and were carried out by different researchers worldwide using different assessment protocols. For example, a number studies on fruit and vegetable waste fail to take into account grains

TABLE 4: Analysis of retail and consumer waste increase/decrease in the USA based on USDA data from [17, 58, 60].

Commodity		Supply/population (S/P)*			Supply/waste (S/W)			Production% increase/decrease		
		1995	2008	2010	1995	2008	2010	1995	2008	2010
Grains	R	17.13	19.55	19.50	0.02	0.12	0.12	2	+10	+10
	C				0.30	0.18	0.19	30	−12	−11
Fruits	R	18.15	22.01	20.76	0.02	0.09	0.09	2	+7	+7
	C				0.22	0.14	0.19	23	−9	−4
Vegetables	R	23.69	36.96	27.09	0.02	0.06	0.08	2	+4	+6
	C				0.24	0.15	0.22	24	−9	−2
Dairy products	R	28.64	27.48	26.80	0.02	0.11	0.11	2	+9	+9
	C				0.30	0.17	0.19	30	−13	−10
Meat/poultry	R	17.82	27.13	17.31	0.01	0.03	0.04	1	+3	+3
	C				0.15	0.23	0.21	15	+8	+6
Fish	R	1.50	1.59	1.55	0.01	0.08	0.08	1	+7	+7
	C				0.15	0.25	0.31	15	+10	+16
Eggs	R	2.97	2.89	3.16	0.02	0.1	0.07	2	+8	+5
	C				0.29	0.15	0.21	29	−14	−8
Nut products	R	0.71	1.04	1.13	0.01	0.06	0.06	1	+5	+5
	C				0.15	0.09	0.09	15	−6	−6

R: retail waste; C: consumer waste.
* Population in 1995 = 266.3 million; in 2008 = 304.06 million; in 2010 = 309.75 million (source: ERS).

and root/tuber wastes. And others have taken into account wastes generated from grains and root/tubers in an attempt to minimise and simplify data collection. Many consumer and retail waste assessments contain very little information about farm practices, processing waste, and wastes resulting from storage and transportation. In spite of their importance, consumers and retailers cannot be considered as the only contributors to waste in the food supply chain. Nevertheless, it is extremely difficult to obtain detailed information from all stakeholders involved in the food supply chain because of business confidentiality considerations. Another limitation arises from the types of measurement procedures used to record and analyse food waste data around the world. In addition, making comparisons is difficult because waste levels can be presented in terms of percentage waste, local currency, and even weight loss. Furthermore, variations can occur between different regions within a country where economic, social, and behavioural reasons may promote specific types of food wastage.

Determining waste levels in a food supply chain often reveals that they are high and costly. For example, in 2008, the United States of America (USA) produced three large waste streams consisting of grain (US$ 34,791 million), vegetables (US$ 103,417 million), and fruits (US$ 62,146 million) at considerable economic cost [13]. Furthermore, each year, the USA produces more than 2.7 million tonnes of fruit and vegetables that are not harvested or remain unsold due to poor crop aesthetics and low market prices [37]. Moreover, most studies only measure or estimate a particular food waste and fail to address any trends in the levels of wastage. To remedy this situation, the United states Department of Agriculture (USDA) carried out a detailed analysis to understand the variation in food waste levels between 1995 and 2010. This data is presented in Table 4 and shows a downward drift in consumer waste compared to retailers waste, although consumers are often blamed for high waste levels. Also, over this period, retail waste has increased for most commodities especially grain products, while consumer waste levels have significantly decreased for some food products such as grain products and fruits. In particular, vegetable waste produced by consumers in 2008 and 2010 was significantly lower than waste levels recorded in 1995.

For developing countries, around 15 to 50% of all fruit and vegetable waste occurs in the postharvest stage [38–40]. For example, in Africa, cassava wastes can be as large as 45% [41] and yam waste levels can reach 50% [42]. In the Philippines, fruit wastes from crops such as papaya can range between 30 and 60% of the total crop [12]. Similarly, around 18 to 40% of all fresh fruits and vegetables go to waste in India every year due to the lack of refrigerated transport and high quality cold storage facilities. This equates to an annual cost to food manufacturers and sellers of around US$ 71,481 million [43]. Unfortunately, much of this data comes from studies carried out almost 40 years ago and because no recent studies have been carried out there is no current assessment of crop waste levels. Thus, there is a critical need for follow-up studies that take into account factors such as technological innovation, population growth, and consumer and marketing trends. It is critical for researchers to document current food waste levels so that stakeholders such as growers, processors, transporters, retailers, and consumers can take steps to address this growing global problem.

The second major food group after fruit and vegetables is grain. Among the grains, rice is recognised as the world's

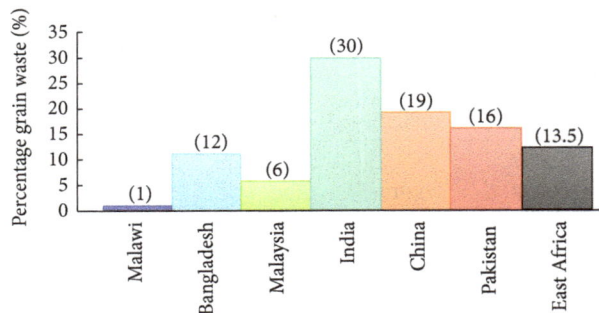

FIGURE 1: Percentage grain waste in selected developing and less developed countries.

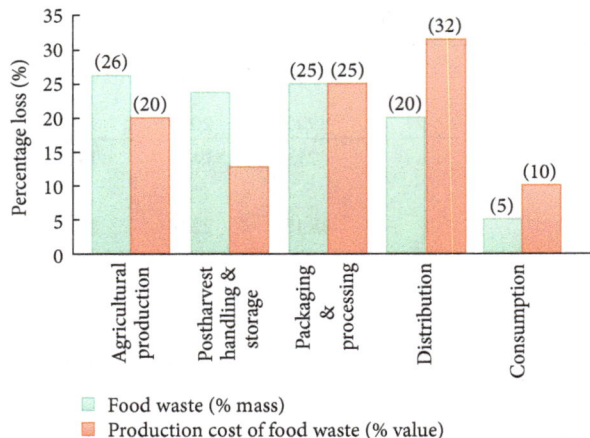

FIGURE 2: Comparison between food waste quantities (%, by mass) and the cost of food waste (%, by value) in each stage of the food supply chain in South Africa [62].

second staple food and on average has a waste level equivalent to 15% of total global production [44, 45]. However, this waste level is not the same for all countries due to variations in climatic zones and various production practices in each respective country. In particular, storage practices in both developing and less developed countries have been extensively studied and reveal significant differences between countries. For example, grain storage waste levels range from less than 1% in Malawi [46, 47] to 12 to 13% in Bangladesh and 3 to 6% in Malaysia [48] as seen in Figure 1. Studies have also shown that grain waste in the Chinese supply chain is 19.0% ± 5.8%, with consumers accounting for the single largest portion of the waste (7.3% ± 4.8%) [49]. And postharvest and preprocessing cereal waste in sub-Saharan Africa was estimated to be around US$ 4 billion. This extremely large cost equates to 13.5% of the total cereal production produced by countries in this region [50].

Inadequate storage capacity, poorly distributed warehouses, lack of adequately designed storage facilities, and inefficient transport and handling management lead to waste levels of around 20 to 30% of India's total grain production [51]. This level of grain waste is estimated to cost around US$ 14 billion each year and is the highest in the region as seen in Figure 1. Amazingly, this level of waste has the potential to provide the minimum annual food requirements of at least 48 million people in India [52]. In Pakistan, grain waste accounts for around 16% of total production, or 3.2 million tonnes annually, and results from inadequate storage infrastructure that permits widespread rodent infestation [53]. Current data indicates that global postharvest crop wastes have direct consequences in terms of food security, malnutrition, and poverty. Except for Malawi, an African country reporting low grain waste levels, other eastern and southern regional African countries have reported waste levels equivalent to around US$ 11 billion or 13.5% of total grain production. Unfortunately, there is very little information available reporting grain wastes in central or West African regions [54]. Most grain waste reports do record total percentage waste for each country but do not give individual crop wastes such as maize, rice, wheat, and barley. Because these reports do not provide individual information on specific grain crops, it is difficult to determine which are more prone to waste. In spite of this, it is evident that policy,

political management, natural calamities, storage, infrastructure facilities, and transportation are the main drivers for producing grain waste in developing countries [55].

Food waste not only costs money, but also consumes other resources such as land, water, energy, and labour. When it comes to water usage, South Africa's (SA) food waste costs become significant since it is one of the world's driest countries. For example, approximately 30% of SA's crop production depends directly on irrigated water, while fruit and vegetable production consumes around 90% of all irrigated water used [56]. The total cost of food waste in SA each year is estimated to be around US$ 5.27 billion and equates to 2.1% of the country's gross domestic product (GDP). Furthermore, agricultural production is more prone to waste than processing, packaging, and consumers. Figure 2 presents a comparison between food waste quantities and the food waste costs for each stage in the SA food supply chain. Interestingly, packaging and processing have similar waste levels and production costs. This suggests that both stages are not cost-effective and rather prone to wastage. While consumer waste levels are relatively low, distribution and infrastructure waste levels are relatively high. The results of this study clearly indicate the severity of waste levels within the SA food supply chain [57].

Grain waste studies have mainly focused on developing countries, with very few studies reporting grain waste in developed countries. A small number of studies conducted in the USA have only investigated grain waste in the retail and consumer stages of the food supply chain as presented in Table 5 [13, 58–60]. It should be noted that in most developed countries grains are considered as livestock feed rather than human food [61]. Thus, there is a crucial need to undertake grain waste studies in developed countries and determine the alternative utilisation of grain and grain wastes as livestock feed. Thus, the lack of reliable food waste data from around the world and the increasing importance of food security mean that significant efforts are needed to fill in the knowledge gaps.

TABLE 5: Amount/percent/value of grain waste in selected world food supply chains.

World zone	Loss amount	Stage of waste	Calculation method	Reference
China	4–6%	Postharvest handling	Average loss in China calculated from data published by several researchers	[211]
	5.7–8.6%	Storage		
	2.2–3.3%	Processing		
	1–1.5%	Distribution		
China	7–10%	Storage	Percentage loss in 2011	[211]
Australia	US$ 435 (grain products)	Consumer	Average annual waste value per person	[19]
Switzerland	62% (grain products)	Production, postharvest handling, processing	Share of losses calculated and estimated in percentage	[8]
	4% (grain products)	Retail		
	32% (grain products)	Household		
Africa	26%	Total supply chain	Percentage of total share	[62]
Sub-Saharan Africa	6%	Production	Percentage (by mass)	[6]
	8%	Postharvest handling and storage		
	3.5%	Processing and packaging		
	2%	Distribution		
	1%	Consumption		
North America	32%	Supply chain	Total weight in lb. (pound) (data collected by USDA in 1995)	[59]
USA	12%	Retail	Estimated total value of food loss in 2008	[13]
	18%	Consumer		

Some studies have just recorded food waste levels, while others have also highlighted methods for managing wastes. There are numerous reports in the literature discussing various recycling and utilisation methods available for processing fruit, vegetable, and grain wastes. The aim of food waste utilisation is to extract the maximum practical benefits and reduce the amount of waste going to landfill [63]. Although there has been extensive information discussing various waste recycling strategies for dealing with agricultural waste, there is very little information assessing the economic benefits of the various waste utilisation methods. At present, there are very few reports available discussing the utilisation of agrowastes on a commercial scale and methods to overcome barriers that currently prevent effective food waste management strategies. All food wastes are a rich source of natural biomolecules and compounds. Fruit and vegetable wastes including peels, stones, and fibres contain a wide range of natural compounds, while grain wastes derived from straw, bagasse, cobs, cotton husk, groundnut husks, and fibrous remnants of forage grasses are mainly composed of useful materials such as cellulose, hemicelluloses, and lignin [64]. Arguably, grain wastes are the most abundant agricultural wastes and the most underutilised [65]. On the whole crop wastes are a valuable source of useful compounds, chemicals, and pharmaceuticals [66]. And currently there is a high demand for pharmaceutical ingredients such as enzymes, solvents, and surfactants all of which can be derived

from crop wastes [67]. Because of the rich source of natural compounds found in crop wastes, the European Union, USA, Canada, Japan, and Malaysia are ambitiously developing and promoting an ecofriendly biologically based market. For example, in 2010, the USA placed a replacement target of 12% on all of its chemical feedstock and by 2030 it is expected that bio-based products will have a market share of around 25% [68]. At present, only a small number of bio-based compounds derived from crop sources have made it into commercial products. Typical examples include (1) succinic acid from crops like sugarcane, maize, rice, barley, and potato [68]; (2) starch based plastic production from cassava, maize, and wheat [69]; (3) surfactants from tropical oil producing grains [70]; (4) fatty acids from coconut and oil palm [71]; (5) polymers, lubricants, adhesives, solvents, and surfactants from rapeseed and sunflower [72]; and (6) lactic acid from carbohydrate containing crops such as cereals, potato, and sugar beet [67]. However, to date, very few products containing compounds and chemicals derived from crop wastes have made it into the commercial marketplace. Estimates of market size, market price, potential bio-based share, potential bio-based production size, potential impact for local producers, potential local employment, and prospects for development are very low and rather poor [73]. Therefore, before large-scale development of bio-based renewable products can take place, more detailed feasibility studies and practical business models are needed. Thus, long-term collaborations between producers, manufacturers, and business are needed to undertake further translational research to bring these new and novel products to the marketplace.

Present research has also shown that grain wastes can be used as a source of bioenergy in the forms of bioethanol, biodiesel, and biogas [74]. For example, bioethanol is currently produced from corn in the USA, European Union, and China [75]. In tropical regions such as in Brazil and Columbia, bioethanol is mainly produced from sugarcane [76]. Unfortunately, because of the constraints imposed by available arable land, there is competition between crops specifically grown for biofuel and those grown for food and feed production [77]. Because of this competition, it is not feasible to increase biofuel production using currently available land and technologies. Consequently, current research has focused on developing more advanced or 2nd-generation biofuel production technologies that use wastes derived from grains, fruits, and vegetables. In the last decade, significant progress has been made in developing chemical processes that can convert agrowastes into ethanol. However, major barriers such as the high cost of pretreatments and inefficient conversion processes have prevented the commercialization of large-scale bioethanol processing facilities [78–80]. Further economic analysis has also identified costs barriers such as feedstock chemicals and capital investment that includes pretreatment facilities, fermenters, and steam generation systems as the main factors restricting large-scale processing facilities [81]. Therefore, to overcome many of these barriers, further research is needed to improve efficiencies in current plant and equipment and to explore and develop new agrowaste conversion technologies.

Research into the generation of biogas from fruit and vegetable wastes has also been carried out. But large-scale commercially viable biogas production is still in its infancy. Currently, municipal wastes are recycled through anaerobic digesters to produce biogas, but agrowastes are yet to be converted using this type of processing facility. The main reasons for this are (1) providing a continuous supply of agrowastes to the facility and (2) developing cost-effective transportation between waste sources and facilities. Thus, without a continuous supply of feedstock, the facility is unable to efficiently deliver a steady flow of biogas. Therefore, the continuous supply of agrowaste essentially becomes a Vehicle Routing Problem (VRP) [82]. VRP is one of the most comprehensively studied problems in transportation literature. However, VRP has not been specifically applied to transporting food wastes produced by a food supply chain. Instead, some studies have considered transporting large amounts of food wastes directly between supply points and processing facilities [83–86]. In the case of crop wastes, they are produced at farms, processing facilities, wholesalers, and retailers and are typically spread over fairly large regional areas including both urban and rural ones. Therefore, there is a need to collect wastes from various dispersed locations and transport them to processing facilities. Thus, collecting and transporting food wastes are fundamentally different from harvesting and shipping agriculture crops to market. The difference arises from waste delivery trucks not receiving a

full load at any one of the locations. For example, a business may only produce a small amount of food waste that does not make a full load. This necessitates the truck to make multiple pickups from other locations before making its delivery to the processing facility. This type of truck routing is a major cost to food waste collection that has not been fully investigated and could limit large-scale crop waste utilisation. Importantly, while the impact of large-scale biogas operations using first- or second-generation biofuels is being debated, there is also considerable interest in developing small-scale biomass processing to produce biofuels. The advantage of small-scale biofuel production plant is that it enables local communities to access a renewable energy source. Small-scale biofuel plants can utilise locally produced food waste and reduce the dependence on fossil fuels and wood resources.

2.2. Livestock, Poultry Meat, and Egg Waste in the Food Supply Chain and Level of Utilisation. Livestock and poultry waste occurs in the early stages of production with animal deaths and animals unsuitable for slaughtering [87]. In the meat industry, the majority of the waste is produced during slaughtering and consists of various nonedible parts that are categorised as byproducts [87, 88]. Meat byproducts consist of bones, tendons, skin, and contents of the gastrointestinal tract, blood, and internal organs. However, these waste parts can vary between each type of animal [89]. Generally, meat products have a relatively short shelf-life ranging between 7 and 26 days [18]. It is for this reason that meat products immediately go to waste if not sold within the labelled expiry date and this is the main reason for wastage at the retail stage. Other reasons for meat product waste include packaging size and date confusion among consumers as detailed in Table 6. Buzby and Hyman in 2012 [13] estimated the total value of meat product waste in the USA at US\$ 83,127 million. Their study found that consumers and retailers were responsible for around 35% and 5%, respectively, of the total waste produced, while the total value of poultry waste was estimated at US\$ 69,100 million, with consumers being responsible for 37% of the total waste [18]. Studies have also revealed a positive growth trend in meat and poultry waste as presented in Table 4. Overall, there has been an increasing trend in meat consumption around the world with the USA recording the largest increase [90]. The increased consumption is around three times as large as the global average; however, at the same time, trends in retail and consumer waste levels are not clearly understood as seen in Table 4 [91]. Furthermore, a study carried out in Canada analysing food waste data between 1961 and 2009 found that red meat accounted for 39.73% of the total waste and poultry waste was estimated to be around 40.74% [92]. However, this analysis did not include bone waste in the slaughterhouse since no data exists. Moreover, it is crucial to note that these percentage wastes only reflect wastage at the consumption stage in the food supply chain and do not take into account farming, processing, and distribution waste data. Similarly, Australian consumers waste around AUS\$ 872.5 million (US\$ 637.5 million) worth of meat and fish every year [19]. Unfortunately, meat and poultry waste has not been studied to the same extent as fruit and vegetable wastes. However, a limited

TABLE 6: Amount/percent/value of meat and poultry waste in the world food supply chains.

World zone	Loss amount	Stage of waste	Calculation method	Reference
UK	7% (meat & fish)	Household		[36]
UK	56% (meat & fish)	Processing industries	Percentage	[36, 93]
14 European countries*	35–42%	Processing industries	Percentage of total share	[94]
	1.4–2.1%	Postharvest handling		
China	2.5–3.7%	Storage	Average loss in China calculated from data published by several researchers	[210]
	1.1%	Processing		
	3%	Distribution		
Australia	US$ 626 (meat & fish)	Consumer	Average annual waste value per person	[19]
Africa	7%	Total supply chain	Percentage of total share	[62]
	15%	Production		
Sub-Saharan Africa	0.7%	Postharvest handling and storage	Percentage (by mass)	[6]
	5%	Processing and packaging		
	7%	Distribution		
	2%	Consumption		
North America	16% (including fish)	Supply chain	Total weight in lb. (pound) (data collected by USDA in 1995)	[59]
USA	5%	Retail	Estimated total value of food loss in 2008	[13]
	35%	Consumer		
Waterloo, Ontario, Canada	6% (including seafood and eggs)	Household	Average of reported food wastage percentages for online survey participants	[213]

*14 European countries: Portugal, Spain, France, Netherlands, Belgium, United Kingdom, Sweden, Finland, Denmark, Germany, Italy, Poland, Hungary, and Greece.

number of studies presented in Table 6 do indicate that most of the waste is produced in the processing of livestock and poultry and the trend is steadily increasing as indicated in Table 4. Inspection of Table 6 reveals that meat processing in the UK accounts for 56% of all wastes [36, 93] and in the grouping of 14 European countries processing wastes vary between 35 and 42% [94]. Unfortunately, there is no data available recording the amount of waste generated during the preslaughtering stage of meat and poultry production.

Like fruit and vegetable wastes, meat and poultry byproducts are also rich in nutritional, medicinal, and pharmaceutical materials [95]. The broad diversity of products has the potential to be used in human food products, animal feeds, fertilizers, and biofuels [96]. Currently, both academic and industry researchers are investigating various methods of adding value to meat and poultry products and make better use of their byproducts. Current research, using the most up-to-date analysis techniques, has been aimed at determining nutritional properties, bioactive molecules, and other useful chemical compounds commonly found in these byproducts. Many of these bioactive molecules and chemical compounds have the potential to be used in fields such as cosmetics and pharmaceuticals [97]. In many countries, slaughterhouse

wastes have already been used to produce cattle and poultry feed, since these wastes are an excellent source of many different types of proteins [95]. Two animal byproducts that have been used without further processing by the fast food industry are tallow and lard. Unfortunately, consumer anxiety in recent years has restricted the use of these byproducts in the fast food industry [98]. In many cases, meat, poultry, and dairy processing wastes have the potential to be recycled and processed into higher value and useful products. But inappropriate use of recycled meat byproducts can create major aesthetic and even health problems. Therefore, most countries have regulatory requirements that limit the use of meat and poultry wastes in the interests of food safety and quality. Also, economic factors have limited the viable use of meat and poultry wastes. For example, at one time, Japanese meat and poultry wastes were extensively used in animal feed until a relatively low priced imported feed concentrate entered the marketplace. And as a result waste usage declined and out of the 20 million tonnes of wastes being produced each year only 3% was used as fertilizer and 5% as animal feed [99]. The remaining large amounts of waste were incinerated or ended up in landfills. In an attempt to reduce the number of enormous landfill sites,

TABLE 7: Amount/percent/value of eggs, milk, and dairy waste in the world food supply chains.

World zone	Loss amount	Stage of waste	Calculation method	Reference
Eggs				
UK	7% (incl. dairy)	Household	-	[36]
Switzerland	18%	Production, postharvest handling, processing	Share of losses calculated and estimated in percentage	[8]
	9%	Retail		
	64%	Household		
North America	31.4%	Supply chain	Total weight in lb. (pound) (data collected by USDA in 1995)	[59]
USA	9%	Retail	Estimated total value of food loss in 2008	[13]
	14%	Consumer		
Milk and dairy				
UK	12%	Food processing industries	Percentage	[36]
14 European countries*	43%–48%	Food processing industries	Percentage of total share	[94]
Australia	US$ 405	Consumer	Average annual waste value per person	[19]
Africa	8%	Total supply chain	Percentage of total share	[62]
Sub-Saharan Africa	6	Production	Percentage (by mass)	[6]
	11	Postharvest handling and storage		
	0.1	Processing and packaging		
	10	Distribution		
	0.1	Consumption		
North America	32.0%	Supply chain	Total weight in lb. (pound) (data collected by USDA in 1995)	[59]
USA	9%	Retail	Estimated total value of food loss in 2008	[13]

*14 European countries: Portugal, Spain, France, Netherlands, Belgium, United Kingdom, Sweden, Finland, Denmark, Germany, Italy, Poland, Hungary, and Greece.

reduce the environmental burden, and prevent gas emissions, the Japanese government introduced a new food-recycling law in May 2001. Unfortunately, just after its introduction, an outbreak of Bovine Spongiform Encephalopathy was reported and created a very negative public response to food recycling. Consequently, public concerns and safety issues have prevented food recycling for human and ruminant consumption [99]. The only other way of processing meat and poultry waste in Japan is via compost production, but to date there has been limited acceptance of this product by farmers.

Eggs are an important food and are extensively used in cooking for the production of a diverse range of food products. In Korea, the annual consumption of eggs was estimated to be around 540,542 tonnes and is expected to increase every year [100]. Because of the extremely large amounts of eggs used worldwide, there are also large amounts of egg wastes produced. For instance, in Switzerland, 18% of all egg wastes occur during production, around 9% occurs in the retail sector, and a massive 64% is produced by consumers as seen in Table 7 [8]. In the North American supply chain, 31.4% of all eggs produced end up as waste [59]. And in the USA around 9% of all egg wastes are produced in the retail sector and consumers produce 14% as seen in Table 7 [13]. Importantly, waste products from both poultry processing and egg production industries must be efficiently dealt with, since growth in both industries largely depends on effective waste management [88]. In the case of egg production, eggs are vulnerable to bacterial attack if the outer shells are not properly and quickly cleaned to remove faecal particles which

contain various microorganisms [101, 102]. In addition, egg waste can also occur during transportation, distribution, and storage if appropriate supportive environment is not supplied. Furthermore, because of the extremely large numbers of eggs used worldwide, approximately 50,000 tonnes of eggshells is produced each year [102]. These eggshells contain high levels of calcium carbonate ($CaCO_3$) that could be used as an alkaline compound to immobilise heavy metals. Therefore, recycling eggshells for the immobilization of heavy metals in wastewater has the potential to significantly reduce environmental pollution [103]. Accordingly, there have been a number studies evaluating eggshells as immobilising agents for heavy metals such as chromium(III) and lead [104–106]. However, to date, the practical use of waste eggshells as immobilising agents is still largely unknown [95].

2.3. Dairy Waste in the Food Supply Chain. The dairy industry, because of its worldwide importance, has been extensively studied to determine its environmental impact. The most important product produced by the dairy industry is raw milk. Raw milk is processed into products such as consumer milk, butter, cheese, yogurt, condensed milk, dried milk (milk powder), and ice cream [107]. In spite of being extensively studied, what is lacking is a comprehensive understanding of waste levels produced throughout the whole dairy industry. The agricultural stage is often reported as the main source of wastes in the life cycle of milk and dairy products [108–110]. However, studies in the UK and Spain have identified the main causes of milk waste coming from poor product quality during the summer period, poor forecasting, packaging mistakes, and breakages occurring at the retail stage [18]. The study also found that poor sales forecasting, slow sales, and cold storage problems during transportation also contributed to the wastage of many dairy products. In addition, cleaning and packaging processes associated with dairy products were also found to significantly contribute to waste levels [111]. Furthermore, dairy product packaging has also been found to significantly contribute to environmental degradation [112]. Generally, waste levels in the dairy industry are quite high. For example, the Mexican milk industry generates between 3.74 and 11.22 million m^3 of waste products each year, which equates to around one to three times the volume of milk produced annually [113]. And, in the case of Denmark, milk and dairy products contribute around 71,000 tonnes of food waste annually [87], while in the grouping of 14 European countries around 43 to 48% of milk and dairy wastes were produced in the processing stage [94]. In North America supply chain wastage was found to be around 32% [59] and USA retailers were found to waste around 9% of all dairy products as seen in Table 7 [13].

Other sources of milk and dairy produce waste result from frequent product changes, but this can be reduced through appropriate product sequencing and more efficient product scheduling [114]. Other methods of reducing milk wastes include capture of fat, protein, and sugars from wastewater produced during milk processing using processes such as evaporation, centrifugation, ultrafiltration, reverse osmosis, and bioconversion. These recapture processes can significantly reduce the amount of milk and dairy wastes being discharged into the environment [5]. Waste reduction can also have a significant impact on product processing efficiency and improved financial returns. For instance, cheese is derived from milk and is widely used as a standalone product and component of many food products around the world. During cheese manufacture, acidified milk is mixed with an enzyme to form solid cheese or casein and the remaining liquid is called whey [115]. The waste whey can have a negative impact if dumped directly into the environment [5]. Today, around 50% of the world's whey production is treated and transformed into various food products. This cost-effective solution adds value to the whey and reduces wastes [116]. Currently, there is a large body of research in the literature that stresses the importance of reducing milk processing waste and wastewater discharge into the environment. However, the amount of milk and dairy wastes being generated throughout the global food supply chain is still largely unknown. Therefore, there is a current need to undertake studies that can identify and document the magnitude of milk and dairy waste occurring throughout the global food supply chain so that proper waste remediation and management steps can be implemented.

3. Aquatic Food Waste in the Food Supply Chain and Level of Waste Utilisation

3.1. Fish Waste in the Food Supply Chain. Historically, fish has always been an import food source and even today it is one of the most traded commodities in international

markets. It was estimated in 2010 that globally around 54.8 million people were engaged in aquaculture and the wider fishing industry [117, 118]. Currently, fish contributes around 16.6% to the total animal protein supply and 6.5% of all proteins consumed by humans worldwide [118]. Fish is highly regarded for its carbohydrates, cholesterol, and low saturated fats. Fish also provides high-value protein and a wide range of essential micronutrients such as vitamins, minerals, and polyunsaturated omega-3 fatty acids. Because of the nutritional and health benefits of fish and other seafoods, the demand is always high and annual consumption is increasing globally. For example, the demand for seafood in Australia has steadily increased over the last three decades [119]. In 2009, Australians on average consumed 25 kg of seafood, compared with 18.8 kg in 1995, 17.3 kg in 1985, and 13.6 kg in 1975. The data indicates that overall seafood consumption in Australia has almost tripled over three decades [118]. In parallel with the increasing global demand for seafood, there are growing concerns about the sustainability and management of the fishing industry. Recent studies have only discussed wastage in general terms and suggest that waste could be as large as US\$ 50 billion each year due to poor management of seafood resources [120, 121]. A recent article by Costello et al. illustrated how fish waste could be reduced in a sustainable way if appropriate management changes were undertaken. The study highlighted that the less studied fisheries have not been closely monitored or assessed, so there is no data recording the amount of waste being produced [122].

Different types and quantities of fish waste are produced throughout the food supply chain, commencing with capture and ending with consumption [123]. Worldwide, around 130 million tonnes of fish waste is produced each year by fisheries and aquaculture. Wastes are produced through by-catch, on-board processing, transport, storage, retailers, and consumers [124]. Fish waste generation begins during wild catching, with by-catch or unintentional catching of marine species being discarded. This problem has been extensively studied and in spite of environmental and business guidelines there is still no effective solution to by-catch waste [125, 126]. It is estimated that globally around 17.9 to 39.5 million tonnes of whole fish is discarded each year by commercial fishing operations [123]. Following capture, processing is the main stage in the food supply chain where most waste occurs. During processing, only the fillets are preserved and the remainder of the fish (up to 66%) is thrown away as seen in Figure 3 [127, 128]. A study by Gavine et al. found that the southeastern Australian seafood industry produced fish waste estimated to be around 20,000 tonnes per year and cost around US\$ 150 per tonne to dispose of in landfill sites [129]. In reality, not only does the waste disposal have a significant cost, but also it has a major environmental impact [130].

Interestingly, in the UK, each tonne of cod purchased by a processor costs about £2,000 (US\$ 3,129) and around 50% of the cod ends up as processing waste. Regrettably, the waste only generates an income of £40 (US\$ 63) as a byproduct and in the worst-case scenario its disposal costs around £60 (US\$ 94). Similarly, only around 43% of shellfish and other fish species are suitable for human consumption

FIGURE 3: During processing, the fillets are considered usable and the remainder is waste.

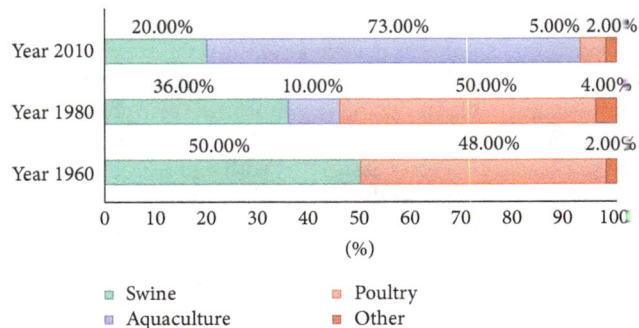

FIGURE 5: Global usage of fishmeal (adapted from World Bank data) [120].

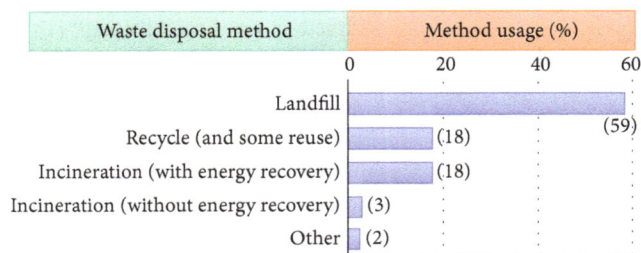

FIGURE 4: Scottish salmon farming waste disposal routes [131].

and the remaining are classified as waste [123]. It has been estimated that in the UK both meat and fish processing were responsible for about 56% of all wastes produced in the food supply chain as shown in Table 6 [36, 93]. A study of a grouping of 14 European countries revealed that fish processing waste could range between 40 and 70% [94]. Thus, it is apparent from these studies that the processing stage is the main contributor to overall waste levels. Research into retail and consumer waste has shown that consumers in the USA are the major contributor to fish waste. Interestingly, the trend in waste by consumers has steadily increased from 16% in 1995 to 31% in 2010 despite having efficient transport and storage facilities [13, 17, 58]. The reasons for the high levels of waste by USA consumers are unknown and need further investigation.

The disposal route for seafood waste is not as straightforward as grains and other crop products. This is because the disposal of seafood wastes involves stricter hygiene, safety, and management of environmental hazards during its disposal and in many cases its disposal is regulated by government organisations. For example, in the UK, landfill costs are much higher for seafood waste disposal because the waste is not categorised as "inactive/inert" waste. Furthermore, regulations regarding the burial or burning of seafood waste are restrictive if there are any alternative utilisation pathways available [123]. However, some fish farming businesses are paying higher landfill costs to dispose of fish wastes compared to other methods of disposal. Currently, around 59% of all fish wastes go to landfill and only around 39% are reused or incinerated as seen in Figure 4 [131].

To reduce the large amount of fish waste produced worldwide, a number of alternative strategies have been developed to add economic value to the wastes. For instance, two different methods, mass transformation and sorting, have been developed to improve the economic value of fish wastes [124]. Mass transformation involves the conversion of fish waste into a single product. Typical examples of transformed fish waste include fishmeal, fish oil, fertilisers, and hydrolysates such as protein hydrolysate. Alternatively, sorting involves utilising various fish body parts such as bones, guts, and fins separately to enhance their economic value. For example, sorting enables the production of specialised products such as liver oil, gelatine, omega-3, protein containing sports food and drinks, calcium, cosmetics, and pharmaceuticals [132]. Wider acceptance and adoption of both methods could lead to significant reductions in wastes going to landfill and reduce the damaging impact of fish wastes on the environment. For example, converting fish wastes into fishmeal has been steadily increasing in recent years with many countries converting their fish wastes using cost-effective reprocessing technologies [118, 133]. In spite of the reprocessing costs associated with converting fish waste into fishmeal, fishmeal's value as a feedstock for aquaculture has offset the reprocessing costs. For example, in Japan, 90% of the ingredients used in fishmeal are derived from fish wastes [134]. Currently, there are only around ten major countries converting fish waste into fishmeal products, that is, Canada, Chile, Denmark, Iceland, Japan, Mexico, Norway, Russian Federation, Thailand, and the USA. However, these countries on average are only using around 25% of their fish wastes to produce fishmeal products [133]. Importantly, fishmeal contains essential amino acids and as a result it is currently the most widely used protein ingredient in aquaculture feeds [135]. Thus, fishmeal usage over a 50-year period (1960 to 2010) reveals its increased use in aquaculture, while its use in both swine and poultry feeds has declined as seen in Figure 5 [136]. One of the contributing factors for this trend was the ban imposed by the European Economic Commission on the use of animal byproducts being used in animal feeds. And similar regulations in the USA have also contributed to the increased usage of fishmeal in aquaculture [137].

In spite of fishmeal being used globally, there has only been limited use of other fish waste byproducts. Fish wastes can also be processed to produce oil, silage, fertiliser, composting matter, and fish protein concentrates [138]. Furthermore, fish wastes are also a rich source of chitin, chitosan, carotenoid pigments, and enzymes that can be used in cosmetics and pharmaceuticals [139]. But, to date, very little has been done to fully develop and commercialise these types of products [123]. However, it should be noted that fish waste processing can be a difficult business in many countries due to problematic issues such as hygiene, safety, and environmental hazards. In addition, the most important factor that any business needs to consider is the economic viability of fish waste processing [140, 141]. For example, large volumes of both solid and liquid wastes are produced after processing Nile perch from Lake Victoria in East Africa. Annually, around 36,000 tonnes of solid waste and approximately 1,838,000 m^3 of produced wastewater containing valuable nutrients are discharged [142]. An investigation of the wastewater revealed that it contained 6,160 mg/L of lipids and 2,000 mg/L of protein [143]. This rich source of lipids and proteins has the potential to produce value-added products through bioconversion. However, current fish waste management in East Africa was found to be inefficient and nonprofitable and was unable to take advantage of the rich source of lipids and proteins present in the wastewater, thus highlighting the need for efficient waste utilisation and waste reduction strategies that can provide viable and profitable options for fish waste processing [142].

A number of aquaculture based industrial studies have examined various types of methods for dealing with seafood waste and its utilisation in Australia [136, 144, 145]. For example, fish wastes are a rich source of essential fatty acids and fish skin-and-bone parts are suitable mineral supplements in fish diets [146]. However, further studies are needed to fully investigate large-scale profitable fish waste processing. On the whole, fish waste processing and utilisation have steadily increased over the years, but several issues restrict its full-scale operation. In particular, environmental issues are major factors preventing large-scale development since fish processing plants can be significant polluters. Obviously, there are good economic and environmental reasons to process fish waste and produce value-added products. But further work is needed to develop effective and efficient methods of processing fish wastes at an economically viable industrial scale with as little environmental impact as possible.

3.2. Aquatic Plant Based Wastes in the Food Supply Chain.
It is interesting to note that the literature in the field often overlooks aquatic plant food wastes. Aquatic plant foods such as algae have been used for both human and animal nutrition for thousands of years. The earliest writings of the ancient Greeks recorded in the *Bellum africanum*, written around 45 B.C., describe the Greeks collecting seaweed from local shorelines and feeding it to their cattle [147]. Many aquatic plants are very rich in protein and are a highly nutritional food that can offer many beneficial advantages as a food supplement as well as having significant medicinal properties

[148–150]. In the search for sources of natural antioxidants, algae and microalgae have been suggested as potentially rich sources. Both algae and microalgae are widely known and consumed in many countries for their advantageous health benefits. In particular, many algae and microalgae are rich sources of polyunsaturated fatty acids that have the potential to reduce the incidence of cardiovascular diseases [151, 152]. In Asian countries like China, Japan, and Korea, the production and consumption of edible aquatic plants had a long tradition. This long-standing tradition has resulted in the widespread incorporation of aquatic plants into the global food supply [153–155]. Rather than just relying on marine capture, currently over 95.5% of the total global production of aquatic plants is supplied by aquaculture [156]. This equates to around 0.44 million tonnes of marine capture and about 12 million tonnes being produced by aquaculture in 2010 as seen in Figure 6.

Studies have shown that the majority of aquaculture production, around 9 million tonnes, was destined for human consumption. Phycocolloids were extracted from the remaining aquatic plants to be used as nutritional supplements in various forms of farm animal and aquaculture feedstock [156, 157]. To date, there has been very little data reported in the literature and wastes levels produced by aquatic plant food industries remain relatively unknown. Likewise, the management of wastes produced during processing remains largely unknown. Therefore, there is a current need to undertake research into aquatic plant food supply chain to determine the current amount of waste, level of utilisation, and management protocols in use.

However, in recent years, research has focused on using microalgae in the production of biodiesel. Microalgae have two major advantages over land based crops. The first is the high growth rate and the second is the high oil content. For example, microalga typically doubles its biomass every 24 hours under normal growing conditions, while the oil content of microalgae can range from 15 to 75% (dry weight) and annually can produce oil from 58,700 up to around 136,900 litres per hectare [158, 159]. Currently, biodiesel production depends on crops such as soybean, rapeseed, canola, sunflower, corn, palm kernels, animal fat, and oils [160]. The biggest hurdle preventing the full-scale production of biodiesel from these crops is land availability [161]. Since the land area needed by microalgae is small compared to oil producing crops, there has been considerable interest in exploring the use of microalgae as an alternative feedstock for the production of biodiesel. The disadvantage of using microalgae for producing biodiesel is the high cost of production and separation that is needed to remove microalgal biomass from the growing media [159, 162]. Another challenge associated with microalgae production in open ponds is contamination from a wide range of naturally occurring algae and bacteria [160]. Similarly, microalgae have also been considered for producing bioethanol. But similar issues encountered for biodiesel production are also prevalent for bioethanol production such as algal biomass separation and contamination [163, 164]. Interestingly, if aquatic plant food wastes proved suitable, they could also be evaluated as a possible feedstock for the production bioenergy products.

(a) Ranking of top captured aquatic plants

(b) Ranking of top farmed aquatic plants

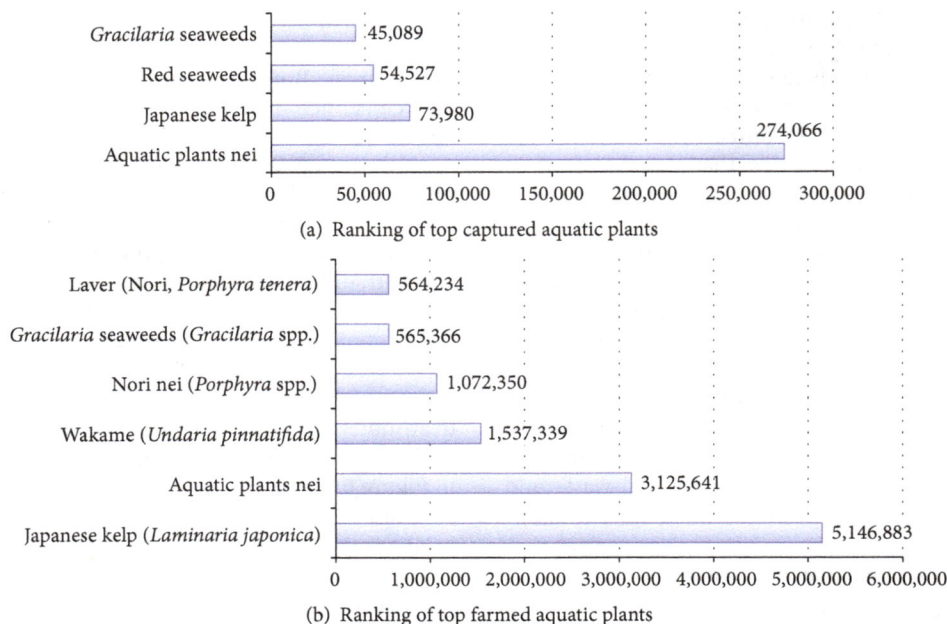

FIGURE 6: Captured and farmed aquatic plant food species in 2010 (data in tonnes) [118, 156].

But this possible application of aquatic plant food wastes needs to be investigated.

4. Discussion

At present, there is very little information in the literature discussing the industrial scale utilisation of food wastes at the local, national, or international level. For example, fruit and vegetable wastes have been extensively studied and reported in the literature. However, food wastes produced and their utilisation in aquaculture, livestock, poultry, and dairy industries are rarely reported and need further research. Much of the currently available food waste data lacks sufficient details and there is even less information discussing waste utilisation in the respective food supply chains. This information is needed before any economic modelling can be done to determine the feasibility of new products and waste transforming facilities needed to produce a commercially successful business outcome. The first step in developing a successful waste utilisation strategy is to assess the type and magnitude of waste [165, 166]. Once waste levels and their location in the food supply chain are known, it is now possible to start developing an effective waste management plan. In developing an effective plan, several important factors need to be considered before successful waste utilisation can be achieved as seen in Figure 7. Ultimately, the main barrier to developing any waste management plan that produces a new product from food waste needs to take into account several strategic factors, for example, new market opportunities, market trends, current market developments, and producing a product that is competitive in the marketplace [167]. Furthermore, each stage of product development needs to be carefully considered. In the case of manufacturing, a

company will need to consider commercial opportunities based on a well-thought-out growth strategy, especially if innovation is a key factor of the product. For packaging and distribution, the product range and associated services will also need to be carefully considered with the view of preventing competitor copying and safeguards to maintain market share. From a governmental perspective, policies may need to be formulated that promote sustainable patterns of consumption and sustainable community lifestyles, foster new job creation strategies, and enhance the economy. For consumers, the combination of diversity, choice, and expectation of high quality produce is a very important issue in their selection process. In summary, any new product produced from a food waste utilisation process that enters the marketplace will need to be both economically and ecologically sustainable. However, at the end of the day, it is consumer acceptance of the new product that is the deciding factor [168].

A recent study by Kummu et al. found that the preferred option for food waste utilisation was to use wastes generated from agriculture and consumers. From a global perspective, their study suggested that 47% of agricultural food wastes and over 86% of consumer wastes could be effectively utilised. The study also found that the biggest improvements in food waste management would occur where the demand for additional food was the least [169]. Therefore, to effectively manage food waste, there needs to be awareness of the benefits of postharvest waste utilisation by farmers, food processors, and government agencies. This awareness is needed so that food waste management capacity can be built up and ultimately lead to improvements in converting wastes into value-added products [54]. Importantly, it is also necessary to fully understand the size of the problem so that there are

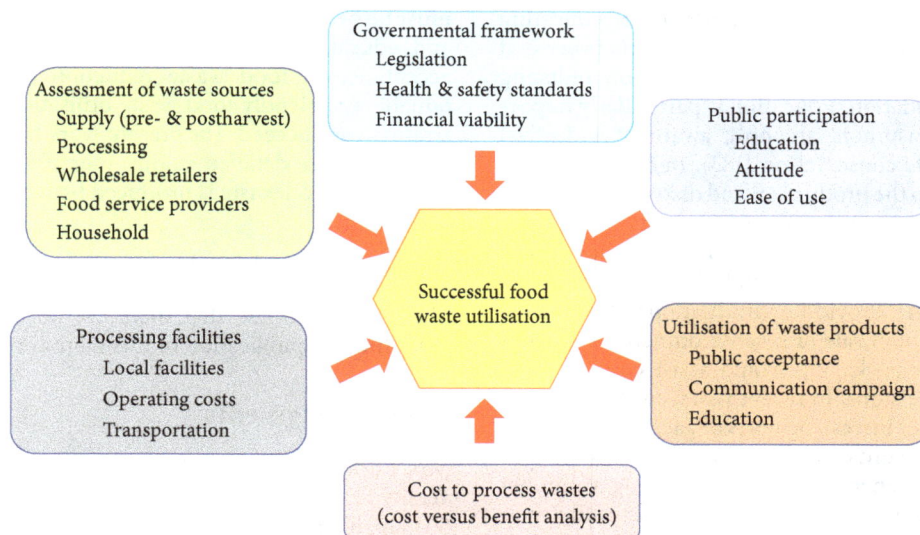

FIGURE 7: Important factors that need to be considered for successful utilisation of a food waste based product.

opportunities to improve food security and reduce poverty using effective waste utilisation strategies. In addition, the reduction of food wastes by effective waste management also reduces environmental degradation and improves economic sustainability of the food supply chain. However, the most important factors that will contribute to the success of any food waste utilisation strategy are its acceptance by the community at large.

Whenever food waste utilisation is debated, it is generally discussed in terms of processing methods, but actual food supply chain losses and their true impacts are often undervalued and underreported [170]. Undervaluing and underreporting are commonly referred to as the "*hidden costs*" of food waste management. Exploring these "*hidden costs*" usually acts as a catalyst for determining the scale of the waste problem, since businesses will only become aware of the problem when it impacts their bottom line. Generally, food related businesses often resolve their waste management problems by keeping their profitability levels high. They usually achieve this goal by reducing energy consumption, reducing raw material usage, and improving recycling activities [171]. Furthermore, businesses will investigate the merits of managing wastes in terms of recovery and value adding as opposed to the cost of disposal [172]. In fact, the disposal cost will have a direct impact on whether a business will go down the recovery and value-adding option or take the waste disposal route [18]. Therefore, food waste management options will often involve a cost versus benefit analysis that ultimately determines businesses profitability. However, because of business confidentiality reasons, food waste management costs are normally not reported. This often leads to partial and unproven estimates of the impact of food waste and makes assessments of waste management strategies difficult [173]. For instance, in many developed countries, the main driver for waste management strategies is government legislation relating to safety, handling of hazardous waste materials, and the environmental impact of the businesses

operational practices. In developing countries, factors such as food type, processing facilities, storage facilities, transport, and even climatic conditions are the principal drivers in food waste management strategies [174]. For example, the drivers for fish, meat, and poultry waste utilisation are health safety and hygiene risks associated with processing the wastes, whereas the economic drivers for fruit and vegetable waste management include microbial spoilage, costs of drying, storage, and shipment of byproducts [175]. Furthermore, these drivers become even more demanding if the food wastes are to be converted into high quality functional compounds [176]. Therefore, waste processing strategies must be optimised to promote production efficiency and cost-effectiveness so that the final products are competitive in the marketplace [177]. Consequently, a cost-and-benefit analysis is of paramount importance before any business adopts a food waste utilisation and management strategy.

The most important factor that needs to be carefully considered when planning to adopt a food waste utilisation strategy that aims to produce value-added products is the consumer. Experience has shown that consumers are often reluctant to accept new products, even when they can see its benefits. Many studies have shown that consumers do not compromise on product quality or performance. This is why consumer behaviour or habit needs to be fully understood when developing and marketing any new product. For example, surveys have consistently shown that consumers are very concerned about the environment and whether new products are ecofriendly [178–180], with consumer queries often focusing on whether environmental guidelines were followed during product manufacture. In Australia, around 62% of all consumer queries involve issues concerning environmental impact [181]. A similar study in Sweden found that customers ranked product taste first, while environmental impact was ranked second [182]. The results of both studies clearly indicate the importance of environmental issues to consumers and how this translates into their purchasing behaviour.

These results also emphasise the importance of educating consumers on the ecofriendly nature of products processed from food wastes and their positive impact on reducing environmental degradation. Education is particularly important since most consumers are only aware of industrial pollution and wildlife conservation [183]. In fact, consumer knowledge relating to the production and distribution of food they purchase and its environmental impact is poor. Thus, consumers need product information so that they can make informed decisions and make ecofriendly based choices when selecting products [24]. Providing information in the form of fact sheets at the point of sale or by environmental indicator labelling on product packaging would assist consumers in making informed decisions. In recent years, consumers have become more aware of increasing costs of gas, electricity, and petrol prices. Accordingly, consumers have been encouraged to reduce their home energy consumption using a number of strategies aimed at improving domestic energy efficiency. Unfortunately, no similar strategies have been aimed at raising the awareness of food waste utilisation. In fact, very few strategies have highlighted the negative environmental impact of dumping food wastes in landfill sites and subsequent greenhouse gas emissions. Thus, consumer education and acceptance of value-added products derived from food wastes will ultimately determine the success of any food waste utilisation and management strategy. Through education, consumers will see the value of these waste derived products and their positive environmental impact. This will ultimately influence consumer behaviour and promote purchasing patterns towards food waste derived value-added products.

5. Conclusion

Today, there is a general absence of detailed information and understanding of the extent of food wastage at different stages of the food value chain from farm to fork. The scale of food waste throughout the food supply chain is complex and can have a significant impact on a number of different fields such as agriculture, food security, economics, waste utilisation and management, environmental conservation, and human health. To resolve food waste problems and promote food waste utilisation strategies in any country will require effective communication and cooperation between all stakeholders. There are a number of hurdles preventing the conversion of food waste to value-added products. These hurdles include developing effective cost/benefit food waste utilisation strategies, developing efficient ecofriendly reprocessing technologies, reducing environmental degradation, and public acceptance. Globally, a number of countries are tackling the problems associated with increasing food waste and food waste utilisation and management. For example, several European countries are promoting utilisation and management strategies such as bioenergy production and regulating landfill costs to discourage waste generation. The key to successful food waste utilisation and management is to develop appropriate ecofriendly reprocessing technologies that can convert all the valuable components present in the waste into valuable products and reduce the amount of waste going to landfill. However, there are many challenges that

must be overcome to achieve this goal. Consumer awareness and education is one such challenge. Without consumer acceptance of food waste reduction approaches, no sustainable ecofriendly food waste utilisation and management strategy can succeed. The present work has also identified the need for more detailed studies identifying where, why, and how much food waste is produced between farm and fork.

Competing Interests

The authors declare that there are no competing interests regarding the publication of this manuscript.

Acknowledgments

Mrs. Purabi Ghosh would like to acknowledge Murdoch University for providing her Ph.D. Scholarship to undertake the Sustainable Food Waste Utilisation and Management studies as part of her Ph.D. project. This work was partly supported by Horticulture Innovation Australia Project Al14003 and Derek Fawcett would like to thank Horticulture Innovation Australia for their research fellowship.

References

[1] N. Mirabella, V. Castellani, and S. Sala, "Current options for the valorization of food manufacturing waste: a review," *Journal of Cleaner Production*, vol. 65, pp. 28–41, 2014.

[2] United Nations Children's Emergency Fund (UNICEF), *Levels and Trends in Child Mortality*, Estimates Developed by the UN Inter-agency Group for Child Mortality Estimation, 2011.

[3] FAO, *The State of Food Insecurity in the World High Food Prices and Food Security—Threats and Opportunities*, Food and Agriculture Organization of the United Nations, Rome, Italy, 2008.

[4] T. Stuart, *Waste: Uncovering the Global Food Scandal*, WW Norton & Company, New York, NY, USA, 2009.

[5] G. T. Kroyer, "Impact of food processing on the environment-an overview," *LWT—Food Science and Technology*, vol. 28, no. 6, pp. 547–552, 1995.

[6] J. Gustavsson, C. Cederberg, U. Sonesson, R. van Otterdijk, and A. Meybeck, *Global Food Losses and Food Waste: Extent, Causes and Prevention*, Food and Agriculture Organization of the United Nations, Rome, Italy, 2011.

[7] "Food industry wastes," in *Introduction: Causes and Challenges of Food Wastage*, M. R. Kosseva and C. Webb, Eds., Academic Press, San Diego, Calif, USA, 2013.

[8] C. Beretta, F. Stoessel, U. Baier, and S. Hellweg, "Quantifying food losses and the potential for reduction in Switzerland," *Waste Management*, vol. 33, no. 3, pp. 764–773, 2013.

[9] P. Gerland, A. E. Raftery, H. Ševčíková et al., "World population stabilization unlikely this century," *Science*, vol. 346, no. 6206, pp. 234–237, 2014.

[10] P. R. Ghosh, S. B. Sharma, Y. T. Haigh, A. Evers, and G. Ho, "An overview of food loss and waste: why does it matter?" *Cosmos*, vol. 11, no. 1, pp. 89–103, 2015.

[11] S. B. Sharma and J. Wightman, *Vision Infinity for Food Security: Some Whys, Why Not's and How's*, Springer Briefs in Agriculture, Springer International, Berlin, Germany, 2015.

[12] B. Redlingshöfer and A. Soyeux, "Food losses and wastage as a sustainability indicator of food and farming systems," in *Proceedings of the Producing and Reproducing Farming Systems: New Modes of Organisation for Sustainable Food Systems of Tomorrow, 10th European IFSA Symposium*, Aarhus, Denmark, 2012.

[13] J. C. Buzby and J. Hyman, "Total and per capita value of food loss in the United States," *Food Policy*, vol. 37, no. 5, pp. 561–570, 2012.

[14] J. Parfitt, M. Barthel, and S. Macnaughton, "Food waste within food supply chains: quantification and potential for change to 2050," *Philosophical Transactions of the Royal Society B: Biological Sciences*, vol. 365, no. 1554, pp. 3065–3081, 2010.

[15] C. S. K. Lin, L. A. Pfaltzgraff, L. Herrero-Davila et al., "Food waste as a valuable resource for the production of chemicals, materials and fuels: current situation and global perspective," *Energy and Environmental Science*, vol. 6, no. 2, pp. 426–464, 2013.

[16] D. Gunders, "Wasted: how America is losing up to 40 percent of its food from farm to fork to landfill," Issue Paper IP:12-06B, Natural Resources Defence Council (NRDC), New York, NY, USA, 2012.

[17] L. S. Kantor, K. Lipton, A. Manchester, and V. Oliveira, "Estimating and addressing America's food losses," *Food Review*, vol. 20, no. 1, pp. 2–12, 1997.

[18] C. Mena, B. Adenso-Diaz, and O. Yurt, "The causes of food waste in the supplier-retailer interface: evidences from the UK and Spain," *Resources, Conservation and Recycling*, vol. 55, no. 6, pp. 648–658, 2011.

[19] D. Baker, J. Fear, and R. Denniss, "What a waste: an analysis of household expenditure on food," Policy Brief 6, Australia Institute, 2009.

[20] A. Baiano, "Recovery of biomolecules from food wastes—a review," *Molecules*, vol. 19, no. 9, pp. 14821–14842, 2014.

[21] M. Gooch, A. Felfel, and N. Marenick, *Food Waste in Canada*, Value Chain Management Centre, 2010.

[22] M. K. Loke and P. Leung, "Quantifying food waste in Hawaii's food supply chain," *Waste Management and Research*, vol. 33, no. 12, pp. 1076–1083, 2015.

[23] R. Ravindran and A. K. Jaiswal, "Exploitation of food industry waste for high-value products," *Trends in Biotechnology*, vol. 34, no. 1, pp. 58–69, 2016.

[24] K.-R. Bräutigam, J. Jörissen, and C. Priefer, "The extent of food waste generation across EU-27: different calculation methods and the reliability of their results," *Waste Management and Research*, vol. 32, no. 8, pp. 683–694, 2014.

[25] A. Jones, "An environmental assessment of food supply chains: a case study on dessert apples," *Environmental Management*, vol. 30, no. 4, pp. 560–576, 2002.

[26] S. Dilas, J. Čanadanović-Brunet, and G. Ćetković, "By-products of fruits processing as a source of phytochemicals," *Chemical Industry and Chemical Engineering Quarterly*, vol. 15, no. 4, pp. 191–202, 2009.

[27] Ó. J. Sánchez and C. A. Cardona, "Trends in biotechnological production of fuel ethanol from different feedstocks," *Bioresource Technology*, vol. 99, no. 13, pp. 5270–5295, 2008.

[28] N. Sarkar, S. K. Ghosh, S. Bannerjee, and K. Aikat, "Bioethanol production from agricultural wastes: an overview," *Renewable Energy*, vol. 37, no. 1, pp. 19–27, 2012.

[29] S.-K. Han and H.-S. Shin, "Biohydrogen production by anaerobic fermentation of food waste," *International Journal of Hydrogen Energy*, vol. 29, no. 6, pp. 569–577, 2004.

[30] R. Zhang, H. M. El-Mashad, K. Hartman et al., "Characterization of food waste as feedstock for anaerobic digestion," *Bioresource Technology*, vol. 98, no. 4, pp. 929–935, 2007.

[31] R. Rathmann, A. Szklo, and R. Schaeffer, "Land use competition for production of food and liquid biofuels: an analysis of the arguments in the current debate," *Renewable Energy*, vol. 35, no. 1, pp. 14–22, 2010.

[32] T. E. Quested, E. Marsh, D. Stunell, and A. D. Parry, "Spaghetti soup: the complex world of food waste behaviours," *Resources, Conservation and Recycling*, vol. 79, pp. 43–51, 2013.

[33] J. Tielens and J. Candel, "Reducing food waste, improving food security?" Food and business knowledge platform, 2014, http://knowledge4food.net/.

[34] G. Kouwenhoven, N. Nalla, R. Vijayender et al., "Creating sustainable businesses by reducing food waste: a value chain framework for eliminating inefficiencies," *International Food and Agribusiness Management Review*, vol. 15, no. 3, pp. 119–137, 2012.

[35] J. Gustavsson and J. Stage, "Retail waste of horticultural products in Sweden," *Resources, Conservation and Recycling*, vol. 55, no. 5, pp. 554–556, 2011.

[36] WRAP, *The Food We Waste*, WRAP, Banbury, UK, 2008.

[37] J. Hirsch and R. Harmanci, *Food Waste: The Next Food Revolution*, Modern Farmer, 2013.

[38] D. G. Coursey and R. H. Booth, "The postharvest phytopathology of perishable tropical produce," *Review of Plant Pathology*, vol. 51, no. 12, pp. 751–765, 1972.

[39] P. Jeffries and M. J. Jeger, "The biological control of postharvest diseases of fruit," *Postharvest News and Information*, vol. 1, no. 5, pp. 365–368, 1990.

[40] K. V. Subrahmanyam, "Post-harvest losses in horticultural crops: an appraisal," *Agricultural Situation in India*, vol. 41, no. 5, pp. 339–343, 1986.

[41] Y. W. Jeon and L. S. Halos, "Addressing R&D for Cassava Postharvest System in West Africa," in *Proceedings of the International Winter Meeting*, The American Society of Agricultural Engineers, Chicago, Ill, USA, 1991.

[42] Z. D. Osunde, "Minimizing postharvest losses in yam (Dioscorea spp.): treatments and techniques," in *Using Food Science and Technology to Improve Nutrition and Promote National Development*, International Union of Food Science & Technology, 2008.

[43] Hindustan Times, "India wastes Rs. 44,000 cr of fruits, vegetables and grains annually," 2013.

[44] L. Liang, "China's post-harvest grain losses and the means of their reduction and elimination," *Jingji Dili (Economic Geography)*, vol. 1, pp. 92–96, 1993.

[45] M. Grolleaud, *Post-Harvest Losses: Discovering the Full Story. Overview of the Phenomenon of Losses during the Post-Harvest System*, Food and Agriculture Organization of the United Nations, Rome, Italy, 2002.

[46] C. D. Singano, B. T. Nkhata, V. Mhango, and M. P. K. J. Theu, "National annual report on larger grain borer monitoring and *Teretrius nigrescens* rearing and releases in Malawi," in *Plant Protection Progress Report for the 2007/2008 Season, Presented at the Department of Agricultural Research Services Planning and Review Meeting, Andrews Hotel, Mangochi, 14–20 September, 2008*, pp. 1–8, 2008.

[47] C. D. Singano, T. Phiri, B. T. Nkhata, V. Mhango, and M. P. K. J. Theu, "National agricultural produce inspection services annual technical report for the period July 2007–June 2008," in

Proceedings of the Department of Agricultural Research Services Planning and Review Meeting, Mangochi, Malawi, September 2008, Plant Protection Progress Report for the 2007/2008 Season.

[48] FAO, *Household Metal Silos: Key Allies in FAO's Fight against Hunger*, Food and Agriculture Organization of the United Nations, Rome, Italy, 2008.

[49] J. Liu, J. Lundqvist, J. Weinberg, and J. Gustafsson, "Food losses and waste in China and their implication for water and land," *Environmental Science & Technology*, vol. 47, no. 18, pp. 10137–10144, 2013.

[50] World Bank, *Missing Food: The Case of Postharvest Grain Losses in Sub-Saharan Africa*, The World Bank, Natural Resources Institute & FAO, Washington, DC, USA, 2011.

[51] Alpha Access, *India Wastes Up to 30% of Annual Food Grain Production Due to Poor Storage Facilities, Claims Study*, Alpha Access Worldwide, Evanston, Ill, USA, 2014, http://www.access-salpha.com/.

[52] K. McConnell, "Post-Harvest Food Loss, Waste Are Focus of New U.S. Institute," IIP Digital, 2012, http://iipdigital.usembassy.gov/st/english/article/2012/09/20120921136390.html#axzz4NatgxlEm.

[53] The Times of India, "India wastes 21 million tonnes of wheat every year," Report, The Times of India, Mumbai, India, 2013.

[54] S. Zorya, N. Morgan, R. Diaz et al., "Missing food: the case of postharvest grain losses in sub-Saharan Africa," Tech. Rep. 60371-AFR, The World Bank, Washington, DC, USA, 2011.

[55] J. Beddington, M. Asaduzzaman, A. Fernandez et al., "Achieving food security in the face of climate change," Final Report, Commission on Sustainable Agriculture and Climate Change, Copenhagen, Denmark, 2012.

[56] DAFF, "South Africa Yearbook 2011/12," Department of Agriculture, Forestry and Fisheries, 2012.

[57] A. J. M. Timmermans, J. Ambuko, W. Belik, and J. Huang, *Food Losses and Waste in the Context of Sustainable Food Systems*, Research Report by the High Level Panel of Experts on Food Security and Nutrition of the Committee on World Food Security and Nutrition, Rome, Italy, 2014.

[58] R. J. Hodges, J. C. Buzby, and B. Bennett, "Postharvest losses and waste in developed and less developed countries: opportunities to improve resource use," *The Journal of Agricultural Science*, vol. 149, no. S1, pp. 37–45, 2011.

[59] A. D. Cuéllar and M. E. Webber, "Wasted food, wasted energy: the embedded energy in food waste in the United States," *Environmental Science & Technology*, vol. 44, no. 16, pp. 6464–6469, 2010.

[60] J. C. Buzby, H. F. Wells, and J. Hyman, *The Estimated Amount, Value, and Calories of Postharvest Food Losses at the Retail and Consumer Levels in the United States*, vol. 121 of *Economic Information Bulletin*, Economic Research Service, United States Department of Agriculture, 2014.

[61] J. S. Sarma, "Cereal feed use in the Third World: past trends and projections to 2000," Research Report 57, International Food Policy Research Institute, Washington, DC, USA, 1986.

[62] A. Nahman and W. de Lange, "Costs of food waste along the value chain: evidence from South Africa," *Waste Management*, vol. 33, no. 11, pp. 2493–2500, 2013.

[63] R. Loehr, *Agricultural Waste Management: Problems, Processes, and Approaches*, Academic Press, New York, NY, USA, 1972.

[64] T. I. N. Ezejiofor, U. E. Enebaku, and C. Ogueke, "Waste to wealth-value recovery from agro-food 2 processing wastes using biotechnology: a review," *British Biotechnology Journal*, vol. 4, no. 4, pp. 418–481, 2014.

[65] R. P. Tengerdy and G. Szakacs, "Bioconversion of lignocellulose in solid substrate fermentation," *Biochemical Engineering Journal*, vol. 13, no. 2-3, pp. 169–179, 2003.

[66] C. M. Galanakis, "Recovery of high added-value components from food wastes: conventional, emerging technologies and commercialized applications," *Trends in Food Science & Technology*, vol. 26, no. 2, pp. 68–87, 2012.

[67] T. M. Carole, J. Pellegrino, and M. D. Paster, "Opportunities in the industrial biobased products industry," *Applied Biochemistry and Biotechnology—Part A Enzyme Engineering and Biotechnology*, vol. 115, no. 1–3, pp. 871–885, 2004.

[68] D. Sengupta and R. W. Pike, *Chemicals from Biomass: Integrating Bioprocesses into Chemical Production Complexes for Sustainable Development*, CRC Press, New York, NY, USA, 2012.

[69] C. Balagopalan, "Cassava utilization in food, feed and industry," in *Cassava: Biology, Production and Utilization*, pp. 301–318, 2002.

[70] D. B. Turley, "The chemical value of biomass," in *Introduction to Chemicals from Biomass*, pp. 21–46, John Wiley & Sons, 2008.

[71] F. Gunstone, *Vegetable Oils in Food Technology: Composition, Properties and Uses*, John Wiley & Sons, New York, NY, USA, 2011.

[72] D. Johansson, *Renewable Raw Materials-A Way to Reduced Greenhouse Gas Emissions for the EU Industry*, DG Enterprise, European Commission, Brussels, Belgium, 2000.

[73] J. W. A. Langeveld, J. Dixon, and J. F. Jaworski, "Development perspectives of the biobased economy: a review," *Crop Science*, vol. 50, supplement 1, pp. S-142–S-151, 2010.

[74] Y. Sun and J. Cheng, "Hydrolysis of lignocellulosic materials for ethanol production: a review," *Bioresource Technology*, vol. 83, no. 1, pp. 1–11, 2002.

[75] J. R. Mielenz, "Ethanol production from biomass: technology and commercialization status," *Current Opinion in Microbiology*, vol. 4, no. 3, pp. 324–329, 2001.

[76] F. Rosillo-Calle and L. A. B. Cortez, "Towards ProAlcool II—a review of the Brazilian bioethanol programme," *Biomass and Bioenergy*, vol. 14, no. 2, pp. 115–124, 1998.

[77] M. von Sivers, G. Zacchi, L. Olsson, and B. Hahn-Hägerdal, "Cost analysis of ethanol production from willow using recombinant *Escherichia coli*," *Biotechnology Progress*, vol. 10, no. 5, pp. 555–560, 1994.

[78] S. Banerjee, S. Mudliar, R. Sen et al., "Commercializing lignocellulosic bioethanol: technology bottlenecks and possible remedies," *Biofuels, Bioproducts and Biorefining*, vol. 4, no. 1, pp. 77–93, 2010.

[79] M. Galbe, P. Sassner, A. Wingren, and G. Zacchi, "Process engineering economics of bioethanol production," *Advances in Biochemical Engineering/Biotechnology*, vol. 108, pp. 303–327, 2007.

[80] O. P. Ward and A. Singh, "Bioethanol technology: developments and perspectives," *Advances in Applied Microbiology*, vol. 51, pp. 53–80, 2002.

[81] P. Sassner, M. Galbe, and G. Zacchi, "Techno-economic evaluation of bioethanol production from three different lignocellulosic materials," *Biomass and Bioenergy*, vol. 32, no. 5, pp. 422–430, 2008.

[82] M. Chen, *Optimization of a supply chain network for bioenergy production from food waste [M.S. thesis]*, University of Illinois at Urbana-Champaign, 2014.

[83] J. Carolan, S. Joshi, and B. E. Dale, "Technical and financial feasibility analysis of distributed bioprocessing using regional biomass pre-processing centers," *Journal of Agricultural & Food Industrial Organization*, vol. 5, no. 2, pp. 1203–1230, 2007.

[84] S. Kang, H. Önal, Y. Ouyang, J. Scheffran, and Ü. D. Tursun, "Optimizing the biofuels infrastructure: transportation networks and biorefinery locations in illinois," in *Handbook of Bioenergy Economics and Policy*, M. Khanna, J. Scheffran, and D. Zilberman, Eds., vol. 33 of *Natural Resource Management and Policy*, pp. 151–173, Springer, New York, NY, USA, 2010.

[85] M. T. Melo, S. Nickel, and F. Saldanha-da-Gama, "Facility location and supply chain management—a review," *European Journal of Operational Research*, vol. 196, no. 2, pp. 401–412, 2009.

[86] N. Parker, "Modelling future biofuel supply chains using spatially explicit infrastructure optimization," Research Report UCD-ITS-RR-11-04, ITS Institute of Transportation Studies, University of Californi, Berkeley, Calif, USA, 2011.

[87] A. Halloran, J. Clement, N. Kornum, C. Bucatariu, and J. Magid, "Addressing food waste reduction in Denmark," *Food Policy*, vol. 49, no. 1, pp. 294–301, 2014.

[88] W. Russ and R. Meyer-Pittroff, "Utilizing waste products from the food production and processing industries," *Critical Reviews in Food Science and Nutrition*, vol. 44, no. 1, pp. 57–62, 2004.

[89] H. Sielaff, *Fleischtechnologie*, Behrs Verlag GmbH & Co, Hamburg, Germany, 1996.

[90] R. Trostle, "Global agricultural supply and demand: factors contributing to the recent increase in food commodity prices," Report, Economic Research Service, United States Department of Agriculture, Washington, DC, USA, 2008.

[91] C. R. Daniel, A. J. Cross, C. Koebnick, and R. Sinha, "Trends in meat consumption in the USA," *Public Health Nutrition*, vol. 14, no. 4, pp. 575–583, 2011.

[92] M. Abdulla, R. C. Martin, M. Gooch, and E. Jovel, "The importance of quantifying food waste in Canada," *Journal of Agriculture, Food Systems, and Community Development*, vol. 3, no. 2, pp. 137–151, 2013.

[93] WRAP, *Waste Arising's in the Supply of Food and Drink to Households in the UK*, WRAP, Banbury, UK, 2010.

[94] L. de Las Fuentes, *AWARENET: Agro-Food Wastes Minimisation and rEduction Network*, European Commission, Brussels, Belgium, 2004.

[95] K. Jayathilakan, K. Sultana, K. Radhakrishna, and A. S. Bawa, "Utilization of byproducts and waste materials from meat, poultry and fish processing industries: a review," *Journal of Food Science and Technology*, vol. 49, no. 3, pp. 278–293, 2012.

[96] H. W. Ockerman and C. L. Hansen, *Animal by-Product Processing & Utilization*, CRC Press, New York, NY, USA, 1999.

[97] F. Toldrá, M.-C. Aristoy, L. Mora, and M. Reig, "Innovations in value-addition of edible meat by-products," *Meat Science*, vol. 92, no. 3, pp. 290–296, 2012.

[98] U. U. Rahman, A. Sahar, and M. A. Khan, "Recovery and utilization of effluents from meat processing industries," *Food Research International*, vol. 65, pp. 322–328, 2014.

[99] T. Kawashima, *The Use of Food Waste as a Protein Source for Animal Feed—Current Status and Technological Development in Japan*, Protein Source of Animal Feed Industry, Food and Agriculture Organization of the United Nations, Rome, Italy, 2004.

[100] MIFAFF, *The Statistics for Consumption of Major Livestock Products*, Ministry for Food, Agriculture, Forestry and Fisheries, Gwacheon, South Korea, 2006.

[101] H. Theron, P. Venter, and J. F. R. Lues, "Bacterial growth on chicken eggs in various storage environments," *Food Research International*, vol. 36, no. 9-10, pp. 969–975, 2003.

[102] K. Palka, "Chemical composition and structure of foods," in *Chemical and Functional Properties of Food Components*, pp. 11–24, CRC Press, Boca Raton, Fla, USA, 2002.

[103] Y. S. Ok, S. S. Lee, W.-T. Jeon, S.-E. Oh, A. R. A. Usman, and D. H. Moon, "Application of eggshell waste for the immobilization of cadmium and lead in a contaminated soil," *Environmental Geochemistry and Health*, vol. 33, no. 1, pp. 31–39, 2011.

[104] C. Arunlertaree, W. Kaewsomboon, A. Kumsopa, P. Pokethitiyook, and P. Panyawathanakit, "Removal of lead from battery manufacturing wastewater by egg shell," *Songklanakarin Journal of Science and Technology*, vol. 29, no. 3, pp. 857–868, 2007.

[105] K. Chojnacka, "Biosorption of Cr(III) ions by eggshells," *Journal of Hazardous Materials*, vol. 121, no. 1–3, pp. 167–173, 2005.

[106] H. J. Park, S. W. Jeong, J. K. Yang, B. G. Kim, and S. M. Lee, "Removal of heavy metals using waste eggshell," *Journal of Environmental Sciences*, vol. 19, no. 12, pp. 1436–1441, 2007.

[107] N. B. Singh, R. Singh, and M. M. Imam, "Waste water management in dairy industry: pollution abatement and preventive attitudes," *International Journal of Science, Environment and Technology*, vol. 3, no. 2, pp. 672–683, 2014.

[108] P. Roy, D. Nei, T. Orikasa et al., "A review of life cycle assessment (LCA) on some food products," *Journal of Food Engineering*, vol. 90, no. 1, pp. 1–10, 2009.

[109] J. Berlin, "Environmental life cycle assessment (LCA) of Swedish semi-hard cheese," *International Dairy Journal*, vol. 12, no. 11, pp. 939–953, 2002.

[110] A. Hospido, M. T. Moreira, and G. Feijoo, "Simplified life cycle assessment of galician milk production," *International Dairy Journal*, vol. 13, no. 10, pp. 783–796, 2003.

[111] J. W. Casey and N. M. Holden, "A systematic description and analysis of GHG emissions resulting from Ireland's milk production using LCA methodology," *DIAS Report, Animal Husbandry*, vol. 61, pp. 219–221, 2004.

[112] U. Sonesson and J. Berlin, "Environmental impact of future milk supply chains in Sweden: a scenario study," *Journal of Cleaner Production*, vol. 11, no. 3, pp. 253–266, 2003.

[113] H. O. Monroy, M. F. Vázquez, J. C. Derramadero, and J. P. Guyot, "Anaerobic-aerobic treatment of cheese wastewater with national technology in Mexico: the case of 'El Sauz'," *Water Science and Technology*, vol. 32, no. 12, pp. 149–156, 1995.

[114] J. Berlin, U. Sonesson, and A.-M. Tillman, "A life cycle based method to minimise environmental impact of dairy production through product sequencing," *Journal of Cleaner Production*, vol. 15, no. 4, pp. 347–356, 2006.

[115] A. J. Mawson, "Bioconversions for whey utilization and waste abatement," *Bioresource Technology*, vol. 47, no. 3, pp. 195–203, 1994.

[116] M. I. González Siso, "The biotechnological utilization of cheese whey: a review," *Bioresource Technology*, vol. 57, no. 1, pp. 1–11, 1996.

[117] K. Cochrane, C. De Young, D. Soto, and T. Bahri, "Climate change implications for fisheries and aquaculture," FAO Fisheries and Aquaculture Technical Paper 530, Food and Agriculture Organization of the United Nations, Rome, Italy, 2009.

[118] FAO, *The State of the World Fisheries and Aquaculture*, Food and Agriculture Organization of the United Nations, Rome, Italy, 2012.

[119] DAFF, "Australia's seafood trade," Updated Report, Australian Government, Department of Agriculture, Canberra, Australia, 2015.

[120] World Bank, "Fish to 2030: prospects for fisheries and aquaculture," Agriculture and Environmental Services Discussion Paper 3, World Bank Group, Washington, DC, USA, 2013.

[121] R. Willman and K. Kelleher, *The Sunken Billions: The Economic Justification for Fisheries Reform*, Agricultural and Rural Development, World Bank Group, Washington, DC, USA, 2010.

[122] C. Costello, D. Ovando, R. Hilborn, S. D. Gaines, O. Deschenes, and S. E. Lester, "Status and solutions for the world's unassessed fisheries," *Science*, vol. 338, no. 6106, pp. 517–520, 2012.

[123] M. Archer, R. Watson, and J. W. Denton, "Fish waste production in the United Kingdom: the quantities produced and opportunities for better utilisation," Seafish Report SR537, Sea Fish Industry Authority, Grimsby, UK, 2001.

[124] M. Sharp and C. Mariojouls, "Waste not, want not: better utilisation of fish waste in the Pacific," *SPC Fisheries Newsletter*, vol. 138, pp. 44–48, 2012.

[125] T. L. Catchpole, C. L. J. Frid, and T. S. Gray, "Discards in North Sea fisheries: causes, consequences and solutions," *Marine Policy*, vol. 29, no. 5, pp. 421–430, 2005.

[126] U. T. Srinivasan, W. W. L. Cheung, R. Watson, and U. R. Sumaila, "Food security implications of global marine catch losses due to overfishing," *Journal of Bioeconomics*, vol. 12, no. 3, pp. 183–200, 2010.

[127] C. Crapo, B. Paust, and J. Babbitt, *Recoveries & Yields from Pacific Fish and Shellfish*, vol. 37 of *Marine Advisory Bulletin Series*, Alaska Sea Grant College Program, University of Alaska, 1993.

[128] I. Knuckey, C. Sinclair, A. Aravind, and W. Ashcroft, "Utilisation of seafood processing waste-challenges and opportunities," in *Proceedings of the 3rd Australian New Zealand Soils Conference*, vol. 2004, SuperSoil, Sydney, Australia, December 2004.

[129] F. M. Gavine, R. M. Gunasekera, G. J. Gooley, and S. S. De Silva, "Value adding to seafood, aquatic and fisheries waste through aquafeed development," Project 1999/424, Department of Natural Resources and Environment, Victoria, Australia, 2001.

[130] C. Jespersen, K. Christiansen, and B. Hummelmose, "Cleaner production assessment in fish processing," United Nations Environmental Program and Danish Environmental Protection Agency, pp. 1–99, Denmark, 2000.

[131] Scottish Aquaculture Research Forum, *Strategic Waste Management and Minimisation in Aquaculture*, SARF Thistle Environmental Partnership, 2008.

[132] A. E. Ghaly, V. V. Ramakrishnan, M. S. Brooks, S. M. Budge, and D. Dave, "Fish processing wastes as a potential source of proteins, amino acids and oils: a critical review," *Journal of Microbial and Biochemical Technology*, vol. 5, no. 4, pp. 107–129, 2013.

[133] J. Shepherd, "Aquaculture: are the criticisms justified? Feeding fish to fish," *World Agriculture*, vol. 3, no. 2, pp. 11–18, 2012.

[134] Trade and Agriculture Directorate, "Fishing for development—green growth in fisheries and aquaculture," Tech. Rep. TAD/FI-(2014)13, Organisation for Economic Co-Operation and Development, Paris, France, 2014.

[135] R. W. Hardy and F. T. Barrows, "Diet formulation and manufacture," *Fish Nutrition*, vol. 3, pp. 505–600, 2002.

[136] R. M. Gunasekera, N. J. Turoczy, S. S. De Silva, F. Gavine, and G. J. Gooley, "An evaluation of the suitability of selected waste products in feeds for three fish species," *Journal of Aquatic Food Product Technology*, vol. 11, no. 1, pp. 57–78, 2002.

[137] S. F. Dominy, "Aquafeed insight 'organic fish: a niche too far," *Feed International*, pp. 35–37, 2000.

[138] B. Wyatt and G. McGourty, "Use of marine by-products on agricultural crops," in *Proceedings of the International By-Products Conference*, Anchorage, Alaska, USA, 1990.

[139] H. Raghuraman, *Extraction of sulfated glycosaminoglycans from mackerel and herring fish waste [M.S. thesis]*, Dalhousie University, Halifax, Canada, 2013.

[140] S. Hwang and C. L. Hansen, "Formation of organic acids and ammonia during acidogenesis of trout-processing wastewater," *Transactions of the ASAE*, vol. 41, no. 1, pp. 151–156, 1998.

[141] Y. P. Kotzamanis, M. N. Alexis, A. Andriopoulou, I. Castritsi-Cathariou, and G. Fotis, "Utilization of waste material resulting from trout processing in gilthead bream (*Sparus aurata* L.) diets," *Aquaculture Research*, vol. 32, no. 1, pp. 288–295, 2001.

[142] R. Gumisiriza, A. Mshandete, T. Manoni et al., "Nile perch fish processing waste along Lake Victoria in East Africa: auditing and characterization," *African Journal of Environmental Science and Technology*, vol. 3, no. 1, pp. 13–20, 2009.

[143] I. S. Arvanitoyannis and A. Kassaveti, "Fish industry waste: treatments, environmental impacts, current and potential uses," *International Journal of Food Science & Technology*, vol. 43, no. 4, pp. 726–745, 2008.

[144] G. L. Allan, S. Parkinson, M. A. Booth et al., "Replacement of fish meal in diets for Australian silver perch, *Bidyanus bidyanus*: I. Digestibility of alternative ingredients," *Aquaculture*, vol. 186, no. 3-4, pp. 293–310, 2000.

[145] C. G. Carter and R. C. Hauler, "Fish meal replacement by plant meals in extruded feeds for Atlantic salmon, *Salmo salar* L.," *Aquaculture*, vol. 185, no. 3-4, pp. 299–311, 2000.

[146] C. K. Rathbone, J. K. Babbitt, F. M. Dong, and R. W. Hardy, "Performance of juvenile Coho salmon *Oncorhynchus kisutch* fed diets containing meals from fish wastes, deboned fish wastes, or skin-and-bone by-product as the protein ingredient," *Journal of the World Aquaculture Society*, vol. 32, no. 1, pp. 21–29, 2001.

[147] F. D. Evans and A. T. Critchley, "Seaweeds for animal production use," *Journal of Applied Phycology*, vol. 26, no. 2, pp. 891–899, 2014.

[148] E. W. Becker, "Micro-algae as a source of protein," *Biotechnology Advances*, vol. 25, no. 2, pp. 207–210, 2007.

[149] T. Fujiwara-Arasaki, N. Mino, and M. Kuroda, "The protein value in human nutrition of edible marine algae in Japan," *Hydrobiologia*, vol. 116-117, no. 1, pp. 513–516, 1984.

[150] R. A. Kay and L. L. Barton, "Microalgae as food and supplement," *Critical Reviews in Food Science & Nutrition*, vol. 30, no. 6, pp. 555–573, 1991.

[151] M. Herrero, E. Ibáñez, J. Señoráns, and A. Cifuentes, "Pressurized liquid extracts from Spirulina platensis microalga: determination of their antioxidant activity and preliminary analysis by micellar electrokinetic chromatography," *Journal of Chromatography A*, vol. 1047, no. 2, pp. 195–203, 2004.

[152] Z. Cohen and A. Vonshak, "Fatty acid composition of Spirulina and spirulina-like cyanobacteria in relation to their chemotaxonomy," *Phytochemistry*, vol. 30, no. 1, pp. 205–206, 1991.

[153] J. Fleurence, M. Morançais, J. Dumay et al., "What are the prospects for using seaweed in human nutrition and for marine animals raised through aquaculture?" *Trends in Food Science & Technology*, vol. 27, no. 1, pp. 57–61, 2012.

[154] P. MacArtain, C. I. R. Gill, M. Brooks, R. Campbell, and I. R. Rowland, "Nutritional value of edible seaweeds," *Nutrition Reviews*, vol. 65, no. 12, pp. 535–543, 2007.

[155] D. J. McHugh, "A guide to the seaweed industry," FAO Fisheries Technical Paper 441, Food and Agriculture Organization of the United Nations, Rome, Italy, 2003.

[156] A. G. J. Tacon and M. Metian, "Fish matters: importance of aquatic foods in human nutrition and global food supply," *Reviews in Fisheries Science*, vol. 21, no. 1, pp. 22–38, 2013.

[157] A. G. J. Tacon and M. Metian, "Fishing for feed or fishing for food: increasing global competition for small pelagic forage fish," *AMBIO*, vol. 38, no. 6, pp. 294–302, 2009.

[158] Y. Chisti, "Biodiesel from microalgae," *Biotechnology Advances*, vol. 25, no. 3, pp. 294–306, 2007.

[159] Y. Chisti, "Biodiesel from microalgae beats bioethanol," *Trends in Biotechnology*, vol. 26, no. 3, pp. 126–131, 2008.

[160] J. J. Cheng and G. R. Timilsina, "Status and barriers of advanced biofuel technologies: a review," *Renewable Energy*, vol. 36, no. 12, pp. 3541–3549, 2011.

[161] S. K. Hoekman, "Biofuels in the U.S.—challenges and opportunities," *Renewable Energy*, vol. 34, no. 1, pp. 14–22, 2009.

[162] P. T. Pienkos and A. L. Darzins, "The promise and challenges of microalgal-derived biofuels," *Biofuels, Bioproducts and Biorefining*, vol. 3, no. 4, pp. 431–440, 2009.

[163] L. Brennan and P. Owende, "Biofuels from microalgae—a review of technologies for production, processing, and extractions of biofuels and co-products," *Renewable and Sustainable Energy Reviews*, vol. 14, no. 2, pp. 557–577, 2010.

[164] R. P. John, G. S. Anisha, K. M. Nampoothiri, and A. Pandey, "Micro and macroalgal biomass: a renewable source for bioethanol," *Bioresource Technology*, vol. 102, no. 1, pp. 186–193, 2011.

[165] H. C. J. Godfray, J. R. Beddington, I. R. Crute et al., "Food security: the challenge of feeding 9 billion people," *Science*, vol. 327, no. 5967, pp. 812–818, 2010.

[166] S. Lundie and G. M. Peters, "Life cycle assessment of food waste management options," *Journal of Cleaner Production*, vol. 13, no. 3, pp. 275–286, 2005.

[167] O. K. Mont, "Clarifying the concept of product-service system," *Journal of Cleaner Production*, vol. 10, no. 3, pp. 237–245, 2002.

[168] J. C. Aurich, C. Fuchs, and C. Wagenknecht, "Life cycle oriented design of technical Product-Service Systems," *Journal of Cleaner Production*, vol. 14, no. 17, pp. 1480–1494, 2006.

[169] M. Kummu, H. De Moel, M. Porkka, S. Siebert, O. Varis, and P. J. Ward, "Lost food, wasted resources: global food supply chain losses and their impacts on freshwater, cropland, and fertiliser use," *Science of the Total Environment*, vol. 438, pp. 477–489, 2012.

[170] S. Binyon, *Reducing and Managing Waste in the Food Industry: Food Industry Sustainability Best Practice Workshop*, Food and Drinks Federation, London, UK, 2007.

[171] K. Hyde, A. Smith, M. Smith, and S. Henningsson, "The challenge of waste minimisation in the food and drink industry: a demonstration project in East Anglia," *Journal of Cleaner Production*, vol. 9, no. 1, pp. 57–64, 2001.

[172] L. Mason, T. Boyle, J. Fyfe, T. Smith, and D. Cordell, *National Food Waste Data Assessment: Final Report*, Department of Sustainability, Environment, Water, Population and Communities, by the Institute for Sustainable Futures, University of Technology Sydney, Sydney, Australia, 2011.

[173] B. Lipinski, C. Hanson, J. Lomax, L. Kitinoja, R. Waite, and T. Searchinger, "Reducing food loss and waste," Tech. Rep., World Resources Institute, Washington, DC, USA, 2013.

[174] M. M. Rutten, "What economic theory tells us about the impacts of reducing food losses and/or waste: implications for research, policy and practice," *Agriculture & Food Security*, vol. 2, no. 1, pp. 2–13, 2013.

[175] A. Schieber, F. C. Stintzing, and R. Carle, "By-products of plant food processing as a source of functional compounds—recent developments," *Trends in Food Science & Technology*, vol. 12, no. 11, pp. 401–413, 2001.

[176] G. Laufenberg, B. Kunz, and M. Nystroem, "Transformation of vegetable waste into value added products: (A) the upgrading concept; (B) practical implementations," *Bioresource Technology*, vol. 87, no. 2, pp. 167–198, 2003.

[177] D. Paul and K. Ohlrogge, "Membrane separation processes for clean production," *Environmental Progress*, vol. 17, no. 3, pp. 137–141, 1998.

[178] H.-K. Bang, A. E. Ellinger, J. Hadjimarcou, and P. A. Traichal, "Consumer concern, knowledge, belief, and attitude toward renewable energy: an application of the reasoned action theory," *Psychology & Marketing*, vol. 17, no. 6, pp. 449–468, 2000.

[179] J. Fien, I. T.-C. P. Ai, D. Yencken, H. Sykes, and D. Treagust, "Youth environmental attitudes in Australia and Brunei: implications for education," *Environmentalist*, vol. 22, no. 3, pp. 205–216, 2002.

[180] A. G. Mertig and R. E. Dunlap, "Environmentalism, new social movements, and the new class: a cross-national investigation," *Rural Sociology*, vol. 66, no. 1, pp. 113–136, 2001.

[181] Australian Bureau of Statistics, *Year Book: Environmental Views and Behaviours (1301.0)*, Australian Bureau of Statistics, Canberra, Australia, 2003.

[182] G. Grankvist and A. Biel, "The importance of beliefs and purchase criteria in the choice of eco-labeled food products," *Journal of Environmental Psychology*, vol. 21, no. 4, pp. 405–410, 2001.

[183] E. Lea and A. Worsley, "Australian consumers' food-related environmental beliefs and behaviours," *Appetite*, vol. 50, no. 2-3, pp. 207–214, 2008.

[184] J. Lundqvist, C. de Fraiture, and D. Molden, *Saving Water: From Field to Fork—Curbing Losses and Wastage in the Food Chain*, SIWI Policy Brief. SIWI, 2008.

[185] UNEP, "Food waste harms climate, water, land and biodiversity-new FAO report," 2013, http://www.unep.org.

[186] C. Nellemann, M. MacDevette, T. Manders et al., "The environmental food crisis—the environment's role in averting future food crises," A UNEP rapid response assessment, United Nations Environment Programme, GRID-Arendal, 2009.

[187] R. Pedreschi, S. Lurie, M. Hertog, B. Nicolaï, J. Mes, and E. Woltering, "Post-harvest proteomics and food security," *Proteomics*, vol. 13, no. 12-13, pp. 1772–1783, 2013.

[188] FAO, *Global Food Losses and Food Waste: Extent, Causes and Prevention*, Food and Agriculture Organization of the United Nations, Rome, Italy, 2011.

[189] C. Pedrick, "Going to waste: missed opportunities in the battle to improve food security," CTA Policy Brief 7, 2012.

[190] FAO, *Partners, Urge Greater Push to Reduce Food Losses and Waste*, Food and Agriculture Organization of the United Nations, Rome, Italy, 2012.

[191] UNEP, *Our Planet*, The Magazine of the United Nations Environment Programme (UNEP), UNEP Division of Communications and Public Information, Nairobi, Kenya, 2013.

[192] CGIAR, "Postharvest loss reduction—a significant focus of CGIAR research," 2013, http://www.cgiar.org/consortium-news.

[193] A. A. Kader, "Increasing food availability by reducing postharvest losses of fresh produce," *Acta Horticulturae*, vol. 682, pp. 2169–2176, 2005.

[194] Y. Waarts, M. M. Eppink, E. B. Oosterkamp et al., "Reducing food waste: obstacles experienced in legislation and regulations," Wageningen, LEI Report 2011-059, 2011.

[195] Eurostat, *Manual on the Economic Accounts for Agriculture and Forestry*, EAA/EAF 97—(Rev. 1.1), Office for Official Publications of the European Communities, Luxembourg City, Luxembourg, 2000.

[196] Delfra, *Report of the Food Industry Sustainability Strategy Champions*, Group on Waste. Department for Environment, Food and Rural Affairs, London, UK, 2007.

[197] V. Monier, V. Escalon, and C. O'Conner, *Preparatory Study on Food Waste Across EU-27*, European Commission, Paris, France, 2011.

[198] N. L. Nemerow, *Zero Pollution for Industry: Waste Minimization through Industrial Complexes*, John Wiley & Sons, New York, NY, USA, 1995.

[199] F. A. Paine and H. Y. Paine, *A Handbook of Food Packaging*, Springer, Boston, Mass, USA, 1992.

[200] F. Schneider, *Wasting Food: An Insistent Behaviour*, Institute of Food Research, Proceeding on Waste: The Social Context, Edmonton, Canada, 2008.

[201] Value Chain Management Centre, *Cut Waste, Grow Profit*, Georges Morris Centre, Guelph, Canada, 2012.

[202] C. Mena, A. Humphries, and T. Y. Choi, "Toward a theory of multi-tier supply chain management," *Journal of Supply Chain Management*, vol. 49, no. 2, pp. 58–77, 2013.

[203] T. Fox and C. Fimeche, *Global Food: Waste Not, Want Not*, Institution of Mechanical Engineers, London, UK, 2013.

[204] G. Wassermann and F. Schneider, "Edibles in household waste," in *Proceedings of the 10th International Waste Management and Landfill Symposium*, Sardinia, Italy, October 2005.

[205] P. Sonigo, J. Bain, A. Tan et al., "Assessment of resource efficiency in the food cycle," Project Report ENV.G4/FRA/2008/0112, European Commission, Brussels, Belgium, 2012.

[206] C. Blume, "Hong Kong Struggles to Cut Food Waste," Voice of America, October 2007, http://www.voanews.com/a/a-13-2007-05-08-voa11-66714397/560009.html.

[207] R. Engström and A. Carlsson-Kanyama, "Food losses in food service institutions examples from Sweden," *Food Policy*, vol. 29, no. 3, pp. 203–213, 2004.

[208] C. Gobel, P. Teitscheid, G. Ritter et al., *Reducing Food Waste-Identification of Causes and Courses of Action in North Rhine-Westphalia*, Abridged Version, Institute for Sustainable Nutrition and Food Production, University of Applied Sciences, Munster, Germany, 2012.

[209] M. Eriksson, I. Strid, and P.-A. Hansson, "Food losses in six Swedish retail stores: wastage of fruit and vegetables in relation to quantities delivered," *Resources, Conservation and Recycling*, vol. 68, pp. 14–20, 2012.

[210] G. Liu, *Food Losses and Food Waste in China*, OECD Publishing, 2014.

[211] T. Zhang, *25 Mt Grain Is Lost due to Inappropriate Storage*, Ministry of Agriculture, 2012 (Chinese).

[212] M. Fehr, M. D. R. Calçado, and D. C. Romão, "The basis of a policy for minimizing and recycling food waste," *Environmental Science & Policy*, vol. 5, no. 3, pp. 247–253, 2002.

[213] U. Schroeder and I. Helena, *Food wastage in the region of Waterloo, Ontario [Thesis]*, University of Waterloo, Waterloo, Canada, 2014, http://hdl.handle.net/10012/8764.

15

Gluten-Free Snacks based on Brown Rice and Amaranth Flour with Incorporation of Cactus Pear Peel Powder: Physical, Nutritional, and Sensorial Properties

Dayanne Vigo Miranda,[1] Meliza Lindsay Rojas,[2] Sandra Pagador,[1] Leslie Lescano,[3] Jesús Sanchez-Gonzalez,[3] and Guillermo Linares [ID][3]

[1]School of Agroindustrial Engineering, Universidad César Vallejo (UCV), AV. Víctor Larco Herrera 13009, Trujillo, Peru
[2]Department of Agri-food Industry, Food and Nutrition (LAN), Luiz de Queiroz College of Agriculture (ESALQ), University of São Paulo (USP), AV. Padua dias 11, Piracicaba, SP, Brazil
[3]School of Agroindustrial Engineering, Universidad Nacional de Trujillo (UNT), Av. Juan Pablo II s/n, Trujillo, Peru

Correspondence should be addressed to Guillermo Linares; glinares@unitru.edu.pe

Academic Editor: Vita Di Stefano

An agroindustrial by-product (cactus pear peel) and whole grains flour (brown rice and amaranth) were used to present a gluten-free snack proposal. The effect of 5% (F1), 7% (F2), and 10% (F3) substitution of brown-rice flour for yellow cactus pear peel powder (*Opuntia ficus-indica*) on the snack physical, sensorial, and nutritional properties was evaluated. In addition, 20% of amaranth flour (*Amaranthus caudatus*) was used for all formulations. As the percentage of substitution increased, the a* value increased, while the L* decreased. The control snacks presented higher hardness, while the snacks with 10% substitution presented a greater crispness. The sensorial properties (overall liking, colour, crispness, and oiliness) reported that the samples containing cactus pear peel powder were the most accepted. The fat content decreased as the substitution percentage increased. The F3 formulation presented the best physical and sensorial properties and when compared with other commercial snack brands, it presented low fat and an adequate protein and fibre content. Therefore, snacks based on brown rice, amaranth, and cactus pear by-product could be considered as a good option of gluten-free product, contributing to reducing the lack of gluten-free products on the markets.

1. Introduction

The food industry has more and more challenges to meet current needs and consumer demands. On one hand, it offers healthy and safe foods with pleasant taste, appearance, and affordable cost. On the other hand, there is an environmental concern, where environment-friendly processes and agroindustrial by-products use are desired. Therefore, food scientists and engineers have the task of providing different alternatives considering the quality of the process, the product, and the special needs of the population, such as diseases that involve an immune response to any dietary compound, such as celiac disease.

Based on studies from United States [1] and Europe, the celiac disease incidence is about 1% of the world population; however, 85% to 90% of this has not been diagnosed nor treated [2]. The pathophysiology of celiac disease involves both the innate and adaptive immune response to dietary gluten. It is characterized by a permanent intolerance for gluten proteins present in dietary wheat, rye, and barley [3, 4]. It was reported that patients (about 50%) present atypical forms of celiac disease related to dermatitis herpetiformis, iron-deficiency anemia, neurologic problems, and others [5]. It was also reported that gluten causes gastrointestinal symptoms in subjects without celiac disease [6], such as irritable bowel syndrome [7]. Therefore, despite the growing number of people with celiac disease and other gluten-sensitivity-related problems [8], there is still a lack of gluten-free products on the market.

Although a gluten-free diet is effective in most patients, many patients find it unsatisfactory. This diet can be burdensome and can limit the quality of life; it is expensive, with bread and pasta substitutes costing substantially more than their gluten-containing counterparts [8]. In fact, some products based on oat, starches, hydrocolloids, gums, pulses ingredients, and fibre [9–11] were used to produce gluten-free products. Therefore, a greater variety of gluten-free products with good physical, nutritional, and sensory characteristics are required, in addition to affordable prices.

One way to produce quality food products with low prices, besides reducing the environmental impact, is the use of agroindustrial by-products. In fruit processing, around 70% of the raw material, such as peel and seeds, is considered waste. Peels are promising fruit by-products because of their high content of insoluble and dietary fibre, pectin, and fructooligosaccharides, as well as phenolic compounds, proteins, minerals, and vitamins. In addition to the favourable physical or nutritional properties, the use of agroindustrial by-products as food additives or supplements has gained increasing interest because their recovery may be economically attractive [12–15].

In this work, the use of an agroindustrial by-product was proposed, that is, the use of cactus pear peels. Cactus pear production using conventional system is around 30-80 tonnes ha^{-1} and intensive system 179-263 tonnes ha^{-1} [16]. Cactus pear peel is the major by-product that represents 38% of fruit weight, resulting from fresh consumption or juice production [17]. The cactus pear peels were used as fibre source; the fibre content of cactus pear peel powder varies from 39% [18] to 64% [12]. On the other hand, there is a growing interest in adding value to Andean crops, so amaranth was used, an Andean grain that presents more than 13% of protein [19]. Additionally, brown-rice flour as the flour basis was used. In fact, the rice flour is gaining popularity as the alternative of wheat flour [20]. Some of the main nutritional compounds of cactus pear peel powder, amaranth flour, and brown-rice flour are presented in Table 1.

Therefore, the objective of the present work was to evaluate the effect of brown-rice flour substitution (5%, 7%, and 10%) with yellow cactus pear peel powder on the physical (instrumental colour, texture, and moisture content), sensory (overall liking and attributes of colour, texture, and oiliness), and nutritional properties (fat, protein, and fibre content) of gluten-free snacks with addition of amaranth flour (20%).

2. Materials and Methods

2.1. Raw Materials

2.1.1. Brown-Rice and Amaranth Flour. The brown rice and amaranth were free of impurities and had low moisture content (amaranth grain: 10.47%; brown rice: 8.21%). The whole grains of rice and amaranth were ground in an electric mill for grains and the resulting product was sieved in a mesh N° 140 (0.105 mm) until obtaining a fine flour. The obtained flour was stored in hermetic containers at room temperature.

2.1.2. Yellow Cactus Pear Peel Powder. Fresh yellow cactus pears (commonly known in Peru as tuna) produced in Santiago de Chuco-La Libertad were obtained from a local market (Trujillo, Peru). The fruits without damage were washed and sanitized. The peels were removed using a knife and cut in pieces of 2 cm^2. The peels presented a pH of 5.73 ± 0.12 and °Brix of 8.90 ± 0.10. Subsequently, the cut peels were dried at 70°C for 24 h using an oven (BOX 1720, MEMMERT, Germany) until reaching 1.84 g water/100g dry matter.

Dried peels were ground using the same grain electric mill and then sieved using a mesh N° 100 (0.149 mm) until obtaining a fine powder. The obtained powder was stored in hermetic containers at room temperature.

2.2. Formulation and Preparation of Snacks. The snacks were prepared according to the formulations shown in Table 2. The control snacks were prepared with 20% of amaranth flour and 80% of brown-rice flour. The amaranth flour percent (20%) was maintained constant for all formulations. Formulations 1, 2, and 3 were prepared by substitution of brown-rice flour with cactus pear peel powder. Substitutions were conducted based on 5%, 7%, and 10% of the weight of the brown-rice flour.

The snacks were prepared for each formulation in the same manner. Firstly, according to the type of formulation (Table 2), the cactus pear peel powder, the amaranth flour, and the brown-rice flour were mixed. For each 100 g of mixture, 5 g of salt, 70 mL of water, and 40 g of butter were added (this quantity provided a texture appropriate to the dough, which allowed future handling without rupture). The mixture was then kneaded to form a smooth dough. The dough was steam precooked for 20 min. Subsequently, the dough was cooled at room temperature, followed by sheeting until reaching a final thickness of 1 mm. The sheets of dough were then cut into circles using a mould of 4 cm diameter. Finally, they were fried in sunflower oil at 170°C for 5 s; the obtained snacks are shown in Figure 1. The snacks were stored in hermetic containers in a cool and dry place for posterior analyses. Four repetitions of each formulation were performed.

2.3. Physical Properties

2.3.1. Instrumental Colour. The instrumental colour of fried snacks was measured on the snack surface according to the method described by Mir et al. [22], using a spectrophotometer (Konica Minolta CM-5, Japan). The CIE (*Commission Internationale d'Eclairage*) colour scale was used, where parameters of L* (lightness), a* (green to red), and b* (blue to yellow) were measured.

2.3.2. Instrumental Texture. The instrumental texture of fried snacks was evaluated according to Alam et al. [23] by a compressive test using a texturometer (TA-HDplus, UK). A stainless-steel spherical probe (P/0.25 s) was inserted at a constant rate of 1 mm/s over a distance of 3 mm until it cracked the snack. The maximum compression force considered as the highest point of the force (N) versus time (s) curve was used for describing the sample texture (in terms of

TABLE 1: Main nutritional compounds of cactus pear peel powder, amaranth flour, and brown rice flour.

Compound (x 100g of product)		Cactus pear peel powder [1][2]	Amaranth flour [3]	Brown rice flour [3]
Protein	g	5.71	13.33	7.23
Total Fat	g	3.33	6.67	2.78
Fibre, total dietary	g	64.15	11.10	4.60
Carbohydrates	g	71.84	66.67	76.48
Total carotenoids	mg	217.11	Nr	nr
Calcium	mg	Nr	178.00	11
Iron	mg	Nr	7.20	1.98

nr: nonreported. Mean values obtained from [1] Elhassaneen and Ragab [18], [2] Diaz-Vela and Totosaus [12], and [3] USDA [21].

TABLE 2: Snacks formulation.

Formulation	Cactus pear peel powder (%)	Amaranth flour (%)	Brown rice flour (%)
Control	-	20	80
Formulation 1 (F1)	5	20	75
Formulation 2 (F2)	7	20	73
Formulation 3 (F3)	10	20	70

FIGURE 1: Snacks obtained with the different used formulations.

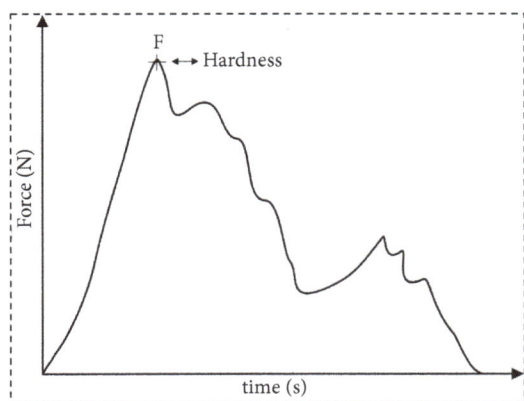

FIGURE 2: Typical curve of force (N) versus time (s) showing the maximum compression force used for describing the snack texture.

hardness) (Figure 2). The analysis was performed in triplicate for each formulation replicate, that is, 12 texture readings for each formulation.

2.4. Moisture Content. The moisture content of the raw materials and snacks was measured by drying the crushed snacks samples at 105°C using a moisture analyzer (MX-50, A&D Company, Tokyo, Japan); for this a HI (high) accuracy mode was selected. The moisture measurement was completed when the drying rate was lower than 0.02%/min.

2.5. Sensorial Properties. The snack sensorial properties were evaluated according to Ho and Abdul Latif [24] and Cruz et al.[25], with some modifications. A total of 97 untrained panellists (63% male and 37% female, age range: 18-25 years) were recruited for the sensory analysis. All participants were informed in an orientation session about the objectives and the use of the scale and signed an ethics consent form. The coded samples were provided to each panellist in a monadic sequential order. For each snack formulation, using a 7-point structured hedonic scale, consumers evaluated the overall liking, the colour (defined as the intensity of colour present on the surface of the snack; it varied from (1) very pale to (7) Very dark), the oiliness (defined as the taste and sensation of oil in snacks; it varied from (1) any oiliness to (7) very intense oiliness), and crispiness (defined as the degree of "crunch" or sound produced when the snack is bitten; it varied from (1) very slight sound to (7) very intense sound).

2.6. Nutritional Properties

2.6.1. Fat Content. The fat percentage (%F) of the snacks was determined using the Soxhlet extraction method, AOAC Method N° 922.06 [26]. After frying, the samples were dried, triturated, and placed inside a thick filter paper thimble, which was placed into the Soxhlet equipment for fat extraction using a combination of analytical grade ethyl ether and

TABLE 3: Instrumental colour parameters in the different formulations of snacks*.

Colour parameters	Formulations			
	Control	F1	F2	F3
L*	50.17 ± 2.22^a	43.29 ± 1.24^b	34.41 ± 0.29^b	32.11 ± 0.53^c
a*	4.55 ± 0.82^a	5.94 ± 0.93^a	6.72 ± 0.42^a	10.37 ± 0.95^b
b*	22.21 ± 1.59^a	23.71 ± 0.64^b	21.80 ± 0.25^a	22.80 ± 0.73^a
ΔE	-	7.35 ± 1.44^a	15.96 ± 1.98^b	19.10 ± 2.23^b

*mean ± standard deviation. Differences among letters indicate significant differences among formulations (α=5%).

petroleum ether (1:1) (Merck, Darmstadt). The determination was performed in triplicate for each formulation replicate.

After having carried out the physical, sensorial, and fat content analyses, the formulation that obtained the best properties and sensory acceptance was selected. For the snacks processed using this formulation, the amounts of protein and fibre were determined. In addition, the fat, protein, and fibre composition of the snacks made in this study was compared with those quantities reported for some brands of commercial snacks. The used brands were *Los cuates* (Karinto S.A.C, Lima, Peru), *Lay's ® classic*, and *Doritos nacho cheese flavoured* (Frito-Lay Inc., Texas, USA) from PepsiCo Inc. Company.

2.6.2. Protein Content. The proteins were determined using the Kjeldahl digestion and distillation method, According to the methodology described by the AOAC method N° 920.87 [27].

2.6.3. Crude Fibre Content. The crude fibre content was determined according to the methodology described by the AOAC method N° 978.10 [28].

2.7. Statistical Analysis. Data were analyzed using the Statgraphics Centurion (StatSoft, USA) software. Analyses were performed through analysis of variance (ANOVA) at 95% confidence ($p < 0.05$), followed by Duncan test to identify significant differences among treatments.

3. Results and Discussion

3.1. Physical Properties

3.1.1. Instrumental Colour. In snacks, the colour is one of the most important attributes judged by the consumer. Visually, as shown in Figure 1, the snacks presented differences in colour; this was evidenced objectively through the instrumental measurement. Table 3 shows the instrumental colour parameters value. The L* parameter (lightness) decreased as the percentage of cactus peel powder increased. Therefore, less lightness was detected in formulations with the highest percentage of cactus pear peel powder (F3). Contrary to the L* values, the a* values increased as the percentage of brown-rice flour substitution increased, the highest a* value being that for the F3 formulation. The values of b* were kept similar in the F2 and F3 samples, with a slight increase in F1 when compared to the control. Lastly, the total colour

difference (ΔE) increased as the percentage of cactus pear powder increased.

In Table 3, it is observed that the lightness value (L*) decreased as the percentage of substitution of brown-rice flour increased. This result makes sense because the cactus pear peel powder used contain pigments, in addition to other reactions that occur with the temperature increase during frying. In fact, the betalain and chlorophyll are pigments of the cactus pear peel powder, which are susceptible to the temperature, pH, and light [29]. These pigments generate compounds of dark brown colouration that decreased the lightness. Additionally, previous studies attributed the changes in L* to reducing sugars content present in the raw material. If this content is low, golden snacks will be obtained; however, if the content of reducing sugars is excessive, it will cause a deep brown colouration in the final product decreasing the lightness [30].

The a* parameter (from green (-) to red (+)) is used to determine the optimal frying point of fried products. In this work, the value of a* increased as the percentage of brown-rice flour substitution was increased (Table 3). That is, the colour of the samples was approximated to a red colour as the percentage of cactus pear powder increased. Meanwhile, in the case of b* parameter (from blue (-) to yellow (+)), for all samples obtained similar and positive results, comparable results were reported by Heredia and Castelló [31]. The obtained results could be due to the used variety of cactus pear, where peel powder was yellow; therefore it conferred to the formulations a yellowish colour.

Finally, all variations in the L*, a*, and b* parameters for each formulation impacted the total colour difference (ΔE) (Table 3). As stated above, both the raw material composition and the reactions that occur during processing contributed to this colour differences. For example, the different chemical reactions such as caramelization (Maillard reaction), nonenzymatic reactions, and structural changes were accelerated by the high temperatures during the frying process [32].

3.1.2. Texture. The texture is another important sensory attribute for the fried snacks acceptance [33]. Usually, the instrumental texture in foods like snacks is determined through a compression test, where the force-displacement curves are obtained (Figure 2). The curves usually have an irregular appearance with several fracture events. This behaviour is observed in crispy foods [34]. The first important drop of force is associated with a major structural breakdown; in this region, the probe mainly deforms the sample [35]. The

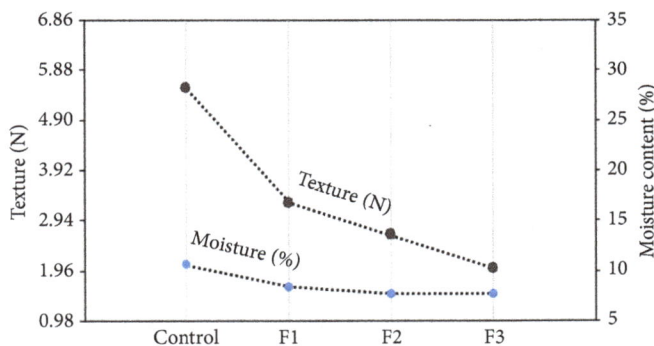

FIGURE 3: Texture (force (N)) and moisture content (%) for each snack formulation.

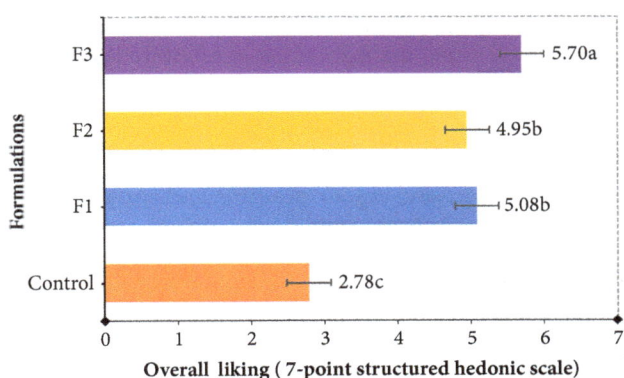

FIGURE 4: Average of the overall liking scores of each formulation, using a 7-point hedonic scale.

texture of the snacks was expressed as the maximum force needed to break the snack.

Significant differences ($p < 0.05$) among the texture of all the samples were observed. The force needed to break the snack decreases as the percentage of substitution of brown-rice flour increases (Figure 3). The control samples obtained the highest value (5.52 ± 0.03 N), while the F3 (10% substitution) samples obtained the lower value of force (2.02 ± 0.02 N). Additionally, it was observed that the moisture content (%) is directly proportional to the strength needed to break the snack (Figure 3).

According to Kita and Lisińska [36], samples with very low solid content and high fat content are less crisp and stickier. This occurred with the control samples, which showed high moisture (Figure 3) and fat content (Table 5). In addition, there are several factors influencing texture properties such as raw material composition, frying temperature, and oil type [37]. In this work, the temperature and time of frying and the oil type were the same; therefore, probably the observed differences on the snack properties were due to their different composition. For example, the control snacks contained a high quantity of brown-rice flour. The brown-rice flour, among others, contains fibre and starch. According to Pacheco-Delahaye [38], during frying, the starch contributes to the crust formation, causing a hard product. As observed,

the control snacks presented the higher hardness value. In contrast, the F3 samples presented lower hardness results, indicating that they were more fragile or crisp.

3.2. Moisture Content. Figure 3 shows the moisture content (%) tendency obtained regarding each formulation. As observed, the force (N) and moisture content (%) showed direct dependency on the snack composition, which decreased as the percentage of brown-rice flour substitution increased. Therefore, the control snacks showed the greater hardness and moisture (%), while the F3 (10% cactus pear peel powder) showed the lowest moisture content with most crispy texture.

Different variables act simultaneously and influence the obtained results [39]. According to van Koerten and Schutyser [40] a lower moisture content results in an increased porosity. The increased porosity decreases the breaking resistance, which in turn results in a crispier behaviour after frying. This could have occurred when the percentage of brown-rice flour substitution increased, where the snacks porosity probably increased, resulting in the observed behaviour.

The observed moisture content could be attributed to the rice flour composition, which contributed to the crust formation. In other products that also contain starch, such as potato chips, the crust formation was reported [41]. Therefore, probably the snacks that contained more quantity of brown-rice flour (as in the case of control snacks) formed rapidly crust increasing the external resistances to water exit. Consequently, the water inside the snack remained trapped, increasing the moisture content.

3.3. Sensorial Properties. The analysis of variance (ANOVA) showed no significant effect on the consumers. However, there was a significant effect for the snacks formulation ($p < 0.05$).

Figure 4 shows the overall liking for each snack formulation. The snacks that were most liked were those prepared using the formulations F1, F2, and F3; these snacks have an average acceptance of more than 5 on a hedonic scale of 7 points. Therefore, the three formulations mentioned above have a great possibility of being accepted by consumers.

TABLE 4: Results of the sensory attributes (colour, crispiness, and oiliness) evaluated for the different snacks formulation*.

Attributes	Formulation			
	Control	F1	F2	F3
Colour	3.15 ± 1.12^a	4.21 ± 1.31^b	5.43 ± 0.89^c	5.79 ± 0.82^d
Crispiness	2.43 ± 1.29^a	4.58 ± 1.26^b	4.9 ± 1.10^c	5.10 ± 1.29^c
Oiliness	4.19 ± 1.47^a	3.7 ± 1.23^b	3.94 ± 1.31^b	2.97 ± 1.19^c

*mean ± standard deviation. Differences among letters indicate significant differences among formulations ($\alpha=5\%$).

TABLE 5: Fat content (%) for the snacks obtained with different formulations*.

Formulations	Fat content (%)
Control	31.75 ± 0.01^a
F1	30.08 ± 0.01^b
F2	30.41 ± 0.01^b
F3	26.89 ± 0.13^c

*mean ± standard deviation. Differences among letters indicate significant differences among formulations ($\alpha=5\%$).

However, among these three formulations, the F3 formulation has the greatest acceptance. In contrast, the sample with the lowest acceptance by consumers was the control sample. Therefore, it can be inferred that the snacks without the addition of cactus pear peel powder presented low acceptance values.

Regarding the specific sensory attributes of snacks (colour, crispiness, and oiliness) (Table 4), the colour attribute scores were among "slightly pale" and "moderately dark" colour. It was observed that the colour score increased as the percentage of cactus pear peel powder increased. That is, the F3 formulation obtained the highest colour intensity punctuation; in fact, it was evidenced instrumentally where these samples showed higher a* and lower L* values. For the crispiness attribute, scores between 2.4 and 5.1 were obtained, which are equivalent in the hedonic scale to "mildly slight" and "slightly intense." In this regard, the snacks prepared with the F3 formulation were the crispiest. Furthermore, significant differences were found in the oiliness attribute, where the snacks of the F3 formulation showed fewer scores, with an average score of 2.97. It means that these snacks presented less taste and an oily sensation. In contrast, the control sample obtained the highest oiliness score of 4.19, equivalent to "slightly intense." The oiliness score agreed with what is reported in Table 5, where the F3 and control samples showed the lowest and highest fat percentage, respectively.

3.4. Nutritional Properties

3.4.1. Fat Content (%). The fat of fried snacks is an important component, which affects their flavour, texture, and shelf life. In the last years, the demand for healthy products, such as low-fat products, has been increasing [42]. Therefore, different studies were performed to obtain low-fat products, for example, by applying pretreatments [25, 43]. In this study, the use of 10% cactus pear peel powder (F3) allowed obtaining

snacks with the lowest fat percentage, while the highest percentage of fat was obtained for control snacks (Table 5).

In addition to the fat content of the raw materials and the fat used in the elaboration of the dough, during deep fat frying, there are coupled heat and mass transfer phenomena. During the mass transfer moisture loss and oil uptake simultaneously occur [25].

The mass transfer (moisture loss and oil uptake) is affected by the food-related properties (such as structure, composition, porosity, vapour pressure, and surface roughness) and by the process conditions (such as temperature, frying time, and type of used oil) [36, 43–48]. In fried products, the highest percentage of oil absorbed is on the surface, where the structure of the formed crust is a key factor [41, 49].

In fact, the control snacks showed the greater hardness and moisture (%) (Figure 3) and a high fat content (Table 5), while F3 (10% cactus pear peel powder) showed the lowest fat and moisture content with most crispy texture. Therefore, the high fat and moisture content observed in the control samples could be attributed to the crust formation. The oil content of the crust was higher than that of the core regions [41]. Probably the control snacks that contained more quantity of brown-rice flour formed rapidly crust storing more oil, thus increasing %F.

3.4.2. Protein and Fibre Content. Under the process conditions, F3 formulation allowed obtaining snacks with the best physical and sensory properties (low fat content, adequate colour and hardness, and greater overall acceptability). Therefore, for this formulation, analyses of proteins and fibre content were performed and compared with other brands typically found in markets (Table 6).

As shown in Table 6, the F3 sample presented lower fat content compared to the other brands; in addition, it contained a slightly higher percentage of proteins (7.22%). This can be due to the protein content of the used raw materials. However, apparently, it had a lower fibre content (2.13%); this is because the fibre content value in this work was reported as crude fibre and the value reported in the label of commercial snacks corresponds at the dietary fibre content. When the crude fibre method is used, the dietary fibre (DF) content is significantly underestimated, since a large part of the hemicellulose and lignin is dissolved, as well as varying

TABLE 6: Nutritional comparison of fried snacks with other brands.

Compound	Composition (%)			
	F3	Los Cuates∗	Lay's® Classic∗	Doritos Nacho cheese flavoured∗
Fat	26.89	36.73	35.71	28.57
Protein	7.22	7.14	7.14	7.14
Fibre	2.13 (CF)	3.57 (DF)	3.57 (DF)	3.57 (DF)

∗Data obtained from nutrition facts reported in the product label. CF: crude fibre; DF: dietary fibre.

amounts of cellulose and all the soluble fibres. The DF values are generally 3 to 5 times higher than the crude fibre [50, 51]. Therefore, the F3 formulation could present a high value of dietary fibre. Although not evidenced, this possibility is supported by the types of flours that were used, both rice and amaranth, which were of an integral type. Further, previous studies reported that cactus pear peel powder has a 64.15% total dietary fibre (Table 1): 33.48% insoluble and 30.67% soluble [12].

4. Conclusions

The substitution of brown-rice flour with cactus pear peel powder directly influenced the physical properties of the snacks. At a higher percentage of substitution, the a∗ colour parameter value increased, while the lightness (L∗) decreased with smaller variations in the b∗ value. In addition, the instrumental texture values decreased as the percentage of substitution increased; the control snacks presented higher hardness values, while the snacks with the higher percentage of substitution (such as F3) had lower hardness values, suggesting a greater crispness. The evaluation of sensorial properties reported that the samples containing cactus pear peel powder were the most accepted. For the specific sensory attributes of colour, texture, and oiliness, it was obtained that the higher percentage of substitution obtained better values of acceptance. The fat content decreased as the substitution percentage increased. Therefore, the F3 formulation presented the best physical and sensorial properties and low fat content. Additionally, when compared with other commercial brands, the F3 snacks elaborated in this study presented an adequate protein and fibre content. Therefore, the elaboration of snacks is a good proposal to exploit the use of agroindustrial by-products and provide gluten-free products.

Conflicts of Interest

The authors declare that there are no conflicts of interest regarding the publication of this paper.

Acknowledgments

The authors are grateful to the Laboratory of Process Engineering, Faculty of Agricultural Sciences, Universidad Nacional de Trujillo, and to Cienciactiva for the M.L. Rojas Ph.D. (CONCYTEC, Peru; Contract 087-2016-FONDECYT) scholarship from "Consejo Nacional de Ciencia, Tecnología e Innovación Tecnológica."

References

[1] A. Rubio-Tapia, J. F. Ludvigsson, T. L. Brantner, J. A. Murray, and J. Everhart, "1037 The Prevalence of Celiac Disease in the United States," *Gastroenterology*, vol. 142, no. 5, pp. S-181–S-182, 2012.

[2] S. Guandalini and A. Assiri, "Celiac disease: A review," *JAMA Pediatrics*, vol. 168, no. 3, pp. 272–278, 2014.

[3] V. M. Wolters and C. Wijmenga, "Genetic background of celiac disease and its clinical implications," *American Journal of Gastroenterology*, vol. 103, no. 1, pp. 190–195, 2008.

[4] C. Catassi, I.-M. Rätsch, E. Fabiani et al., "Coeliac disease in the year 2000: exploring the iceberg," *The Lancet*, vol. 343, no. 8891, pp. 200–203, 1994.

[5] A. Fasano and C. Catassi, "Current approaches to diagnosis and treatment of celiac disease: an evolving spectrum," *Gastroenterology*, vol. 120, no. 3, pp. 636–651, 2001.

[6] J. R. Biesiekierski, E. D. Newnham, P. M. Irving et al., "Gluten Causes gastrointestinal symptoms in subjects without celiac disease: a double-blind placebo-controlled trial," *American Journal of Gastroenterology*, vol. 106, no. 3, pp. 508–514, 2011.

[7] E. F. Verdu, D. Armstrong, and J. A. Murray, "Between celiac disease and irritable bowel syndrome: the ' no man's land' of gluten sensitivity," *American Journal of Gastroenterology*, vol. 104, no. 6, pp. 1587–1594, 2009.

[8] B. Lebwohl, J. F. Ludvigsson, and P. H. R. Green, "Celiac disease and non-celiac gluten sensitivity," *BMJ*, vol. 351, Article ID h4347, 2015.

[9] E. Gallagher, T. R. Gormley, and E. K. Arendt, "Recent advances in the formulation of gluten-free cereal-based products," *Trends in Food Science & Technology*, vol. 15, no. 3-4, pp. 143–152, 2004.

[10] M. Peräaho, P. Collin, K. Kaukinen, L. Kekkonen, S. Miettinen, and M. Mäki, "Oats can diversify a gluten-free diet in celiac disease and dermatitis herpetiformis," *Journal of the Academy of Nutrition and Dietetics*, vol. 104, no. 7, pp. 1148–1150, 2004.

[11] J. J. Han, J. A. M. Janz, and M. Gerlat, "Development of gluten-free cracker snacks using pulse flours and fractions," *Food Research International*, vol. 43, no. 2, pp. 627–633, 2010.

[12] J. Diaz-Vela, A. Totosaus, A. E. Cruz-Guerrero, and M. De Lourdes Pérez-Chabela, "In vitro evaluation of the fermentation of added-value agroindustrial by-products: Cactus pear (Opuntia ficus-indica L.) peel and pineapple (Ananas comosus) peel as functional ingredients," *International Journal of Food Science & Technology*, vol. 48, no. 7, pp. 1460–1467, 2013.

[13] A. Nawirska and M. Kwaśniewska, "Dietary fibre fractions from fruit and vegetable processing waste," *Food Chemistry*, vol. 91, no. 2, pp. 221–225, 2005.

[14] V. Oreopoulou and C. Tzia, "Utilization of plant by-products for the recovery of proteins, dietary fibers, antioxidants, and colorants," *Utilization of By-Products and Treatment of Waste in the Food Industry*, pp. 209–232, 2007.

[15] A. Schieber, F. C. Stintzing, and R. Carle, "By-products of plant food processing as a source of functional compounds—recent developments," *Trends in Food Science & Technology*, vol. 12, no. 11, pp. 401–413, 2001.

[16] FAO, Agro-industrial utilization of cactus pear. 2013, Food and Agriculture Organization of the United Nations: Rome.

[17] M. Namir, K. Elzahar, M. F. Ramadan, and K. Allaf, "Cactus pear peel snacks prepared by instant pressure drop texturing: Effect of process variables on bioactive compounds and functional properties," *Journal of Food Measurement and Characterization*, vol. 11, no. 2, pp. 388–400, 2017.

[18] Y. Elhassaneen, "Improvement of Bioactive Compounds Content and Antioxidant Properties in Crackers with the Incorporation of Prickly Pear and Potato Peels Powder," *International Journal of Nutrition and Food Sciences*, vol. 5, no. 1, p. 53, 2016.

[19] R. Repo-Carrasco-Valencia, J. K. Hellström, J.-M. Pihlava, and P. H. Mattila, "Flavonoids and other phenolic compounds in Andean indigenous grains: quinoa (*Chenopodium quinoa*), kañiwa (*Chenopodium pallidicaule*) and kiwicha (*Amaranthus caudatus*)," *Food Chemistry*, vol. 120, no. 1, pp. 128–133, 2010.

[20] G. We et al., *Development of rice flour-based puffing snack for early childhood*, Food Engineering Progress, 2010.

[21] USDA, USDA Branded Food Products Database,.

[22] S. A. Mir, S. J. D. Bosco, and M. A. Shah, "Technological and nutritional properties of gluten-free snacks based on brown rice and chestnut flour," *Journal of the Saudi Society of Agricultural Sciences*, 2017.

[23] M. S. Alam, S. Pathania, and A. Sharma, "Optimization of the extrusion process for development of high fibre soybean-rice ready-to-eat snacks using carrot pomace and cauliflower trimmings," *LWT- Food Science and Technology*, vol. 74, pp. 135–144, 2016.

[24] L.-H. Ho and N. W. b. Abdul Latif, "Nutritional composition, physical properties, and sensory evaluation of cookies prepared from wheat flour and pitaya (Hylocereus undatus) peel flour blends," *Cogent Food Agriculture*, vol. 2, no. 1, Article ID 1136369, 2016.

[25] G. Cruz, J. P. Cruz-Tirado, K. Delgado et al., "Impact of pre-drying and frying time on physical properties and sensorial acceptability of fried potato chips," *Journal of Food Science and Technology*, vol. 55, no. 1, pp. 138–144, 2018.

[26] AOAC, Association of Official Analytical Chemists, in Official methods of analysis: Proximate Analysis and Calculations Crude Fat (CF) Flour. Method No 922.06. 2006: Gaithersburg, MD.

[27] AOAC, Association of Official Analytical Chemists, in Official Methods of Analysis: Official Method for Protein. Method No. 920.87. 1995: Washington DC.

[28] AOAC, Association of Official Analytical Chemists, in Official methods of analysis: Proximate Analysis and Calculations Crude Fiber. Method No 978.10. 2006: Gaithersburg, MD.

[29] U. Osuna-Martínez, J. Reyes-Esparza, L. Rodríguez-Fragoso, and U. Osuna-Martínez, "Cactus (Opuntia ficus-indica): a review on its antioxidants properties and potential pharmacological use in chronic diseases," *Natural Products Chemistry & Research*, 2014.

[30] B. Altunakar, S. Sahin, and G. Sumnu, "Functionality of batters containing different starch types for deep-fat frying of chicken nuggets," *European Food Research and Technology*, vol. 218, no. 4, pp. 318–322, 2004.

[31] A. Heredia, M. L. Castelló, A. Argüelles, and A. Andrés, "Evolution of mechanical and optical properties of French fries obtained by hot air-frying," *LWT- Food Science and Technology*, vol. 57, no. 2, pp. 755–760, 2014.

[32] P. J. Fellows, *Food processing technology: principles and practice*, Elsevier, 2009.

[33] M. Taniwaki, N. Sakurai, and H. Kato, "Texture measurement of potato chips using a novel analysis technique for acoustic vibration measurements," *Food Research International*, vol. 43, no. 3, pp. 814–818, 2010.

[34] J. Chen, C. Karlsson, and M. Povey, "Acoustic envelope detector for crispness assessment of biscuits," *Journal of Texture Studies*, vol. 36, no. 2, pp. 139–156, 2005.

[35] A. Salvador, P. Varela, T. Sanz, and S. M. Fiszman, "Understanding potato chips crispy texture by simultaneous fracture and acoustic measurements, and sensory analysis," *LWT—Food Science and Technology*, vol. 42, no. 3, pp. 763–767, 2009.

[36] A. Kita, G. Lisińska, and G. Gołubowska, "The effects of oils and frying temperatures on the texture and fat content of potato crisps," *Food Chemistry*, vol. 102, no. 1, pp. 1–5, 2007.

[37] A. Kita, G. Lisińska, and M. Powolny, "The influence of frying medium degradation on fat uptake and texture of French fries," *Journal of the Science of Food and Agriculture*, vol. 85, no. 7, pp. 1113–1118, 2005.

[38] E. Pacheco-Delahaye, "Evaluación nutricional de hojuelas fritas y estudio de la digestibilidad del almidón de plátano verde (Musa spp.)," *Rev. Fac. Agron*, vol. 28, pp. 42–48, 2002.

[39] F. Pedreschi and P. Moyano, "Oil uptake and texture development in fried potato slices," *Journal of Food Engineering*, vol. 70, no. 4, pp. 557–563, 2005.

[40] K. N. van Koerten, M. A. I. Schutyser, D. Somsen, and R. M. Boom, "Crust morphology and crispness development during deep-fat frying of potato," *Food Research International*, vol. 78, pp. 336–342, 2015.

[41] J. Rahimi, P. Adewale, M. Ngadi, K. Agyare, and B. Koehler, "Changes in the textural and thermal properties of batter coated fried potato strips during post frying holding," *Food and Bioproducts Processing*, vol. 102, pp. 136–143, 2017.

[42] S. Mazurek, R. Szostak, and A. Kita, "Application of infrared reflection and Raman spectroscopy for quantitative determination of fat in potato chips," *Journal of Molecular Structure*, vol. 1126, pp. 213–218, 2016.

[43] A. O. Oladejo, H. Ma, W. Qu et al., "Effects of ultrasound pretreatments on the kinetics of moisture loss and oil uptake during deep fat frying of sweet potato (Ipomea batatas)," *Innovative Food Science and Emerging Technologies*, vol. 43, pp. 7–17, 2017.

[44] M. C. Moreno, C. A. Brown, and P. Bouchon, "Effect of food surface roughness on oil uptake by deep-fat fried products," *Journal of Food Engineering*, vol. 101, no. 2, pp. 179–186, 2010.

[45] J. Yang, A. Martin, S. Richardson, and C.-H. Wu, "Microstructure investigation and its effects on moisture sorption in fried potato chips," *Journal of Food Engineering*, vol. 214, pp. 117–128, 2017.

[46] N. Z. Ngobese, T. S. Workneh, and M. Siwela, "Effect of low-temperature long-time and high-temperature short-time blanching and frying treatments on the French fry quality of six

Irish potato cultivars," *Journal of Food Science and Technology*, vol. 54, no. 2, pp. 507–517, 2017.

[47] T. M. Millin, I. G. Medina-Meza, B. C. Walters, K. C. Huber, B. A. Rasco, and G. M. Ganjyal, "Frying Oil Temperature: Impact on Physical and Structural Properties of French Fries During the Par and Finish Frying Processes," *Food and Bioprocess Technology*, vol. 9, no. 12, pp. 2080–2091, 2016.

[48] E. Karacabey, M. S. Turan, Ş. G. Özçelik, C. Baltacıoğlu, and E. Küçüköner, "Optimisation of pre-drying and deep-fat-frying conditions for production of low-fat fried carrot slices," *Journal of the Science of Food and Agriculture*, vol. 96, no. 13, pp. 4603–4612, 2016.

[49] E. J. Pinthus, P. Weinberg, and I. S. Saguy, "Oil Uptake in Deep Fat Frying as Affected by Porosity," *Journal of Food Science*, vol. 60, no. 4, pp. 767–769, 1995.

[50] N. Pak, "Análisis de fibra dietética," in *Produccion y manejo de datos de composicion quimica de alimentos en nutricion*, C. Moron, I. Zacarias, and., and S. d. Pablo, Eds., FAO. Dirección de Alimentación y Nutrició, Santiago, 1997.

[51] J. L. Slavin, "Dietary fiber: classification, chemical analyses, and food sources." *Journal of the Academy of Nutrition and Dietetics*, vol. 87, no. 9, pp. 1164–1171, 1987.

16

Roots and Tuber Crops as Functional Foods: A Review on Phytochemical Constituents and their Potential Health Benefits

Anoma Chandrasekara and Thamilini Josheph Kumar

Department of Applied Nutrition, Wayamba University of Sri Lanka, Makandura, 60170 Gonawila, Sri Lanka

Correspondence should be addressed to Anoma Chandrasekara; anomapriyan@yahoo.ca

Academic Editor: Jose M. Prieto

Starchy roots and tuber crops play a pivotal role in the human diet. There are number of roots and tubers which make an extensive biodiversity even within the same geographical location. Thus, they add variety to the diet in addition to offering numerous desirable nutritional and health benefits such as antioxidative, hypoglycemic, hypocholesterolemic, antimicrobial, and immunomodulatory activities. A number of bioactive constituents such as phenolic compounds, saponins, bioactive proteins, glycoalkaloids, and phytic acids are responsible for the observed effects. Many starchy tuber crops, except the common potatoes, sweet potatoes, and cassava, are not yet fully explored for their nutritional and health benefits. In Asian countries, some edible tubers are also used as traditional medicinal. A variety of foods can be prepared using tubers and they may also be used in industrial applications. Processing may affect the bioactivities of constituent compounds. Tubers have an immense potential as functional foods and nutraceutical ingredients to be explored in disease risk reduction and wellness.

1. Introduction

Starchy root and tuber crops are second only in importance to cereals as global sources of carbohydrates. They provide a substantial part of the world's food supply and are also an important source of animal feed and processed products for human consumption and industrial use. Starchy roots and tubers are plants which store edible starch material in subterranean stems, roots, rhizomes, corms, and tubers and are originated from diversified botanical sources. Potatoes and yams are tubers, whereas taro and cocoyams are derived from corms, underground stems, and swollen hypocotyls. Cassava and sweet potatoes are storage roots and canna and arrowroots are edible rhizomes. All these crops can be propagated by vegetative parts and these include tubers (potatoes and yams), stem cuttings (cassava), vine cuttings (sweet potatoes), and side shoots, stolons, or corm heads (taro and cocoyam).

The contribution of roots and tubers to the energy supply in different populations varies with the country. The relative importance of these crops is evident through their annual global production which is approximately 836 million tonnes [1]. Asia is the main producer followed by Africa, Europe, and America. Asian and African regions produced 43 and 33%, respectively, of the global production of roots and tubers [1]. A number of species and varieties are consumed but cassava, potatoes, and sweet potatoes consist of 90% global production of root and tuber crops [1].

Nutritionally, roots and tubers have a great potential to provide economical sources of dietary energy, in the form of carbohydrates (Table 1). The energy from tubers is about one-third of that of an equivalent weight of rice or wheat due to high moisture content of tubers. However, high yields of roots and tubers give more energy per land unit per day compared to cereal grains [2]. In general the protein content of roots and tubers is low ranging from 1 to 2% on a dry weight basis [2]. Potatoes and yams contain high amounts of proteins among other tubers. Sulphur-containing amino acids, namely, methionine and cystine, are the limiting ones in root crop proteins. Cassava, sweet potatoes, potatoes, and yam contain some vitamin C and yellow varieties of sweet potatoes, yam, and cassava contain β-carotene. Taro is a good

TABLE 1: Nutritional composition of selected tuber crops.

Nutrients (per 100 g)	Potatoes		Sweet potatoes, raw	Cassava, raw	Yam, raw
	White flesh and skin, raw	Red flesh and skin, raw			
Proximate composition					
Energy (kcal)	69.0	70	86.0	160.0	118.0
Protein (g)	1.7	1.9	1.6	1.4	1.5
Total lipid (fat) (g)	0.1	0.1	0.1	0.3	0.2
Carbohydrate, by difference (g)	15.7	15.9	20.1	38.1	27.9
Fibre, total dietary (g)	2.4	1.7	3.0	1.8	4.1
Sugars, total (g)g	1.2	1.3	4.2	1.7	0.5
Minerals					
Calcium, Ca (mg)	9	10	30	16	17
Magnesium, Mg (mg)	21	22	25	21	21
Potassium, K (mg)	407	455	337	271	816
Phosphorus, P (mg)	62	61	47	27	55
Sodium, Na (mg)m	16	18	55	14	9
Vitamins					
Total ascorbic acid (mg)	19.70	8.60	2.40	20.60	17.10
Thiamin (mg)	0.07	0.08	0.08	0.09	0.11
Riboflavin (mg)	0.03	0.03	0.06	0.05	0.03
Niacin (mg)	1.07	1.15	0.56	0.85	0.55
Vitamin B-6 (mg)	0.203	0.170	0.209	0.088	0.293
Folate (μg-DFE)	18	18	11	27	23
Vitamin E (mg)	0.01	0.01	0.26	0.19	0.35
Vitamin K (μg)	1.6	2.9	1.8	1.9	2.3
Vitamin A (IU)IU	8	7	14187	13	138

Source: USDA [105].

source of potassium. Roots and tubers are deficient in most other vitamins and minerals but contain significant amounts of dietary fibre [2]. Similar to other crops, nutritional value of roots and tubers varies with variety, location, soil type, and agricultural practices, among others.

The burden of noncommunicable diseases (NCDs) increases globally in both developed and developing countries and plays a pivotal role as the major cause of death. Oxidative stress which would be harbored by both endogenous and exogenous factors contributes immensely to the etiology of NCDs as well as the aging process. The association between plant food intake and reduced NCDs episodes has been the main focus of a number of scientific investigations in the recent past. Furthermore, identification of specific plant constituents which convey health benefits is of much interest. Foods of plant origin consist of a wide range of nonnutrient phytochemicals. They are synthesized as secondary metabolites and serve a wide range of ecological roles in home plants [3]. Tubers and root crops are significant sources of a number of compounds, namely, saponins, phenolic compounds, glycoalkaloids, phytic acids, carotenoids, and ascorbic acid. Several bioactivities, namely, antioxidant, immunomodulatory, antimicrobial, antidiabetic, antiobesity, and hypocholesterolemic activities, among others, are reported for tubers and root crops.

This review focuses on the bioactivities of phytochemicals and their distribution in starchy roots and tuber crops.

Furthermore, the effect of processing on bioactive compounds of roots and tubers is also discussed.

2. Roots and Tuber Crops

Plants producing starchy roots, tubers, rhizomes, corms, and stems are important to nutrition and health. They play an essential role in the diet of populations in developing countries in addition to their usage for animal feed and for manufacturing starch, alcohol, and fermented foods and beverages.

Roots and tuber crops are important cultivated staple energy sources, second to cereals, generally in tropical regions in the world. They include potatoes, cassava, sweet potatoes, yams, and aroids belonging to different botanical families but are grouped together as all types produce underground food. An important agronomic advantage of root and tuber crops as staple foods is their favourable adaptation to diverse soil and environmental conditions and a variety of farming systems with minimum agricultural inputs. In addition, variations in the growth pattern and adopting cultural practices make roots and tubers specific in production systems. However, roots and tuber crops are bulky in nature with high moisture content of 60–90% leading them to be associated with high transportation cost, short shelf life, and limited market margin in developing countries even where they are mainly

TABLE 2: Different types of tuber crops commonly consumed in world.

	Botanical name	Family	Common name
Potatoes	Solanum tuberosum	Solanaceae	
Country potato Hausa potato	Solenostemon rotundifolius	Lamiaceae (mint family)	Innala, ratala (Sri Lanka)
Cannas	Canna edulis	Cannaceae	Buthsarana (Sri Lanka)
	Maranta arundinacea L.	Marantaceae	Arrow root Hulankeeriya (Sri Lanka) Aru aru, arawak (India)
Taro	Xanthosoma sagittifolium	Araceae	Kiriala (Sri Lanka) Keladi (Malaysia) Phueak (Thailand) Khoai mon (Vietnam) Sato-imo (Japan)
Yam	Dioscorea alata	Dioscoreaceae	Purple yam; greater yam Guyana; water yam Winged yam Raja ala (Sri Lanka) Ube (Philippines)
Sweet potatoes	Ipomoea batatas	Convolvulaceae	Camote; batata Shakarkand
Cassava	Manihot esculenta	Euphorbiaceae	Yuxco; mogo; manioc mandioca; kamoteng kahoy
Elephant foot yam	Amorphophallus paeoniifolius	Araceae	White pot giant arum; stink lily

cultivated. Table 2 presents commonly consumed starchy tuber and root crops worldwide.

2.1. Potatoes (Solanum tuberosum). Potato is currently the fourth most important food crop in the world after maize, wheat, and rice, with a production of 368 million tonnes [1]. It ranks the third after rice and wheat in terms of consumption. Potato is a crop of highland origin and has been domesticated in the high Andes of South America and has become a major food crop in the cool highland areas of South America, Asia, and Central and Eastern Africa [4].

In developed countries potatoes play a pivotal role in the diet compared to those of the developing ones. The energy intake from potatoes by an individual in developed and developing countries was 130 and 41 kcal/day, respectively [5]. Potatoes provide significant amounts of carbohydrates, potassium, and ascorbic acid in the diet [6]. Furthermore, they contribute to 10% of the total folate intake in some European countries, such as Netherlands, Norway, and Finland [7]. In addition, ascorbic acid present in potatoes protects folates from oxidative breakdown [8]. About 50% of the recommended dietary allowance (RDA) of vitamin A may be provided by 250 g of genetically carotenoid enriched potatoes [9]. Potatoes have several secondary metabolites which demonstrated antioxidant as well as other bioactivities [10].

2.2. Sweet Potatoes (Ipomoea batatas L.). The origin of sweet potato is Central America, but at present it is widely grown in many tropical and subtropical countries in different ecological regions. It is the seventh largest food crop, grown in tropical, subtropical, and warm temperate regions in the

world [11]. Sweet potato can be grown all around the year under suitable climatic conditions and complete crop loss under adverse climatic conditions is rare; thus it is considered as an "insurance crop." The crop is particularly important in Southeast Asia, Oceania, and Latin America regions and China claims about 90% of total world production. Sweet potatoes are considered as a typical food security crop for disadvantaged populations as the crop can be harvested little by little over a long period of time. National Aeronautics and Space Administration (NASA) has selected sweet potatoes as a candidate crop to be grown and incorporated into the menus for astronauts on space missions due to their unique features and nutritional value [12]. The consumption of 125 g orange fleshed sweet potatoes, rich in carotenoids, improves vitamin A status of children, especially in developing countries [13]. In addition, sweet potatoes are rich in dietary fibre, minerals, vitamins, and bioactive compounds such as phenolic acids and anthocyanins, which also contribute to the color of the flesh.

2.3. Cassava (Manihot esculenta). Cassava is the most widely cultivated root crop in the tropics and because of long growth season (8–24 months), its production is limited to the tropical and subtropical regions in the world. Cassava is a perennial shrub belonging to the family Euphorbiaceae. The genus *Manihot* comprises 98 species and *M. esculenta* is the most widely cultivated member [14]. Cassava originated in South America and subsequently was distributed to tropical and subtropical regions of Africa and Asia [15]. Cassava plays an important role as staple for more than 500 million people in the world due to its high carbohydrate content [15]. A number of bioactive compounds, namely, cyanogenic glucosides such as linamarin and lotaustralin, noncyanogenic glucosides, hydroxycoumarins such as scopoletin, terperoids, and flavonoids, are reported in cassava roots [15–17].

2.4. Yams (Dioscorea sp.).

Yam is a member of the monocotyledonous family Dioscoreaceae and is a staple food in West Africa, Southeast Asia, and the Caribbean regions [18]. Yam is consumed as raw yam, cooked soup, and powder or flour in food preparations. Yam tubers have various bioactive components, namely, mucin, dioscin, dioscorin, allantoin, choline, polyphenols, diosgenin, and vitamins such as carotenoids and tocopherols [19, 20]. Mucilage of yam tuber contains soluble glycoprotein and dietary fibre. Several studies have shown hypoglycemic, antimicrobial, and antioxidant activities of yam extracts [21, 22]. Yams may stimulate the proliferation of gastric epithelial cells and enhance digestive enzyme activities in the small intestine [23].

2.5. Aroids.

Aroids are tuber or underground stem bearing plants belonging to the family Araceae. There are several edible tubers/stems such as taro (Colocasia), giant taro (Alocasia), tannia or yautia (Xanthosoma), elephant foot yam (Amorphophallus), and swamp taro (Cyrtosperma). The origin of tannia is South America and the Caribbean regions [4]. Colocasia, originating in India and Southeast Asia, is a staple food in many islands of the South Pacific, such as Tonga and Western Samoa, and in Papua New Guinea. Furthermore, taro is the most widely cultivated crop in Asia, Africa, and Pacific as well as Caribbean Islands.

2.6. Minor Tuber Crops

2.6.1. Canna.

Canna is rhizomatous type tuber which is widely distributed throughout the tropics and subtropics. The genus Canna belongs to the family Cannaceae. The edible types of Canna edulis originated in the Andean region or Peruvian coast and extended from Venezuela to northern Chile, in South America. It is commercially cultivated in Australia for the production of starch.

2.6.2. Arrowroot.

Maranta arundinacea L. (West Indian arrowroot) is cultivated for its edible rhizomes. It belongs to Marantaceae and is believed to have originated in the Northwestern part of South America. Arrowroot has been widely distributed throughout the tropical countries like India, Sri Lanka, Indonesia, the Philippines, and Australia and West Indies.

2.7. Bioactive Compounds in Tuber Crops.

Bioactive compounds in plants are secondary metabolites having pharmacological or toxicological effects in humans and animals. Secondary metabolites are produced within the plants besides the primary biosynthesis associated with growth and development. These compounds perform several essential functions in plants, including protection from undesirable effects, attraction of pollinators, or signaling of essential functions.

2.7.1. Phenolic Compounds.

Phenolic compounds have an aromatic ring with one or more hydroxyl groups and act as antioxidants. They are derived from biosynthetic precursors such as pyruvate, acetate, a few amino acids, acetyl-CoA, and malonyl-CoA following the pentose phosphate, shikimate, and phenylpropanoid metabolism pathways. In plants, phenylalanine and to a lesser extent tyrosine are the two major amino acids involved in the synthesis of phenolic compounds [3]. Major groups of phenolic compounds abundantly found in plants are simple phenolics, phenolic acids, flavonoids, coumarins, stilbenes, tannins, lignans, and lignins. The quantity of phenolic compounds present in a given species of plant material varies with a number of factors such as cultivar, environmental conditions, cultural practices, postharvest practices, processing conditions, and storage [3]. Two classes of phenolic acids, hydroxybenzoic acids and hydroxycinnamic acids, are found in plant materials. Compounds with a phenyl ring (C_6) and a C_3 side chain are known as phenylpropanoids and serve as precursors for the synthesis of other phenolic compounds. Flavonoids are synthesized by condensation of a phenylpropanoid compound with three molecules of malonyl coenzyme A. The phenolics present in tubers render several health benefits, namely, antibacterial, anti-inflammatory, and antimutagenic activities, among others.

2.7.2. Saponins and Sapogenins.

Saponins are high molecular weight glycosides consisting of a sugar moiety linked to a triterpene or steroid aglycone. The aglycone portion of the saponin molecule is called sapogenin. Depending on the type of sapogenin present saponins are divided into three groups, namely, triterpene glycosides, steroid glycosides, and steroid alkaloid glycosides. Saponins having a steroid structure are precursors for the chemical synthesis of birth control pills (with progesterone and estrogen), similar hormones, and corticosteroids [24]. According to recent findings steroidal saponins could be a novel class of prebiotics to lactic acid bacteria and are effective candidates for treating fungal and yeast infections in humans and animals [25].

2.7.3. Bioactive Proteins.

The protein contents of roots and tuber crops are variable. The global contribution of proteins from roots and tubers in the diet is less than 3%. However, in African countries, this contribution may vary from 5 to 15% [4].

Dioscorin is the main storage protein found in tropical Dioscorea yams. It accounts for 90% of water extractable soluble proteins in a majority of Dioscorea species. Dioscorin was reported to have carbonic anhydrase and trypsin inhibitor activities [26]. Furthermore, dehydroascorbate reductase and monodehydroascorbate reductase activities of dioscorin in the presence of glutathione have been reported [27]. Dioscorin from fresh yam (Dioscorea batatas) exhibited DPPH radical scavenging activity [28] and showed beneficial effects in lowering blood pressure [19, 29]. In addition, dioscorin demonstrated angiotensin converting enzyme (ACE) inhibitory and antihypertensive activities on spontaneously hypertensive rats [29, 30]. The dioscorin from yam exhibited carbonic anhydrase, trypsin inhibitor, dehydroascorbate reductase (DHA), and monodehydroascorbate reductase (MDA) activities and immunomodulatory activities [18, 26, 27].

Sporamin is a soluble protein and is the main storage protein in sweet potato roots and accounts for about 60–80% of its total proteins [31]. The sporamin of sweet potatoes is initially known as ipomoein. It is a nonglycoprotein without glycan and is stored in the vacuole in the monomeric form. Sporamin is initially produced as preprosporamin, which is synthesized by the membrane-bound polysome in the endoplasmic reticulum (ER) [32]. Sporamin is a trypsin inhibitor with a Kunitz-type trypsin inhibitory activity which has potential application in the transgenic insect resistant plants [33]. Furthermore, sporamin showed various antioxidant activities related to stress tolerance, such as DHA and MDA reductase activities [34].

2.7.4. Glycoalkaloids.

Glycoalkaloids are an important class of phytochemicals found in many species of the genera *Solanum* and *Veratrum* [35]. Alkaloids are nitrogen-containing secondary metabolites found mainly in several higher plants and microorganisms and animals [36]. The skeleton of most alkaloids is derived from amino acids and moieties from other pathways, such as those originating from terpenoids. The primary function of alkaloids in plants is acting as phytotoxins, antibactericides, insecticides, and fungicides and as feeding deterrents to insects, herbivorous mammals, and mollusks [37]. There are two main glycoalkaloids in commercial potatoes. These include α-chaconine and α-solanine which are glycosylated derivatives of the aglycone solonidine. Wild potatoes (*Solanum chacoense*) and egg plants contain the glycoalkaloid solasonine. The major glycoalkaloid reported in tomatoes is α-tomatine which is a glycosylated derivative of aglycone tomatidine. Steroidal alkaloids and their glycosides present in several species of *Solanum* are known to possess a variety of biological activities such as antitumour, antifungal, teratogenic, antiviral, and antiestrogenic activities. Certain glycoalkaloids are used as anticancer agents [38, 39]. The steroidal alkaloid glycosides showed cytotoxic activity against various tumour cell lines [40].

2.7.5. Carotenoids.

Carotenoids are among the most widespread natural pigments with yellow, orange, and red colors in plants. The carotenes are hydrocarbons soluble in nonpolar solvents such as hexane and petroleum ether. The oxygenated derivatives of carotenes, xanthophylls, dissolve better in polar solvents such as alcohols [41]. The majority of carotenoids are unsaturated tetraterpenes with the same basic C 40 isoprenoid skeleton resulting from the joining of eight isoprene units in a head-to-tail manner with the exception of the tail-to-tail connection at the centre. Carotenoids play important biological roles in living organisms. In photosynthetic systems of higher plants, algae, and phototrophic bacteria, carotenoids participate in a variety of photochemical reactions [42]. Carotenoids, either isolated from natural sources or chemically synthesized, have been widely utilized, due to their distinctive coloring properties,

as natural nontoxic colorants in manufactured foods, drinks, and cosmetics. Carotenoids possess numerous bioactivities and play important roles in human health and nutrition, including provitamin A activity, antioxidant activity, regulation of gene expression, and induction of cell-to-cell communication [43], which are involved in a myriad of health beneficial effects. It has been demonstrated that zeaxanthin and lutein are stable throughout artificial digestion, whereas β-carotene and all-trans lycopene are degraded in the jejunal and ileal compartments. Among the isomers, the stability of 5-*cis* lycopene is superior to that of all-trans lycopene and 9-*cis* lycopene [44]. Yellow varieties of sweet potatoes and yams are good sources of carotenoids.

2.7.6. Ascorbic Acid.

Ascorbic acid, also known as vitamin C, is a water-soluble vitamin. It naturally occurs in plant tissues, primarily in fruits and vegetables. Ascorbic acid occurs in considerable quantities in several root crops. However, the level could be reduced during cooking of roots unless skins and cooking water are utilized. Root crops, if carefully prepared, can make a significant contribution to the vitamin C content of the diet. As reported by the 1983 Nutritional Food Survey Committee, potatoes serve as a principal source of vitamin C in British diets, providing 19.4% of the total requirement [2]. In general yams contain 6–10 mg of vitamin C/100 g and may vary up to 21 mg/100 g. In addition, the vitamin C content of potatoes is very similar to those of sweet potatoes and cassava. The concentration of ascorbic acid varies with the species, location, crop year, maturity at harvest, soil, and nitrogen and phosphate fertilizers [2].

3. Bioactivities of Phytochemicals in Roots and Tubers

3.1. Antioxidant Activity.

Accumulating research evidences demonstrate that oxidative stress plays a major role in the development of several chronic diseases such as different types of cancer, cardiovascular diseases, arthritis, diabetes, autoimmune and neurodegenerative disorders, and aging. Though internal antioxidant defense systems, either enzymes (superoxide dismutase, catalase, and glutathione peroxidase) or other compounds (lipoic acid, uric acid, ascorbic acid, α-tocopherol, and glutathione), are available in the body, external sources of antioxidants are needed, as internal defense system may get overwhelmed by excessive exposure to oxidative stress. A number of studies have reported the antioxidant activities of several roots and tuber crops.

Methanolic extract of potatoes demonstrated high phenolic content and strong antioxidant activity as determined by 2,2-diphenyl-1-picrylhydrazyl (DPPH) radical scavenging activity [45]. Authors further showed that the total phenolic content (TPC) ranged from 16.6 to 32 mg gallic acid equivalents (GAE)/100 g dry sample and EC 50 of DPPH radical scavenging activity was 94 mg/mL (dry matter).

Several authors reported that the peels of sweet potato possessed a potent wound healing effect, which appears to

TABLE 3: Percentage of 2,2-diphenyl 1-1-picrylhydrazyl (DPPH) radical scavenging activity, flavonoids, and phenolic content of selected tuber crops.

	% DPPH inhibition	Flavonoids	Total phenolics
St. Vincent yam (*Dioscorea alata*)	18.9 ± 0.56	390.65 ± 40.63	16.03 ± 0.79
Water yam (*Dioscorea alata*)	95.83 ± 0.21	410.52 ± 20.22	13.10 ± 1.03
Coco yam (*Xanthosoma* sp.)	12.59 ± 0.66	145.31 ± 5.61	9.39 ± 0.68
Sweet potatoes (*Ipomoea batatas*)	28.01 ± 1.34	165.34 ± 5.81	4.37 ± 0.27
Potatoes (*Solanum tuberosum*)	20.47 ± 1.38	85.21 ± 4.32	18.26 ± 1.35
Yellow yam (*Dioscorea cayenensis*)	13.55 ± 0.52	150.67 ± 30.34	3.43 ± 0.19

Source: Dilworth et al. [49].

be related to the free radical scavenging activity of the phytoconstituents and their ability in lipid oxidation inhibition [46, 47]. The healing effect of sweet potato fibre for burns or decubital wounds in a rat model was demonstrated and the reduction in size and changes in the quality of the wounds were observed [48]. They further found that the rats treated with sweet potato fibre covering had reduced wound area compared to those of the control. Petroleum ether extract of sweet potato had shown significant closure of scar area for complete epithelialization compared to the control [46].

The methanolic extracts of the peels and peel bandage of sweet potato roots were screened for wound healing effect by excision and incision wound models on Wistar rats [47]. They further showed that hydroxyproline content was found to be significantly increased in the test group compared to that of wounded control group. Increased hydroxyproline content leads to enhancement of collagen synthesis which improves wound healing. In addition, the content of malondialdehyde decreased in test groups compared to that of wounded control group, indicating lipid oxidation inhibitory effect of sweet potato peels [47]. Water yam (*Dioscorea alata*) was reported to possess the highest DPPH radical scavenging activity of 96% among different selected tuber crops such as sweet potato, potato, coco yam, and other *Dioscorea* yams (Table 3; [49]).

Cornago et al. [50] showed the TPC and antioxidant activities of two major Philippine yams of *Ube* (purple yam) and *Tugui* (lesser yam). Purple yam (*Dioscorea alata*) and lesser yam (*Dioscorea esculenta*) had a TPC which ranged from 69.9 to 421.8 mg GAE/100 g dry weight. The purple yam variety *Daking* had the highest TPC and antioxidant activities as measured by DPPH radical scavenging activity, reducing power and ferrous ion chelating capacity, whereas varieties *Sampero* and *Kimabajo* showed the lowest TPC and antioxidant activities.

Hsu et al. [51] studied the antioxidant activity of water and ethanolic extracts of yam peel on tert-butyl hydroperoxide (t-BHP) induced oxidative stress in mouse liver cells (Hepa 1–6 and FL83B). Ethanolic extracts of yam peel exhibited a better protective effect on t-BHP treated cells compared to that of water extracts. Furthermore, it was observed that ethanolic extract increased catalase activity, whereas water extract decreased it. According to Chen and Lin [52] heating affected the TPC, antioxidant capacity, and the stability

of dioscorin of various yam tubers. Raw yams contained higher TPC than their cooked counterparts. Furthermore, the DPPH radical scavenging activities declined with increasing temperature. TPC and dioscorin content of yam cultivars (*Dioscorea alata* L. var. Tainung number 2) and keelung yam (*D. japonica* Thunb. var. pseudojaponica (Hay.) Yamam) correlated with DPPH radical scavenging activity and ferrous ion chelating effect [52]. Phytochemicals of yams seem to enhance the activities of endogenous antioxidant enzymes. The administration of yams decreased the levels of γ-glutamyl transpeptidase (GGT), low density lipoprotein, and triacylglycerol in serum of rats in which hepatic fibrosis was induced by carbon tetrachloride [53]. Treatment of rats with yams increased the antioxidant activities of hepatic enzymes, namely, glutathione peroxidase and superoxide dismutase [53].

Extracts of flavonoids and flavones from potatoes showed high scavenging activities toward oxygen radicals. Potatoes scavenged 94% of hydroxyl radicals [54]. The major phenolic compound reported in potato was chlorogenic acid which constituted more than 90% of its phenolics. The range of hydrophilic oxygen radical absorbance capacity (ORAC) of 200 to 1400 μmol trolox equivalents/100 g fresh weight and the lipophilic ORAC range of 5 to 30 nmol alpha-tocopherol equivalents/100 g fresh weight were reported for potatoes [55].

Using a rat model, it was shown that ethanolic extracts of purple fleshed potato flakes had effective free radical scavenging activity and inhibition of linoleic acid oxidation [56]. Further potato extracts enhanced hepatic manganese superoxide dismutase (SOD), Cu/Zn-SOD, and glutathione peroxidase (GSH-Px) activities as well as mRNA expression, suggesting a reduced hepatic lipid peroxidation and an improved antioxidant potential.

Recently, Ji et al. [57] reported the contents of phenolic compounds and glycoalkaloids of 20 potato clones and their antioxidant, cholesterol uptake and neuroprotective activities *in vitro*. Peels of purple and red pigmented potato clones showed higher phenolic content compared to those of yellow and unpigmented clones [57]. Chlorogenic acid (50–70%) and anthocyanins, namely, pelargonidin and petunidin glycosides, were identified as major phenolic compounds present in potatoes. In addition, major glycoalkaloids present in potatoes α-chaconine and α-solanine and their contents were reduced by granulation process. Peels of potato clones showed the highest DPPH radical scavenging activity followed by flesh and granules [57].

A few studies have reported the antioxidant activities of cassava roots. A recent study [58] showed that the antioxidant activities of organically grown cassava tubers were higher than those of mineral-base fertilized roots. They found that TPC and flavonoid content (FC) were significantly higher for organic cassava compared to those of cassava grown with inorganic fertilizers.

3.2. Antiulcerative Activities.
The antiulcerative activity of sweet potato roots was studied in a rat model [59]. The extract of sweet potatoes did not show any toxic or deleterious effects by oral route up to 2000 mg/kg. Furthermore, superoxide dismutase (SOD), catalase (CAT), glutathione peroxidase (GPx), and glutathione reductase (GR) activities were significantly elevated by administration of tuber extracts in treated rats, indicating the ability of restoring enzyme activities compared to the control. Kim et al. [60] showed that butanol fraction of sweet potato could be a better source for treating gastric ulcers induced by excessive alcohol intake.

3.3. Anticancer Activities.
Cancer is a leading cause of death worldwide, and it is mostly related to unhealthy food habits and lifestyle. It is important to find ways to reduce and prevent the risk of cancer through dietary components, which are present in plant foods. Cancer is a multistage disease condition and tapping at any initial stage could help attenuate the disease condition. Root and tuber phytochemicals have demonstrated anticancer effects in several types of carcinoma cell lines and animal models.

Huang et al. [61] showed that aqueous extract of sweet potatoes had higher antiproliferative activity than that of ethanol extracts. Cell proliferation was analyzed at 48 h after human lymphoma NB4 cells which had been cultured with several concentrations of extracts, 0, 25, 50, 100, 200, 400, 800, or 1000 μg of dry matter/mL, in the media using the microculture tetrazolium treatment (MTT) assay. The phytochemicals present in sweet potato roots may exert a significant effect on antioxidant and anticancer activities. Furthermore, antioxidant activity is directly related to the phenolics and flavonoid contents of the sweet potato extracts. An additive role of phytochemicals was also found, which may contribute significantly to the potent antioxidant activity and antiproliferative activity in vitro [61]. Furthermore, two anthocyanin pigments, namely, 3-(6,6$'$-caffeylferulylsophoroside)-5-glucoside of cyanidin (YGM-3) and peonidin (YGM-6), purified from purple sweet potatoes effectively inhibited the reverse mutation induced by mutagenic pyrolysates of tryptophan (Trp-P-1, Trp-P-2) and imidazoquinoline (IQ) in the presence of rat liver microsomal activation systems [62].

A greater inhibition of carcinogenesis was shown by red pigmented cultivar of potatoes compared to a white Russet Burbank cultivar in breast cancer induced rats [63]. The red cultivar of potatoes was reported to have high levels of anthocyanins and chlorogenic acid derivatives.

According to Madiwale et al. [64], purple fleshed potato showed higher potential in suppressing proliferation and elevated apoptosis of HT-29 human colon cancer cell lines compared with white fleshed potato. It was interesting to note that storage of potatoes affected their antioxidant and anticancer activities and TPC. The extracts from both fresh and stored potatoes inhibited cancer cell proliferation and elevated apoptosis, but anticancer effects were higher in fresh potatoes than in stored tubers. The study further demonstrated that storage duration had a strong positive correlation with antioxidant activity and percentage of viable cancer cells and a negative correlation existed with apoptosis induction. These findings further elaborated that antioxidant activity and phenolic content of potatoes increased with storage, but antiproliferative and proapoptotic activities were decreased [64].

In addition to phenolic compounds, saponins present in roots and tubers play a pivotal role as anticancer/antitumour agents. There are several groups of saponins, namely, cycloartanes, ammaranes, oleananes, lupanes, and steroids, demonstrating strong antitumour effect on different types of cancers. For instance, cycloartane saponins possess antitumour properties in human colon cancer cells and tumour xenografts [65]. They downregulated expression of the hepatocellular carcinoma (HCC) tumour marker α-fetoprotein and suppressed HepG2 cell growth by inducing apoptosis and modulating an extracellular signal-regulated protein kinase- (ERK-) independent NF-κB signaling pathway [65]. Furthermore, oleananes saponins exerted their antitumour effect through various pathways, such as anticancer, antimetastasis, immunostimulation, and chemoprevention pathways [65].

Several studies have shown that glycoalkaloids such as α-chaconine and α-solanine present in tubers are potential anticarcinogenic agents [66]. Glycoalkaloids showed antiproliferative activities against human colon (HT-29) and liver (HepG2) cancer cells as assessed by the MTT assay [66].

Wang et al. [67] reported that the aqueous extract of yam (Dioscorea alata) inhibited the H_2O_2-$CuSO_4$ induced damage of calf thymus DNA and protected human lymphoblastoid cells from $CuSO_4$ induced DNA damage. Extract of water yam contains a homogenous compound with a single copper-binding site and also is a good natural, safe (redox inactive) copper chelator. In addition to phenolic compounds saponins and mucilage polysaccharides present in yams are responsible for this activity. Furthermore, water-soluble mucilage polysaccharides are the most important copper chelators in extract of water yam. Thus Dioscorea alata aqueous extracts could serve as potential agents in the management of copper-mediated oxidative disorders and diabetes [67].

3.4. Immunomodulatory Activities.
Purified dioscorin from yam tubers showed immunomodulatory activities in vitro [18]. The effects of dioscorin on native BALB/c mice spleen cell proliferation were assayed by MTT assay. It was found that dioscorin stimulated RAW 264.7 cells to produce nitric oxide (NO), in the absence of lipopolysaccharide (LPS) contamination. Yam dioscorin exhibited immunomodulatory activities by the innate immunity which is a nonspecific immune system which comprises the cells and mechanisms that defend the host from infection by other organisms in

a nonspecific manner. Dioscorin was reported to stimulate cytokine production and to enhance phagocytosis. Furthermore, the released cytokines may act synergistically with phytohemagglutinin (PHA) which is a lectin found in plants that stimulate the proliferation of splenocytes [18].

Several studies have demonstrated the immune activity of yam mucopolysaccharides (YMP). *In vitro* cytotoxic activity of mouse splenocyte against leukemia cell was increased in the presence of YMP of *Dioscorea batatas* at 10 μg/mL [68]. Furthermore, the production of IFN-γ was significantly increased in the YMP treated splenocytes, suggesting their capability of inducing cell-mediated immune responses. In addition, YMP at a concentration of 50 μg/mL increased uptaking capacity and lysosomal phosphatase activity of peritoneal macrophages [68].

Dioscorea phytocompounds enhanced murine splenocyte proliferation *ex vivo* and improved regeneration of bone marrow cells *in vivo* [69]. Mice which were fed with a *Dioscorea* extract recovered damaged bone marrow progenitor cell populations that had been depleted by large doses of 5-fluorouracil (5-FU). Furthermore, they reported that the phytocompound(s) responsible for these bioactivities had a high molecular weight (≥100 kDa) and were most likely polysaccharides. They postulated those high-molecular-weight polysaccharides in DsCE-II act on specific target cell types in the GI tract (dendritic cells, intestinal epithelial cells, and T-cells) to mediate a cascade of immunoregulatory activities leading to the recovery of damaged cell populations following 5-FU or other chemical insults in the bone marrow, spleen, or other immune cell systems [69].

Dioscorea tuber mucilage from Taiwanese yams (*Dioscorea Japonica* Thunb var.) showed significant effects on the innate immunity and adaptive immunity on BALB/c mice through oral administration [70]. In addition, it was found that the specific antibodies rapidly responded against foreign proteins (or antigens) in the presence of yam mucilage. Mucilage from these yam varieties exhibited a stimulatory effect on phagocytic activity by granulocyte and monocyte (*ex vivo*), on peritoneal macrophages, and on the RAW 264.7 cells (*in vivo*) of mice [70].

Yams (*Dioscorea esculenta*) showed anti-inflammatory activity on carrageenan induced oedema in the right hind paw of Wistar rats [71]. However, this activity was short lived as it was quickly removed from the system after reaching the peak within 2 hours. Phytochemical screening of *D. esculenta* confirmed the presence of saponins, β-sitosterol, stigmasterol, cardiac glycosides, fats, starch, and diosgenin, which could be responsible for the observed activity [71]. Diosgenin contained in Chinese yam was an immunoactive steroidal saponin which also showed prebiotic effect. Diosgenin had also beneficial effects on the growth of enteric lactic acid bacteria [25].

The impact of the consumption of pigmented potatoes on oxidative stress and inflammatory damages has been demonstrated in humans [72]. Participants were given white, yellow (high concentrations in phenolic acids and carotenoids), or purple fleshed (high content of anthocyanin and phenolic acids) potato once per day in a randomized 6-week trial

which reported good compliance. The results showed that the consumption of pigmented potatoes was responsible for elevated antioxidant status and reduced inflammation and DNA damage, which was observed through the reduction of inflammatory cytokine and C-reactive protein concentrations [72]. Methanolic extracts of *A. campanulatus* (Suran) tuber also showed immunomodulatory activity [73]. Authors indicated that the presence of steroids and flavonoids in *A. campanulatus* (Suran) tuber may be responsible for the observed immunomodulatory activity [73].

3.5. Antimicrobial Activity. Yam varieties with their phenolic compounds are potential agents with antimicrobial efficacy. Sonibare and Abegunde [74] reported that the methanolic extracts of *Dioscorea* yams (*Dioscorea dumetorum* and *Dioscorea hirtiflora*) showed antioxidant and antimicrobial activities. Antimicrobial activity was determined by the agar diffusion method (for bacteria) and pour plate method (for fungi). Nonedible *D. dumetorum* showed the highest *in vitro* antibacterial activity against *Proteus mirabilis*. The methanolic extracts from *D. hirtiflora* demonstrated antimicrobial activity against all tested organisms, namely, *Staphylococcus aureus*, *E coli*, *Bacillus subtilis*, *Proteus mirabilis*, *Salmonella typhi*, *Candida albicans*, *Aspergillus niger*, and *Penicillium chrysogenum*.

3.6. Hypoglycemic Activities. Diabetes mellitus is a chronic disorder marked by elevated levels of glucose in the blood and life-threatening complications that can untimely lead to death. Extracts of sweet potato peels have shown reduced plasma glucose levels of diabetic patients [75]. The extract of white skinned sweet potatoes (WSSP) reduced hyperinsulinemia in Zucker fatty rats by 23, 26, 60, and 50%, after 3, 4, 6, and 8 weeks, respectively, starting the oral administration of WSSP similar to troglitazone (insulin sensitizer) [76]. In addition, blood triacylglycerol (TG), free fatty acid (FFA), and lactate levels were also lowered by the oral administration of WSSP. In histological examinations of the pancreas of Zucker fatty rats, remarkable regranulation of pancreatic islet B-cells was observed in the WSSP and troglitazone groups after 8 weeks of treatment. Based on these findings it was concluded that WSSP was likely to improve the abnormal glucose and lipid metabolism in insulin-resistant diabetes mellitus. In support of these observations, ingestion of 4 g of caiapo, the extract of WSSP, per day for 6 weeks reduced fasting blood glucose and total as well as low density lipoprotein (LDL) cholesterol in male Caucasian type 2 diabetic patients who were previously treated by diet alone [75]. However, significant changes were not observed between low dose of caiapo and placebo. The improvement of insulin sensitivity, as determined by the frequently sampled intravenous glucose tolerance test (FSIGT), indicated that caiapo extracts showed beneficial effects via reducing insulin resistance. Later Ludvik et al. [77] confirmed the beneficial effects of caiapo on glucose and serum cholesterol levels in type 2 diabetic patients treated by diet alone for 3 months after administration for the study. Improved fasting blood glucose levels and glucose levels during an OGTT and in the postprandial state as well as improvement in long-term glucose control were also observed as expressed by the significant decrease in HbA1c [77].

Ethanolic extract of tubers of *Dioscorea alata* showed an antidiabetic activity in alloxan-induced diabetic rats [78]. Diabetic rats with administered yam extract exhibited significantly lower creatinine levels which could be a result of improved renal function by reduced plasma glucose level and subsequent glycosylation of renal basement membranes. Several bioactive compounds, including phenolics, were identified in the ethanolic extract of *D. alata*. These include hydro-Q9 chromene, γ-tocopherol, α-tocopherol, feruloyl glycerol, dioscorin, cyanidin-3-glucoside, catechin, procyanidin, cyanidin, peonidin 3-gentiobioside, and alatanins A, B, and C [78].

3.7. Hypocholesterolemic Activity. Cardiovascular diseases are among the leading causes of death worldwide. It is well known that diet plays an important role in the regulation of cholesterol homeostasis. External agents possessing anticholesterolemic activities continuously show beneficial effects on risk reduction and management of the disease conditions.

Diosgenin, a steroidal saponin of yam (*Dioscorea*), demonstrated antioxidative and hypolipidemic effects *in vivo* [79]. Rats fed with a high-cholesterol diet were supplemented with either 0.1 or 0.5% diosgenin for 6 weeks. The lipid profile of the plasma and liver, lipid peroxidation and antioxidant enzyme activities in the plasma, erythrocyte, and gene expression of antioxidative enzymes in the liver and the oxidative DNA damage in lymphocytes were measured. Diosgenin showed pancreatic lipase inhibitory activity, protective effect of liver under high-cholesterol diet, reduced total cholesterol level, and protection against the oxidative damaging effects of polyunsaturated fatty acids [79]. Steroidal saponins of yams are used for industrial drug processing. Saponins, such as dioscin and gracilin, and prosapogenins of dioscin have long been identified from yam. The content of dioscin was about 2.7% (w/w). Diosgenin content was about 0.004 and 0.12–0.48% in cultivated yam and wild yam, respectively. The antihypercholesterolemic effect of yam saponin is related to its inhibitory activity against cholesterol absorption [80].

The effect of yam diosgenin on hypercholesterolemia had been also reported by Cayen and Dvornik [81]. Hypercholesterolemic rats fed with yam (*Dioscorea*) showed that diosgenin had decreased cholesterol absorption, increased hepatic cholesterol synthesis, and increased biliary cholesterol secretion without affecting serum cholesterol level. In agreement with this finding, several studies showed that diosgenin, in some *Dioscorea*, could enhance fecal bile acid secretion and decrease intestinal cholesterol absorption [82, 83]. The relative contribution of biliary secretion and intestinal absorption of cholesterol in diosgenin-stimulated fecal cholesterol excretion were studied using wild-type (WT) and Niemann-Pick C1-Like 1 (NPC1L1) knockout (LIKO) mice. NPC1L1 was recently identified as an essential protein for intestinal cholesterol absorption [84]. Diosgenin significantly increased biliary cholesterol and hepatic expression of cholesterol synthetic genes in both WT and LIKO mice. In addition diosgenin stimulation of fecal cholesterol excretion was primarily attributable to its impact on hepatic cholesterol metabolism rather than NPC1L1-dependent intestinal cholesterol absorption [85].

Native protein of dioscorin purified from *D. alata* (cv. Tainong number 1) (TN1-dioscorin) and its peptic hydrolysates presented ACE inhibitory activities in a dose-dependent manner [29]. According to kinetic analysis, dioscorin showed mixed noncompetitive inhibition against ACE. Dioscorin from *Dioscorea* might be beneficial in controlling high blood pressure [85].

Chen et al. [23] reported the effects of Taiwan's yam (*Dioscorea alata* cv. Tainung number 2) on mucosal hydrolase activities and lipid metabolism in male Balb/c mice. High level of Tainung number 2 yam in the diet (50% w/w) reduced plasma and hepatic cholesterol levels and increased fecal steroid excretions in mice model. This could be due to the loss of bile acid in the enterohepatic cycle to fecal excretion [23]. They further suggested that the increased viscosity of the digest and the thickness of the unstirred layer in the small intestine caused by Tainung number 2 yam fibre (or/and mucilage) decreased the absorption of fat, cholesterol, and bile acid. Short term (3-week) consumption of 25% Tainung number 2 yam in the diet could reduce the atherogenic index but not total cholesterol level in nonhypercholesterolemic mice. Furthermore, additional dietary yam (50% yam diet) could consistently exert hypocholesterolemic effects in these mice [23]. However, Diosgenin was not elucidated in Tainung number 2 used in this study [23]. Thus, authors suggested that diosgenin might not be involved in the cholesterol lowering effect of Tainung number 2 yam. Dietary fibre and viscous mucilage could be active components for the beneficial cholesterol lowering effects of yam. Furthermore, short term consumption (3-week) of 25% uncooked keelung yam effectively reduced total blood cholesterol levels and the atherogenic index in mice. Authors indicated that the active components for the lipid lowering effects may be attributed to dietary fibre, mucilage, plant sterols, or synergism of these active components [23].

3.8. Hormonal Activities. Yam (*Dioscorea*) has the ability to reduce the risk of cancer and cardiovascular diseases in postmenopausal women [86]. It was shown that the levels of serum estrogen and sex hormone binding globulin (SHBG) increased significantly after subjects had been on a yam diet for 30 days. Furthermore, three serum hormone parameters measured, namely, estrogen, estradiol, and SHBG, did not change in those who were fed with sweet potatoes as the control. The risk of breast cancer which increased by estrogens might be balanced by the elevated SHB and the ratio of estrogen plus estradiol to SHBG. Authors further showed that high SHBG levels had a protective effect against the occurrence of type 2 diabetes mellitus and coronary heart diseases in women [86].

Chronic administration of *Dioscorea* may enhance bone strength and provide insight into the role of *Dioscorea* in bone remodeling and osteoporosis during the menopause [87]. Administration of *Dioscorea* to ovariectomised rats decreased the porosity effect on bones and increased the ultimate force of bones. The changes in biochemical and physiological functions seen in these animals were similar to those in menopausal women [87].

3.9. Antiobesity. Differentiation and proliferation inhibitory activity of sporamin of sweet potato roots in 3T3-L1 preadipocytes were reported [84]. Sporamin did not exhibit any cytotoxic activity toward the model cell line 3T3-L1 preadipocytes which have frequently been used to study differentiation of adipocytes *in vitro*. Concentration range of 0.025 to 1 mg/mL sporamin showed antidifferentiation and antiproliferation effects on 3T3-L1 cells similarly to 0.02 mg/mL berberine. Berberine is a traditional Chinese medicine used as an antimicrobial and antitumour agent [84]. Hwang et al. [88] demonstrated that purple sweet potato is a potential agent, which can be used for the prevention of obesity. Anthocyanin fractions of purple sweet potato inhibited hepatic lipid accumulation through the induction of adenosine monophosphate activated protein kinase (AMPK) signaling pathways. AMPK plays an important role in the regulation of lipid synthesis in metabolic tissues [88]. Anthocyanin dose of 200 mg/kg of body weight per day reduced weight gain and hepatic triacylglycerol accumulation and improved serum lipid parameters in mice fed for 4 weeks with purple sweet potatoes. Furthermore, anthocyanin administration increased the phosphorylation of AMPK and acetyl coenzyme A carboxylase (ACC) in the liver and HepG2 hepatocytes. These authors further suggested that anthocyanin fraction may improve high fat diet induced fatty liver disease and regulate hepatic lipid metabolism [88].

4. Effect of Processing on Bioactivities of Roots and Tubers

Tuber crops are processed in a number of ways before they are consumed and these include hydrothermal treatments such as boiling, frying, baking and roasting, dehydration, and fermentation. Depending on the region and country different types of foods are prepared at domestic and industrial levels.

Major proportions of cassava produced in Africa and Latin America are processed into fermented foods and food additives such as organic acids and monosodium glutamate. Lactic acid bacteria and yeasts are two major groups of microorganisms used for cassava fermentation. Those fermented foods from cassava include *gari*, *fufu*, *lafun*, *chickwanghe*, *agbelima*, *attieke*, and *kivunde* in Africa, tape in Asia, and "cheese" bread and "coated peanut" in Latin America [89]. Sweet potatoes also may be fermented into soy sauce, vinegar, lactojuices, lactopickles, and *sochu* (an alcoholic drink produced in Japan), and yams may be fermented into fermented flour. Furthermore, yams are used in a number of foods. Greater yams (*Dioscorea alata* L.), commonly known as *ube* in the Philippines, are utilized in sweetened food delicacies due to their attractive violet color and unique taste. *Ube* is used in a number of native delicacies such as *halaya* (yam pudding with milk), *sagobe* (with parboiled chocomilk and glutinous rice balls), *puto* (rice cake), *halo-halo*, *hopia*,

and different types of rice cake using glutinous rice such as *suman*, *sapin-sapin*, *bitso*, and *bibingka*. It can also be used as an ingredient for flavour and/or filling in ice cream, tarts, bread, and cakes. Different food processing conditions alter the content of phytochemicals as well as their bioactivities.

Processing conditions, such as peeling, drying, and sulphite treatment, could change physiochemical properties and nutritional quality of sweet potato flour [90]. Sweet potato flour is generally used to enhance characteristics of food products through color, flavour, and natural sweetness and is supplemented nutrients. It also serves as substitute for cereal flours, which particularly contain gluten which is not suited for individuals diagnosed with celiac disease [91]. Peeled and unpeeled sweet potato flour with or without sulphite treatment showed higher browning indices at 55°C and decreased with increasing drying temperatures [90]. Furthermore, β-carotene content of all flours decreased with increasing temperature. Total phenolic content decreased at a higher drying temperature for peeled and unpeeled sweet potato flour without sulphite treatment. Unpeeled sweet potato flour had higher phenolic content which was contributed by the tuber skin [92]. Sweet potato flour with sulphite treatment showed higher phenolics and ascorbic acid contents and this could be due to the inactivation of polyphenol oxidase by sulphites. Sulphite reacts with quinines and inhibits the polyphenol oxidase activity and depletes oxygen content [93].

Shih et al. [94] demonstrated the physiochemical and physiological properties and bioactivities of sweet potato dry products made from two varieties, yellow (variety: Tainong 57) and orange (variety: Tainong 66), using different processing methods such as freeze-drying, hot air-drying, and extrusion. Antioxidant contents were different between 70% methanolic extracts of yellow and orange sweet potatoes. In addition, differences of antioxidant activity were observed among freeze-dried, hot air-dried, and extruded samples [94]. The freeze-dried samples of orange potatoes showed high phenolics, β-carotene and anthocyanin contents, and free radical scavenging activities compared to those of yellow sweet potatoes. The extrusion process significantly increased the DPPH radical scavenging activity and TPC, whereas anthocyanin and β-carotene contents were decreased. The increase in the level of phenolic acids after extrusion could be due to the release of the bound phenolic acids and their derivatives from the cell walls of the plant matter [94]. The DPPH radical scavenging activity was higher for hot air-dried samples compared to those of freeze-dried yellow sweet potatoes but this trend was opposite for orange sweet potatoes [94]. They also reported that the methanolic extracts of sweet potatoes had immunomodulatory activities. Methanolic extracts of sweet potatoes increased the proliferation of the lectin concanavalin A (Con A) stimulated splenocytes of BALB/c mice in a concentration dependent manner. Freeze-dried sweet potatoes had the highest mitogenic index compared to that of hot air-drying and extrusion processing samples of both yellow and orange sweet potatoes. Phenolic compounds, anthocyanins, and β-carotene could be responsible compounds for the observed bioactivities [94, 95].

The retention of β-carotene decreased with the duration of boiling, steaming, and microwave cooking [96]. It

TABLE 4: Content of phenolic compounds during various processing methods of purple yam (*Dioscorea alata*).

	Fresh yam	Blanched	Frozen	Vacuum frying
Total phenolic content (mg gallic acid equivalent/100 g)	478 ± 11.62	306 ± 7.7	293 ± 8.93	204 ± 5.49
Total anthocyanin content (mg/100 g)	31.0 ± 2.35	12.6 ± 1.65	12.6 ± 1.54	7.92 ± 1.03
Sinapic acid (mg/100 g)	135 ± 5.27	93.1 ± 7.53	91.73 ± 5.71	55.5 ± 3.32
Ferulic acid (mg/100 g)	31.3 ± 1.08	15.94 ± 1.15	15.77 ± 1.05	9.48 ± 1.11

Source: Fang et al. [103].

was found that 50-minute boiling reduced one-third of β-carotene content. Furthermore, steaming reduced higher content of β-carotene compared to that of boiling. It was noted that microwave cooking had the highest loss. Authors suggested that long cooking time leads to high reductions in β-carotene content of sweet potatoes [96].

Different cooking methods affect the phenolic contents of potatoes [97]. A significant loss in phenolic contents of peeled and unpeeled potatoes was observed during boiling and baking. Furthermore, boiling caused loss of protocatechuic and caffeoylquinic acids of peeled potatoes by 86 and 26%, respectively. In addition, microwave cooking resulted in loss of 50 to 83% of protocatechuic acid and 27 to 64% of caffeoylquinic acids in peeled potatoes. Interestingly the losses were reported to be lower in unpeeled potatoes than those in peeled potatoes [97]. The reported losses in phenolic contents could be due to a combination of degradation caused by leaching into water, heat, and polyphenol oxidation [10, 98]. Potatoes cooked with peel had a high amount of total phenolics in the cortex and internal tissues [99]. This has been attributed to the migration of phenolics from the peel into both cortex and internal tissues of the tuber [10]. Navarre et al. [100] also showed the effects of cooking (microwave cooking, boiling, and steaming) on the phenolic contents of potatoes and the results showed that none of these methods decreased the content of chlorogenic, cryptochlorogenic, and neochlorogenic acids. Furthermore, it was found that flavonols (rutin) remained during cooking; however, there was an increase in phenolic contents of stir-fried potatoes. Different processing methods decreased the anthocyanin content due to the effect of high temperature, enzyme activity, change in pH, and presence of metallic ions and proteins [101, 102]. Another study showed that anthocyanin content in raw potatoes was higher than that of potato chips and French fries which were processed by deep frying [103].

Fang et al. [103] identified the major phenolic compounds in Chinese purple yam and their changes during vacuum frying (Table 4). They demonstrated that 40 and 63% of anthocyanin remained after blanching and vacuum frying, respectively [103]. Furthermore, hydroxycinnamic acids, namely, sinapic acids and ferulic acids, showed higher stability than that of anthocyanins and freezing did not influence the phenolic contents due to short freezing time of 20 hours [103].

Different dehydration methods used to produce yam flours could affect their antioxidant activities [104]. Freeze-dried yam flours had the strongest DPPH radical scavenging activities compared to tjose of hot air-dried or drum-dried yam flours. Chen and Lin [52] also reported the negative

impact of temperature on the content of phenolic compounds and dioscorin and antioxidant activities of yam cultivars. Chen et al. [104] further showed the effect of pH on phenolic contents, antioxidant activity, and stability of dioscorin in various yam tubers. The TPC observed of yam varieties were the highest at pH 5 and gradual decrease was observed with increased pH. The DPPH radical scavenging activity showed a similar pattern to those of phenolic contents of yams at pH 5, but ferrous ion chelating capacity was found to be high for all yams at pH 8 [104].

5. Conclusions

Roots and tubers are important diet components for humans and add variety to it. In addition to the main role as an energy contributor, they provide a number of desirable nutritional and health benefits such as antioxidative, hypoglycemic, hypocholesterolemic, antimicrobial, and immunomodulatory activities. A variety of foods can be prepared using tubers and type and usage vary with the country and region. Processing affects the bioactivities of constituent compounds. Tubers may serve as functional foods and nutraceutical ingredients to attenuate noncommunicable chronic diseases and to maintain wellness.

Competing Interests

The authors declare that they have no competing interests.

Acknowledgments

This research was supported by the Research Grant Scheme of Wayamba University of Sri Lanka through a grant (SRHDC/RP/04/13-09) to Anoma Chandrasekara.

References

[1] FAOSTAT, 2013, http://faostat3.fao.org.

[2] Food and Agriculture Organization (FAO), *Roots, Tubers, Plantains and Bananas in Human Nutrition*, vol. 24 of *Food and Nutrition Series*, Food and Agriculture Organization, Rome, Italy, 1990.

[3] M. Naczk and F. Shahidi, "Phenolics in cereals, fruits and vegetables: occurrence, extraction and analysis," *Journal of Pharmaceutical and Biomedical Analysis*, vol. 41, pp. 1523–1542, 2006.

[4] FAO, *Production Year Book*, vol. 53, Food and Agriculture Organization, Rome, Italy, 1999.

[5] B. Burlingame, B. Mouillé, and R. Charrondière, "Nutrients, bioactive non-nutrients and anti-nutrients in potatoes," *Journal*

of Food Composition and Analysis, vol. 22, no. 6, pp. 494–502, 2009.

[6] A. L. Hale, L. Reddivari, M. N. Nzaramba, J. B. Bamberg, and J. C. Miller Jr., "Interspecific variability for antioxidant activity and phenolic content among *Solanum* species," *American Journal of Potato Research*, vol. 85, no. 5, pp. 332–341, 2008.

[7] D. A. Navarre, A. Goyer, and R. Shakya, "Nutritional value of potatoes; Vitamin, phyto-nutrient and mineral content," in *Advances in Potatoes Chemistry and Technology*, J. Singh and L. Kaur, Eds., Elsevier, Amsterdam, The Netherlands, 2009.

[8] H. McNulty and K. Pentieva, "Folate bioavailability," *Proceedings of the Nutrition Society*, vol. 63, no. 4, pp. 529–536, 2004.

[9] G. Diretto, S. Al-Babili, R. Tavazza, V. Papacchioli, P. Beyer, and G. Giuliano, "Metabolic engineering of potato carotenoid content through tuber-specific overexpression of a bacterial mini-pathway," *PLoS ONE*, vol. 2, no. 4, article e350, 2007.

[10] R. Ezekiel, N. Singh, S. Sharma, and A. Kaur, "Beneficial phytochemicals in potato-a review," *Food Research International*, vol. 50, no. 2, pp. 487–496, 2013.

[11] G. J. Scott, "Transforming traditional food crops: product development for roots and tubers," in *Product Development for Root and Tuber Crops*, vol. 1, pp. 3–20, Asia International Potato Center, 1992.

[12] A. C. Bovell-Benjamin, "Sweet potato: a review of its past, present, and future role in human nutrition," *Advances in Food and Nutrition Research*, vol. 52, pp. 1–59, 2007.

[13] P. J. van Jaarsveld, D. W. Marais, E. Harmse, P. Nestel, and D. B. Rodriguez-Amaya, "Retention of β-carotene in boiled, mashed orange-fleshed sweet potato," *Journal of Food Composition and Analysis*, vol. 19, no. 4, pp. 321–329, 2006.

[14] N. M. A. Nassar, D. Y. C. Hashimoto, and S. D. C. Fernandes, "Wild *Manihot* species: botanical aspects, geographic distribution and economic value," *Genetics and Molecular Research*, vol. 7, no. 1, pp. 16–28, 2008.

[15] I. S. Blagbrough, S. A. L. Bayoumi, M. G. Rowan, and J. R. Beeching, "Cassava: an appraisal of its phytochemistry and its biotechnological prospects—review," *Phytochemistry*, vol. 71, no. 17-18, pp. 1940–1951, 2010.

[16] H. Prawat, C. Mahidol, S. Ruchirawat et al., "Cyanogenic and non-cyanogenic glycosides from *Manihot esculenta*," *Phytochemistry*, vol. 40, no. 4, pp. 1167–1173, 1995.

[17] K. Reilly, R. Gómez-Vásquez, H. Buschmann, J. Tohme, and J. R. Beeching, "Oxidative stress responses during cassava postharvest physiological deterioration," *Plant Molecular Biology*, vol. 56, no. 4, pp. 625–641, 2004.

[18] Y.-W. Liu, H.-F. Shang, C.-K. Wang, F.-L. Hsu, and W.-C. Hou, "Immunomodulatory activity of dioscorin, the storage protein of yam (*Dioscorea alata* cv. Tainong No. 1) tuber," *Food and Chemical Toxicology*, vol. 45, no. 11, pp. 2312–2318, 2007.

[19] M. M. Iwu, C. O. Okunji, G. O. Ohiaeri, P. Akah, D. Corley, and M. S. Tempesta, "Hypoglycaemic activity of dioscoretine from tubers of *Dioscorea dumetorum* in normal and alloxan diabetic rabbits," *Planta Medica*, vol. 56, no. 3, pp. 264–267, 1990.

[20] M. R. Bhandari, T. Kasai, and J. Kawabata, "Nutritional evaluation of wild yam (*Dioscorea* spp.) tubers of Nepal," *Food Chemistry*, vol. 82, no. 4, pp. 619–623, 2003.

[21] J. E. Kelmanson, A. K. Jäger, and J. van Staden, "Zulu medicinal plants with antibacterial activity," *Journal of Ethnopharmacology*, vol. 69, no. 3, pp. 241–246, 2000.

[22] Y.-C. Chan, C.-K. Hsu, M.-F. Wang, and T.-Y. Su, "A diet containing yam reduces the cognitive deterioration and brain lipid peroxidation in mice with senescence accelerated," *International Journal of Food Science and Technology*, vol. 39, no. 1, pp. 99–107, 2004.

[23] H.-L. Chen, C.-H. Wang, C.-T. Chang, and T.-C. Wang, "Effects of Taiwanese yam (*Dioscorea japonica* Thunb var. pseudojaponica Yamamoto) on upper gut function and lipid metabolism in Balb/c mice," *Nutrition*, vol. 19, no. 7-8, pp. 646–651, 2003.

[24] I. Podolak, A. Galanty, and D. Sobolewska, "Saponins as cytotoxic agents: a review," *Phytochemistry Reviews*, vol. 9, no. 3, pp. 425–474, 2010.

[25] C.-H. Huang, J.-Y. Cheng, M.-C. Deng, C.-H. Chou, and T.-R. Jan, "Prebiotic effect of diosgenin, an immunoactive steroidal sapogenin of the Chinese yam," *Food Chemistry*, vol. 132, no. 1, pp. 428–432, 2012.

[26] W.-C. Hou and Y.-H. Lin, "Dioscorins, the major tuber storage proteins of yam (*Dioscorea batatas* Decne), with dehydroascorbate reductase and monodehydroascorbate reductase activities," *Plant Science*, vol. 149, no. 2, pp. 151–156, 1999.

[27] W.-C. Hou, H.-J. Chen, and Y.-H. Lin, "Dioscorins from different *Dioscorea* species all exhibit both carbonic anhydrase and trypsin inhibitor activities," *Botanical Bulletin of Academia Sinica*, vol. 41, no. 3, pp. 191–196, 2000.

[28] W.-C. Hou, M.-H. Lee, H.-J. Chen et al., "Antioxidant activities of dioscorin, the storage protein of yam (*Dioscorea batatas* Decne) tuber," *Journal of Agricultural and Food Chemistry*, vol. 49, no. 10, pp. 4956–4960, 2001.

[29] F.-L. Hsu, Y.-H. Lin, M.-H. Lee, C.-L. Lin, and W.-C. Hou, "Both dioscorin, the tuber storage protein of yam (*Dioscorea alata* cv. Tainong No. 1), and its peptic hydrolysates exhibited angiotensin converting enzyme inhibitory activities," *Journal of Agricultural and Food Chemistry*, vol. 50, no. 21, pp. 6109–6113, 2002.

[30] J.-Y. Lin, S. Lu, Y.-L. Liou, and H.-L. Liou, "Antioxidant and hypolipidaemic effects of a novel yam-boxthorn noodle in an in vivo murine model," *Food Chemistry*, vol. 94, no. 3, pp. 377–384, 2006.

[31] P. R. Shewry, "Tuber storage proteins," *Annals of Botany*, vol. 91, no. 7, pp. 755–769, 2003.

[32] R. Senthilkumar and K.-W. Yeh, "Multiple biological functions of sporamin related to stress tolerance in sweet potato (*Ipomoea batatas* Lam)," *Biotechnology Advances*, vol. 30, no. 6, pp. 1309–1317, 2012.

[33] K.-W. Yeh, J.-C. Chen, M.-I. Lin, Y.-M. Chen, and C.-Y. Lin, "Functional activity of sporamin from sweet potato (*Ipomoea batatas* Lam.): a tuber storage protein with trypsin inhibitory activity," *Plant Molecular Biology*, vol. 33, no. 3, pp. 565–570, 1997.

[34] W.-C. Hou and Y.-H. Lin, "Dehydroascorbate reductase and monodehydroascorbate reductase activities of trypsin inhibitors, the major sweet potato (*Ipomoea batatas* [L.] Lam) root storage protein," *Plant Science*, vol. 128, no. 2, pp. 151–158, 1997.

[35] M. F. B. Dale, D. W. Griffiths, H. Bain, and D. Todd, "Glycoalkaloid increase in *Solarium tuberosum* on exposure to light," *Annals of Applied Biology*, vol. 123, no. 2, pp. 411–418, 1993.

[36] I. Ginzberg, J. G. Tokuhisa, and R. E. Veilleux, "Potato steroidal glycoalkaloids: biosynthesis and genetic manipulation," *Potato Research*, vol. 52, no. 1, pp. 1–15, 2009.

[37] M. Friedman, "Potato glycoalkaloids and metabolites: roles in the plant and in the diet," *Journal of Agricultural and Food Chemistry*, vol. 54, no. 23, pp. 8655–8681, 2006.

[38] K.-W. Kuo, S.-H. Hsu, Y.-P. Li et al., "Anticancer activity evaluation of the Solanum glycoalkaloid solamargine: triggering apoptosis in human hepatoma cells," *Biochemical Pharmacology*, vol. 60, no. 12, pp. 1865–1873, 2000.

[39] L.-F. Liu, C.-H. Liang, L.-Y. Shiu, W.-L. Lin, C.-C. Lin, and K.-W. Kuo, "Action of solamargine on human lung cancer cells-enhancement of the susceptibility of cancer cells to TNFs," *FEBS Letters*, vol. 577, no. 1-2, pp. 67–74, 2004.

[40] T. Ikeda, H. Tsumagari, T. Honbu, and T. Nohara, "Cytotoxic activity of steroidal glycosides from solanum plants," *Biological and Pharmaceutical Bulletin*, vol. 26, no. 8, pp. 1198–1201, 2003.

[41] W. Stahl and H. Sies, "Lycopene: a biologically important carotenoid for humans?" *Archives of Biochemistry and Biophysics*, vol. 336, no. 1, pp. 1–9, 1996.

[42] N. I. Krinsky, "Actions of carotenoids in biological systems," *Annual Review of Nutrition*, vol. 13, pp. 561–587, 1993.

[43] H. Tapiero, D. M. Townsend, and K. D. Tew, "The role of carotenoids in the prevention of human pathologies," *Biomedicine and Pharmacotherapy*, vol. 58, no. 2, pp. 100–110, 2004.

[44] S. Blanquet-Diot, MahaSoufi, M. Rambeau, E. Rock, and M. Alric, "Digestive stability of xanthophylls exceeds that of carotenes as studied in a dynamic in vitro gastrointestinal system," *Journal of Nutrition*, vol. 139, no. 5, pp. 876–883, 2009.

[45] F. Hesam, G. R. Balali, and R. T. Tehrani, "Evaluation of antioxidant activity of three common potato (Solanum tuberosum) cultivars in Iran ," *Avicenna Journal of Phytomedicine*, vol. 2, no. 2, pp. 79–85, 2012.

[46] R. Chimkode, M. B. Patil, and S. S. Jalalpure, "Wound healing activity of tuberous root extracts of *Ipomoea batatas*," *Advances in Pharmacology and Toxicology*, vol. 10, pp. 69–72, 2009.

[47] V. Panda and M. Sonkamble, "Anti-ulcer activity of *Ipomoea batatas* tubers (sweet potato)," *Functional Foods in Health and Disease*, vol. 2, no. 3, pp. 48–61, 2011.

[48] T. Suzuki, H. Tada, E. Sato, and Y. Sagae, "Application of sweet potato fiber to skin wound in rat," *Biological and Pharmaceutical Bulletin*, vol. 19, no. 7, pp. 977–983, 1996.

[49] L. Dilworth, K. Brown, R. Wright, M. Oliver, S. Hall, and H. Asemota, "Antioxidants, minerals and bioactive compounds in tropical staples," *African Journal of Food Science and Technology*, vol. 3, no. 4, pp. 90–98, 2012.

[50] D. F. Cornago, R. G. O. Rumbaoa, and I. M. Geronimo, "Philippine Yam (*Dioscorea* spp.) tubers phenolic content and antioxidant capacity," *Philippine Journal of Science*, vol. 140, no. 2, pp. 145–152, 2011.

[51] C.-K. Hsu, J.-Y. Yeh, and J.-H. Wei, "Protective effects of the crude extracts from yam (*Dioscorea alata*) peel on tert-butylhydroperoxide-induced oxidative stress in mouse liver cells," *Food Chemistry*, vol. 126, no. 2, pp. 429–434, 2011.

[52] Y.-T. Chen and K.-W. Lin, "Effects of heating temperature on the total phenolic compound, antioxidative ability and the stability of dioscorin of various yam cultivars," *Food Chemistry*, vol. 101, no. 3, pp. 955–963, 2007.

[53] Y.-C. Chan, S.-C. Chang, S.-Y. Liu, H.-L. Yang, Y.-C. Hseu, and J.-W. Liao, "Beneficial effects of yam on carbon tetrachloride-induced hepatic fibrosis in rats," *Journal of the Science of Food and Agriculture*, vol. 90, no. 1, pp. 161–167, 2010.

[54] Y. H. Chu and C. L. Chang, "Flavonoid content of several vegetables and their antioxidant activity," *Journal of the Science of Food and Agriculture*, vol. 80, no. 5, pp. 561–566, 2000.

[55] C. R. Brown, "Breeding for phytonutrient enhancement of potato," *American Journal of Potato Research*, vol. 85, no. 4, pp. 298–307, 2008.

[56] K.-H. Han, K.-I. Shimada, M. Sekikawa, and M. Fukushima, "Anthocyanin-rich red potato flakes affect serum lipid peroxidation and hepatic SOD mRNA level in rats," *Bioscience, Biotechnology and Biochemistry*, vol. 71, no. 5, pp. 1356–1359, 2007.

[57] X. Ji, L. Rivers, Z. Zielinski et al., "Quantitative analysis of phenolic components and glycoalkaloids from 20 potato clones and in vitro evaluation of antioxidant, cholesterol uptake, and neuroprotective activities," *Food Chemistry*, vol. 133, no. 4, pp. 1177–1187, 2012.

[58] N. F. Omar, S. A. Hassan, U. K. Yusoff, N. A. P. Abdullah, P. E. M. Wahab, and U. R. Sinniah, "Phenolics, flavonoids, antioxidant activity and cyanogenic glycosides of organic and mineral-base fertilized cassava tubers," *Molecules*, vol. 17, no. 3, pp. 2378–2387, 2012.

[59] V. Panda and M. Sonkamble, "Phytochemical constituents and pharmacological activities of *Ipomoea batatas* l. (Lam)—a review," *International Journal of Research in Phytochemistry & Pharmacology*, vol. 2, no. 1, pp. 25–34, 2012.

[60] J. J. Kim, C. W. Kim, D. S. Park et al., "Effects of sweet potato fractions on alcoholic hangover and gastric ulcer," *Laboratory Animal Research*, vol. 24, pp. 209–216, 2008.

[61] D.-J. Huang, C.-D. Lin, H.-J. Chen, and Y.-H. Lin, "Antioxidant and antiproliferative activities of sweet potato (*Ipomoea batatas* [L.] Lam "Tainong 57") constituents," *Botanical Bulletin of Academia Sinica*, vol. 45, no. 3, pp. 179–186, 2004.

[62] M. Yoshimoto, S. Okuno, M. Yoshinaga, O. Yamakawa, M. Yamaguchi, and J. Yamada, "Antimutagenicity of sweet potato (*Ipomoea batatas*) roots," *Bioscience, Biotechnology and Biochemistry*, vol. 63, no. 3, pp. 537–541, 1999.

[63] M. D. Thompson, H. J. Thompson, J. N. McGinley et al., "Functional food characteristics of potato cultivars (*Solanum tuberosum* L.): phytochemical composition and inhibition of 1-methyl-1-nitrosourea induced breast cancer in rats," *Journal of Food Composition and Analysis*, vol. 22, no. 6, pp. 571–576, 2009.

[64] G. P. Madiwale, L. Reddivari, D. G. Holm, and J. Vanamala, "Storage elevates phenolic content and antioxidant activity but suppresses antiproliferative and pro-apoptotic properties of colored-flesh potatoes against human colon cancer cell lines," *Journal of Agricultural and Food Chemistry*, vol. 59, no. 15, pp. 8155–8166, 2011.

[65] K. K.-W. Auyeung, P.-C. Law, and J. K.-S. Ko, "*Astragalus* saponins induce apoptosis via an ERK-independent NF-κB signaling pathway in the human hepatocellular HepG2 cell line," *International Journal of Molecular Medicine*, vol. 23, no. 2, pp. 189–196, 2009.

[66] K.-R. Lee, N. Kozukue, J.-S. Han et al., "Glycoalkaloids and metabolites inhibit the growth of human colon (HT29) and liver (HepG2) cancer cells," *Journal of Agricultural and Food Chemistry*, vol. 52, no. 10, pp. 2832–2839, 2004.

[67] T.-S. Wang, C.-K. Lii, Y.-C. Huang, J.-Y. Chang, and F.-Y. Yang, "Anticlastogenic effect of aqueous extract from water yam (*Dioscorea alata* L.)," *Journal of Medicinal Plant Research*, vol. 5, no. 26, pp. 6192–6202, 2011.

[68] E. M. Choi, S. J. Koo, and J.-K. Hwang, "Immune cell stimulating activity of mucopolysaccharide isolated from yam (*Dioscorea batatas*)," *Journal of Ethnopharmacology*, vol. 91, no. 1, pp. 1–6, 2004.

[69] P.-F. Su, C.-J. Li, C.-C. Hsu et al., "Dioscorea phytocompounds enhance murine splenocyte proliferation ex vivo and improve regeneration of bone marrow cells in vivo," *Evidence-Based*

Complementary and Alternative Medicine, vol. 2011, Article ID 731308, 11 pages, 2011.

[70] H.-F. Shang, H.-C. Cheng, H.-J. Liang, H.-Y. Liu, S.-Y. Liu, and W.-C. Hou, "Immunostimulatory activities of yam tuber mucilages," *Botanical Studies*, vol. 48, no. 1, pp. 63–70, 2007.

[71] J. O. Olayemi and E. O. Ajaiyeoba, "Anti-inflammatory studies of yam (*Dioscorea esculenta*) extract on Wistar rats," *African Journal of Biotechnology*, vol. 6, no. 16, pp. 1913–1915, 2007.

[72] K. L. Kaspar, J. S. Park, C. R. Brown, B. D. MAthison, D. A. Navarre, and B. P. Chew, "Pigmented potato consumption alters oxidative stress and inflammatory damage in men," *Journal of Nutrition*, vol. 141, no. 1, pp. 108–111, 2011.

[73] A. S. Tripathi, V. Chitra, N. W. Sheikh, D. S. Mohale, and A. P. Dewani, "Immunomodulatory activity of the methanol extract of *Amorphophallus campanulatus* (Araceae) Tuber," *Tropical Journal of Pharmaceutical Research*, vol. 9, no. 5, pp. 451–454, 2010.

[74] M. A. Sonibare and R. B. Abegunde, "*In vitro* antimicrobial and antioxidant analysis of *Dioscorea dumetorum* (Kunth) Pax and *Dioscorea hirtiflora* (Linn.) and their bioactive metabolites from Nigeria," *Journal of Applied Biosciences*, vol. 51, pp. 3583–3590, 2012.

[75] B. H. Ludvik, K. Mahdjoobian, W. Waldhaeusl et al., "The effect of *Ipomoea batatas* (Caiapo) on glucose metabolism and serum cholesterol in patients with type 2 diabetes: a randomized study," *Diabetes Care*, vol. 25, no. 1, pp. 239–240, 2002.

[76] S. Kusano and H. Abe, "Antidiabetic activity of white skinned sweet potato (*Ipomoea batatas* L.) in obese Zucker fatty rats," *Biological and Pharmaceutical Bulletin*, vol. 23, no. 1, pp. 23–26, 2000.

[77] B. Ludvik, B. Neuffer, and G. Pacini, "Efficacy of *Ipomoea batatas* (Caiapo) on diabetes control in type 2 diabetic subjects treated with diet," *Diabetes Care*, vol. 27, no. 2, pp. 436–440, 2004.

[78] V. Maithili, S. P. Dhanabal, S. Mahendran, and R. Vadivelan, "Antidiabetic activity of ethanolic extract of tubers of *Dioscorea alata* in alloxan induced diabetic rats," *Indian Journal of Pharmacology*, vol. 43, no. 4, pp. 455–459, 2011.

[79] I. S. Son, J. H. Kim, H. Y. Sohn, K. H. Son, J.-S. Kim, and C.-S. Kwon, "Antioxidative and hypolipidemic effects of diosgenin, a steroidal saponin of yam (*Dioscorea* spp.), on high-cholesterol fed rats," *Bioscience, Biotechnology and Biochemistry*, vol. 71, no. 12, pp. 3063–3071, 2007.

[80] H. Y. Ma, Z. T. Zhao, L. J. Wang, Y. Wang, Q. L. Zhou, and B. X. Wang, "Comparative study on anti-hypercholesterolemia activity of diosgenin and total saponin of *Dioscorea panthaica*," *China Journal of Chinese Materia Medica*, vol. 27, no. 7, pp. 528–531, 2002.

[81] M. N. Cayen and D. Dvornik, "Effect of diosgenin on lipid metabolism in rats," *Journal of Lipid Research*, vol. 20, no. 2, pp. 162–174, 1979.

[82] A. Thewles, R. A. Parslow, and R. Coleman, "Effect of diosgenin on biliary cholesterol transport in the rat," *Biochemical Journal*, vol. 291, no. 3, pp. 793–798, 1993.

[83] K. Uchida, H. Takse, Y. Nomura, K. Takeda, N. Takeuchi, and Y. Ishikawa, "Effects of diosgenin and B-sisterol on bile acids," *The Journal of Lipid Research*, vol. 25, pp. 236–245, 1984.

[84] X. Zhi-Dong, L. Peng-Gao, and M. Tai-Hua, "The differentiation- and proliferation-inhibitory effects of sporamin from sweet potato in 3T3-L1 preadipocytes," *Agricultural Sciences in China*, vol. 8, no. 6, pp. 671–677, 2009.

[85] R. E. Temel, J. M. Brown, Y. Ma et al., "Diosgenin stimulation of fecal cholesterol excretion in mice is not NPC1L1 dependent," *Journal of Lipid Research*, vol. 50, no. 5, pp. 915–923, 2009.

[86] W.-H. Wu, L.-Y. Liu, C.-J. Chung, H.-J. Jou, and T.-A. Wang, "Estrogenic effect of yam ingestion in healthy postmenopausal women," *Journal of the American College of Nutrition*, vol. 24, no. 4, pp. 235–243, 2005.

[87] J.-H. Chen, J. S.-S. Wu, H.-C. Lin et al., "Dioscorea improves the morphometric and mechanical properties of bone in ovariectomised rats," *Journal of the Science of Food and Agriculture*, vol. 88, no. 15, pp. 2700–2706, 2008.

[88] Y. P. Hwang, J. H. Choi, E. H. Han et al., "Purple sweet potato anthocyanins attenuate hepatic lipid accumulation through activating adenosine monophosphate–activated protein kinase in human HepG2 cells and obese mice," *Nutrition Research*, vol. 31, no. 12, pp. 896–906, 2011.

[89] R. C. Ray and P. S. Sivakumar, "Traditional and novel fermented foods and beverages from tropical root and tuber crops: review," *International Journal of Food Science and Technology*, vol. 44, no. 6, pp. 1073–1087, 2009.

[90] M. Ahmed, M. S. Akter, and J.-B. Eun, "Peeling, drying temperatures, and sulphite-treatment affect physicochemical properties and nutritional quality of sweet potato flour," *Food Chemistry*, vol. 121, no. 1, pp. 112–118, 2010.

[91] L. Caperuto, J. Amaya-Farfan, and C. Camargo, "Performance of quinoa (*Chenopodium quinoa* wild) flour in the manufacture of gluten-free spaghetti," *Journal of the Science of Food and Agriculture*, vol. 81, pp. 95–101, 2000.

[92] N. I. Mondy and B. Y. Gosselin, "Effect of peeling on total phenols, total glycoalkaloids, discoloration and flavor of cooked potatoes," *Journal of Food Science*, vol. 53, no. 3, pp. 756–759, 1988.

[93] G. M. Sapers, P. H. Cooke, A. E. Heidel, S. T. Martin, and R. L. Miller, "Structural changes related to texture of pre-peeled potatoes," *Journal of Food Science*, vol. 62, no. 4, pp. 797–803, 1997.

[94] M.-C. Shih, C.-C. Kuo, and W. Chiang, "Effects of drying and extrusion on colour, chemical composition, antioxidant activities and mitogenic response of spleen lymphocytes of sweet potatoes," *Food Chemistry*, vol. 117, no. 1, pp. 114–121, 2009.

[95] M. Kampa, V.-I. Alexaki, G. Notas et al., "Antiproliferative and apoptotic effects of selective phenolic acids on T47D human breast cancer cells: potential mechanisms of action," *Breast Cancer Research*, vol. 6, no. 2, pp. R63–74, 2004.

[96] X. Wu, C. Sun, L. Yang, G. Zeng, Z. Liu, and Y. Li, "β-Carotene content in sweet potato varieties from China and the effect of preparation on β-carotene retention in the Yanshu No. 5," *Innovative Food Science & Emerging Technologies*, vol. 9, no. 4, pp. 581–586, 2008.

[97] A. A. Barba, A. Calabretti, M. d'Amore, A. L. Piccinelli, and L. Rastrelli, "Phenolic constituents levels in cv. Agria potato under microwave processing," *LWT—Food Science and Technology*, vol. 41, no. 10, pp. 1919–1926, 2008.

[98] M. Takenaka, K. Nanayama, S. Isobe, and M. Murata, "Changes in caffeic acid derivatives in sweet potato (*Ipomoea batatas* L.) during cooking and processing," *Bioscience, Biotechnology and Biochemistry*, vol. 70, no. 1, pp. 172–177, 2006.

[99] N. I. Mondy and B. Gosselin, "Effect of irradiation on discoloration, phenols and lipids of potatoes," *Journal of Food Science*, vol. 54, no. 4, pp. 982–984, 1989.

[100] D. A. Navarre, R. Shakya, J. Holden, and S. Kumar, "The effect of different cooking methods on phenolics and vitamin C in

developmentally young potato tubers," *American Journal of Potato Research*, vol. 87, no. 4, pp. 350–359, 2010.

[101] M. Rein, *Copigmentation reactions and color stability of berry anthocyanins [Dissertation]*, University of Helsinki, Helsinki, Finland, 2005.

[102] A. Patras, N. P. Brunton, C. O'Donnell, and B. K. Tiwari, "Effect of thermal processing on anthocyanin stability in foods; mechanisms and kinetics of degradation," *Trends in Food Science and Technology*, vol. 21, no. 1, pp. 3–11, 2010.

[103] Z. Fang, D. Wu, D. Yü, X. Ye, D. Liu, and J. Chen, "Phenolic compounds in Chinese purple yam and changes during vacuum frying," *Food Chemistry*, vol. 128, no. 4, pp. 943–948, 2011.

[104] Y.-T. Chen, W.-T. Kao, and K.-W. Lin, "Effects of pH on the total phenolic compound, antioxidative ability and the stability of dioscorin of various yam cultivars," *Food Chemistry*, vol. 107, no. 1, pp. 250–257, 2008.

[105] USDA NAL, 2015, https://fnic.nal.usda.gov/food-composition.

Porous Crumb Structure of Leavened Baked Products

H. A. Rathnayake ⓘ,[1] **S. B. Navaratne,**[1] **and C. M. Navaratne**[2]

[1]*Department of Food Science and Technology, Faculty of Applied Sciences, University of Sri Jayewardenepura,
 Gangodawila, Nugegoda, Sri Lanka*
[2]*Department of Agricultural Engineering, Faculty of Agriculture, University of Ruhuna, Mapalana, Kamburupitiya, Sri Lanka*

Correspondence should be addressed to H. A. Rathnayake; heshani@sci.sjp.ac.lk

Academic Editor: Salam A. Ibrahim

Quality evaluation of the porous crumb structure of leavened baked goods, especially bread, has become a vast study area of which various research studies have been carried out up to date. Here is a brief review focusing on those studies with six main parts including porous crumb structure development, crumb cellular structure analysis, application of fractal dimension for evaluating crumb cellular structure, mechanical and sensorial properties of crumb structure, changes of porous crumb structure with staling, and modifications to obtain a well-developed porous crumb structure and retard staling. Development of the porous crumb structure mainly depends on dough ingredients and processing conditions. Hence, certain modifications for those factors (incorporating food hydrocolloids, emulsifiers, improvers, etc.) have been conducted by cereal sciences for obtaining well-developed porous crumb structure and retard staling. Several image analysis methods are available for analyzing microstructural features of porous crumb structure, which can directly affect the mechanical and sensorial properties of the final product. A product with a well-developed porous crumb structure may contain the property of higher gas retention capacity which results in a product with increased volume and reduced crumb hardness with appealing sensorial properties.

1. Introduction

Leavened baked products include a wide range of food products, such as bread, buns, and cakes that are commonly consumed throughout the world over the past 150 years [1]. Even though foods have similar chemical composition, they may exhibit different mechanical behavior and sensorial properties depending on their cellular structure [2, 3]. Hence, the quality parameters of leavened baked products are mainly related to crumb's mechanical and sensorial properties that may influence consumer purchase [3, 4].

Crumb grain has been defined in literature as the exposed cell structure of crumb when a leavened baked product is sliced [5–7] which can be generally seen as a two-phase soft cellular solid, consisting of a solid phase apparent in the cell wall structure and a fluid phase made up of air [8, 9]. According to materials science approach, solid cellular materials can be categorized basically as open or closed cell foam. Open cell structured porous food materials consist of pores that are connected to each other through an interconnected

network [10, 11], which is comparatively softer than closed cell foam structures [11]. The cell foam that does not have interconnected pores is considered as closed cell forms [10, 11] and contains higher compressive strength due to the dense structures [11].

Due to the complex mechanical behavior of crumb structure [8, 12–15], close examination of different slices can reveal considerable variation in the cell characteristics even within a single sample [13]. Hence, vast range of research studies have been carried out through decades for understanding the structure and properties of the crumb structure with regard to the mechanical and sensorial quality of the final product [2, 5, 12, 16–18].

The purpose of this review is to identify and summarize the literature that covers the characteristics and development of the porous structure of dough and how it can affect the physical and sensorial quality of the final product. The most commonly concerned properties include product volume [19, 20], texture [4, 6, 8], and cellular structural properties. Further, this review covers some literature of physicochemical changes of porous crumb structure during storage period (crumb staling) and certain research studies carried out with a view to improve the porous crumb structure and retard the staling process.

2. Porous Crumb Structure Development

Development of porous crumb structure mainly depends on dough ingredients, processing conditions [5, 8, 11, 12, 19, 21], yeast activity, fermentation temperature, and gas bubble formation [21, 22].

The basic ingredients that are used for a leavened baked product are flour, water, leavening agent (either a chemical leavening agent like $NaHCO_3$ or biological leavening agent like yeast), NaCl [8, 10], sugar, and shortening. There are number of processes to convert the ingredients into a well-developed porous structure, where the main processing steps involve kneading, fermentation, proofing, and baking.

Water and flour are the most significant ingredients which may affect considerably the texture and crumb properties [1].

Wheat flour is the most commonly used flour type for leavened baked products made up of a mixture of two groups of proteins named gliadins and glutenins [23, 24]. During mixing and hydration, these two proteins combine together and form a viscoelastic gluten network that can retain leavened gas during fermentation and baking [23, 25–27]. The starch associated with this gluten network (rather the moistened starch) becomes gelatinized during heating and form a semirigid structure to the product along with the coagulated gluten (gluten protein gets denatured during heating and protein-protein crosslinking occurs via formation of disulfide bond) [11, 28, 29]. Additionally, Rouillé et al. [30] stated that the soluble fraction of wheat flour affects both loaf volume and crumb fineness in an opposite way. According to the research conducted by He and Hoseney [31], selection of flour with better protein quality may result in a product with better porous structure with uniform sized gas cells. Poorly built gluten network may fail to retain leavened gas that results in a product with lower loaf volume.

Approximately 50% water results in a finely textured, light crumb, and a dough prepared with a higher water percentages may result in a coarser crumb with more carbon dioxide (CO_2) [1].

Two forms of yeast are being used for leavened baked products naming, moist pressed cakes and dehydrated granules both of which consist of billions of living cells of Saccharomyces cerevisiae [14, 28]. When wheat flour was rehydrated, the yeast begins to metabolize and ferment producing CO_2 as a by-product (the mechanism of yeast fermentation will be described in a latter part of this review). When using a wheat flour type with a little gluten development ability (E.g., cake flour), baking powder (chemical leavening agents) may be used. If a biological leavening agent is used, gas evolving rate becomes high and leavened gas will largely escape from the batter. Hence the gas cells may overexpand and may lead to collapse resulting in a coarse-grain structure with lowered volume [28].

Brooker [32] had mentioned that the addition of small amount of shortening to the dough may lead to improve loaf volume and results in finer and more uniform crumb structure within cell walls. Further, Brooker [32] has found that the addition of crystal fat is far better than addition of oil. When adding shortening to the dough mix, fat crystals are ejected from shortenings during mixing, become enveloped by a fat (crystal)–water interface, and are able to stabilise large numbers of small air bubbles by adsorbing to their surface. During baking, air bubbles can expand without rupturing because of extra interfacial material provided by adsorbed fat crystals when they melt which result in a product with fine crumb structure [28, 32].

Sugar may act as a tenderizer, sweetener, and additional fermentable substrate. And also sugar has moisture retaining properties of baked goods [28]. Additionally, sugar has the ability to increase starch gelatinization and protein denaturation temperatures, which may lead to improve air bubble expansion during baking [11, 33].

Kneading process may lead the ingredients to get mixed homogeneously, to absorb water by hydrophilic groups of flour protein molecules, for the development of gluten protein, and to build up a viscoelastic structure and entrainment of air into the dough mass [8, 12, 14, 23, 25–28]. Certain researches have mentioned that the nuclei for gas cell development can be generated during the mixing process within the air phase of the dough [8, 12, 14, 23].

During fermentation, the yeast cells utilize the carbohydrates in the absence of oxygen (since dough making is reported as an anaerobic process) to produce energy, alcohol (ethanol), and CO_2 as the end-products [12, 34, 35] via a series of intermediate stages, in which many enzymes take part. Apart from that, fermentation process is also important for formation of flavor substances [14, 35].

The generated CO_2 may partly dissolve within the liquid phase and diffuses to the nuclei generated during mixing stage due to concentration gradient of gas [8, 12, 31, 36–38] that causes the modifications of the dough structure causing physicochemical changes in gluten network and other proteins giving the characteristic porosity of the porous crumb [22, 37]. When CO_2 defused to the nuclei in the liquid phase, the nuclei may expand into gas cells [8, 31, 39] and the density of the dough can be reduced [8] while increasing the pressure a little [31]. He and Hoseney [31] had stated that the pressure inside the gas cells could be slightly greater than that of the atmospheric pressure which had been reported as 1.01 atm. This small pressure increment (0.01atm) had occurred due to the result of surface tension at the gas-dough interface and the viscous resistance of the dough to expansion.

The process of carbohydrate fermentation is scientifically known as the tricarboxylic acid cycle (TCA) [35]. Baker's yeast can ferment all the main types of sugars in dough including glucose, fructose, saccharose (sucrose), and maltose [14, 35]. Glucose and fructose become fermented immediately. After almost all the built-up fructose and glucose have been depleted, sucrose is first converted to glucose and fructose by the enzyme amylase [28, 35, 40]. This latter process occurs very rapidly, and a few minutes after mixing the dough, all sucrose molecules have been converted into glucose and fructose. Maltose molecules may be hydrolyzed to two molecules of glucose with the help of the yeast enzyme maltase [14, 35, 41]. The simplified equation of dough fermentation can be indicated as mentioned in the following equation.

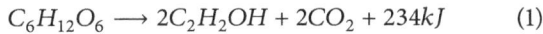

$$C_6H_{12}O_6 \longrightarrow 2C_2H_2OH + 2CO_2 + 234kJ \qquad (1)$$

The amount of carbon dioxide produced in dough by sugar fermentation may amount to about 70% of the theoretical amount indicated by the chemical equation. This can be explained by the fact that part of the sugar is used for energy and reproduction of the yeast cells within the dough [35]. Dough expansibility during fermentation can be mainly determined by viscoelasticity. The viscous components in the dough mass allow gas cells to expand to equalize the pressure, whereas the dough elastic components provide the relevant strength to prevent the dough from overexpansion and collapsing [31].

Heat and mass transfer phenomena are taking place simultaneously during bread baking which causes physical, chemical, and structural transformation [1] including water evaporation, volume expansion, starch gelatinization, and protein denaturation, settling the porous structure [1] which leads to setting the final bread crumb structure within the oven [8]. Usually, the protein denaturation and starch gelatinization occur during the temperature interval of 60–85°C and contribute to the change from dough to crumb [1]. With the temperature increment within the oven, thermal expansion of vapor occurs and the saturation pressure of water within the dough get increased. This causes the loaf to be expanded (oven spring). According to Hayman, Hoseney, and Faubion [42], the loaf expansion occurs by increasing the product volume during the first 6-8 min in baking, creating a high strain within the dough that can compress the heat set cellular structure of the outer reagents of the product [1, 8, 31, 43]. As a result of that, the outer cells can be elongated with their long axes parallel to the crust planes [8]. CO_2 also plays an important role in the expansion of bubbles during baking by releasing from the dough when the bubble walls start to break under pressure, making the porous structure more continuous and open to the outside of the bread [1].

When analyzing the crumb structure, several factors can be taken into account and the most common factors considered in most researches are crumb appearance, product volume [8, 44–50], resilience of the product [8], crumb color [8, 9, 16, 38], consumer appeal [2, 8, 11, 16, 51–53], physical texture of the product [7–9, 46, 54–57], taste [8, 9], compactness and uniformity of the crumb grains [58], size, shape, uniformity, and wall thickness of crumb cells or pores [2, 7, 8, 16, 17, 30, 58, 59].

3. Evaluating Crumb Cellular Structure

Recently, image analysis (IA) has been used as a quantitative tool that provides directly interpretable data for reliably assessing the microstructural features of crumb and its relationship with the crumb mechanical and sensorial properties [1, 2, 7, 10, 41, 44] of the final product as well as the crumb structure evaluation during fermentation and baking [22]. The most common features that can be analyzed using digital image analysis can be considered mainly as cell size, cell size distribution, number of cells per unit area, cell wall thickness, void fraction (porosity), shape factor, and number of missing cell walls [2, 8, 13, 16, 17, 30, 41, 59].

Image analysis involves several steps including image acquisition, image preprocessing, image segmentation, feature extraction, and classification [38]. Three elements are reported to be required to acquire the digital image of cut surface of a porous crumb including a source of illumination, the specimen, and an image sensing device [60].

There are certain methods that have been used for image acquisition, among which light microscopy and electron microscopy [2, 61] had been recorded as the most convenient imaging techniques applied to food structure analysis [2]. Apart from that, digital scanners and conventional photography [5, 21, 22, 38, 58] had been commonly used to capture two-dimensional (2D) high-resolution images of porous crumb structure and have been recommended as fast, convenient, economically feasible, and robust methods that provide good accuracy by acting independently from external light [3, 5, 38, 58]. There are some more advanced high-resolution techniques also available for the purpose of providing quantitative information of the porous crumb structure [21, 22], such as scanning electron microscopy [5, 14, 21, 22], X-ray computed tomography [2, 3, 10, 11, 14, 21, 22, 38, 62, 63], and magnetic resonance imaging [3, 5, 11, 14, 21, 22, 38, 64, 65].

X-ray computerized microtomography (X-MCT) had been the high spatial resolution version of the computed axial tomography routinely used for medical diagnosis and been applied for porous crumb structure analysis in several studies [2, 10, 11, 62, 63, 66]. Figure 1 represents an example of X-ray microtomography 2D reconstructed cross section images of cake samples from the research done by Sozer et al. [11]. X-MCT provides the ability to obtain three-dimensional (3D) representation of the inside structure of a sample from a set of projection measurements recorded from a certain number of points of view and examine their textural characteristics [2, 10, 11]. Mathieu et al. [62] had mentioned that the kinetics of bubble growth and foam setting in dough during fermentation and proofing can be determined by this method. The main disadvantage of X-MCT is the poor intrinsic contrast of low-density materials or porous crumb structures [2].

Several research studies have been carried out up to date regarding porous crumb structure evaluation by using magnetic resonance imaging (MRI) [21, 64, 65, 71]. Among them, Wagner et al. [65] had used a spacious MRI oven compatible with a low-field MRI scanner (0.2 T) to monitor bread loaf fermentation and baking. Figure 2 shows normalized magnetic resonance images of bread taken within that oven during baking. Bajd and Serša [21] have proved that the use of high-field MRI scanners with magnetic resonance microscopy (MRM) can overcome the resolution problem obtained in low-field MRI experiments (because even though those low-field methods have good temporal resolution and image quality, they are lacking in spatial resolution). The advantages of applying MRI in image analysis have been recorded in literature as noninvasiveness, ability to determine precise moisture content, and containing a comparatively high spatial resolution [14, 21, 65].

FIGURE 1: X-ray microtomography 2D reconstructed cross section images of cake samples [11].

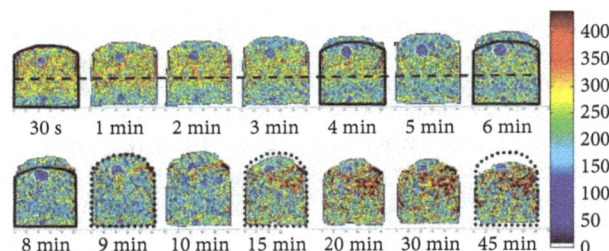

FIGURE 2: Normalized magnetic resonance images (MRI) of bread acquired during baking. That is, during baking, the cold colors convert to warm colors corresponding to low to high signal intensities, respectively [65].

3.1. Image Segmentation. Image segmentation can be considered as a method that separates object(s) of interest within an image from its background, typically yielding a binary image [5, 8, 16]. This segmentation process is important in crumb structure analysis to accurately segregate the gas and solid phases and define the distribution of cell and cell wall sizes [7, 8]. The most common way of segmenting images had been recognized as thresholding and edge detection [5, 7, 8, 30].

Thresholding can be either subjective (chosen manually) or objective. The objective method is based on statistical techniques referred to as clustering and it is considered as one of the most commonly used techniques for optimal thresholding [5, 8, 30]. Literature stated that the using of single threshold value for image segmentation of porous crumb structure may lead to under- and overestimation of cell sizes due to nonuniform structure of the bread crumb [8]. Hence more sophisticated thresholding techniques such as multiple thresholding had been recommended. Scanlon and Zghal [8] had described an example for this phenomenon as a local segmentation where there is application of neighborhood of pixels to detect individual objects within an image or individual crumb cells using gray level threshold for each crumb cell. Figure 3 shows an example for digital image segmentation. The scanned images of porous crumb structure can be threshold and analyzed by different scientific image analysis software which has been developed by numerous researchers utilizing multiple algorithms from cell segmentation techniques that aim at determining the cell size and shape distribution [5, 13, 67]. An example list of software is mentioned in Table 1.

Before introducing scientific image analysis software, image segmentation has been done following cluster analysis method which was commonly known as the K-means algorithm which was mostly used for classifying digitized images into cells and background [3, 7]. In general, the algorithm groups a set of data that contains M observations described by N variables or features into K clusters [7]. This

algorithm adapts the gray level thresholding of each bread slice image, depending on the overall brightness of the crumb image and the distribution of constituent pixel gray levels, both of which can be affected by the crumb structure itself [44]. The corresponding segmented images appear to provide an accurate binary representation of the complex cellular structure seen in the original gray-scale images. Figure 4 represents a comparison of an original and segmented image with K-means algorithm.

Crumb fineness or cell density is the value determined by the total number of cells detected over total surveyed area. And the ratio between the numbers of cells lower than 1 mm to the number of cells higher than 1 mm in diameter gives the cell uniformity which is reported to be directly correlated to fineness [3, 5, 30, 67, 73]. Cell density can strongly influence the mechanical properties of bread crumb [67]. Che Pa et al. [5] mentioned that higher value of crumb fineness indicates a finer crumb structure.

Cell wall thickness (μm) is determined on the cubic subvolumes ($100 \times 100 \times 100$ pixels) randomly extracted from the considered whole volume [2]. According to Scanlon and Zghal [8], thickness of the cell wall depends on the differences in starch content (e.g., thinner cell wall may result due to availability of lesser amount of starch granules) and moisture content of the dough mass. Further, Scanlon and Zghal [8] described that thinner cell walls may cause greater mechanical strength and greater deflection at break (i.e., the flexibility of thinner cell walls is higher compared to thicker cell walls) and also cause crumbs to be softer [47].

When the crumb structure rises during proofing, defects in the cell walls (missing due to coalescence or ruptured cell walls) and variability in cell wall distribution are some factors that must be considered in microstructure analysis of the porous crumb structure [8]. This has been considered in literature as the missing cell walls. Zghal, Scanlon, and Sapirstein [20] had derived an equation (2) to calculate the number of missing cell walls by calculating the theoretical

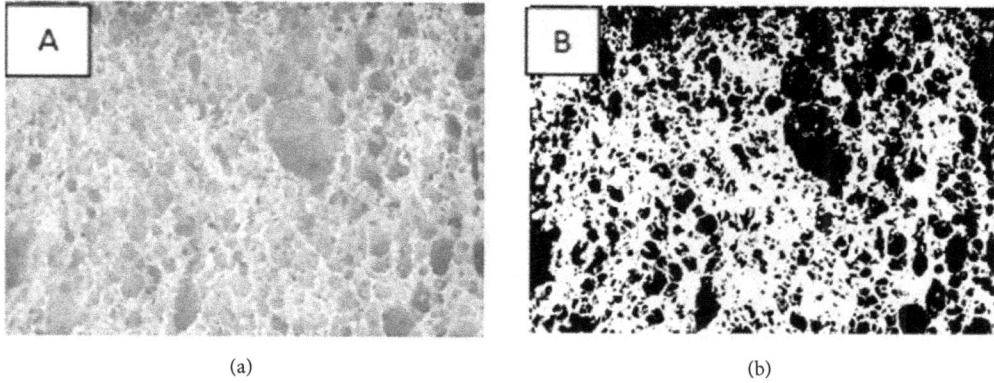

(a)

(b)

FIGURE 3: Original (2D) Gray Image (a) and segmented (b) gas cells where black pixels represent bubbles and white pixels represent the porous structure [18].

TABLE 1: List of examples of image processing software.

Image processing software	Detail	References
ImageJ	version 1.29, Natl. Inst. of Health, Bethseda, Md., U.S.A	Lassoued et al., [3]
		Bajd and Serša, [21]
		Tlapale-Valdivia et al., [38]
		Pérez-Nieto et al., [17]
	http://rsb.info.nih.gov/ij or https://imagej.net/	Curic et al., [48]
		Scheuer et al., [67]
SigmaScan®Pro	Version:5.50.4522.1800IC by Drug discovery Online	Romano et al., [12]
		Tlapale-Valdiviaet al., [38]
	http://www.sigmaplot.co.uk/products/sigmascan/sigmascan.php	Angioloni and Collar [16]
MATLAB	The MathWorks Inc.,	Gonzales-Barron and Butler [13]
		Bajd and Serša, [21]
	Natick, MA, USA	Shehzad et al.,[22]
		Rouillé et al.,[30]
	https://in.mathworks.com/	Verdú et al., [18]
		Eduardo, Svanberg and Ahrné [68]
Gebäckanalyse	Ver 1.3c 1997/98 program (Hochschule Ostwestfalen Lippe, Germany	Onyango, Unbehend and Lindhauer [69]
Labview	Vision Assistant 2009, National Instruments, USA	Che Pa et al., [5]
UTHSCSA ImageTool programme	Version 2.0, University of Texas Health Science Centre, San Antonio, Texas	Skendi et al., [70]

number of cells/cm^2 and the number of cells/cm^2 determined by image analysis.

$$N_x = N_0 \left(1 - \left[\frac{(1/\rho_x)^{2/3} - (1/\rho_0)^{2/3}}{(1/\rho_0)^{2/3}} \right] \right) \quad (2)$$

N_x represents the theoretical number of cells at the time t_x. Zghal, Scanlon, and Sapirstein [20] have considered t_x as 35 minutes from the beginning of proofing time. N_o means the number of cells/cm^2 at the highest density of the sample at t_o time measured by digital image analysis (DIA) and ρ_x and ρ_0 are densities of the product at the time t_x and t_o, respectively.

Then the number of missing cell walls can then be calculated from the following equation:

$$N_{mcw} = N_x - N_f \quad (3)$$

N_{mcw} represents the number of missing cell walls, and N_f represents the number of cells/cm^2 of the final product determined by DIA.

Crumb porosity (void fraction) had been expressed as the mean value of the total cell to total area ratio on each slice of the considered volume. A higher void fraction suggests an increase of the number of larger cells (>1 mm diameter) and consequently a decrease in the degree of cell uniformity [5].

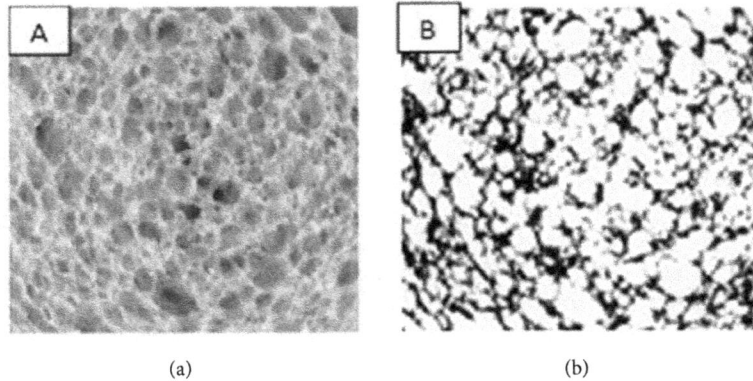

FIGURE 4: Original 2D Gray Image (a) and segmented (b) gas cells using K-means algorithm [44].

According to Zghal, Scanlon, and Sapirstein [44], the values of the mean cell area and void fraction have to be multiplied by a corrective factor of 1.5, because they found that, on average, the cell volume based on observation of cell size of the surface of a slice is 41% lower than the actual volume, assuming that the cells are spherical in shape and sectioned randomly to give cells with a uniform size distribution [5].

Relative density of bread crumb is a dominant physical characteristic which can affect the elastic properties and mechanical strength representing the 3D structure of cellular solids [11, 20] and is defined as the fraction of voxels segmented as cell walls. It is comparable to the ratio $\rho*/\rho_s$, where $\rho*$ represents the density of the crumb and ρ_s is the density of the material of the cell walls [3]. Zghal, Scanlon, and Sapirstein [44] have proved that, with the increment of the proofing time, gas cell coalescence may occur which will lead to nonuniformity in relative density due to missing cell walls which can weaken the crumb strength.

According to that, a product with well-developed porous crumb structure should have a high porosity and fine, regular gas cell structure [9, 74].

Apart from the undoubted advantages, Falcone et al. [2] have stated that there are some problems that occurred when imaging techniques are used on such food materials. One of them is that most of the imaging techniques require a sample preparation that can produce artifacts (e.g., especially when image analyzing using light microscopy and electron microscopy) which has to be considered to avoid wrong conclusions in microstructure investigation. Further, some imaging techniques are more expensive because they require sophisticated equipment; those can only be applied on foods having a high commercial value. Farrera-Rebollo [58] had also reviled certain problems arising in image analysis. For example, there are certain differences in the results of different image analysis methods (such as scanning resolution) even for similar products. Che Pa et al. [5] have mentioned that it is difficult to accurately determine the structure of the porous crumb structure due to the lack of uniformity in cell distribution and the higher variation in gas cell size. Mathieu et al. [62] and Lassoued et al. [3] have also mentioned the complex nature and of porous structure and difficulty in cell segmentation of 2D images and exact identification of the

relationship between microstructure and mechanical properties. Hence, researches had focused more on overcoming those disadvantages in microstructure analysis of foods with complex cellular structure.

4. Application of Fractal Dimension for Evaluating Crumb Cellular Structure

Visual textures are generally formed by the interaction of light with a rough surface. In a digital image of a surface, information is stored as an array of pixels with different intensities or gray levels. Therefore, the local variation of brightness from one pixel to the next (or within a small region) is often called texture [75].

Image texture analysis also called texture feature is a region where there is a descriptive approach that provides a measure of properties such as smoothness, coarseness, and regularity [3, 8, 75]. Fractal dimension (FD) provides a numerical descriptor of the morphology of objects with complex and irregular structures and it is reported to be applied to explain changes in the structure of food materials during or as a consequence of processing [17, 58]. FD can be evaluated using the Box Counting Method (BCM) [13, 17, 75], Differential Fractal Brownian Motion Method (FBMM) [13, 75], the Frequency Domain Method (FDM) [13, 75], morphological fractal (M) [13], mass fractal method (MF), and spectral dimension or random walks method (RW) [13]. All these studies show that image texture analysis has potential for determining some cellular structural features, while avoiding thresholding and cell segmentation of 2D images [3].

Fractal Brownian Motion Method (FBMM) is based on the average absolute difference of pixel intensities and is an example of a statistical fractal that can be described by the Hurst coefficient [13]. In Frequency Domain Method, the fast Fourier transforms (FFT) are taken in the horizontal and vertical directions and then the FD is taken from the average value of the vertical and horizontal fractal dimensions FFT_d. The mass fractal dimension, MF_d, is mostly used to describe the heterogeneity and space-filling ability of an object which can be estimated from the negative slope of the logarithmic plot of number of pores, in m number of pixels versus log m [3].

Pérez-Nieto et al. [17] have clearly described the method of calculating the mean fractal dimension of the perimeter of the pores ($\overline{FD}sc$) using the results obtained from ImageJ analysis and stated that resulting in a higher mean fractal dimension may indicate a more convoluted or jagged pore structure resulting in a product with a rough fractal texture. Equation (4) represents how to calculate the FDsc value and afterwards, the mean fractal dimension ($\overline{FD}sc$) can be calculated by (5).

$$FDsc = 2\frac{\log(P/4)}{\log A} \qquad (4)$$

$$\overline{FD}sc = \frac{\sum_{i=1}^{n}(FDsc)\,i}{n} \qquad (5)$$

n represents the number of cells (objects), FDsc represents fractal dimension of the perimeter of a single cell, and P and A represent the individual cell perimeter in pixel and individual cell area in pixel, respectively.

Additionally Pérez-Nieto et al. [17] have calculated the fractural dimension of the crumb structure using the Shifting Differential Box Counting Method (FD_{SDBC}) using ImageJ software which corresponds to the 2D gray level crumb images by the slope of the least-squares linear regression of the log (box count) versus log (box size) plot and by (6), where "N" is the number of boxes and "r" is the length of the side of box. Higher values of FD_{SDBC} represent more complex or rougher gray level crumb images, while low values of FD_{SDBC} can be associated with simpler or smoother images. The same test was described by Quevedo et al. [75] using Matlab 5.0 software using different food materials. Gonzales-Barron and Butler [13] have described relative differential Box Counting Method (RDBC) for calculating FD.

$$FD_{SDBC} = \frac{\log(N)}{\log(1/r)} \qquad (6)$$

5. Mechanical and Sensorial Properties of the Porous Crumb Structure

Crumb mechanical properties can vary microscopically and macroscopically, where the microscopic variations may occur due to the volume fraction of the granules that can be determined by the finite thickness of the cell wall. And the macroscopic variations may occur due to the variation of crumb moisture content across the product that may represent the differences in the degree of melting of starch granules [8].

Yeast fermented dough is considered to have complex mechanical behaviors, and the dimensions and physical properties of the dough may change with the time [8, 12–15]. Lack of homogeneity in the crumb cell distribution and development of complex stress combination during crumb mechanical testing are also named as the reasons for the complex mechanical behavior of bread crumbs [8]. Furthermore, invasive, continuous measurements on dough are generally not adequate as they may provoke dough collapse [12]. Hence,

the choice of the most appropriate analytical procedure is thus crucial for the full comprehension of the underlying mechanisms of leavening and crumb structure development.

5.1. Texture Analysis. Apart from the crumb grain appearance, physical texture is also an important quality to determine the porous structure of baked products where the texture is related to the geometric and mechanical properties of the product, which heavily depends on its cellular structure [5, 8, 9, 13, 44, 55, 76] such as cell wall thickness, cell size, and uniformity [5, 8, 44] which has been defined as the cellular structure of a crumb of a slice of a product [8].

Texture profile analysis had been created as an imitative test that resembles what goes on in the human mouth and is a parameter to determine the human perception of the product's texture and how it behaved when handled and eaten. Furthermore, it incorporates all the attributes (mechanical, geometric, and surface) of the food, suggesting that the experience of texture is one of many stimuli working together in combination [77]. The most commonly considered attributes of leavened baked products include hardness, springiness, adhesiveness, chewiness, gumminess, and cohesiveness [46, 54].

Hardness (g) is measured from the peak force on first compression and is defined as the force required for biting bread samples. Springiness (mm) is calculated from the distance of the sample recovered after the first compression. Adhesiveness (mJ) represents the work necessary to overcome the attractive forces between the surface of food and that of the sensor during the first and second compression cycle. Cohesiveness is a characteristic of mastication that can be calculated by the ratio of the active work done under the second cycle area to the first cycle area. More cohesive dough may result in a product with higher specific volume and softer texture. Gumminess (g) depends on hardness and cohesiveness which presents the density that persists throughout chewing. Gumminess of the crumb also depends on the tenacity and extensibility of the dough and the flour protein content. Chewiness (mJ) depends on gumminess and springiness which describes how long it takes to chew a food sample to the consistency suitable for swallowing [7, 46, 54–57]. Figure 5 represents the compression curve of force versus time and the summary of obtaining the major texture parameters.

5.2. Sensory Evaluation. Crumb mechanical properties as well as consumer acceptability can also be determined by the application of sensory evaluation [2], because how the crumb feels by touch or in the mouth is greatly influenced by the size or the structure of the crumb cells. As an example, crumb with finer, thin-walled, uniformly sized cells yields a softer and more elastic texture than crumb with coarse and thick-walled cell structure [11, 16].

Crumb appearance, aroma, texture, taste, and degree of satisfaction can be named as the main parameters tested in sensory evaluation [48, 52]. Appearance induces the palatability and the consumer acceptability of the product. Aroma and taste of the product represents the presence of many volatile and nonvolatile components, whereas the nonvolatile compounds contribute mainly to the taste and the volatiles

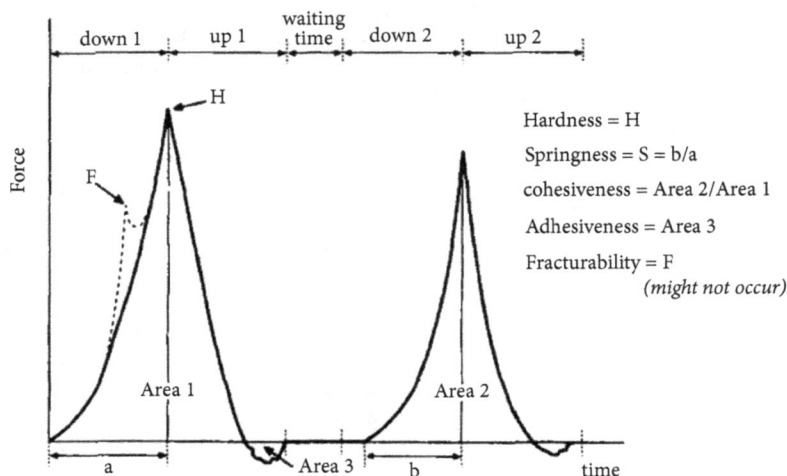

FIGURE 5: Instrumental texture profile analysis obtained with a TA-XT2 Texturometer [72].

influence both taste and aroma [52]. According to Schieberle [53], the amount of flavor compounds formed in bread can be affected by yeast amount and activity, fermentation and baking time, and temperature [51, 52].

When evaluating the texture of the bread crumb in sensory evaluation, improved elasticity, softness, and sponginess are the main parameters considered to determine the quality of the product.

5.3. Dough pH. The degree of acidity is considered as a parameter that determines the physical state of the gluten, influences the growth and the activity of the yeast, and controls the growth of many other microorganisms. According to Miller, Graf, and Hoseney [78], carbon dioxide dissolves in the aqueous dough phase until it becomes saturated (especially in the early stages of the fermentation) and after that, gaseous carbon dioxide diffuses into bubbles or into the atmosphere. The dissolved carbon dioxide reacts with water to form carbonic acid which imparts the acidic pH of the dough. With the fermentation process progressing, the pH of the dough get reduced [38, 49] and should be within the range of 5.2-6.0 [79].

5.4. Product Volume. Product volume (cm^3) is considered as an important bread characteristic since it provides a quantitative measurement of baking performance [1, 55] and leavened gas retention capability within the dough mass [37, 43, 69]. The desirable loaf volume of yeast fermented products is achieved only if the dough provides a favorable environment for yeast growth and gas generation. At the same time, it also represents a strong gluten matrix that is capable of maximum gas retention [12]. Zghal, Scanlon, and Sapirstein [44] and Tlapale-Valdivia et al. [38] had stated that the product volume can be affected by the dough mixing and proofing time.

Rapeseed displacement method is the most commonly used method for the determination of product volume [44–50]. Apart from that, there are some other methods also available for determination of gas production of a fermented

dough such as the oven rise recorder, alveograph method, and pressure meter methods [12, 14].

5.5. Other Common Physical Properties. Apart from the aforementioned mechanical and sensorial properties that are most commonly carried out to determine the properties of porous crumb structure, some other parameters were also described in literature for evaluating the mechanical properties of porous crumb structure. Among them, specific volume (gcm^{-3}) is an important visual characteristic for leavened baked products which could strongly influence the consumer's choice when evaluating product quality [80] and is also reported to have an effect on certain mechanical properties of the crumb structure such as crumb hardness and relative elasticity [8]. Bulk density (gcm^{-3}) is mostly used to describe the density of the cellular solid [8]. The bulk density can be affected by the particle size and the density of flour or flour blends [81] and also by crumb porosity [51].

Determination of crumb permeability utilizing Darcy's law is considered as a simple tool for assessing crumb texture along with specific volume [10]. Compression testing using stress-strain pots and calculating Young's moduli by the slope of the stress-strain curve is another method mentioned in literature for evaluating porous crumb structure. Certain researches mention Young's module for characterizing crumb properties [8, 10, 20, 62] where density is the most highly correlated parameter [20]. Further, Zghal, Scanlon, and Sapirstein [20] stated that Young's modulus can be positively correlated with density, crumb brightness, $cells/cm^2$, and uniformity of cell wall thickness and negatively correlated with void fraction, cell wall thickness, mean cell area, and number of missing cell walls.

Crumb color can be determined by chrommameter method ($L*,a*,b*$ values) [16, 38, 70] or computerized image analyses such as Photoshop system, which may facilitate not only a methodology for measurements of uneven coloration, but also it can be applied for the assessment of many other attributes of whole appearance as well [16]. Tlapale-Valdivia et al. [38] have mentioned that the crumb color can be mainly affected by both kneading and fermentation procedures and also by the fineness and homogeneity of the crumb grain [13].

6. Crumb Staling

Staling has been defined as a term that indicates decreasing consumer acceptance of bakery products caused by changes in crumb properties, except those resulting from the action of spoilage organisms [29, 82]. The most widely used indicator for staling can be considered as the increment of crumb firmness which is the parameter that is most commonly identified by the customers [9, 16, 70, 83].

Usually, leavened baked products begin to undergo deteriorative changes commencing with removal from the oven [82, 84]. The mechanism of crumb staling is more complex, more important, and less understood [1, 29]. Most of the sources identify the retrogradation (increasing crystallinity caused by cross linkages of starch molecules) of amylopectin as the main reason for staling [1, 37, 48, 82, 85–88]. Apart from the increment of starch crystallinity and opacity, increase in crumb firmness, changes in flavor, decrease in water absorption capacity, amount of starch and enzyme susceptibility of starch, and changes in X-ray diffraction pattern scan can also occur due to staling [1, 29, 48].

Moisture migrations can be also involved in staling process [11, 29, 82, 87]. This has been described in literature as the reversal in the location of water; therein water migrated to starch from gluten during baking and may return back to gluten proteins during storage [29, 82]. And also water may migrate from the crumb to the crust when the evaporation from the crust is prevented. This may cause the crust to gradually increase the leatheriness and remain soft while reducing the total moisture content in the crumb [29, 82, 89]. Usually, moisture in bread crumb acts as a plasticizer, where the reduction of crumb moisture content due to crumb staling can lead to formation of hydrogen bonds among the starch polymers or between starch and the proteins yielding increased crumb firmness [89, 90].

The staling rate can be affected by product size, moisture content, production process (e.g., baking time and temperature [1, 90]), packaging [89], and the storage temperature [29, 82]. As an example, lower storage temperatures (such as −1, 10, and 21°C) can accelerate starch recrystallization of the crumb compared with the storage at a warm room temperature (such as 32 and 43°C) [29, 82]. According to Slade and Levine [91] 4°C (refrigerator temperature) is the single optimum temperature that balances nucleation and crystallization and that the melting temperature involved implicates amylopectin as the polymer crystallizing. The effect of baking temperature on bread staling has been stated by Giovanelli, Peri, and Borri [90]. According to that, bread baked at lower temperatures (e.g., under slight vacuum to achieve crumb cooking at temperatures < 100°C) stales at a slower rate in terms of starch retrogradation [90].

Staling rate or crumb firming can be influenced by crumb structure, which is closely related to the gluten content, degree of starch gelatinization, and moisture redistribution

[88]. This crumb firming kinetics can be applied to modified Avrami model (7) [85, 87, 92] based on starch retrogradation [87] where θ represents the recrystallization still to occur and T_∞, T_0, and T_t represent the bread hardness at ∞ time, bread hardness at zero time, and bread hardness at t time, respectively. "k" is a rate constant (constant time to compare bread hardening rate) that represents the parameters characterizing the crystallization process, and "n" is the Avrami exponent that is related to the crystal shape, way of crystallites nucleation, their subsequent growth, and the time dependence of the nucleation process [85, 87, 92]. Crumb firming kinetics may strongly depend on both "k" and "n". As an example, crumbs with low "n" and/or "k" and/or "T_∞" values may indicate slow crumb firming kinetics [92].

$$\theta = \frac{T_\infty - T_t}{T_\infty - T_0} = e^{-kt^n} \tag{7}$$

Additionally, the degree of staleness can be analyzed by different methods [29]:

(i) Compression stress-strain curves and then determining Young's module

(ii) Thermal analysis (including differential scanning calorimetry (DSC) and differential thermal analysis (DTA), thermogravimetric analysis (TGA), thermomechanical analysis (TMA), and dynamic mechanical analysis (DMA)) that provide basic information on starch retrogradation. As an example, DSC can measure the enthalpy associated with amylopectin recrystallization and monitors the progressive magnitude of staling endotherm [83]

(iii) Infrared spectroscopy (Fourier transform infrared (FTIR) spectroscopy and near-infrared (NIR) spectroscopy)

(iv) Magnetic resonance imaging (MRI)

(v) X-ray crystallography to measure the crystalline nature of the starch in the system

(vi) Microscopy (transmitted light and polarized light microscopy can be used to monitor changes in starch granules from crumb before and after staling and confocal laser scanning microscopy to investigate the changes in starch granules in crumb during staling and electron microscope)

(vii) Sensory evaluation

Several studies have shown that the total increment in crumb firmness may depend on bread specific volume, where higher crumb firmness may result in a product with lower specific volume [87]. According to study done by Błaszczak et al. [88], some properties like hardness and gumminess have increased while elasticity, cohesiveness, and volume recovery coefficient had decreased during the time of storage of bread samples. Additionally Błaszczak et al. [88] have showed that the crumb pores may become smaller and round during staling. Nussinovitch et al. [50] had proved that the crumb loses its elastic properties over time due to crumb staling. According to their research, the greatest loss occurs during the first 24 hours after baking and the loss had been drastic beyond that point. At a lower deformation, the damage can be slight or moderate and thus some recoverable work can be observed and its relative change during storage had been observed.

Literature states several studies that have been focused on the application of different additives and enzymes for extending the shelf-life of the leavened baked products by retarding the staling process [83]. Incorporation of surfactants such as monoglyceride, sodium stearoyl lactylate, and sucrose esters has been proved as a source that retards staling effect by retarding amylopectin retrogradation and by blocking moisture migration from gluten to starch [29, 86, 93]. Additionally, enzymes like α-amylases, hydrolases of non-starch polysaccharides [29, 83, 85, 88], proteases and lipases [29], emulsifiers like sodium/calcium stearoyl lactylate, mono/diglyceride [83], and hydrocolloids/gums [29, 68, 83] also have been applied to retard staling process (more details are described under "methods to improve porous crumb structure development and retard crumb staling").

Freezing and frozen storage conditions can also have great influence on the quality and shelf-life of leavened baked products [1, 83, 94, 95]. The production of frozen dough requires flour with higher gluten quality than that used in conventional bread making processes as well as freeze-tolerant yeasts [83, 95], because, during freezing and frozen storage, the number of viable yeast cells decreases [83, 95, 96] and a reducing compound named glutathione is released as a consequence. This compound can break down the disulphide bonds among proteins leading to a weakening effect on the gluten. The weakening of the gluten network leads to increase in the proofing time as well as reduction in the oven spring and the dough resistance to stress conditions. This can result in the crumb to be lower in volume [48, 83, 95, 96] and coarse in texture with large and nonuniform air cells [95]. Additionally, this process is also reported to increase the fermentation time as well [48, 95, 96]. Rosell and Gómez [95] have described the applications of certain additives for improving the quality of bread prepared from frozen dough. According to that, application of additives such as gluten, emulsifiers, and hydrocolloids can improve the product volume and can improve the stability during frozen storage due to the water retention capacity of hydrocolloids.

Certain resent studies have moved from frozen dough to partially baked bread, called part-baked bread or prebaked bread [83, 95]. Many researches have mentioned that freezing prebaked bread is a better method to improve the shelf-life of the bread while keeping the bread quality similar to fresh products [48, 83, 94, 95]. According to Novotni et al. [89], partially baked bread has longer oxidative stability than fully baked frozen bread. One of the major problems of the part-baked and frozen bakery product is crust flaking. This relates to mechanical damage that occurs due to the intense thermomechanical shock during chilling–freezing and final baking [48]. According to the research conducted by Bácenas and Rosell [94], moisture content of the prebaked breads has been reduced while increasing the hardness with the increment of frozen storage. In addition, long times of frozen storage had been shown to be associated with greater aging rates. The physical damage of the prebaked bread during

freezing has been caused by the progressive growing of the ice crystals which has been reported to be the main responsible cause for the quality loss and the greater speed rate of aging.

7. Methods to Improve the Porous Crumb Structure Development and Retard Crumb Staling

According to Seibel [97] and Jongh [98], pregelatinized flour and/or emulsifiers when working with composite flour with or without wheat flour, monoglycerides (0.5 - 1.0%), calcium and sodium stearoyl lactate (CSL and SSL) at a dose of 0.5-1.0% (flour basis), and binding agents (CMC, guar gum, carob gum, pregelatinized potato starch) can be applied to obtain better leavening and porous crumb structure and antistaling properties.

Food hydrocolloids or gums can be included in leavened baked products for diverse purpose and act as binding agents, such as gluten substitutes, to improve texture, to slow down the starch retrogradation, to increase moisture retention, and to extend the overall quality of the product throughout the storage period by decreasing the moisture loss consequently retarding the crumb hardening [23, 57, 68, 69]. As described by Collar et al. [57], the aforementioned qualities can be obtained even by using small quantities (<1% (w/w) in flour) of hydrocolloids.

Certain examples can be given for the application of several hydrocolloids in leavened baked products. Among them hydroxypropylmethylcellulose (HPMC) and xanthan gum are the most widely used hydrocolloids to obtain better crumb properties (especially in gluten-free products) [80]. Additionally, guar gum, carboxymethylcellulose (CMC), locust bean gum, and agarose can also be used to improve the porous crumb structure of gluten-free bread formulas [99]. Anton and Artfield [23] and Bhol and Bosco [51] have cited Gan et al. [100] for their research findings regarding the application of HPMC, CMC, and guar gum in 50:50 wheat flour: rice flour formulation. According to that, HPMC at 1.7% and CMC at 0.4% have produced better bread characteristics than the application of guar gum at 0.7%. HPMC can preferentially bind to starch granules [68] and result in a rigid yet well porous and soft crumb texture with higher volume, improved sensory characteristics, and an extended shelf-life by providing the necessary viscosity of the dough to trap leavened gas during fermentation [23] and also retarding starch retrogradation by inhibiting the migration of water through interaction with starch molecules [29, 68]. Additionally, Eduardo, Svanberg, and Ahrné [68] have cited the findings of Shittu, Aminu, and Abulude [101] as the application of xanthan gum (1% from the composite flour weight) can improve dough handling properties, loaf specific volume, and crumb softness when incorporated into breads with composite cassava-wheat formulations.

Emulsifiers, pentosans, enzymes, or combinations of these can also be used as binding agents to improve the porous crumb structure and to obtain antistaling effects [69]. Emulsifiers are widely used in commercial bread formulas [68] to strengthen the dough that mainly interacts with gluten proteins, to improve gas retention capacity, and to soften the crumb as well as to retard staling effect [1, 68, 95]. The research carried out by Onyango, Unbehend, and Lindhauer [69] resulted in the fact that crumb hardness as well as staling rate can be decreased with the increase of the emulsifier concentration (better results have been observed for application of emulsifiers in 2.4% w/w flour wet basis than 0.4% w/w flour wet basis). Several examples for the emulsifiers incorporated in bread formulation include monoglycerides, sodium stearoyl lactylate (SSL), and diacetyl tartaric ester methyl (DATEM) [95].

The most frequently used enzymes to obtain better crumb structure are the α-amylases from different origins (cereal, fungal, and bacterial) to increase product volume and to improve crumb grain properties, crust, and crumb color and also have a contribution to flavor development [29, 85, 88]. Błaszczak et al. [88] have found that the addition of fungal (0.0014 g/100 g of flour) and bacterial α-amylases (0.03 g/100 g of flour) can have a substantial effect on starch behavior during dough fermentation, bread baking, and staling. In particular, addition of these enzymes had shown improved structural changes in the starch–protein matrix showing an antistaling mechanism [57, 88]. Hydrolases of non-starch polysaccharides (such as cellulose, xylanase, β-glucanase) are considered as another group of enzymes that can also be applied to improve properties of porous crumb structure [29, 85]. According to a research carried out by Haros, Rosell and Benedito, [85], addition of hydrolases of non-starch polysaccharides (the remaining enzyme activity in each flour as follows: cellulase treated flour (CEL) 85.5 mU/g of flour, xylanase treated flour (XYL) 2.9 mU/g of flour, and β-glucanase treated flour (GLUC) 1.2 mU/g of flour) had resulted in a product with lower crumb hardness, gumminess, and chewiness and reduced staling effect by reducing the initial crumb firmness and the kinetics of the firming process during storage. Skendi et al. [70] had found that bread prepared with barley β-glucan isolates (1.00 x10^5, BG-100; lower molecular weight and 2.03 x10^5, BG-200; higher molecular weight) resulted in a product with increased specific volume and reduced crumb firmness (especially the BG-200 sample).

Naturally extracted food gelling sources can also be used to improve the porous crumb structure of leavened products. According to the study done by Navaratne [79], bread samples containing *Davulkurudu* leaf (DKL) extract (the extraction of 10:100 leaves:water ratio had been incorporated until the dough moisture content reached to 58%) had given a better and well-developed porous crumb structure with reduced bulk density and better sensory attributes. Apart from that, staling process has also been reported to be retarded by 6 to 8 hours.

According to Różyło et al. [102], the enrichment with algal protein can also improve the rheological properties of dough, increase the gas retention capacity, and thereby lead to increase in product volume (4% per flour basis). Apart from that, the addition of algae had shown a positive influence on increasing elasticity and reducing the hardness of the crumb and reducing the degree of staling of gluten-free leavened products, which could have been found as a result of the presence of natural hydrocolloids in algae.

Garimella Purna et al. [37] had incorporated waxy wheat flour (15%, 30%, and 45% flour basis) into bread formulation and obtained a product with more open and porous structure (due to excessive swelling of waxy starch), high product volume, and softer texture (due to the combination of less amylose and more soluble starch from amylopectin). But the method had not retarded the staling process (due to rapid retrogradation of amylose in the initial stages of cooling and slow retrogradation of amylopectin for further firming the crumb). Additionally, they have mentioned that a significant post-bake shrinkage can be occurred in formulations with higher levels (>30%) of waxy wheat flour.

Researchers have found that the dough kneading and proofing have a vital effect on the quality of the crumb structure [44]. Poorly kneaded product had resulted in lower crumb brightness, fewer cells/cm^2, thicker cell walls, and larger gas cells when compared with optimally mixed dough. Overkneading also resulted in lower quality product. Zghal, Scanlon, and Sapirstein [44] found that overproofing may lead to causing reduced cell wall thickness, increased cell size, and higher void fraction causing the crumb to be harder. And also it may lead to an increase in cell coalescence that results in losing gas cell walls.

8. Conclusion

Dough ingredients and processing conditions have a vital effect on the development of porous crumb structure of leavened foods. Certain modifications of those factors such as the addition of certain additives like hydrocolloids/gums, enzymes, and emulsifiers can impact the properties of porous crumb structure and crumb staling. Porous crumb structure is a heterogeneous complex structure and hence the crumb microstructure is strongly affecting the mechanical and sensorial properties of the final product. Well-developed porous crumb structure has the ability of retaining more leavened gas resulting in a product with increased volume and reduced crumb hardness.

Conflicts of Interest

The authors declare that they do not have any conflicts of interest regarding the publication of this paper.

Acknowledgments

The authors wish to offer their gratitude towards all the academic and nonacademic staff of the Department of Food Science and Technology, University of Sri Jayewardenepura for supplying laboratory facilities and National Science Foundation, Sri Lanka, for funding the project. This work was supported by the National Science Foundation, Sri Lanka (Grant no. TG/2017/Tech-D/03).

References

[1] A. Mondal and A. K. Datta, "Bread baking - A review," *Journal of Food Engineering*, vol. 86, no. 4, pp. 465–474, 2008.

[2] P. M. Falcone, A. Baiano, F. Zanini et al., "A Novel Approach to the Study of Bread Porous Structure: Phase-contrast X-Ray Microtomography," *Journal of Food Science*, vol. 69, no. 1, pp. FEP38–FEP43, 2004.

[3] N. Lassoued, P. Babin, G. Della Valle, M.-F. Devaux, and A.-L. Réguerre, "Granulometry of bread crumb grain: Contributions of 2D and 3D image analysis at different scale," *Food Research International*, vol. 40, no. 8, pp. 1087–1097, 2007.

[4] E. J. Pyler, *BAKING Science & Technology volume II*, Sosland, Kansas, 3rd edition, 1988.

[5] N. F. Che Pa, N. L. Chin, Y. A. Yusof, and N. Abdul Aziz, "Measurement of bread crumb texture via imaging of its characteristics," *Journal of Food, Agriculture and Environment (JFAE)*, vol. 11, no. 2, pp. 48–55, 2013.

[6] P. W. Kamman, "Factors affecting the grain and texture of white bread," *Bak Dig*, vol. 44, no. 2, pp. 34–38, 1970.

[7] H. D. Sapirstein, R. Roller, and W. Bushuk, "Instrumental measurement of bread crumb grain by digital image analysis," *Cereal Chem*, vol. 71, no. 4, pp. 383–391, 1994.

[8] M. G. Scanlon and M. C. Zghal, "Bread properties and crumb structure," *Food Research International*, vol. 34, no. 10, pp. 841–864, 2001.

[9] J. Korczyk-Szabó and M. Lacko-Bartošová, "Crumb texture of spelt bread," *Journal of Central European Agriculture*, vol. 14, no. 4, pp. 1326–1335, 2013.

[10] S. Wang, P. Austin, and S. Bell, "It's a maze: The pore structure of bread crumbs," *Journal of Cereal Science*, vol. 54, no. 2, pp. 203–210, 2011.

[11] N. Sozer, H. Dogan, and J. L. Kokini, "Textural properties and their correlation to cell structure in porous food materials," *Journal of Agricultural and Food Chemistry*, vol. 59, no. 5, pp. 1498–1507, 2011.

[12] A. Romano, G. Toraldo, S. Cavella, and P. Masi, "Description of leavening of bread dough with mathematical modelling," *Journal of Food Engineering*, vol. 83, no. 2 pp. 142–148, 2007.

[13] U. Gonzales-Barron and F. Butler, "Fractal texture analysis of bread crumb digital images," *European Food Research and Technology*, vol. 226, no. 4, pp. 721–729, 2008.

[14] A. Ali, A. Shehzad, M. R. Khan, M. A. Shabbir, and M. R. Amjid, "Yeast, its types and role in fermentation during bread making process-A Review," *Pakistan Journal of Food Sciences*, vol. 22, no. 3s, pp. 171–179, 2012.

[15] A. H. Bloksma, "Dough structure, dough rheology and baking quality," *Cereal Food World*, vol. 1990, no. 35, pp. 237–244, 1990.

[16] A. Angioloni and C. Collar, "Bread crumb quality assessment: A plural physical approach," *European Food Research and Technology*, vol. 229, no. 1, pp. 21–30, 2009.

[17] A. Pérez-Nieto, J. J. Chanona-Pérez, R. R. Farrera-Rebollo, G. F. Gutiérrez-López, L. Alamilla-Beltrán, and G. Calderón-Domínguez, "Image analysis of structural changes in dough during baking," *LWT- Food Science and Technology*, vol. 43, no. 3, pp. 535–543, 2010.

[18] S. Verdú, E. Ivorra, A. J. Sánchez, J. M. Barat, and R. Grau, "Relationship between fermentation behavior, measured with a 3D vision Structured Light technique, and the internal structure of bread," *Journal of Food Engineering*, vol. 146, pp. 227–233, 2015.

[19] A. C. Ballesteros López, A. J. Guimarães Pereira, and R. G. Junqueira, "Flour mixture of rice flour, corn and cassava starch in the production of gluten-free white bread," *Brazilian Archives of Biology and Technology*, vol. 47, no. 1, pp. 63–70, 2004.

[20] M. C. Zghal, M. G. Scanlon, and H. D. Sapirstein, "Cellular structure of bread crumb and its influence on mechanical properties," *Journal of Cereal Science*, vol. 36, no. 2, pp. 167–176, 2002.

[21] F. Bajd and I. Serša, "Continuous monitoring of dough fermentation and bread baking by magnetic resonance microscopy," *Magnetic Resonance Imaging*, vol. 29, no. 3, pp. 434–442, 2011.

[22] A. Shehzad, H. Chiron, G. Della Valle, K. Kansou, A. Ndiaye, and A. L. Réguerre, "Porosity and stability of bread dough during proofing determined by video image analysis for different compositions and mixing conditions," *Food Research International*, vol. 43, no. 8, pp. 1999–2005, 2010.

[23] A. A. Anton and S. D. Artfield, "Hydrocolloids in gluten-free breads: A review," *International Journal of Food Sciences and Nutrition*, vol. 59, no. 1, pp. 11–23, 2009.

[24] S. Hamada, N. Aoki, and Y. Suzuki, "Effects of water soaking on bread-making quality of brown rice flour," *Food Science and Technology Research*, vol. 18, no. 1, pp. 25–30, 2012.

[25] E. Nwanekezi, "Composite Flours for Baked Products and Possible Challenges – A Review," *Nigerian Food Journal*, vol. 31, no. 2, pp. 8–17, 2013.

[26] A. Lazaridou, D. Duta, M. Papageorgiou, N. Belc, and C. G. Biliaderis, "Effects of hydrocolloids on dough rheology and bread quality parameters in gluten-free formulations," *Journal of Food Engineering*, vol. 79, no. 3, pp. 1033–1047, 2007.

[27] H. Wieser, "Chemistry of gluten proteins," *Food Microbiology*, vol. 24, no. 2, pp. 115–119, 2007.

[28] N. N. Potter and J. H. Hotchkiss, *Food Science. Fifth*, Chapman and Hall, Inc., New York, NY, USA, 2007, Food Science. Fifth.

[29] J. A. Gray and J. N. Bemiller, "Bread staling: Molecular basis and control," *Comprehensive Reviews in Food Science and Food Safety*, vol. 2, no. 1, pp. 1–21, 2003.

[30] J. Rouillé, G. Della Valle, M. F. Devaux, D. Marion, and L. Dubreil, "French bread loaf volume variations and digital image analysis of crumb grain changes induced by the minor components of wheat flour," *Cereal Chemistry*, vol. 82, no. 1, pp. 20–27, 2005.

[31] H. He and R. Hoseney, "Factors controlling Gas Retention in Nonheated Doughs," *Cereal Chem*, vol. 69, no. 1, pp. 1–6, 1992.

[32] B. E. Brooker, "The role of fat in the stabilisation of gas cells in bread dough," *Journal of Cereal Science*, vol. 24, no. 3, pp. 187–198, 1996.

[33] S. S. Kim and C. E. Walker, "Effects of sugars and emulsifiers on starch gelatinization evaluated by differential scanning calorimetry," *Cereal Chemistry Journal*, vol. 69, pp. 447–452, 1992.

[34] W. F. Petersen, *Pressure fermentation of dough for bread and similar bakery products [Internet]*. US1923880A, 1930, https://patents.google.com/patent/US1923880.

[35] C. A. Stear, *Handbook of Breadmaking Technology - C. A. Stear - Google Books*, Elsevier Science Publishers LTD, New York, NY, USA, 1990.

[36] N. Chamberlain and T. H. Collins, "The Chorleywood bread process: the roles of oxygen and nitrogen," *Bak Dig*, vol. 53, no. 1, pp. 18-19, 1979.

[37] S. K. Garimella Purna, R. A. Miller, P. A. Seib, R. A. Graybosch, and Y.-C. Shi, "Volume, texture, and molecular mechanism behind the collapse of bread made with different levels of hard waxy wheat flours," *Journal of Cereal Science*, vol. 54, no. 1, pp. 37–43, 2011.

[38] A. D. Tlapale-Valdivia, J. Chanona-Pérez, R. Mora-Escobedo, R. R. Farrera-Rebollo, G. F. Gutiérrez-López, and G. Calderón, "Dough and crumb grain changes during mixing and fermentation and their relation with extension properties and bread quality of yeasted sweet dough," *International Journal of Food Science & Technology*, vol. 45, no. 3, pp. 530–539, 2010.

[39] E. Chiotellis and G. M. Campbell, "Modelling the Evolution of the Bubble Size Distribution," *Trans IChemE*, vol. 81, no. Part c, pp. 194–206, 2003.

[40] S. Sahlström, W. Park, and D. R. Shelton, "Factors Influencing Yeast Fermentation and the Effect of LMW Sugars and Yeast Fermentation on Hearth Bread Quality," *Cereal Chemistry*, vol. 81, no. 3, pp. 328–335, 2004.

[41] Y. Sasano, Y. Haitani, I. Ohtsu, J. Shima, and H. Takagi, "Proline accumulation in baker's yeast enhances high-sucrose stress tolerance and fermentation ability in sweet dough," *International Journal of Food Microbiology*, vol. 152, no. 1-2, pp. 40–43, 2012.

[42] D. Hayman, R. C. Hoseney, and J. M. Faubion, "Effect of pressure (crust formation) on bread crumb grain development," *Cereal Chemistry*, vol. 75, no. 5, pp. 581–584, 1998.

[43] J. A. Delcour, S. Vanhamel, and R. C. Hoseney, "Physicochemical and functional properties of rye nonstarch polysaccharides. II. Impact of a fraction containing water-soluble pentosans and proteins on gluten starch loaf volumes," *Cereal Chemistry*, vol. 68, pp. 72–76, 1991.

[44] M. C. Zghal, M. G. Scanlon, and H. D. Sapirstein, "Prediction of bread crumb density by digital image analysis," *Cereal Chemistry*, vol. 76, no. 5, pp. 734–742, 1999.

[45] "AACC International. Standard 10-05," in *Approved Methods of American Association of Cereal Chemists*, 10 th. St. Paul, MN: American Association of Cereal Chemists, 2000.

[46] I. Švec and M. Hrušková, "Image data of crumb structure of bread from flour of czech spring wheat cultivars," *Czech Journal of Food Sciences*, vol. 22, no. No. 4, pp. 133–142, 2011.

[47] F. Cornejo and C. M. Rosell, "Influence of germination time of brown rice in relation to flour and gluten free bread quality," *Journal of Food Science and Technology*, vol. 52, no. 10, pp. 6591–6598, 2015.

[48] D. Curic, D. Novotni, D. Skevin et al., "Design of a quality index for the objective evaluation of bread quality: Application to wheat breads using selected bake off technology for bread making," *Food Research International*, vol. 41, no. 7, pp. 714–719, 2008.

[49] K. S. Aplevicz, P. J. Ogliari, and E. S. Sant'Anna, "Influence of fermentation time on characteristics of sourdough bread," *Brazilian Journal of Pharmaceutical Sciences*, vol. 49, no. 2, pp. 233–239, 2013.

[50] a. Nussinovitch, M. steffens, P. Chinachoti, and M. Peleg, "Effect of strain level and storage time on the recoverable work of compressed bread crumb," *Journal of Texture Studies*, vol. 23, no. 1, pp. 13–24, 1992.

[51] S. Bhol and S. J. D. Bosco, *Quality Characteristics of Yeast Leavened Bread*, Pondicherry University, 2014.

[52] M. Sabovics, E. Straumite, and R. Galoburda, "The influence of baking temperature on the quality of triticale bread," in *Foodbalt*, Latvia, 2014, http://llufb.llu.lv/conference/foodbalt/2014/FoodBalt_Proceedings_2014-228-233.pdf.

[53] P. Schieberle, "Intense aroma compounds - useful tools to monitor the influence of processing and storage on bread aroma," *Advances in Food Sciences*, vol. 18, no. 5/6, pp. 237–244, 1996.

[54] A. Meterei, E. Syendrei, and A. Fekete, "Prediction of Bread crumb texture quality," in *Proceedings of International Conference On Agricultural Engineering and Food*, Montpelier, France, 2004.

[55] S. Maktouf, K. B. Jeddou, C. Moulis, H. Hajji, M. Remaud-Simeon, and R. Ellouz-Ghorbel, "Evaluation of dough rheological properties and bread texture of pearl millet-wheat flour mix," *Journal of Food Science and Technology*, vol. 53, no. 4, pp. 2061–2066, 2016.

[56] K. Kaack, L. Pedersen, H. N. Laerke, and A. Meyer, "New potato fibre for improvement of texture and colour of wheat bread," *European Food Research and Technology*, vol. 224, no. 2, pp. 199–207, 2006.

[57] C. Collar, P. Andreu, J. C. Martínez, and E. Armero, "Optimization of hydrocolloid addition to improve wheat bread dough functionality: a response surface methodology study," *Food Hydrocolloids*, vol. 13, no. 6, pp. 467–475, 1999.

[58] R. R. Farrera-Rebollo, M. P. de la Salgado-Cruz, J. Chanona-Pérez, G. F. Gutiérrez-López, L. Alamilla-Beltrán, and G. Calderón-Domínguez, "Evaluation of Image Analysis Tools for Characterization of Sweet Bread Crumb Structure," *Food and Bioprocess Technology*, vol. 5, no. 2, pp. 474–484, 2012.

[59] I. Y. Zayas, "Digital image texture analysis for bread crumb grain evaluation," *Cereal Foods World*, vol. 38, no. 10, pp. 760–766, 1993.

[60] J. Chan and B. Batchelor, "Machine Vision for the Food Industry," in *Food Process Monitoring Systems*, A. C. Pinder and G. Godfrey, Eds., Chapman and Hall, 1st edition, 1993.

[61] M. Ferrando and W. Spiess, "Review: Confocal scanning laser microscopy. A powerful tool in food science Revision: Microscopía láser confocal de barrido. Una potente herramienta en la ciencia de los alimentos," *Food Science and Technology International*, vol. 6, no. 4, pp. 267–284, 2016.

[62] V. Mathieu, A.-F. Monnet, S. Jourdren, M. Panouillé, C. Chappard, and I. Souchon, "Kinetics of bread crumb hydration as related to porous microstructure," *Food & Function*, vol. 7, no. 8, pp. 3577–3589, 2016.

[63] K. S. Lim and M. Barigou, "X-ray micro-computed tomography of cellular food products," *Food Research International*, vol. 37, no. 10, pp. 1001–1012, 2004.

[64] S.-W. Hong, Z.-Y. Yan, M. S. Otterburn, and M. J. McCarthy, "Magnetic resonance imaging (MRI) of a cookie in comparison with time- lapse photographic analysis (TLPA) during baking process," *Magnetic Resonance Imaging*, vol. 14, no. 7-8, pp. 923–927, 1996.

[65] M. J. Wagner, M. Loubat, A. Sommier et al., "MRI study of bread baking: experimental device and MRI signal analysis," *International Journal of Food Science & Technology*, vol. 43, no. 6, pp. 1129–1139, 2008.

[66] P. Babin, G. Della Valle, R. Dendievel, N. Lassoued, and L. Salvo, "Mechanical properties of bread crumbs from tomography based Finite Element simulations," *Journal of Materials Science*, vol. 40, no. 22, pp. 5867–5873, 2005.

[67] P. M. Scheuer, J. A. S. Ferreira, B. Mattioni, M. Z. de Miranda, and A. de Francisco, "Optimization of image analysis techniques for quality assessment of whole-wheat breads made with fat replacer," *Food Science and Technology*, vol. 35, no. 1, pp. 133–142, 2015.

[68] M. Eduardo, U. Svanberg, and L. Ahrné, "Effect of hydrocolloids and emulsifiers on baking quality of composite cassava-maize-wheat breads," *International Journal of Food Science*, vol. 2014, 2014.

[69] C. Onyango, G. Unbehend, and M. G. Lindhauer, "Effect of cellulose-derivatives and emulsifiers on creep-recovery and crumb properties of gluten-free bread prepared from sorghum and gelatinised cassava starch," *Food Research International*, vol. 42, no. 8, pp. 949–955, 2009.

[70] A. Skendi, C. G. Biliaderis, M. Papageorgiou, and M. S. Izydorczyk, "Effects of two barley β-glucan isolates on wheat flour dough and bread properties," *Food Chemistry*, vol. 119, no. 3, pp. 1159–1167, 2010.

[71] F. De Guio, M. Musse, H. Benoit-Cattin, T. Lucas, and A. Davenel, "Magnetic resonance imaging method based on magnetic susceptibility effects to estimate bubble size in alveolar products: application to bread dough during proving," *Magnetic Resonance Imaging*, vol. 27, no. 4, pp. 577–585, 2009.

[72] S. M. Fiszman, M. Pons, and M. H. Damásio, "New parameters for instrumental texture profile analysis: Instantaneous and retarded recoverable springiness," *Journal of Texture Studies*, vol. 29, no. 5, pp. 499–508, 1998.

[73] A. Pérez-Nieto, J. Chanona-Pérez, R. Farrera-Rebollo, G. Gutiérrez-López, L. Alamilla-Beltrán, and G. Calderón-Domínguez, "Image analysis of structural changes in dough during baking," *LWT- Food Science and Technology*, vol. 43, no. 3, pp. 535–543, 2010.

[74] N. Lassoued, J. Delarue, B. Launay, and C. Michon, "Baked product texture: Correlations between instrumental and sensory characterization using Flash Profile," *Journal of Cereal Science*, vol. 48, no. 1, pp. 133–143, 2008.

[75] R. Quevedo, L.-G. Carlos, J. M. Aguilera, and L. Cadoche, "Description of food surfaces and microstructural changes using fractal image texture analysis," *Journal of Food Engineering*, vol. 53, no. 4, pp. 361–371, 2002.

[76] J. Tan, H. Zhang, and X. Gao, "SEM image processing for food structure analysis," *Journal of Texture Studies*, vol. 28, no. 6, pp. 657–672, 1997.

[77] A. Rosenthal, "Relation Between Instrumental and Sensory Measures of Food Texture," in *Food Texture: Measurement and Perception*, Aspen Publishers, Gaithersburg, 1999.

[78] R. A. Miller, E. Graf, and R. C. Hoseney, "Leavened Dough pH Determination by an Improved Method," *Journal of Food Science*, vol. 59, no. 5, pp. 1086-1087, 1994.

[79] S. B. Nawaratne, "Exploring Possibility of Using Davulkurundu (Neolitsea involurate) Leaf Extract to Improve Leavening Action and Crumb Structure of Wheat Bread," *Tropical Agricultural Research and Extension*, vol. 10, pp. 67–73, 2007.

[80] A.-S. Hager and E. K. Arendt, "Influence of hydroxypropylmethylcellulose (HPMC), xanthan gum and their combination on loaf specific volume, crumb hardness and crumb grain characteristics of gluten-free breads based on rice, maize, teff and buckwheat," *Food Hydrocolloids*, vol. 32, no. 1, pp. 195–203, 2013.

[81] A. A. Adebowale, M. T. Adegoke, S. A. Sanni, M. O. Adegunwa, and G. O. Fetuga, "Functional properties and biscuit making potentials of sorghum-wheat flour composite," *American Journal of Food Technology*, vol. 7, no. 6, pp. 372–379, 2012.

[82] Y. H. Hui, H. Corke, and I. De Leyn, *Bakery Products: Science and Technology*, Blackwell Publishing, Iowa, 2008.

[83] M. E. Barcenasa, M. Harosb, B. Carmen, and C. M. Rosell, "Effect of freezing and frozen storage on the staling of," *Food Research International*, vol. 36, pp. 863–869, 2003.

[84] B. D'Appolonia and M. Morad, "Bread Staling," *Cereal Foods World*, vol. 58, no. 3, pp. 186–190, 1891.

[85] M. Haros, C. M. Rosell, and C. Benedito, "Effect of different carbohydrases on fresh bread texture and bread staling," *European Food Research and Technology*, vol. 215, no. 5, pp. 425–430, 2002.

[86] P. A. Rao, A. Nussinovitch, and P. Chinachoti, "Effects of selected surfactants on amylopectin recrystallization and on recoverability of bread crumb during storage," *Cereal chemistry*, vol. 69, pp. 613–618, 1992.

[87] E. Armero and C. Collar, "Crumb firming kinetics of wheat breads with anti-staling additives," *Journal of Cereal Science*, vol. 28, no. 2, pp. 165–174, 1998.

[88] W. Błaszczak, J. Sadowska, C. M. Rosell, and J. Fornal, "Structural changes in the wheat dough and bread with the addition of alpha-amylases," *European Food Research and Technology*, vol. 219, no. 4, pp. 348–354, 2004.

[89] D. Novotni, D. Ćurić, K. Galić et al., "Influence of frozen storage and packaging on oxidative stability and texture of bread produced by different processes," *LWT- Food Science and Technology*, vol. 44, no. 3, pp. 643–649, 2011.

[90] G. Giovanelli, C. Peri, and V. Borri, "Effects of baking temperature on crumb-staling kinetics," *Cereal Chemistry*, vol. 74, no. 6, pp. 710–714, 1997.

[91] L. Slade and H. Levine, "Recent advances in starch retrogradation," in *Industrial Polysaccharides-The Impact of Biotechnology and Advanced Methodologies*, S. Stivala, V. Crescenzi, and I. Dea, Eds., pp. 387–430, Gordon and Breach, New York, NY, USA, 1987.

[92] S. Hejri-Zarifi, Z. Ahmadian-Kouchaksaraei, A. Pourfarzad, and M. H. H. Khodaparast, "Dough performance, quality and shelf life of flat bread supplemented with fractions of germinated date seed," *Journal of Food Science and Technology*, vol. 51, no. 12, pp. 3776–3784, 2014.

[93] W. Pisesookbunterng and B. DAppolonia, "Staling Studies. I. Effect of Surfactanta on Moisture Migrartion from Crumb to Crust and Firmness Value of Bread Crumb," *Cereal Chem*, vol. 60, no. 4, pp. 298–300, 1983.

[94] M. E. Bárcenas and C. M. Rosell, "Effect of frozen storage time on the bread crumb and aging of par-baked bread," *Food Chemistry*, vol. 95, no. 3, pp. 438–445, 2006.

[95] C. M. Rosell and M. Gómez, "Frozen dough and partially baked bread: An update," *Food Reviews International*, vol. 23, no. 3, pp. 303–319, 2007.

[96] P. D. Ribotta, A. E. León, and M. C. Añón, "Effect of freezing and frozen storage of doughs on bread quality," *Journal of Agricultural and Food Chemistry*, vol. 49, no. 2, pp. 913–918, 2001.

[97] W. Seibel, "16 Composite Flours," *Docstock*, pp. 193–198, 2011, https://muehlenchemie.de/downloads-future-of-flours/fof_kap_16.pdf.

[98] G. Jongh, "The Formation of Dough and Bread Structures. I. The Ability of Starch to Form Structures, and the Improving Effect of Glyceryl Monostearate," *Cereal Chem*, vol. 38, pp. 140–152, 1961.

[99] K. Wang, F. Lu, Z. Li, L. Zhao, and C. Han, "Recent developments in gluten-free bread baking approaches: a review," *Journal of Food Science and Technology*, vol. 37, no. supp 1, pp. 1–9, 2017, http://www.scielo.br/scielo.php?script=sci_arttext&pid=S0101-20612017000500001&lng=en&tlng=en.

[100] J. Gan, L. G. B. Rafael, and S. D. Cato L, "Evaluation of the potential of different rice flours in bakery formulations," in *Proceedings of the Conference C 2001. P of the 51st ACC*, M.

Wooton, I. L. Batey, and C. W. Wrigley, Eds., pp. 309–312, Royal Australian Chemical Institute, New South Wales, Australia, 2001.

[101] T. A. Shittu, R. A. Aminu, and E. O. Abulude, "Functional effects of xanthan gum on composite cassava-wheat dough and bread," *Food Hydrocolloids*, vol. 23, no. 8, pp. 2254–2260, 2009.

[102] R. Różyło, W. Hameed Hassoon, U. Gawlik-Dziki, M. Siastała, and D. Dziki, "Study on the physical and antioxidant properties of gluten-free bread with brown algae," *CYTA - Journal of Food*, vol. 15, no. 2, pp. 196–203, 2017.

18

Effect of Water Activity and Packaging Material on the Quality of Dehydrated Taro (*Colocasia esculenta* (L.) Schott) Slices during Accelerated Storage

A. R. Sloan, M. L. Dunn, L. K. Jefferies, O. A. Pike, Sarah E. Nielsen Barrows, and F. M. Steele

Department of Nutrition, Dietetics and Food Science, Brigham Young University, S-221 ESC, Provo, UT 84602, USA

Correspondence should be addressed to F. M. Steele; frost_steele@byu.edu

Academic Editor: Melvin Pascall

The quality of dehydrated taro slices in accelerated storage (45°C and 75% RH) was determined as a function of initial water activity (a_w) and package type. Color, rehydration capacity, thiamin content, and α-tocopherol content were monitored during 34 weeks of storage in polyethylene and foil laminate packaging at initial storage a_w of 0.35 to 0.71. Initial a_w at or below 0.54 resulted in less browning and higher rehydration capacity, but not in significantly higher α-tocopherol retention. Foil laminate pouches resulted in a higher rehydration capacity and increased thiamin retention compared to polyethylene bags. Type of packaging had no effect on the color of the samples. Product stability was highest when stored in foil laminate pouches at $0.4 a_w$. Sensory panels were held to determine the acceptability of rehydrated taro slices using samples representative of the taro used in the analytical tests. A hedonic test on rehydrated taro's acceptability was conducted in Fiji, with panelists rating the product an average of 7.2 ± 1.5 on a discrete 9-point scale. Using a modified Weibull analysis (with 50% probability of product failure), it was determined that the shelf life of dehydrated taro stored at 45°C was 38.3 weeks.

1. Practical Applications

Taro (*Colocasia esculenta* (L.) Schott) is a tuber grown in the tropics and subtropics. Though it contains high levels vitamins such as α-tocopherol and thiamine, it is a highly perishable root, resulting in high postharvest losses. In areas such as the Pacific Islands, where taro is a staple crop, this loss of nutrients is significant in the overall diet. Though some research has been done on the various ways taro can be preserved, little if any work has been reported on the effect of water activity (a_w) and packaging type on dehydrated taro slices' quality and acceptability. The purpose of this research was to expand our understanding of the influence that packaging type and initial a_w have on stored, dehydrated taro's nutritional, physical, and sensory characteristics so as to increase shelf life, thereby reducing postharvest losses.

2. Introduction

Taro (*Colocasia esculenta* (L.) Schott) is a tuber crop grown throughout the tropics and subtropics. The total world production of taro in 2012 was about 9.97 million tons from an area of 1.32 million hectares, mainly in Africa, Asia, and Oceania [1]. Taro can be categorized into two main groups: the dasheen type, which has a large central corm (a thick, rounded stem base), with a few small side cormels, and the eddoe type, which has a small corm with large cormels. Raw taro is about three-quarters water with its main component being carbohydrate in the form of starch. In the Pacific Islands, taro (dasheen type) is an important staple food and plays a critical role in household, community, and national food security [2].

The major limitation to the use of taro is the scarcity of preservation methods, which leads to high postharvest losses. Taro corms can be stored at ambient tropical conditions for about 2 weeks before becoming unfit for human consumption due to postharvest rot [3–5]. This leads to heavy losses in developing countries where corms are stored at ambient temperatures, often at least 23.9°C [6]. In fact, Passam [5] reported average postharvest losses of taro of about 50% after six weeks of ambient storage. Research aimed at developing storable processed forms of taro would greatly contribute to minimizing postharvest losses of this staple crop [2].

Though some research has been done on the rehydratability and color characteristics of dehydrated taro slices during extended storage, little has been done on the effect of a_w. A high temperature short time (HTST) pneumatic drying pretreatment prior to conventional hot air drying has been used to produce shelf-stable taro pieces with good rehydration and color characteristics [7]. In another study, Moy et al. [8] stored dehydrated taro slices in polyethylene (PE) bags at 21, 38, and 60°C for up to 6 months, finding that enzyme activity decreased and some browning occurred at 38 and 60°C. In another study, guava- and papaya-taro flakes stored at 38°C for 24 weeks maintained acceptable flavor; however, the ascorbic acid content and color of the samples were unstable during storage [9].

The impact of packaging on the color characteristics of dehydrated taro is unknown. However, in other food systems, it appears that packaging that maintains a low oxygen environment is most likely to preserve color. For example, in a study by Henríquez et al. [10], dried apple peel powder was stored in high-density polyethylene (HDPE) and metalized films of high barrier (MFHB), which is less permeable to oxygen than the HDPE. The MFHB packaging preserved most of the phenolic compounds, exhibited less moisture increase, and contributed to a longer shelf life. However, changes in packaging may not always have this effect on color. In a study on the color preservation of dehydrated green bell peppers by Sigge et al. [11], packaging type did not seem to affect color. Samples were stored in laminated pouches of different oxygen and moisture transmission rates; color changes due to storage temperature were significant, but changes due to packaging type were insignificant. Because of the variability of color changes due to packaging types in different food systems, it would be useful to specifically test the effects of packaging type on the color characteristics of taro.

Though color is certainly an important sensory characteristic and can also indicate the retention of some pigment nutrients, it is not always a good predictor of shelf life [12]. Therefore, shelf life should be studied separately.

An important aspect of shelf life is nutritional quality. Nutrient degradation during storage is typically a factor of storage temperature, pH, exposure to oxygen, porosity, light, and the presence of organic chemicals [13]. Taro is a good source of several vitamins, most notably α-tocopherol and thiamine, supplying 2.93 mg α-tocopherol and 0.107 mg thiamine for every 100 g of cooked taro [14]. This is 15% of the USDA recommended Daily Value for α-tocopherol and 7% of the recommended Daily Value for thiamine [15]. Due to the lack of previous research on the presence and stability of these vitamins in dried taro, a study of the stability of α-tocopherol and thiamine in dehydrated taro may be helpful.

Alpha-tocopherol is a fairly unstable, fat-soluble vitamin. It is degraded by storage and heat treatment, though it is stable at high temperatures if no oxygen is present [16]. Thiamine is a more stable molecule, but it can be heat sensitive, and interaction with SO_2 can completely destroy thiamine [13].

Typically, the biggest concerns to thiamine degradation in food systems are time of heating, pH, and storage time [17].

In addition to nutritional aspects, some sensory aspects of dried taro products have also been investigated. In a consumer taste panel conducted in Fiji, 73% of panelists indicated they would consume dehydrated taro slices as a part of their regular diet [18]. Jayaraman et al. [7] stored HTST-dried taro pieces packed in paper-aluminum foil-polyethylene laminate pouches for 1 year under ambient conditions and found that the taro pieces retained both sensory acceptability and rehydration characteristics.

The Weibull distribution has been widely used in many shelf life studies. Hough and Garitta [19] name it the most statistically and experimentally sound methodology for shelf life estimations. It has been used on a variety of products, from coffee to baked products [20]. Because it is determined by the consumers' acceptance or rejection a product, it is most appropriate for studies in which the overall acceptability of a product is to be determined.

There exists a close relationship between a_w and many reactions involved in food stability such as microbial growth, enzymatic hydrolysis, oxidation, nonenzymatic browning, pigment loss, and nutrient loss [21]. Packaging with a poor moisture barrier allows the product to reabsorb moisture, which may lead to microbial growth, discoloration, and deterioration in flavor [22]. The aim of this project was to determine the effects of a_w and packaging material on physical, nutritional, and sensory properties of dehydrated taro slices during accelerated storage.

3. Materials and Methods

3.1. Experimental Design. An incomplete $4 \times 2 \times 4$ factorial design was used with four levels of a_w (0.71, 0.54, 0.40, and 0.35), two levels of packaging material (foil laminate and polyethylene), and four levels of storage time (0, 13, 26, and 34 weeks). The amount of product required for a complete $4 \times 2 \times 4$ factorial design exceeded the resources available due to limitations on the amount of product that could be dried during the available time. The design was incomplete in that investigating only one packaging material at a_w levels of 0.35 and 0.71 (as described below) allowed for the evaluation of a broader a_w range while still allowing a packaging material effect to be measured at a_w levels of 0.40 and 0.54.

3.2. Sample Preparation. Fresh corms of *Colocasia esculenta* (L.) Schott were obtained from a local market at Suva, Fiji. Corms were washed, peeled, and sliced with an electric food slicer (Model 213-A, Krups Gmbh, Solingen, Germany) into 0.5 cm thick slices for drying. Samples were randomly assigned to one of the four target a_w treatment levels (0.80, 0.70, 0.60, and 0.50). Drying was done using a forced-air tray dryer (Model FD-60, Nesco, Two Rivers, Wis., USA) at 57°C. During drying, the a_w of the samples was monitored by periodically testing the a_w of the largest taro slice in the batch using an Aqualab CX-2 water activity meter (Decagon Devices Inc., Pullman, Wash., USA) until the samples reached a high (0.80), intermediate (0.70), low (0.60), or very low

TABLE 1: Initial and final a_w of dehydrated taro slices stored at 45°C and 75% RH for 34 weeks[a].

Package type	Storage time (wk)	
	0	34
Polyethylene	0.35 ± 0.01	0.64 ± 0.06
Polyethylene	0.40 ± 0.004	0.65 ± 0.04
Polyethylene	0.54 ± 0.01	0.66 ± 0.03
Foil laminate	0.40 ± 0.004	0.56 ± 0.02
Foil laminate	0.54 ± 0.01	0.66 ± 0.02
Foil laminate	0.71 ± 0.01	0.73 ± 0.02

[a]Mean of two replicates ± standard deviation.

(0.50) target a_w level. The samples were then placed in sealed PE bags for each a_w treatment level and transported under ambient conditions to the USA for evaluation.

In USA, samples were left in the PE bags and moisture was allowed to equilibrate at room temperature for 3-4 weeks. At equilibration, the a_w of three random samples from each bag was measured and the mean value for each bag was recorded. The resulting means (0.71, 0.54, 0.40, and 0.35) were designated as the actual a_w treatment levels. These treatment values were lower than the target levels because the samples were dried until the largest slice reached the target level to ensure that the majority of the taro slices were at the target a_w or below. Therefore, many of the smaller slices were below the target value at the time of drying, leading to lower final a_w values after equilibration.

After equilibration, samples at each of the 4 a_w levels were split into portions such that eight portions of approximately 276 g each could be randomly assigned to each of two types of packaging: PE bags (B01062WA, Nasco International, Fort Atkinson, Wis., USA) without oxygen absorbers and foil laminate (FL) pouches (Cal Pac 1500, Basaw Manufacturing, Inc., North Hollywood, Calif., USA) with oxygen absorbers (Ageless Z-Series, Mitsubishi Gas Chemical America Inc., New York, NY, USA). The FL pouches represent the most stability-enhancing packaging. The PE bags had a headspace oxygen level equal to atmospheric oxygen whereas FL pouches with oxygen absorbers provided a low (<0.01 ppm) headspace oxygen level, as measured using a 3500-Series Headspace Oxygen Analyzer (Illinois Instruments Inc., Johnsburg, Ill., USA). The PE bags (57 μm thickness) were composed of a blend of low-density polyethylene and linear low-density polyethylene while the FL pouches (177 μm thickness) were composed of layers of polyethylene, aluminum foil, and polyester. Both the PE bags and the FL pouches were sealed with a heat sealer (Model AIE-305 AI, American International Electric Inc., Whittier, Calif., USA). The samples with a_w of 0.35 were packaged only in the PE bags while the $0.71 a_w$ samples were packaged only in FL pouches due to restrictions of product available. Each portion was packed separately, two portions for each of the four storage times. Packages for each treatment combination were stored in the dark in a controlled atmosphere chamber at an accelerated storage temperature of 45°C. The relative humidity (RH) of the storage chambers was set at 75%, simulating tropical humidity. Two portions of 138 g each were sampled from each treatment combination at 0, 13,

26, and 34 weeks for physical measurements and at 0 and 34 weeks for nutritional measurements.

At the conclusion of the 34 weeks of storage, random samples were taken from each treatment combination to measure any changes in a_w of the samples (Table 1). Only the treatments with initial a_w of 0.54 were compared when investigating headspace oxygen as a possible result of packaging differences because they had equilibrated to approximately the same a_w level (0.66).

3.3. Physical Analyses

Water Activity. Water activity was determined by the chilled mirror technique using an Aqualab CX-2 water activity meter (Decagon Devices Inc., Pullman, Wash., USA).

Color. CIE L^* (0 to 100, dark to light), a^* (±, green/red), and b^* (±, blue/yellow) color values were obtained by measuring reflectance with a Hunterlab Colorflex Spectrophotometer (Hunter Associates Laboratory, Reston, Va., USA) using a 64 mm glass sample cup. The mean values of triplicate measurements were reported.

Rehydration. Rehydration characteristics were determined on triplicate samples by immersing a taro slice (4–12 g) in 250 mL of distilled water at room temperature for 24 hours. The slice was drained for 2 min on a number 4 sieve (Fisher Scientific International Inc., Hampton, NH, USA) prior to weighing. Rehydration capacity was defined as the weight ratio between the water absorbed by the sample and the weight of the dehydrated sample [23].

3.4. Nutritional Analyses.
Vitamin determinations were performed in a randomized order under subdued light. All chemical standards and enzymes were obtained from Sigma-Aldrich (St. Louis, MO, USA).

Thiamin. Thiamin was extracted in duplicate using the method of Ndaw and others [24]. Samples were ground using a coffee grinder (Model E160B, Hamilton Beach/Proctor-Silex, Inc., Glen Allen, VA, USA). Five grams of taro powder was placed in a 125 mL flask with 50 mL of 0.05 M sodium acetate that had been adjusted to pH 4.5. Ten mg papain, 10 mg acid phosphatase, and 10 mg α-amylase were added and mixed well. Samples were incubated at 37°C for 18 h. Thiamin was converted to thiochrome for quantification using an Agilent Model 1100 high performance liquid chromatograph (HPLC) (Agilent Technologies, Palo Alto, Calif., USA) and a Luna 5μ C18 (2), 150 × 4.6 mm reverse phase column (Phenomenex, Inc., Torrence, Calif., USA). Separations were accomplished with the following HPLC conditions: mobile phase of methanol in 0.05 M sodium acetate (30 : 70, v/v); 23°C; 1 mL/min flow rate; 10 μL injection volume; fluorescence detector, at an excitation wavelength of 366 nm and an emission wavelength of 435 nm. Data was quantified using an external standard.

α-Tocopherol. α-Tocopherol content was measured in triplicate using the method of Peterson and Qureshi [25] with

modifications. Samples were ground using a coffee grinder (Model E160B, Hamilton Beach/Proctor-Silex, Inc., Glen Allen, VA, USA). Samples of taro powder weighing 1 g were extracted with 10 mL methanol, sonicated for 3 min, and centrifuged for 10 min. The supernatant was evaporated on a rotary evaporator (Model R-215, Buchi Labortechnik AG, Flawil, Switzerland) at 40°C and the residue was dissolved in hexane. Determinations were done using an Agilent Model 1100 HPLC (Agilent Technologies, Palo Alto, Calif., USA). The following HPLC conditions were used: mobile phase of hexane with 0.2% isopropanol; 23°C; 1 mL/min flow rate; 20 μL injection volume; normal phase μ Prisol column (Waters Corp., Milford, Mass., USA), fluorescence detector, at an excitation wavelength of 295 nm and an emission wavelength of 330 nm. Data was quantified using an external standard.

3.5. Sensory Tests. Two sensory tests were performed. First, preliminary hedonic testing was performed on rehydrated taro slices in Fiji to determine general acceptability of the product. Second, a Weibull analysis of rehydrated taro stored in the US was conducted to determine shelf life of the product.

For the preliminary hedonic testing of rehydrated taro slices, eighty-six panelists (44 males, 25 of which were under 20 and 52 women, 25 of which were under 20) were recruited. These panelists were regular taro consumers who liked taro. Fresh taro of the same variety used in the study and of an approximate moisture content of 67% was purchased from the local market, peeled, and sliced in approximately 1 cm thick slices. The taro slices (about 8–12 g each) were dehydrated in a convection solar dryer, as described in Russon et al. [26], until they snapped when pressure was applied. Dried slices were then held for at least one week in FL pouches (Cal Pac 1500, Basaw Manufacturing, Inc., North Hollywood, Calif., USA) at ambient temperature, which is often at least 24°C, before sensory tests were conducted [6]. For the sensory test, the taro slices were rehydrated to as close to the original moisture content as possible (±3%) by boiling in 0.4% brine solution until the taro began to crack at the edges, fall apart, and turn gray, all of which are typical characteristics of taro doneness. Panelists seated at individual tables were presented a one-slice sample of taro to eat with their hands, as is typical in traditional consumption of taro. To analyze this product, panelists were given paper ballots to fill out. Panelists evaluated appearance, aroma, flavor, texture, and overall liking using a discrete 9-point hedonic scale where 9 = like extremely, 5 = neither like nor dislike, 1 = dislike extremely. The question about overall liking was placed at the end of the questionnaire to obtain a response that allowed time for panelists to consider all aspects of sensory quality, as has been done previously [27]. Acceptance was further determined by asking panelists if they would eat the sample as part of their regular diet and if they would eat it in an "emergency situation" in which there was a shortage of food.

Subsequent sensory analysis was conducted at Brigham Young University Sensory Laboratory (Provo, UT, USA) to determine shelf life. An abbreviated Weibull Hazard sensory analysis was conducted, modeled after that used by Cardelli and Labuza [28]. The end of shelf life was defined as the time at which 50% consumers found the product unacceptable.

Polynesians, including Fijians, who liked taro and consumed it occasionally were recruited as panelists from a database of campus and local communities. Both genders, across ages 18 to ≥60 years were represented. The university Institutional Review Board approved the study and panelists provided their informed consent. Panelists were monetarily compensated for their time.

Six panels were held after 0, 15, 26, 32.5, 39, and 42 weeks of storage. Beginning at week 26, tests were performed more frequently to ensure that the limit for shelf life could be determined accurately. The Weibull Hazard Method as described in Cardelli and Labuza [28] was used to determine the time of failure (end of shelf life) for the product and as a model for the method of conducting the panel. The first panel consisted of three panelists. The number of panelists was increased by $C = 1$ at each panel until 50% of the panelists deemed the samples unacceptable, as was done in other studies [19, 28, 29]. After that, the number of panelists was increased by $C + U$, with U = number of unacceptable responses for the previous test time, ending with a total of 15 panelists.

Dehydrated taro slices stored at 45°C in PE bags were used for the sensory panels conducted. They were representative of the taro used in the determination of a_w, rehydration capacity, α-tocopherol content, and thiamine content. For each panel, slices of taro were soaked in distilled water at refrigerated temperatures for 15 hours and then boiled in a 0.4% brine solution until the taro began to crack at the edges, fall apart, and turn gray, indicating doneness. One slice of taro (8–12 g, 0.3–0.5 cm thick, and 5–8 cm diameter) was served to each panelist on a 15.24 cm diameter Styrofoam plate labeled with a three-digit blinding code. The sample was received through bread box-style compartments in isolated booths under normal 17 Watt fluorescent lighting. Panelists were instructed to use a bite of unsalted cracker and a sip of bottled water to cleanse the palate before tasting. Questions were presented one at a time on a computer screen and data was collected using Compusense®5 (version 4.6) software (Compusense Inc., Guelph, Ontario, Canada). Untrained panelists were asked to evaluate the samples as either "acceptable" or "unacceptable", and to indicate whether or not they would consume the sample as part of their regular diet by answering either "yes" or "no."

3.6. Statistical Analysis. Nutrient content data was analyzed using PROC MIXED of SAS (SAS Inst. Inc., Cary, NC, USA). Significant differences among treatment means were determined using Fisher's LSD for all pairwise comparisons with a significance level of 0.05.

Mean hedonic scores for the sensory analysis performed in Fiji were determined by averaging the 86 scores for the various aspects evaluated and calculating the standard deviations using a standard formula. Shelf life of taro slices was determined by a modified Weibull Hazard analysis, as presented in Cardelli and Labuza [28]. Hazard values were

TABLE 2: Changes in L^* (0 to 100, dark to light) color values of dehydrated taro slices stored at different a_w levels and in different packaging materials[1].

a_w level	Package type	Storage time (wk)			
		0	13	26	34
0.35	Polyethylene	74.32^{abA}	79.22^{aA}	68.23^{bAB}	73.90^{abA}
0.40	Polyethylene	70.61^{aA}	72.57^{aA}	72.07^{aAB}	68.86^{aA}
0.54	Polyethylene	70.90^{aA}	72.11^{aA}	76.68^{aAB}	74.53^{aA}
0.40	Foil laminate	78.73^{aA}	70.51^{aA}	73.13^{aAB}	74.65^{aA}
0.54	Foil laminate	73.83^{aA}	70.41^{aA}	78.65^{aA}	73.26^{aA}
0.71	Foil laminate	69.49^{aA}	72.41^{aA}	67.65^{aB}	76.70^{aA}

[1] Mean of two replicates.
a-b Means within a row with different small letters are significantly different ($P < 0.05$).
A-B Means within a column with different capital letters are significantly different ($P < 0.05$).

TABLE 3: Changes in a^* ($\pm$, green/red) color values of dehydrated taro slices stored at different a_w levels and in different packaging materials[1].

a_w level	Package type	Storage time (wk)			
		0	13	26	34
0.35	Polyethylene	2.08^{bA}	2.76^{abA}	3.93^{aAB}	3.46^{abA}
0.40	Polyethylene	2.95^{aA}	3.10^{aA}	3.88^{aAB}	4.08^{aA}
0.54	Polyethylene	1.63^{cA}	2.29^{bcA}	3.25^{abB}	3.95^{aA}
0.40	Foil laminate	0.77^{bB}	3.02^{aA}	3.76^{aAB}	3.15^{aA}
0.54	Foil laminate	1.93^{bA}	3.20^{abA}	2.77^{abB}	3.67^{aA}
0.71	Foil laminate	2.71^{bA}	3.68^{abA}	5.17^{aA}	4.57^{aA}

[1] Mean of two replicates.
a-c Means within a row with different small letters are significantly different ($P < 0.05$).
A-B Means within a column with different capital letters are significantly different ($P < 0.05$).

TABLE 4: Changes in b^* ($\pm$, blue/yellow) color values of dehydrated taro slices stored at different a_w levels and in different packaging materials[1].

a_w level	Package type	Storage time (wk)			
		0	13	26	34
0.35	Polyethylene	11.58^{cB}	13.61^{bcB}	17.09^{aB}	15.64^{abB}
0.40	Polyethylene	13.21^{aA}	13.56^{aB}	16.41^{aB}	15.60^{aB}
0.54	Polyethylene	11.70^{bA}	13.72^{abB}	15.20^{aB}	16.51^{aB}
0.40	Foil laminate	12.05^{bA}	15.34^{abAB}	17.35^{aB}	16.82^{aB}
0.54	Foil laminate	11.93^{bA}	15.79^{aAB}	16.75^{aB}	18.49^{aAB}
0.71	Foil laminate	12.59^{cA}	17.72^{bA}	21.52^{aA}	21.01^{abA}

[1] Mean of two replicates.
a-c Means within a row with different small letters are significantly different ($P < 0.05$).
A-B Means within a column with different capital letters are significantly different ($P < 0.05$).

calculated as described in Gacula and Singh [29]. The log cumulative hazard was then plotted against time. The end of shelf life was determined to be the time for 50% probability that an untrained tester would grade the samples as being unacceptable. This probability corresponds to a cumulative hazard of 69.3.

4. Results and Discussion

4.1. Physical Quality

Color. The L^* (lightness) values ranged from 67.65 to 78.65 and showed no significant effects due to storage time, a_w, or package type (Table 2). a^* (redness) and b^* (yellowness) values showed significant effects from storage time and a_w. a^* and b^* values increased significantly with storage time (Tables 3 and 4). In the FL pouches, a^* and b^* values generally increased with increasing a_w. This effect was not present in the PE bags, possibly because the samples equilibrated to the same a_w level or because of the greater headspace oxygen compared to the FL pouches. Nonenzymatic browning rate generally increases with increasing a_w, reaches a maximum at a_w of 0.3–0.7, depending on the type of food, and decreases with a further increase in a_w [30]. Most enzymatic reactions are slowed at a_w < 0.8, which may explain the relatively light appearance of the dried taro, in as much as all a_w values studied were below this value. Water may accelerate both enzymatic and nonenzymatic browning by enhancing mobility of the reactants. On the other hand, an increase in water content may decrease browning rate by diluting the reactive components [31].

With respect to packaging material, the FL pouches did not result in less color change of the samples compared to the PE bags. However, if the FL packaging is utilized with lower initial a_w, the dehydrated taro slices will likely experience less color change during storage.

Rehydration. Storage time, a_w, and packaging material had significant effects on rehydration capacity. Rehydration capacity decreased significantly with storage time (Figure 1).

Most of the decrease in rehydration capacity occurred during the first thirteen weeks of storage. Sanjuán et al. [32, 33] observed similar results with dehydrated broccoli stored at 40°C. They noticed a sharp decrease in rehydration capacity occurring by 15 weeks of storage followed by a negligible decrease from that time forward. In the FL pouches, a_w was inversely correlated with rehydration capacity with lower a_w samples having higher rehydration capacities. The initial a_w of the samples in the PE bags had no effect on the rehydration capacity, again, likely due to the equilibrating of a_w levels to a relatively high level.

4.2. Nutritional Quality

Thiamin. Thiamin content is shown in Table 5. Results were adjusted to account for a recovery rate average of 85%. Storage time, a_w, and packaging material had significant effects on thiamin content ($P < 0.01$). The initial thiamin content of the dried samples ranged within 2.03–2.54 μg/g and did not show significant differences due to initial a_w levels prior to packaging. The thiamin content decreased significantly to 0.14–1.38 μg/g [5.48–68.2% retention] after 34 weeks of storage. Retention of thiamin was highest in the FL pouches at

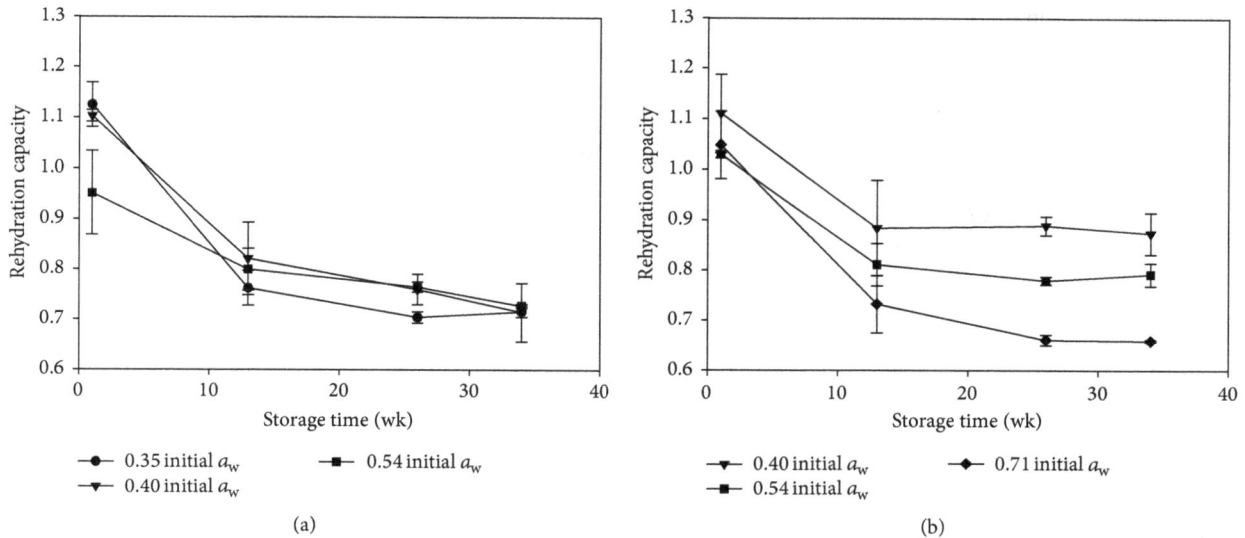

FIGURE 1: Dimensionless rehydration capacity of dehydrated taro slices dried to different a_w stored in (a) polyethylene bags and (b) foil laminate pouches at 45°C and 75% RH for 34 weeks ($n = 3$).

$0.4a_w$. Both a_w and packaging material had significant effects on thiamin retention ($P < 0.01$). Degradation of thiamin generally increased with increasing a_w. Dennison et al. [34] found similar results in a model system at 45°C; however, at temperatures ≤ 37°C thiamin was found to be quite stable regardless of storage temperature or a_w. Thiamin retention was significantly higher in FL pouches. This could be due to the lower water vapor permeability which may have caused the samples to reabsorb moisture at a much slower rate than the samples in the PE bags. The higher thiamin retention could also have possibly been related to the exclusion of oxygen in the FL pouches.

α-Tocopherol. α-Tocopherol content is shown in Table 6. Results were adjusted to account for a recovery rate of 77% at 0 weeks storage and 84% at 34 weeks storage. Results are reported as μg α-tocopherol per g of dry sample. Storage time, a_w, and packaging material had significant effects on α-tocopherol content. The initial α-tocopherol content of the dried samples ranged within 1.59–9.97 μg/g and was significantly lower at lower a_w levels, suggesting that it was susceptible to degradation during the prolonged drying required to obtain lower a_w levels. The α-tocopherol content generally decreased to values ranging within 1.30–7.59 μg/g after 34 weeks of storage with the exception of the FL pouches at $0.40a_w$ which increased in α-tocopherol content. The results for that treatment combination are likely due to sampling variation within the treatment. The percent retention data was too variable to detect any differences in retention. α-Tocopherol content was generally higher in samples with higher initial a_w. In contrast, Widicus et al. [35] found that α-tocopherol was less stable with increasing a_w and increasing molar ratio of oxygen in a dehydrated model food system at 37°C. If there was increased stability in taro at lower a_w levels it was secondary in importance to the effect of

TABLE 5: Thiamin content of dehydrated taro slices stored for 34 weeks at different a_w levels and in different packaging materials[1].

a_w level	Package type	Initial content (μg/g)	Final content (μg/g)	Retention (%)
0.35	Polyethylene	2.51[aA]	0.45[bC]	18.01[C]
0.40	Polyethylene	2.54[aA]	0.14[bC]	5.48[D]
0.54	Polyethylene	2.47[aA]	0.17[bC]	6.82[D]
0.40	Foil laminate	2.03[aB]	1.38[bA]	68.20[A]
0.54	Foil laminate	2.54[aA]	0.95[bB]	37.46[B]
0.71	Foil laminate	2.16[aAB]	0.34[bC]	15.66[C]

[1]Mean of two replicates.
[a-b]Means within a row with different small letters are significantly different ($P < 0.05$).
[A-D]Means within a column with different capital letters are significantly different ($P < 0.05$).

extended drying times on the initial content of α-tocopherol. α-Tocopherol content after storage was generally higher in the FL pouches. The exclusion of oxygen in the FL pouches may have slowed the rate of degradation of α-tocopherol.

4.3. Sensory Analysis. In the hedonic sensory test in Fiji, panelists rated the overall acceptability of the rehydrated taro at 7.2 ± 1.5 on a hedonic scale (Table 7). They also rated the appearance at 6.9 ± 1.9, the aroma at 7.0 ± 1.9, the flavor at 7.1 ± 1.8, and the texture at 7.0 ± 1.7. These results are congruent with "like moderately," which is indicated by a score of 7.0 on the hedonic scale used. Of the 86 panelists, 73% indicated that they would accept this product into their regular diet, and 97% also said that they would use it in emergencies where there was a shortage of food. These results are consistent with results obtained by Rowe et al. [18], who

TABLE 6: α-Tocopherol content of dehydrated taro slices stored for 34 weeks at different a_w and in different packaging materials[1].

a_w level	Package type	Initial content (μg/g)	Final content (μg/g)	Retention (%)
0.35	Polyethylene	1.59[aC]	1.30[aC]	82.06[B]
0.40	Polyethylene	3.02[aC]	2.03[aC]	66.86[B]
0.54	Polyethylene	6.69[aB]	2.90[bBC]	44.64[B]
0.40	Foil laminate	2.79[bC]	4.45[aB]	160.57[A]
0.54	Foil laminate	9.97[aA]	7.59[bA]	77.59[B]
0.71	Foil laminate	8.83[aA]	6.64[bA]	75.23[B]

[1] Mean of three replicates.

[a-b] Means within a row with different small letters are significantly different ($P < 0.05$).

[A-C] Means within a column with different capital letters are significantly different ($P < 0.05$).

TABLE 7: Mean hedonic scores and percent acceptance from Fijian cultural acceptability sensory panel for dehydrated taro slices[a].

Appearance	6.9 ± 1.9
Aroma	7.0 ± 1.9
Flavor	7.1 ± 1.8
Texture	7.0 ± 1.7
Overall acceptability	7.2 ± 1.5
Regular diet acceptance	73%
Emergency use acceptance	97%

[a] Mean of 86 scores ± standard deviation.

TABLE 8: Weibull sensory data for dehydrated taro slices stored at 45°C in PE bags.

Weeks	Acceptability													
0	+	+	+											
15	+	+	+	+										
26	−	+	+	+	−									
32.5	−	+	−	+	+	−								
39	−	+	−	+	+	+	+	+	−	−				
42	+	−	+	−	−	+	−	−	+	−	−	−	−	+

+: acceptable sample as assessed by an untrained panelist.

−: unacceptable sample as assessed by an untrained panelist.

conducted a consumer taste panel in Fiji, finding that 73% of panelists would consume dehydrated taro slices as part of their regular diet.

For the subsequent sensory shelf life tests in the US, the ratings of each individual panelist are shown in Table 8. The product was deemed acceptable by 100% of the panelists for each panel until week 26 when some panelists began to deem it unacceptable. Weibull Hazard rankings were assigned for the failures in weeks 26 through 42, as shown in Table 9. Shelf life was determined to be 38.5 weeks as shown in Figure 2. In comparison, fresh taro's shelf life in tropical climates is typically only about 2 weeks [3–5]. The extension of shelf life of an acceptable taro product to 38.5 weeks is a substantial improvement.

TABLE 9: Weibull hazard ranking table for dehydrated taro slices stored at 45°C in PE bags.

Rank	Weeks	H value[a]	$\sum H$
19	26	5.26	5.26
18	26	5.56	10.8
17	32.5	5.88	16.7
16	32.5	6.25	23
15	32.5	6.67	29.6
14	39	7.14	36.8
13	39	7.69	44.5
12	39	8.33	52.8
11	39	9.09	61.9
10	42	10.0	71.9
9	42	11.1	83
8	42	12.5	95.5
7	42	14.3	110
6	42	16.7	126
5	42	20.0	146
4	42	25.0	171
3	42	33.3	205
2	42	50.0	255
1	42	100.0	355

[a] H = hazard value = 100/rank.

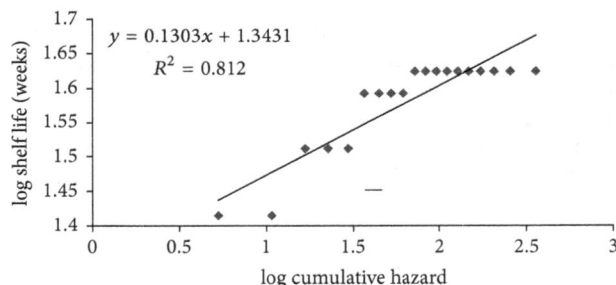

$y = 0.1303x + 1.3431$
$R^2 = 0.812$

Shelf life calculation:
69.3 cumulative hazard = 50% probability of failure
log 69.3 = 1.84
log shelf life = 0.1303(1.84) + 1.3431 = 1.58
Shelf life = 38.3 weeks

FIGURE 2: Weibull hazard plot for dehydrated taro slices stored at 45°C in PE.

As discussed earlier, PE's permeability to oxygen and moisture was associated with higher end a_w levels and greater headspace oxygen during storage. This higher a_w was associated with greater browning, lower rehydration capacity, and lower vitamin retention. This corresponds to Fennema's [21] observations which show the highest rate of browning, oxidation, and other nonenzymatic reactions in an a_w range of 0.6 to 0.8 of many food products. In the Weibull analysis conducted, increased browning and lower rehydration capacity could have had an impact on sensory quality, as they tend to indicate that a food is less fresh.

A sorption isotherm at 25°C, GAB model (data not shown), was run on the fresh taro to determine the effect of hysteresis on potential physical changes in the starch granules during dehydration and subsequent rehydration as described by Nurtama and Lin [36]; however, minimal hysteresis was observed which would support the high degree of acceptance in both the initial sensory test and the Weibull sensory data (Tables 7 and 8).

5. Conclusions

The quality of dehydrated taro slices in accelerated storage was investigated as a function of a_w and package type. Samples stored in low oxygen packaging (FL pouches) exhibited less a_w fluctuation during storage. Also, samples stored in FL pouches at lower a_w exhibited less browning and higher rehydration capacity. Thiamin retention was significantly higher in FL packaging and at lower a_w. Alpha-tocopherol content was also higher in FL pouches, though it was likely susceptible to degradation during drying. It appears that packaging capable of maintaining a low a_w, such as FL pouches, optimizes the physical quality of dehydrated taro slices during storage. In a hedonic test of dehydrated taro, panelists indicated that they liked product moderately, and in a Weibull analysis of dehydrated taro stored in PE bags, the shelf life was determined to be 38.5 weeks. If FL pouches were used instead of PE bags, the a_w may have remained lower, thus improving these two sensory impacting aspects. Thus, storing the dehydrated taro in FL pouches would likely increase the shelf life of this product even further. However, even this extension of shelf life through simple dehydration is not insignificant, especially as nutritional quality can be retained through the appropriate use of packaging and a_w level. Maintaining the nutritional and sensory attributes of taro can improve its functionality as an emergency or secondary significant food source in the Pacific Islands and elsewhere.

Competing Interests

The authors declare that they have no competing interests.

References

[1] FAO, *FAOSTAT Statistical Database*, 2013, http://faostat3.fao .org.

[2] I. Onwueme, *Taro Cultivation in Asia and the Pacific*, RAP Publication 1999/16, FAO, Bangkok, Thailand, 1999, http://www.fao.org/docrep/005/ac450e/ac450e00.htm.

[3] D. E. Gollifer and R. H. Booth, "Storage losses of taro corms in the British Solomon Islands Protectorate," *Annals of Applied Biology*, vol. 73, no. 3, pp. 349–356, 1973.

[4] S. Chandra, "Handling, storage, and processing of root crops in Fiji," in *Small-Scale Processing and Storage of Tropical Root Crops*, D. L. Plucknett, Ed., pp. 53–63, Westview Press, Boulder, Colo, USA, 1979.

[5] H. C. Passam, "Experiments on the storage of eddoes and tannias (Colocasia and Xanthosoma spp.) under tropical ambient conditions," *Tropical Science*, vol. 24, pp. 39–46, 1982.

[6] Fiji Meteorological Services, The Climate of Fiji, 2006, http://www.met.gov.fj/ClimateofFiji.pdf.

[7] K. S. Jayaraman, V. K. Gopinathan, P. Pitchamuthu, and P. K. Vijayaraghavan, "The preparation of quick-cooking dehydrated vegetables by high temperature short time pneumatic drying," *International Journal of Food Science & Technology*, vol. 17, no. 6, pp. 669–678, 1982.

[8] J. H. Moy, N. T. S. Wang, and T. O. M. Yama, "Dehydration and processing problems of taro," *Journal of Food Science*, vol. 42, no. 4, pp. 917–920, 1977.

[9] W. K. Nip, "Development and storage stability of drum-dried guava- and papaya-taro flakes," *Journal of Food Science*, vol. 44, no. 1, pp. 222–225, 1979.

[10] C. Henríquez, A. Córdova, M. Lutz, and J. Saavedra, "Storage stability test of apple peel powder using two packaging materials: high-density polyethylene and metalized films of high barrier," *Industrial Crops and Products*, vol. 45, pp. 121–127, 2013.

[11] G. O. Sigge, C. F. Hansmann, and E. Joubert, "Effect of storage conditions, packaging material and metabisulphite treatment on the color of dehydrated green bell peppers (*Capsicum annuum* L.)," *Journal of Food Quality*, vol. 24, no. 3, pp. 205–218, 2001.

[12] S. Devahastin and C. Niamnuy, "Modelling quality changes of fruits and vegetables during drying: a review," *International Journal of Food Science and Technology*, vol. 45, no. 9, pp. 1755–1767, 2010.

[13] S. S. Sablani, "Drying of fruits and vegetables: retention of nutritional/functional quality," *Drying Technology*, vol. 24, no. 2, pp. 123–135, 2006.

[14] US Department of Agriculture, *USDA Nutrient Database for Standard Reference. Release 26*, 2004, http://ndb.nal.usda.gov/ndb/foods/show/3540.

[15] FDA 2013, Guidance for Industry: A Food Labeling Guide 14, Appendix F: Calculate the Percent Daily Value for the Appropriate Nutrients, http://www.fda.gov/Food/GuidanceRegulation/GuidanceDocumentsRegulatoryInformation/LabelingNutrition/ucm064928.htm.

[16] C. M. Sabliov, C. Fronczek, C. E. Astete, M. Khachaturyan, L. Khachatryan, and C. Leonardi, "Effects of temperature and UV light on degradation of α-tocopherol in free and dissolved form," *JAOCS: Journal of the American Oil Chemists' Society*, vol. 86, no. 9, pp. 895–902, 2009.

[17] B. K. Dwivedi and R. G. Arnold, "Chemistry of thiamine degradation: mechanisms of thiamine degradation in a model system," *Journal of Food Science*, vol. 37, no. 6, pp. 886–888, 1972.

[18] J. P. Rowe, A. R. Sloan, and F. M. Steele, "Long-term preservation of cassava, taro, and breadfruit by solar-drying in the Pacific Islands area," in *IFT Annual Meeting Book of Abstracts*, p. 201, Institute of Food Technologists, Chicago, Ill, USA, 2004.

[19] G. Hough and L. Garitta, "Methodology for sensory shelf-life estimation: a review," *Journal of Sensory Studies*, vol. 27, no. 3, pp. 137–147, 2012.

[20] A. Giménez, F. Ares, and G. Ares, "Sensory shelf-life estimation: a review of current methodological approaches," *Food Research International*, vol. 49, no. 1, pp. 311–325, 2012.

[21] D. S. Reid and O. R. Fennema, "Molecular mobility and food stability," in *Food Chemistry*, S. Damodaran, K. L. Parkin, and O. R. Fennema, Eds., pp. 46–65, CRC Press, Boca Raton, Fla, USA, 2008.

[22] W. F. A. Horner, "Drying: chemical changes," in *Encyclopedia of Food Science, Food Technology and Nutrition*, R. Macrae, R. K. Robinson, and M. J. Sadler, Eds., pp. 1485–1489, Academic Press, New York, NY, USA, 1993.

[23] T. D. Durance and J. H. Wang, "Energy consumption, density, and rehydration rate of vacuum microwave- and hot-air convection- dehydrated tomatoes," *Journal of Food Science*, vol. 67, no. 6, pp. 2212–2216, 2002.

[24] S. Ndaw, M. Bergaentzlé, D. Aoudé-Werner, and C. Hasselmann, "Extraction procedures for the liquid chromatographic determination of thiamin, riboflavin and vitamin B6 in foodstuffs," *Food Chemistry*, vol. 71, no. 1, pp. 129–138, 2000.

[25] D. M. Peterson and A. A. Qureshi, "Genotype and environment effects on tocols of barley and oats," *Cereal Chemistry*, vol. 70, pp. 157–162, 1993.

[26] J. K. Russon, M. L. Dunn, and F. M. Steele, "Optimization of a convective air flow solar food dryer," *International Journal of Food Engineering*, vol. 5, no. 1, article 8, 2009.

[27] M. McEwan, L. V. Ogden, and O. A. Pike, "Effects of long-term storage on sensory and nutritional quality of rolled oats," *Journal of Food Science*, vol. 70, no. 7, pp. S453–S458, 2005.

[28] C. Cardelli and T. P. Labuza, "Application of weibull hazard analysis to the determination of the shelf life of roasted and ground coffee," *LWT—Food Science and Technology*, vol. 34, no. 5, pp. 273–278, 2001.

[29] M. C. Gacula and J. Singh, *Statistical Methods in Food and Consumer Research*, Academic Press, New York, NY, USA, 1984.

[30] K. Eichner, "The influence of water content on non-enzymic browning reactions in dehydrated foods and model systems and the inhibition of fat oxidation by browning intermediates," in *Water Relations of Foods*, R. B. Duckworth, Ed., pp. 417–434, Academic Press, New York, NY, USA, 1975.

[31] T. P. Labuza and M. Saltmarch, "The nonenzymatic browning reaction as affected by water in foods," in *Water Activity: Influences on Food Quality*, L. B. Rockland and G. F. Stewart, Eds., pp. 605–650, Academic Press, New York, NY, USA, 1981.

[32] N. Sanjuán, J. Benedito, J. Bon, and A. Mulet, "Changes in the quality of dehydrated broccoli stems during storage," *Journal of the Science of Food and Agriculture*, vol. 80, no. 11, pp. 1589–1594, 2000.

[33] N. Sanjuán, J. Bon, G. Clemente, and A. Mulet, "Changes in the quality of dehydrated broccoli florets during storage," *Journal of Food Engineering*, vol. 62, no. 1, pp. 15–21, 2004.

[34] D. Dennison, J. Kirk, J. Bach, P. Kokoczka, and D. Heldman, "Storage stability of thiamin and riboflavin in a dehydrated food system," *Journal of Food Processing and Preservation*, vol. 1, no. 1, pp. 43–54, 1977.

[35] W. A. Widicus, J. R. Kirk, and J. F. Gregory, "Storage stability of α-tocopherol in a dehydrated model food system containing no fat," *Journal of Food Science*, vol. 45, no. 4, pp. 1015–1018, 1980.

[36] B. Nurtama and J. Lin, "Moisture sorption isotherm characteristics of taro flour," *World Journal of Dairy & Food Sciences*, vol. 5, pp. 1–6, 2010.

19

Comparative Analysis of Nutritional and Bioactive Properties of Aerial Parts of Snake Gourd (*Trichosanthes cucumerina* Linn.)

Ruvini Liyanage,[1] **Harshani Nadeeshani,**[2] **Chathuni Jayathilake,**[1] **Rizliya Visvanathan,**[1] **and Swarna Wimalasiri**[2]

[1]*Division of Nutritional Biochemistry, National Institute of Fundamental Studies, Hantana Road, Kandy, Sri Lanka*
[2]*Department of Food Science and Technology, Faculty of Agriculture, University of Peradeniya, Peradeniya, Sri Lanka*

Correspondence should be addressed to Ruvini Liyanage; ruvini@ifs.ac.lk

Academic Editor: Salam A. Ibrahim

The present investigation was carried out to determine the nutritional and functional properties of *T. cucumerina*. Water extracts of freeze dried flowers, fruits, and leaves of *T. cucumerina* were evaluated for their total phenolic content (TPC), total flavonoid content (TFC), antioxidant activity, α-amylase inhibitory activity, and fiber and mineral contents. Antioxidant activity, TPC, and TFC were significantly higher ($P \leq 0.05$) in leaves than in flowers and fruits. A significant linear correlation was observed between the TPC, TFC, and antioxidant activities of plant extracts. Although, leaves and flower samples showed a significantly higher ($P \leq 0.05$) amylase inhibitory activity than the fruit samples, the overall amylase inhibition was low in all three parts of *T. cucumerina*. Soluble and insoluble dietary fiber contents were significantly higher ($P \leq 0.05$) in fruits than in flowers and leaves. Ca and K contents were significantly higher ($P \leq 0.05$) in leaf followed by fruit and flower and Mg, Fe, and Zn contents were significantly higher ($P \leq 0.05$) in leaves followed by flowers and fruits. In conclusion, *T. cucumerina* can be considered as a nourishing food commodity which possesses high nutritional and functional benefits for human health.

1. Introduction

Trichosanthes cucumerina is a well-known plant, the fruit of which is mainly consumed as a vegetable. The plant is commonly called snake gourd, viper gourd, snake tomato, or long tomato in many countries. It is an annual climber belonging to the family Cucurbitaceae and commonly grown in Asian countries including Sri Lanka, India, Malaysia, Peninsula, and Philippines [1]. The fruit is usually consumed as a vegetable due to its high nutritional value. The plant is a rich source of functional constituents other than its basic nutrients such as flavonoids, carotenoids, phenolic acids, and soluble and insoluble dietary fibers and essential minerals, which makes the plant pharmacologically and therapeutically active [2, 3]. The plant contains proteins, fat, fiber, carbohydrates, minerals, and vitamins A and E in high levels. The predominant mineral elements are potassium (121.6 mg/100 g) and phosphorus (135 mg/100 g) and also sodium, magnesium, and zinc are found in fairly high amounts [4].

In ancient medicine *T. cucumerina* was used for treating headache, alopecia, fever, abdominal tumors, bilious, boils, acute colic diarrhea, haematuria, and skin allergy. It has a prominent place in medicinal systems like *Ayurveda* and *Siddha* [2]. The whole plant including roots, leaves, fruits, and seeds is reported to show medicinal properties such as antidiabetic, antibacterial, anti-inflammatory, anthelmintic, antifebrile, gastroprotective, and antioxidant activity [1, 2, 5–7]. In Sri Lanka, it is the aerial parts of *T. cucumerina* that are used in the traditional medicinal system for treating disease conditions. *T. cucumerina* is one of the major ingredients in several polyherbal preparations that are prescribed in Sri Lanka for the control of Diabetes Mellitus [1, 6]. In a study done by Arawwawala et al. [1], the hot water extract of aerial parts of *T. cucumerina* was found to significantly reduce the blood glucose levels and improve the glucose tolerance of normoglycemic and STZ-induced diabetic rats. In addition to this, Arawwawala et al. have done several studies on Sri Lankan *T. cucumerina* and have reported it to show antioxidant [8], anti-inflammatory [9], antimicrobial [10], and gastroprotective [11] properties. However, to date, there are no detailed studies on the micronutrient content and functional properties of aerial parts of Sri Lankan *T. cucumerina*. Therefore, the present study was conducted to determine the antioxidant potential, total phenolic and total flavonoid content, α-amylase inhibitory activity, soluble and insoluble dietary fiber content, and the essential mineral element content of *T. cucumerina* grown in Sri Lanka.

2. Material and Methods

2.1. Plant Samples. Fresh snake gourd fruits, leaves, and flowers belonging to the variety TA-2 were collected from Horticultural Crop Research and Development Institute (HORDI), Gannoruwa, Sri Lanka, in September 2014. Fruits and leaves were harvested three months after cultivation while fruits were collected two weeks after fruit set. All the chemicals and solvents used were of analytical grade.

2.2. Preparation of Plant Extracts for Antioxidant, Total Phenolic, and Total Flavonoid Assays. Nearly 50 g of fresh and disease-free snake gourd leaves, flowers, and fruits were washed with distilled water and sliced. The pieces were freeze-dried, ground into fine powder, and stored at $-20°C$ till further analysis. Freeze dried samples (0.1 g) were homogenized at 5000 rpm for 10 minutes in 7.5 mL of deionized water. The homogenized plant samples were centrifuged at 7000 rpm for 15 minutes in a microcentrifuge (5340R, Germany). The supernatant was filtered through Whatman® 42 filter paper. The extracts were appropriately diluted with distilled water and used for analysis.

2.3. Total Phenolic Content (TPC). The total phenolic content (TPC) of the extracts was determined colorimetrically as described by Samatha et al. [12] with slight modifications. The reaction mixture was prepared by mixing 50 μL of sample extract, 105 μL of 10% Folin-Ciocalteu's reagent dissolved in deionized water, 80 μL of sodium carbonate (Na_2CO_3, 7.5%, w/v), and 15 μL of deionized water. After 3 minutes, Na_2CO_3 was added and the mixture was incubated for 30 minutes at room temperature. The absorbance was taken using UV-Visible microplate spectrophotometer (Multiskan®, 2011) at 760 nm. Results were expressed in terms of milligrams of Gallic acid equivalents (mg of GAE) per gram dry weight (g DW). All tests were conducted in triplicate.

2.4. Total Flavonoid Content (TFC). The procedure described by Agbo et al. [13] was followed with slight modifications. A volume of 0.5 mL of aqueous extract was taken in to a test tube and 2 mL of distilled water was added followed by the addition of 0.15 mL of sodium nitrite ($NaNO_2$, 5%, w/v) and allowed to stand for 6 min. Afterward, 0.15 mL of aluminum trichloride ($AlCl_3$, 10%, w/v) was added and incubated for 6 min, followed by the addition of 2 mL of sodium hydroxide (NaOH, 4%, w/v), and the volume was made up to 5 mL with distilled water. After 15 min of incubation, absorbance was

measured at 510 nm using UV-Visible microplate spectrophotometer. Results were expressed in terms of mg of catechin equivalents (CE) per gram of extract.

2.5. Antioxidant Capacity

2.5.1. DPPH (2,2-Diphenyl-1-picrylhydrazyl Radical Scavenging Activity). The DPPH assay was performed according to the method described by Sanjeevkumar et al. [14] with modifications. DPPH solution (100 μL) was added to different volumes (0, 30, 60, 90, 120, and 150 μL) of aqueous sample extract and diluted with distilled water until the volume reached 250 μL and was allowed to stand for 30 minutes in dark at room temperature. The absorbance was read at 517 nm. The sample concentration providing 50% inhibition (IC_{50}) was calculated.

2.5.2. ABTS (2,2'-Azino-bis(3-ethylbenzothiazoline-6-sulphonic Acid)) Radical Scavenging Activity. The total antioxidant capacity of the extracts was determined using ABTS radical. The ABTS radical was generated by reacting ABTS (1.25 mM) with potassium persulphate (PP) (2.0 mM), which acts as the radical generator. From each extract 50 μL was mixed with 150 μL of ABTS radical solution and absorbance was read over 6 minutes at 1-minute interval at 734 nm. The radical scavenging activity after lapse of 6 minutes was calculated as percentage of ABTS discolouration [15]. Results were presented as mM of Trolox equivalents (mM of TE) per gram dry weight (g DW).

2.5.3. Ferric Reducing Antioxidant Power (FRAP) Assay. The ability to reduce ferric ions was measured using the method described by Shukla et al. [16] with some modifications. The FRAP reagent was prepared by mixing 300 mM sodium acetate buffer (pH 3.6), 10.0 mM (tripyridyl triazine) TPTZ, and 20.0 mM $FeCl_3·6H_2O$ solution in a ratio of 10:1:1, respectively. FRAP reagent (150 μL) was added to 100 μL of sample extract and the reaction mixture was incubated at 37°C for 30 min. The absorbance was measured at 593 nm. Fresh working solutions of $FeSO_4$ were used for the standard curve. The results were expressed as mM Fe^{2+} equivalents per gram of sample.

2.6. Potent α-Amylase Inhibitory Activity

2.6.1. Preparation of Plant Extracts for α-Amylase Inhibitory Assays. Freeze dried snake gourd sample (0.1 g) was homogenized at 5,000 rpm for 15 minutes in 7.5 mL of 1% Dimethyl Sulphoxide (DMSO). The homogenized plant samples were centrifuged at 7,000 rpm for 15 minutes in a microcentrifuge. The supernatant was used for enzyme assays.

2.6.2. DNSA (3,5-Dinitrosalicylic Acid) α-Amylase Inhibitory Assay. The assay was performed according to Sudha et al. [17] with slight modifications. The assay mixture consisted of 50 μL of α-amylase from porcine pancreas (A3176 SIGMA) in phosphate buffer (PBS) (0.02 M; 6.7 mM NaCl; pH 6.9) and 50 μL plant extracts in different concentrations ranging from to 25 mg/mL. The content was incubated for 30 minutes at room temperature followed by the addition of 100 μL of 1% soluble starch (S2004 SIGMA) and incubated for another 3 minutes at room temperature. The reaction was terminated by adding 100 μL DNSA color reagent and placed in 85°C water bath for 15 minutes. After cooling, the sample mixture was diluted with 900 μL of distilled water and absorbance was measured at 540 nm. The control samples were prepared without plant extracts. The % inhibition was calculated and the results were expressed in terms of IC_{50} value. Acarbose was used as the standard inhibitor.

2.6.3. Starch-Iodine α-Amylase Inhibitory Assay. Starch-io-dine assay was carried out as described by Sudha et al. [17] with slight modifications. Assay reaction was initiated by adding 40 μL of PBS, 40 μL of plant extracts in different concentrations (5–25 mg/mL), and 40 μL of enzyme in PBS into microplate wells. After 10 minutes of incubation at 37°C, 40 μL of 0.3% soluble starch solution was added and again incubated for another 15 minutes at 37°C. To stop the reaction 20 μL of 1 M HCl was added, followed by the addition of 100 μL of iodine solution (5 mM I_2 and 5 mM KI). Immediately after addition the absorbance was taken at 620 nm.

2.6.4. Starch Hydrolase α-Amylase Inhibitory Assay. The inhibitory activities of plant extracts were quantified based on turbidity measurements according to previous work done by Liu et al. [18] with slight modifications. Assay was initiated by adding 100 μL of PBS, 40 μL of enzyme solution, and 40 μL of plant inhibitor solution. The reaction mixture was incubated for 10 minutes at 37°C and thereafter 100 μL of 1% starch solution was added and the mixture was again incubated for another 15 minutes at 37°C. The turbidity change was monitored at 660 nm. The % inhibition was calculated.

2.7. Determination of Mineral Content by Microwave Assisted Digestion. This analysis was performed according to the method described by Negi et al. [19] with slight modifications. From each freeze dried plant sample, 0.2 g was digested with 5 mL of HNO_3 (69%) and 1 mL of H_2O_2 (30%) in a microwave digestion system and diluted to 50 mL with deionized distilled water. A blank digest was carried out in the same way. The digestion condition in microwave digestion unit was programmed as 15 min ramping time, 10 min holding time or digestion time at 180°C, and 15 min cooling time. Calibration curves were prepared for each mineral using 1000 ppm stock solutions of relevant standard (Fe, Zn, Ca, Mg, and K). Necessary dilutions were done in order to get absorbance values. Mineral concentration was given as mg/100 g fresh weight (FW).

2.8. Determination of Insoluble and Soluble Dietary Fiber. This analysis was performed according to the method described by Shin [20] with minor modifications. Duplicates of 1 g sample were weighed and suspended in phosphate buffer (0.08 M) and digested sequentially with heat-stable α-amylase (A3306 Sigma) (pH 6; 100°C; 30 min), protease

(P3910 Sigma) (pH 7.5; 60°C; 30 min), and amyloglucosidase (A9913 Sigma) (pH 4–4.6; 60°C; 30 min) to remove protein and starch. Enzyme digestates were filtered through glass fritted crucibles. Crucibles containing insoluble dietary fiber were rinsed with dilute ethanol followed by acetone and dried overnight in a 105°C oven and weighed to nearest 0.1 mg. Filtrates plus washings were mixed with 4 volumes of 95% ethanol to precipitate materials that were soluble in the digestates. After 1 h, precipitates were filtered through fritted crucibles. One of each set of duplicate insoluble fiber residues and soluble fiber residues was ashed in a muffle furnace at 525°C for 5 h. Another set of residues were used to determine protein as Kjeldahl nitrogen. Soluble or insoluble dietary fiber residues were obtained through calculation. Total dietary fiber was calculated as the sum of soluble and insoluble dietary fiber.

2.9. Statistical Analysis. Data were analyzed using the SAS statistical software version 9.1.3 (SAS Institute Inc., Cary, NC). Results were calculated and expressed as mean ± standard deviation (SD) of 3 independent analyses. P values of ≤0.05 were considered to be significant.

3. Results

3.1. Total Phenolic and Total Flavonoid Content. Table 1 lists the total phenolic content and the total flavonoid content of *T. cucumerina* aerial parts analyzed in the present study. The differences between the extracts for total phenolic content were significant ($P \leq 0.05$). Leaf samples of *T. cucumerina* showed the highest phenolic content whereas the fruit samples showed the least. As in TPC values, TFC values also differ in the same manner in aerial parts of *T. cucumerina* (Table 1). Leaf samples of *T. cucumerina* showed the highest flavonoid content and the fruit samples showed the lowest phenolic content.

3.2. Antioxidant Capacity. In the DPPH assay (Table 1), leaves sample had the lowest IC_{50} value which indicates that it possessed the highest antioxidant activity ($P \leq 0.05$) among all three samples. Fruit extract showed the least significant antioxidant activity ($P \leq 0.05$) by giving the highest IC_{50} value as 10.83 ± 0.7 mg/mL. According to the ABTS assay results (Table 1), leaves had significantly higher ($P \leq 0.05$) antioxidant activity followed by the flower and fruit extract (Figure 1). The antioxidant data in FRAP assay also vary in the same manner. Leaves samples of *T. cucumerina* showed the highest antioxidant activity while the fruit samples showed the least activity.

According to statistical data, the antioxidant activity of *T. cucumerina* was in accordance with their amount of total phenolic and flavonoid contents (Table 2). Total phenolic and flavonoid contents showed strong positive correlation with the ABTS and FRAP assay values and negative correlation with IC_{50} values of DPPH assay.

3.3. Potent α-Amylase Inhibitory Activity. Results indicate that the aerial parts of *T. cucumerina* possess α-amylase

TABLE 1: Total phenolic content and total flavonoid content and antioxidant capacity.

Plant part (Dry weight)	TPC (mg GAE/g)	TFC (mg CE/g)	DPPH IC$_{50}$ (mg/mL)	ABTS (μmol TE/g) 6th minute	FRAP (mM Fe^{2+}/g⁻)
Fruits	4.64 ± 0.3^c	0.77 ± 0.1^c	10.83 ± 0.7^a	24.05 ± 0.8^c	0.40 ± 0.01^b
Leaves	27.39 ± 1.2^a	6.05 ± 0.1^a	3.08 ± 0.2^b	235.71 ± 8.5^a	4.24 ± 0.08^a
Flowers	19.97 ± 1.6^b	4.46 ± 0.1^b	4.16 ± 0.1^b	159.19 ± 2.4^b	4.08 ± 0.11^a

Means followed by the same letter within each column are not significantly different at $P \leq 0.05$, according to the least significant difference test.

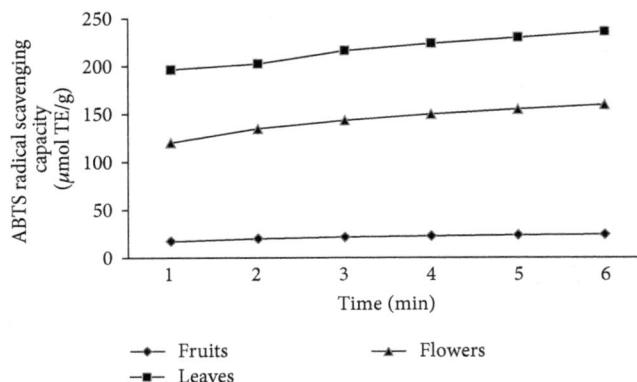

FIGURE 1: Changing of radical scavenging activity of ABTS assay with time.

TABLE 2: Correlation of antioxidant capacity and total phenolic and total flavonoid contents.

Characteristics	TPC	TFC	DPPH	ABTS	FRAP
TPC					
TFC	0.999				
DPPH	−0.961	−0.971			
ABTS	0.998	0.995	0.945		
FRAP	0.919	0.933	0.991	0.896	

TABLE 3: Acarbose positive control for α-amylase inhibitory assays

Inhibitory method	IC$_{50}$ (μg/mL)
Starch-iodine method	80
Turbidity method	83
DNSA method	78

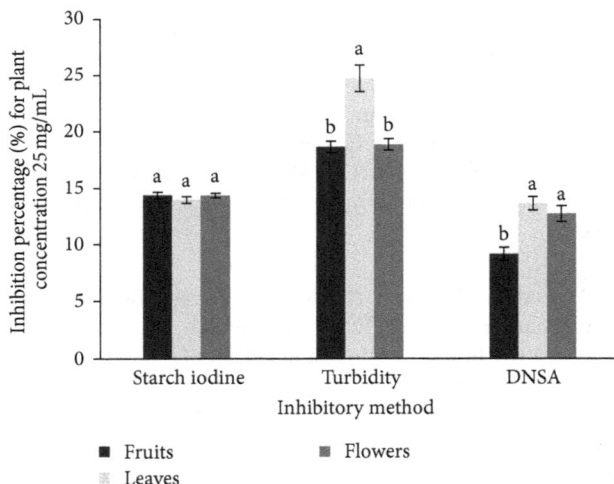

FIGURE 2: Potent α-amylase inhibitory activity values for different assays. Means followed by the same letter(s) (a and b) within a cluster are not significantly different at $P > 0.05$.

inhibitory activity but not in significantly high level. According to the three amylase inhibitory assays, none of the parts of *T. cucumerina* gave an IC$_{50}$ value within the concentration of 25 mg/mL (Figure 2). Acarbose was used as the positive control (Table 3).

3.4. Mineral Element Content. Potassium was recorded as the highest ($P \leq 0.05$) mineral element followed by calcium

and magnesium. Iron and zinc were recorded as the least. Among all three parts, flower samples had the highest mineral element content while the lowest was recorded in the fruit samples, except for calcium and potassium ($P \leq 0.05$). On the contrary, fruit samples were significantly high in potassium ($P \leq 0.05$) (Table 4).

3.5. Insoluble and Soluble Dietary Fiber. Insoluble dietary fiber (IDF) content of all three parts was higher than soluble dietary fiber (SDF). When considering the total dietary fiber (TDF) content, fruit samples had significantly higher fiber content (31.76 ± 3.7 g/100 g) followed by flowers and leaves (21.63 ± 1.71 g/100 g and 9.95 ± 1.25 g/100 g, resp.) (Table 5). The statistical variation was the same as in TDF for both SDF and IDF.

4. Discussion

Snake gourd contains a rich variety of nutrients, vitamins, and minerals that are essential for human health, including significant levels of dietary fiber, a small amount of calories, and high levels of protein. These medicinal plants are reported to possess different pharmacological properties due to the presence of phytochemicals such as alkaloids, flavonoids, glycosides, tannins, and steroids. It is well-known that antioxidative properties of plants help to protect the body from free radical damage in cell structures, nucleic acids, lipids, proteins, and other body components and thereby reduce the risk and progression of many acute and chronic

TABLE 4: Mineral content by microwave assisted digestion.

Plant part	Mineral elements (mg/100 g dry weight basis)				
	Fe	Zn	Ca	Mg	K
Fruits	4.81 ± 0.09^c	1.21 ± 0.05^{bc}	2452.31 ± 25.10^a	1179.50 ± 17.27^c	2472.77 ± 16.94^{ab}
Leaves	11.72 ± 0.48^{ab}	2.81 ± 0.35^{bc}	1450.23 ± 33.25^{bc}	2146.12 ± 74.88^b	2367.89 ± 22.45^{ab}
Flowers	13.87 ± 0.14^{ab}	7.13 ± 0.05^a	1351.88 ± 24.47^{bc}	2533.75 ± 21.59^a	1379.07 ± 17.10^c

Means followed by the same letter(s) within each column are not significantly different at $P \leq 0.05$, according to the least significant difference test.

TABLE 5: Insoluble and soluble dietary fiber.

Plant part	Insoluble dietary fiber (g/100 g)	Soluble dietary fiber (g/100 g)	Total dietary fiber (g/100 g)
Leaves	9.17 ± 0.91^c	0.78 ± 0.02^c	9.95 ± 1.25^c
Flowers	17.49 ± 1.78^b	4.14 ± 0.50^b	21.63 ± 1.71^b
Fruits	19.68 ± 1.50^a	12.07 ± 1.45^a	31.76 ± 3.70^a

Means followed by the same letter(s) within each column are not significantly different at $P \leq 0.05$, according to the least significant difference test.

diseases like cancer, cardiovascular diseases, diabetes, and other metabolic syndromes [21]. The results of the present study on screening the phenolic and flavonoid contents and also the total antioxidant capacity confirmed that all three parts of T. cucumerina possess antioxidative properties. These results are in agreement with those reported by Choudhary et al. [22] in India, where the leaves possessed the highest total phenolic content while fruits possessed the lowest. As in TPC values, TFC values also differ in the same manner in aerial parts of T. cucumerina. According to Ademosun et al. [23], the total flavonoid content of T. cucumerina was significantly ($P \leq 0.05$) higher than that of a tomato variety of Lycopersicon esculentum Mill. var. esculentum.

Antioxidants are substances that have the ability to neutralize free radicals. In a normal healthy human body, the scavenging of prooxidants such as reactive oxygen species (ROS) and reactive nitrogen species (RNS) is effectively regulated by various antioxidant defense systems. With the exposure to adverse physicochemical, environmental, or pathological conditions, the regularly maintained balance is shifted in favor of prooxidants. This results in oxidative stress which is highly responsible for adverse health conditions including autoimmune destruction in the human body [24]. Hence, assessing the antioxidant capacities in various foods or herbs is important to elevate the antioxidant levels and their action in the body. In the present study, the antioxidant capacity was measured using DPPH, ABTS, and FRAP assays and all three extracts were found to show potential antioxidant activity. In a study done by Choudhary et al. [22], the DPPH radical scavenging activity was higher for the leaf extract than the fruit extract. In another study, Singh and Prakash [25] reported the IC_{50} values for leaves and fruit as $278.29 \pm 2.98 \, \mu g/mL$ and $260.30 \pm 2.23 \, \mu g/mL$, respectively. The antioxidant activities of T. cucumerina well correlated with the amount of total phenolic and flavonoid contents. Several reports have shown a close relationship between total phenolic content and antioxidant activities [22].

Natural plants or their products are related in retarding the absorption of glucose by inhibiting starch hydrolyzing enzymes, such as pancreatic amylase and glucosidase. The inhibition of these enzymes delays carbohydrate digestion thereby reducing the glucose releasing rate and consequently lowers the increase in postprandial plasma glucose. Several indigenous medicinal plants have a high potential in inhibiting α-amylase activity [26]. Though the present findings did not give significant enzyme inhibition activity for aerial parts of T. cucumerina, the findings of Ademosun et al. [23] have shown a higher inhibitory effect on α-amylase activity with an IC_{50} value of $2.15 \pm 0.09 \, mg/mL$.

Dietary minerals and trace elements are chemical substances required by living organisms in addition to carbon, hydrogen, nitrogen, and oxygen that are present in nearly all organic molecules. These elements perform various functions, including building of bones and cell structures, regulating the body's pH, carrying charge, and driving chemical reactions. Minerals and trace elements are usually obtained through the diet. Though food sources of animal origin provide most essential minerals, some plants are also considered as rich sources of minerals. Several other researchers have reported positive results for mineral element analysis of T. cucumerina. According to Ilelaboye and Pikuda [27], T. cucumerina seeds contain iron (187 mg/kg DM), zinc (37.5 mg/kg DM), calcium (1643 mg/kg DM), magnesium (1896 mg/kg DM), and potassium (8704 mg/kg DM).

Most of the foods with a low or medium glycemic index are reported to contain considerable amounts of dietary fiber [28, 29]. There are two different types of fiber, soluble and insoluble. Both are important for health, digestion, and preventing diseases. The increased satiety associated with high fiber foods may result in reduction of appetite due to decreased rate of digestion and absorption [30]. According to Braaten et al. [31], among the two types of dietary fiber, the SDF fraction has shown more promising effect on lowering postprandial blood glucose values. The study done by Osuagwu and Edeoga [32] also indicated positive results for total fiber content of T. cucumerina leaves as 12 g/100 g similar to that of the present study. However, Hettiaratchi et al. [33] had reported that incorporating T. cucumerina fresh fruit into a salad meal did not increase the total fiber content in comparison to the control meal incorporated with *Lasia spinosa* rhizome.

The results of this study suggest that *T. cucumerina* is rich in phenolic compounds and has good antioxidant activity. The plant can also be considered as a good source of essential mineral elements and rich in both soluble and insoluble dietary fiber. Several studies have also proven that *T. cucumerina* possesses so many functional health benefits [1–11]. Thus, *T. cucumerina* can be considered as a natural source of functional nutrients which may yield many beneficial effects to the human body.

5. Conclusion

From the present study it is concluded that the water extract of *Trichosanthes cucumerina*, fruits, leaves, and flowers possessed significant amounts of beneficial compounds. The highest concentration of phenolic compounds, flavonoids, and antioxidant capacity was found in the leaves and least in the fruits. The α-amylase inhibitory activity of the plant was not significant. Furthermore, the mineral element content of the leaves and flowers is considerably higher than that of the fruit, while the fiber content is significantly high in the fruit compared to the leaves and flowers.

Competing Interests

The authors declare that there is no conflict of interests regarding the publication of this paper.

Acknowledgments

The research was funded by National Institute of Fundamental Studies, Kandy, Sri Lanka.

References

[1] M. Arawwawala, I. Thabrew, and L. Arambewela, "Antidiabetic activity of *Trichosanthes cucumerina* in normal and streptozotocin–induced diabetic rats," *International Journal of Biological and Chemical Sciences*, vol. 3, no. 2, pp. 287–296, 2009.

[2] S. Sandhya, K. R. Vinod, J. C. Sekhar, R. Aradhana, and V. S. Nath, "An updated review on *Tricosanthes cucumerina* L," *International Journal of Pharmaceutical Sciences Review and Research*, vol. 1, no. 2, pp. 56–60, 2010.

[3] A. A. Yusuf, O. M. Folarin, and F. O. Bamiro, "Chemical composition and functional properties of snake gourd (*Trichosanthes cucumerina*) seed flour," *The Nigerian Food Journal*, vol. 25, no. 1, pp. 36–45, 2007.

[4] O. A. Ojiako and C. U. Igwe, "The nutritive, anti-nutritive and hepatotoxic properties of *Trichosanthes anguina* (snake tomato) fruits from Nigeria," *Pakistan Journal of Nutrition*, vol. 7, no. 1, pp. 85–89, 2008.

[5] H. Kirana and B. P. Srinivasan, "*Trichosanthes cucumerina* Linn. improves glucose tolerance and tissue glycogen in non insulin dependent diabetes mellitus induced rats," *Indian Journal of Pharmacology*, vol. 40, no. 3, pp. 103–106, 2008.

[6] L. D. A. M. Arawwawala, M. I. Thabrew, and L. S. R. Arambewela, "A review of the pharmacological properties of *Trichosanthes cucumerina* Linn of Sri Lankan origin," *Unique Journal of Pharmaceutical and Biological Science*, vol. 1, no. 1, pp. 3–6, 2013.

[7] T. O. Ajiboye, S. A. Akinpelu, H. F. Muritala et al., "Trchosanthes cucumerina fruit extenuates dyslipidemia, protein oxidation, lipid peroxidation and DNA fragmentation in the liver of high-fat diet-fed rats," *Journal of Food Biochemistry*, vol. 38, no. 5, pp. 480–490, 2014.

[8] M. Arawwawala, I. Thabrew, and L. Arambewela, "In vitro and in vivo evaluation of antioxidant activity of *Trichosanthes cucumerina* aerial parts," *Acta Biologica Hungarica*, vol. 62, no. 3, pp. 235–243, 2011.

[9] M. Arawwawala, I. Thabrew, L. Arambewela, and S. Handunnetti, "Anti-inflammatory activity of *Trichosanthes cucumerina* Linn. in rats," *Journal of Ethnopharmacology*, vol. 131, no. 3, pp. 538–543, 2010.

[10] L. D. A. M. Arawwawala, "Antibacterial activity of *Trichosanthes cucumerina* Linn. extracts," *International Journal of Pharmaceutical & Biological Archive*, vol. 2, no. 2, 2011.

[11] L. D. A. M. Arawwawala, M. I. Thabrew, and L. S. R. Arambewela, "Gastroprotective activity of *Trichosanthes cucumerina* in rats," *Journal of Ethnopharmacology*, vol. 127, no. 3, pp. 750–754 2010.

[12] T. Samatha, R. Shyamsundarachary, P. Srinivas, and N. R. Swamy, "Quantification of total phenolic and total flavonoid contents in extracts of *Oroxylum indicum* L.Kurz," *Asian Journal of Pharmaceutical and Clinical Research*, vol. 5, no. 4, pp. 177–179, 2012.

[13] M. O. Agbo, P. F. Uzor, U. N. Akazie-Nneji, C. U. Eze-Odurukwe, U. B. Ogbatue, and E. C. Mbaoji, "Antioxidant, total phenolic and flavonoid content of selected Nigerian medicinal plants," *Dhaka University Journal of Pharmaceutical Sciences*, vol. 14, no. 1, pp. 35–41, 2015.

[14] C. B. Sanjeevkumar, R. L. Londonkar, U. M. Kattegouda, and N. K. A. Tukappa, "Screening of in vitro antioxidant activity of chloroform extracts of *Bryonopsis laciniosa* fruits," *International Journal of Current Microbiology and Applied Sciences*, vol. 5, no. 3, pp. 590–597, 2016.

[15] J. Al-humaidi, "Phytochemical screening, total phenolic and antioxidant activity of crude and fractionated extracts of cynomorium coccineum growing in Saudi Arabia," *European Journal of Medicinal Plants*, vol. 11, no. 4, pp. 1–9, 2016.

[16] A. Shukla, R. Tyagi, S. Vats, and R. K. Shukla, "Total phenolic content, antioxidant activity and phytochemical screening of hydroalcoholic extract of *Casearia tomentosa* leaves," *Journal of Chemical and Pharmaceutical Research*, vol. 8, no. 1, pp. 136–141, 2016.

[17] P. Sudha, S. S. Zinjarde, S. Y. Bhargava, and A. R. Kumar, "Potent α-amylase inhibitory activity of Indian Ayurvedic medicinal plants," *BMC Complementary and Alternative Medicine*, vol. 11, article 5, 2011.

[18] T. Liu, L. Song, H. Wang, and D. Huang, "A high-throughput assay for quantification of starch hydrolase inhibition based on turbidity measurement," *Journal of Agricultural and Food Chemistry*, vol. 59, no. 18, pp. 9756–9762, 2011.

[19] J. S. Negi, V. K. Bisht, A. K. Bhandari, and R. C. Sundriyal, "Determination of mineral contents of *Digitalis purpurea* L. and *Digitalis lanata* Ehrh," *Journal of Soil Science and Plant Nutrition*, vol. 12, no. 3, pp. 463–469, 2012.

Comparative Analysis of Nutritional and Bioactive Properties of Aerial Parts of Snake Gourd...

195

[20] D. Shin, "Analysis of dietary insoluble and soluble fiber contents in school meal," *Nutrition Research and Practice*, vol. 6, no. 1, pp. 28–34, 2012.

[21] F. Alhakmani, S. A. Khan, and A. Ahmad, "Determination of total phenol, *in-vitro* antioxidant and anti-inflammatory activity of seeds and fruits of Zizyphus spina-christi grown in Oman," *Asian Pacific Journal of Tropical Biomedicine*, vol. 4, pp. S656–S660, 2014.

[22] S. Choudhary, B. S. Tanwer, and R. Vijayvergia, "Total phenolics, flavonoids and antioxidant activity of *Tricosanthes cucumerena* Linn," *Drug Invention Today*, vol. 4, no. 5, pp. 368–370, 2012.

[23] O. Ademosun, G. Oboh, T. M. Adewuni, A. J. Akinyemi, and T. A. Olasehinde, "Antioxidative properties and inhibition of key enzymes linked to type-2 diabetes by snake tomato (*Tricosanthes cucumerina*) and two tomato (*Lycopersicon esculentum*) varieties," *African Journal of Pharmacy and Pharmacology*, vol. 7, no. 33, pp. 2358–2365, 2013.

[24] T. P. A. Devasagayam, J. C. Tilak, K. K. Boloor, K. S. Sane, S. S. Ghaskadbi, and R. D. Lele, "Free radicals and antioxidants in human health: current status and future prospects," *Journal of Association of Physicians of India*, vol. 52, pp. 794–804, 2004.

[25] S. K. Singh and V. Prakash, "Screening of antioxidant activity and phytochemicals strength of some herbal plants," *International Journal of Pharmacy and Pharmaceutical Sciences*, vol. 5, no. 3, pp. 296–300, 2013.

[26] I. G. Tamil, B. Dineshkumar, M. Nandhakumar, M. Senthilkumar, and A. Mitra, "In vitro study on α-amylase inhibitory activity of an Indian medicinal plant, *Phyllanthus amarus*," *Indian Journal of Pharmacology*, vol. 42, no. 5, pp. 280–282, 2010.

[27] N. O. A. Ilelaboye and O. O. Pikuda, "Determination of minerals and anti-nutritional factors of some lesser-known crop seeds," *Pakistan Journal of Nutrition*, vol. 8, no. 10, pp. 1652–1656, 2009.

[28] M. O. Weickert and A. F. H. Pfeiffer, "Metabolic effects of dietary fiber consumption and prevention of diabetes," *Journal of Nutrition*, vol. 138, no. 3, pp. 439–442, 2008.

[29] J. R. Perry and W. Ying, "A review of physiological effects of soluble and insoluble dietary fibers," *Journal of Nutrition & Food Sciences*, vol. 6, article 476, 2016.

[30] L. Chambers, K. McCrickerd, and M. R. Yeomans, "Optimising foods for satiety," *Trends in Food Science and Technology*, vol. 41, no. 2, pp. 149–160, 2015.

[31] J. T. Braaten, P. J. Wood, F. W. Scott, K. D. Riedel, L. M. Poste, and M. W. Collins, "Oat gum lowers glucose and insulin after an oral glucose load," *The American Journal of Clinical Nutrition*, vol. 53, no. 6, pp. 1425–1430, 1991.

[32] A. N. Osuagwu and H. O. Edeoga, "Nutritional properties of the leaf, seed and pericarp of the fruit of four *Cucurbitaceae* species from South-East Nigeria," *IOSR Journal of Agriculture and Veterinary Science*, vol. 7, no. 9, pp. 41–44, 2014.

[33] U. P. K. Hettiaratchi, S. Ekanayake, and J. Welihinda, "Sri Lankan rice mixed meals: effect on glycaemic index and contribution to daily dietary fibre requirement," *Malaysian Journal of Nutrition*, vol. 17, no. 1, pp. 97–104, 2011.

20

Drying Rate and Product Quality Evaluation of Roselle (*Hibiscus sabdariffa* L.) Calyces Extract Dried with Foaming Agent under Different Temperatures

Mohamad Djaeni ⓘ, Andri Cahyo Kumoro, Setia Budi Sasongko, and Febiani Dwi Utari

Department of Chemical Engineering, Faculty of Engineering, Diponegoro University, Jl. Prof. H. Soedarto, SH, Tembalang, Semarang 50275, Indonesia

Correspondence should be addressed to Mohamad Djaeni; moh.djaeni@live.undip.ac.id

Academic Editor: Marie Walsh

The utilisation of roselle (*Hibiscus sabdariffa* L.) calyx as a source of anthocyanins has been explored through intensive investigations. Due to its perishable property, the transformation of roselle calyces into dried extract without reducing their quality is highly challenging. The aim of this work was to study the effect of air temperatures and relative humidity on the kinetics and product quality during drying of roselle extract foamed with ovalbumin and glycerol monostearate (GMS). The results showed that foam mat drying increased the drying rate significantly and retained the antioxidant activity and colour of roselle calyces extract. Shorter drying time was achieved when higher air temperature and/or lower relative humidity was used. Foam mat drying produced dried brilliant red roselle calyces extract with better antioxidant activity and colour qualities when compared with nonfoam mat drying. The results showed the potential for retaining the roselle calyces extract quality under suggested drying conditions.

1. Introduction

Roselle (*Hibiscus sabdariffa* L.) is commercially cultivated in some countries like India, Indonesia, Malaysia, Sudan, Egypt, and Mexico [1]. The roselle calyx is brilliant red in colour due to the existence of anthocyanins, such as cyanidin-3-sambubioside, cyanidin-3-glucoside, and delphinidin-3-glucoside [2]. The calyx is usually used to prepare jam, jelly, cakes, ice cream, preserves, and herbal beverage [3]. The consumption of roselle calyx tea has been reported to promote health benefits, which mainly functions as an antioxidant [4]. The relationship between antioxidant activity and anthocyanin of roselle calyx has been reported in the literatures [5, 6]. The total anthocyanin content of roselle calyces is 2.52 g/100 g expressed as delphinidin-3-glucoside [7]. Besides, the roselle extracts also contain ascorbic acid. Wang et al. observed that anthocyanins have many times more antioxidant activity than ascorbic acid [8]. Therefore, the antioxidant activity of roselle extract is predominantly contributed by anthocyanins. However, the use of anthocyanins

in food products experienced problems related to their instability during processing and storage caused by direct exposure to heat, oxygen, and light [9]. Mazza and Miniati observed that thermal degradation of anthocyanin of roselle extract was fast at temperatures above 100°C [10].

Microencapsulation technique is one of the methods to maintain anthocyanins stability by entrapping them inside a coating material to reduce direct interactions with external factors, such as temperature, light, moisture, and oxygen. Although spray drying and freeze drying are the most common microencapsulation methods applied in the food and pharmaceutical industries, freeze drying is 30 to 50 times more costly than that of spray drying [11]. However, extreme moisture loss during spray drying may trigger shrinking and deformation of dried particles [12].

Foam mat drying is carried out by transforming liquid and semisolid materials into stable foam by incorporation of air and a foaming agent. It is a good option to shorten drying time and to retain product quality [13]. Ovalbumin is usually chosen as a foaming agent due to the ability of its proteins to generate a dense film around the air bubbles, reducing the surface tension instability and retaining the entrapped air [14]. In addition, glycerol monostearate (GMS) and methyl cellulose (MC) are the other common foaming agents [13, 15]. With the agents, the surface area of product becomes greater for mass and heat transfer. These conditions allow higher drying rates [13].

The foam mat drying can be a potential option for roselle extract drying. However, the application cannot be straightforward. As far as literature survey is being done, no studies have been carried out to investigate the degradation of the anthocyanins, colour, and antioxidant activity as a function of temperature, relative humidity, and time during foam mat drying of roselle extract. This study aimed to study the effect of air temperatures and relative humidity on the kinetics and product quality during foam drying of roselle extract. As indicators, the drying time and the quality of roselle extract in terms of antioxidant activity and colour were evaluated. The results were also compared with conventional convective drying without foaming agent.

2. Materials and Methods

2.1. Plant Materials and Chemicals. Traditionally dried brilliant red roselle (*Hibiscus sabdariffa* Linn.) calyces, with average size and with no bruises, were obtained from an herbal medicine market in Solo, Indonesia. 2,2-Diphenyl-1-picrylhydrazyl (DPPH) and ethanol were of analytical grade and purchased from Merck (Darmstadt, Germany).

2.2. Roselle Extract Preparation. The calyces were ground using a blender (Panasonic MX-895 M, Japan) and sieved carefully to obtain powder with 0.25 mm in size. The roselle was extracted using water as solvent as previously suggested by Chumsri et al. [16]. One hundred grams of roselle calyces powder was mixed with water in a beaker glass at a mass to volume ratio (gmL^{-1}) of 1 : 10 at 50°C under continuous agitation for 1 hour. The extract was then filtered through a nylon filter to obtain brilliant red colour liquid extract.

2.3. Foam Drying Process. Foam drying was used to reduce the moisture content of the roselle extract [13]. The drying system is depicted in Figure 1. The roselle extract (100 ml) was mixed with 4.23 grams of ovalbumin as the foaming agent and 0.423 grams of GMS as foam stabiliser. After complete mixing, the solution was whipped in a domestic mixer (360 W power) at maximum speed for 20 min to provide the mechanical incorporation of air (the density of foam was 17.73 kg/m^3). After whipping, the foam was placed on aluminium beds (length 20 cm, width 15 cm and thickness of 0.2 cm). The beds were then put in the dryer which used ambient air as drying medium. Before entering the dryer, the ambient air with relative humidity of 80%, temperature range of 30–32°C, and superficial velocity of 0.22 m s^{-1} was heated up to certain inlet drying temperature (40°C, 50°C, and 60°C) using an electric heater equipped with thermo-controller. So, the relative humidity of ambient air decreased corresponding to the inlet drying temperatures. Because of

water releases during drying process, the foam thickness is usually reduced to about 5% of its initial condition. The moisture loss was determined gravimetrically by weighing the samples every 10 min on an electronic scale. In addition, antioxidant activity and colour of the dried extract were also determined accordingly. The experiments were terminated after 120-minute drying times. As a comparison, the roselle extract was also dried without addition of foaming agent and stabiliser (control sample).

2.4. Determination of Moisture Content. The moisture content in roselle extract-ovalbumin-glycerol monostearate mixture was measured referring to AOAC method [17]. Briefly, 1 gram of dried extract sample was heated at the temperature of 105°C in the oven until a constant weight was obtained. In this case, the initial moisture content in the extract before drying was 95% (wet basis) or 19 grams of moisture per gram of dry extract (dry basis).

2.5. Modelling of Foam Drying. Referring to thin layer drying model, the correlation between moisture ratios with time can be represented by the following [18]:

$$MR = \frac{(M_t - M_e)}{(M_0 - M_e)}, \quad (1)$$

where M_t was the moisture content at time t, M_0 was the initial moisture content, and M_e was the equilibrium moisture content of the foam, all of them in dry basis (g g^{-1}). The equilibrium moisture content in roselle extract with the foaming agent was estimated by considering the composition of the mixture and both equilibrium moistures in roselle and ovalbumin as expressed by [19]

$$M_e = \sum x_i M_{ei}, \quad (2)$$

where x was the mass fraction of roselle and ovalbumin, i referred to components (roselle and ovalbumin), and M_e was the equilibrium moisture content (g water per g dry solid). The equilibrium moisture of roselle and ovalbumin was influenced by relative humidity or water activity (Aw) and temperature [19, 20]. Langová et al. suggested that the equilibrium moisture of roselle can be extended as follows [19]:

$$M_e = A^{Aw-B} + C,$$
$$A = 1568.6e^{-0.072T},$$
$$B = -0.0076T + 0.4687,$$
$$C = -0.1898T + 13.49, \quad (3)$$

where A, B, and C were the constant in equilibrium moisture content equation and T was the drying temperatures (40°C, 50°C, and 60°C). Using the same principles, Christ et al. suggested that the equilibrium moisture of ovalbumin can be satisfactorily predicted using [20]

$$M_e = \exp^{\{-0.2387-(0.0035(T+273))+[1.5933\exp(A_w)]\}}. \quad (4)$$

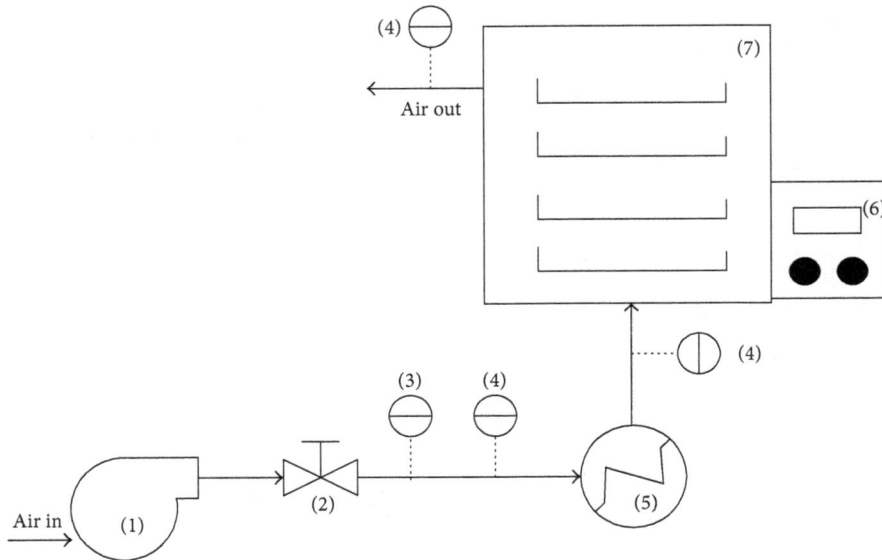

FIGURE 1: The schematic overview of the tray drying equipment ((1) blower; (2) valve; (3) temperature and relative humidity sensor (T-RH); (4) temperature and velocity sensor (T-V); (5) heater; (6) thermocontroller; (7) drying chamber).

The drying rate (DR) was estimated based on the moisture ratio (dimensionless moisture) reduction as a function of time as suggested by Franco et al. [18].

$$DR = \frac{(M_{t+dt} - M_t)}{dt},\tag{5}$$

where M_{t+dt} and M_t were the moisture contents at times $t+dt$ and t, respectively. Based on literature surveys, the thin layer model (Newton model) accuracy has been proven to predict the rate of foam drying process [21, 22]. According to Newton model, the drying time for rosella extract can be estimated by

$$t = -\frac{\ln(MR)}{k_d}.\tag{6}$$

Here, k_d was the drying rate constant (minute^{-1}). The dependency of k_d on temperature (T) can be correlated using Arrhenius like equation as shown by

$$k_d = k_0 \exp^{-E_a/R(T+273)},\tag{7}$$

where k_0 was the preexponential factor (minute^{-1}), E_a was the activation energy (J mol^{-1}), and R was the gas constant (8.32 J mol^{-1} K^{-1}).

The mass diffusion coefficient can be calculated using the analytical solution for a flat plate, which was previously used by Franco et al. [18].

$$MR = \left(\frac{8}{\pi^2}\right)\exp^{(-D_{if}\pi^2 t)/\tau^2},\tag{8}$$

where D_{if} was the mass diffusion coefficient (m^2 min^{-1}), t was the drying time (minute), and τ was the thickness of the foam (m).

2.6. Mass and Heat Transfer Analysis. The mass and heat convective transfer coefficients, h and h_m, were determined by (9) and (10) using dimensionless numbers obtained from (11) to (14) which were previously used by Franco et al. [18].

$$h = \frac{Nu\, k_o}{d},\tag{9}$$

$$h_m = \frac{Sh\, D_{WA}}{d},\tag{10}$$

$$Re = \frac{\rho_o v_o}{\mu_o},\tag{11}$$

$$Nu = 0.664\, Re^{1/2} Pr^{1/3},\tag{12}$$

$$Sc = \frac{\mu_o}{\rho_o D_{WA}},\tag{13}$$

$$Sh = 0.664\, Re^{1/2} Sc^{1/3},\tag{14}$$

$$D_{WA} = 1.87 \times 10^{-10}\frac{(T+273)^{2.072}}{P},\tag{15}$$

where Re, Nu, Sc, and Sh were the Reynold, Nusselt, Schmidt, and Sherwood number, respectively, whereas k_o, P, T, ρ_o, v_o, and μ_o were, respectively, the thermal conductivity (Wm^{-1}K^{-1}), pressure (atm), temperature (°C), density (kgm^{-3}), velocity (ms^{-1}), and viscosity (Pa s) of air. In addition, d was the material length (m) and D_{WA} was the diffusivity of water in air (m^{2}s^{-1}).

2.7. DPPH Free Radical-Scavenging Activity Assay. The capacity of roselle extracts to inhibit the sequestration activity of 2,2-diphenyl-1-picrylhydrazyl (DPPH) free radical was observed according to the method of Brand-Williams et al. [23]. Precisely, 100 μL of ethanolic extract solutions was added to 1.4 mL DPPH radical ethanolic solution (10^{-4} mol L^{-1}) in a test tube. The content of the test tube was shaken vigorously and let to react at room temperature for 30 minutes in the dark. The absorbance at 517 nm was then measured against a blank (100 μL of ethanol in 1.4 mL of DPPH radical solution) using a spectrophotometer. The antioxidant activity was calculated as percentage inhibition (%) of DPPH radical formation, as follows:

$$I\,(\%) = \left[\frac{A_0 - A_s}{A_s} \right] 100\%, \tag{16}$$

where A_0 was absorbance of control blank and A_s was absorbance of sample extract. All determinations were conducted in triplicate.

2.8. Colour Measurement. The colour changes of dried roselle extract were observed by measuring L, a, and b values using a Chroma Meter (CR-300, Minolta Co., Ltd., Osaka, Japan) as previously used by Kumar et al. [6]. The total colour degradation during drying process was evaluated based on the changes of total colour ratio (TCR) as suggested by Ahmed et al. [24] and the value was calculated using

$$\text{TCR} = \frac{L_s a_s b_s}{L_o a_o b_o}, \tag{17}$$

where L_s, a_s, and b_s were the measured on a sample at a given time, while L_o, a_o, and b_o were measured on raw roselle extract (before the drying process). All determinations were conducted in triplicate.

2.9. Total Anthocyanin Content Measurement. The total anthocyanin content was observed with the method referring to Anuar et al. [25]. The samples (fresh and dried roselle extract) were diluted in 0.025 M KCl buffer (pH 1) and 0.4 M CH3COONa (pH 4.5). The absorbance of the solution was measured at 520 nm and 700 nm using UV-Vis Spectrophotometer (UVmini-1240, Shimadzu, Japan). The total anthocyanin content (TAC), as cyanidin-3-glucoside was calculated using the following [25]:

$$\text{TAC}\,(\text{mg/L}) = \frac{A \times \text{MW} \times \text{DF} \times 1000}{\varepsilon \times d}, \tag{18}$$

where A was the absorbance, MW was molecular weight of anthocyanin rounding 449.2 g/mol, DF was the dilution factor, ε was the molar absorptivity (26900), and d was the cuvette length (1 cm). Absorbance was calculated using [25]

$$A = (A_{520} - A_{700})_{\text{pH1}} - (A_{520} - A_{700})_{\text{pH4.5}}. \tag{19}$$

2.10. Kinetics of Antioxidant Activity and Colour Degradation. Kinetics of antioxidant activity and colour degradations were assumed to follow the first-order reaction. First-order reaction approach has been successfully used in describing the colour degradation of food products [24]. Since the antioxidant activity was linearly correlated to anthocyanins

content, then the antioxidant activity degradation should obey the following equation:

$$\ln \left(\frac{I_t}{I_0} \right) = -k_a t. \tag{20}$$

Similarly, the colour degradation can be described by

$$\ln \left(\frac{\text{TCR}_t}{\text{TCR}_0} \right) = -k_c t. \tag{21}$$

k_a and k_c were the constant rate of antioxidant activity and colour degradations (minute^{-1}). I_0 and I_t were antioxidant activity of the sample at time 0 and t, while TCR$_0$ and TCR$_t$ were total colour ratio at time 0 and t. All the constants were obtained numerically with the help of Excel solver.

3. Results and Discussion

3.1. Effect of the Presence of Foaming Agent. The roselle extract was dried with foaming agent and without foaming agent (nonfoam drying) at 40°C and 60°C. This step was expected to obtain dried product with moisture content of 0.11 gram of moisture per gram of dried product. The drying curves in the form of moisture ratio versus time were depicted in Figure 2. In all cases, the foaming agent played important role in speeding up water evaporation. As seen in Figure 2, the moisture ratio in product with foam reduced drastically. As expected, the foam drying had faster moisture evaporation leading to shorter drying time than that of nonfoam drying. During foam drying, the total surface area that was necessary for mass and heat transfer increased drastically as the honeycomb structure was formed through the destruction of the surface area of product into a plenty of smaller units. With this larger surface area, the moisture evaporation rate increased significantly [13]. Even, the drying rates of roselle extract with foam was 3 times higher than that of without foam (see Figure 3). The results were more significant compared with foam mat drying applied to the other food products using foam. Rajkumar et al. found a reduction of about 35 minutes to 40 minutes drying time during foam drying of mango pulp [21]. Meanwhile, Kandasamy et al. spent a half of drying time usually required for nonfoam drying of papaya pulp [15]. With shorter drying time or higher drying rate, foam mat drying offered a promising potential to save energy consumption.

3.2. Effect of Air Temperature and Relative Humidity. Table 1 presented the fact that the use of higher temperature shortened the drying time of roselle extract without foam. Similarly, Table 2 also showed that the drying time was shorter when higher temperatures were applied on foam drying. This behaviour can be explained by the increased drying rate caused by larger temperature gradient between drying air and foam which promote faster water evaporation rate [13, 15, 21].

Tables 1 and 2 showed that, at all temperatures studied, the use of low air relative humidity significantly reduced the drying time of both foam and nonfoam drying. With low air relative humidity, the driving force for moisture transfer

TABLE 1: Predicted drying time of roselle extract without foam at various air conditions.

Aw/RH	Drying time (minutes)					
	30°C	40°C	50°C	60°C	70°C	80°C
0.0	572.6	410.4	256.8	214.9	153.6	116.9
0.2	587.2	419.7	261.9	218.4	155.4	117.7
0.4	656.0	451.7	275.9	226.0	158.5	118.8
0.6	nd*	nd*	377.0	247.2	164.5	120.5
0.8	nd*	nd*	nd*	nd*	178.6	123.2

nd*: not dried (equilibrium moisture of the product was upper 10% wet basis).

TABLE 2: Predicted drying time of roselle extract with foam at various air conditions.

Aw/RH	Drying time (minutes)					
	30°C	40°C	50°C	60°C	70°C	80°C
0.0	163.0	131.6	93.3	84.9	66.9	55.5
0.2	167.6	134.9	95.4	86.5	67.8	56.0
0.4	184.6	145.3	101.0	90.0	69.7	56.9
0.6	nd*	nd*	187.7	101.5	73.9	58.8
0.8	nd*	nd*	nd*	nd*	99.1	63.6

nd*: not dried (equilibrium moisture of the product was upper 10% wet basis).

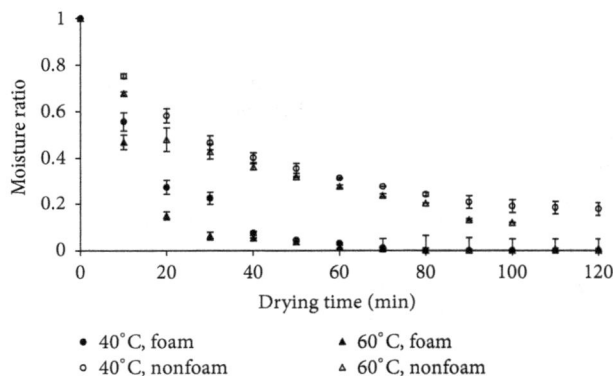

FIGURE 2: Moisture ratio of foam and nonfoam roselle extract drying at different air temperatures.

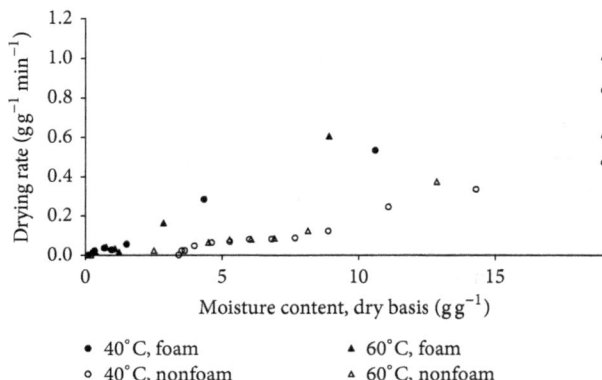

FIGURE 3: Drying rate of foam and nonfoam roselle extract drying at different air temperatures.

was larger due to the lower equilibrium moisture content of the product. Hence, the roselle extract can be dried to even lower moisture content. Djaeni et al. reported that the use of low air relative humidity reduced the drying time of carrageenan significantly, especially for drying at low or medium temperatures (40 to 60°C) [26].

The drying rate constants were estimated using Newton model (see (5) and (6)), and the results were tabulated in Table 3. The model agreed well with the experimental data as indicated by high values of R^2. This simple empirical model was chosen due to its simplicity and versatility for application in various drying experiments with different materials [21, 22]. The higher drying rate constants were achieved at higher temperature for both foam and nonfoam dryings. Ambekar et al. observed that the drying rate constant at 70°C was 2 times higher than at 50°C during foam drying of passion fruit pulp [27].

Arrhenius like model was used to describe the correlation between the drying rate constant (k_d) and temperatures (see (7)). The value of Arrhenius constants were listed in Table 3. Using these constants, the drying rate and drying time for roselle extract at the desired drying temperatures can be well predicted. Based on the mass and heat convective transfer coefficients shown in Table 4, the Sherwood (Sh) number confirmed that mass convective flow was superior over the moisture diffusion and that moisture diffusion coefficients were significantly depending on the air temperature. Since the foam had a high porosity, it was expected that the internal motion of moisture was slower than its evaporation into the flowing air. Further, the Nusselt number (Nu) also confirmed that the convective heat transfer was faster than thermal diffusion because the high porosity of the food might hinder the heat conduction. Similar result was reported by Franco et al. in foam drying of yacon juice [18].

TABLE 3: The constants of roselle extract drying at different temperature.

T (°C)	Nonfoam					Foam				
	k_d (min^{-1})	$D_{if} \times 10^9$ (m^2 min^{-1})	R^2	$k_o \times 10^{-2}$	Ea (kJ mol^{-1})	k_d (min^{-1})	$D_{if} \times 10^9$ (m^2 min^{-1})	R^2	$k_o \times 10^{-2}$	Ea (kJ mol^{-1})
40	0.017	3.13	0.90			0.052	8.56	0.92		
50	0.026	4.28	0.86	2.38	29.77	0.071	11.69	0.93	3.12	19.92
60	0.030	4.94	0.87			0.076	12.51	0.87		

TABLE 4: Heat and mass transfer coefficients.

T (°C)	$D_{wa} \times 10^{-5}$ (m^{2}s^{-1})	Re*	Sc*	Sh*	k (W m^{-1}K^{-1})	Nu*	h (W m^{-1}K^{-1})	$h_m \times 10^{-4}$ (ms^{-1})
40	2.77	25.63	113483.61	2.81	27.15	3.01	408.05	3.903
50	2.96	24.62	83114.76	2.70	28.80	2.94	409.19	3.996
60	3.15	22.99	77646.68	2.55	28.59	2.84	406.49	4.028

Re*: Reynold number, Sc*: Schmidt number, Sh*: Sherwood number, and Nu*: Nusselt number.

TABLE 5: The Arrhenius parameters for antioxidant activity degradation.

T (°C)	Nonfoam				Foam			
	k_a (min^{-1})	R^2	$k_o \times 10^6$	Ea (kJ mol^{-1})	k_a (min^{-1})	R^2	$k_o \times 10^1$	Ea (kJ mol^{-1})
40	0.00094	0.85			0.0009	0.93		
50	0.0025	0.94	5.9	58.42	0.001	0.97	3.1	27.34
60	0.0036	0.90			0.0017	0.89		

3.3. Antioxidant Activity Degradation.

In this work, the antioxidant activity of roselle extract was expressed as percentage of inhibition ($I\%$) to DPPH solution [23]. Degradation of antioxidant activity of dried roselle extract obtained from foam and nonfoam dryings at different air temperature is depicted in Figure 4. The dried roselle extracts experienced significant reduction of percentage of inhibition ($I\%$) at both higher air temperature and longer drying time for all drying methods. In general, the antioxidant activity of roselle extracts obtained from foam drying was slightly higher than that of nonfoam drying indicating their higher total antioxidant content. Since the antioxidant activity of dried roselle extracts was mainly contributed by anthocyanins, which were sensitive to heat, then lower percentage of inhibition ($I\%$) of dried roselle extract might be due to its lower anthocyanins content [6].

The antioxidant degradation rate constants k_a at various temperatures were calculated using (20), while their dependence on temperature was estimated using Arrhenius model (analog with (7)). The results were presented in Table 5. The higher value of k_o indicates that antioxidant degradation rate for nonfoam drying was faster than that of foam drying.

3.4. Colour Degradation.

The dried roselle extracts obtained from nonfoam drying were all darker than those obtained from foam drying at the same drying conditions. With longer drying time, the sugar contained in the roselle extract may experience caramelisation and lead to giving browning effect on nonfoam drying products [28]. The stable foam available

TABLE 6: The total Anthocyanin content (TAC) of roselle extract drying with foam.

Variable	TAC (mg/L)
Mixture of roselle extract with foaming agent	83.49 ± 3.34
Dried product, 40°C	58.71 ± 1.52
Dried product, 50°C	46.85 ± 1.25
Dried product, 60°C	41.76 ± 0.47

along the foam drying promoted faster moisture evaporation rate and resulted in shorter drying time, by which the product quality can be maintained [29].

Colour degradation in the form or TCR_t/TCR_0 was observed during the drying process as seen in Figure 5. Increase in drying temperatures led to reduction in the colour quality of the roselle extract during the drying process as indicated by darker colour. These phenomena can be explained by the formation of brown pigment during drying. In addition, the browning process also increased with an increase in drying temperature [30]. Reyes and Cisneros-Zevallos also reported that the degradation rate constant of colour of anthocyanins increased with temperatures as can be seen in the case of aqueous extracts of purple and red-flesh potatoes. The constant rate of the colour degradation at 50°C was 9.5 times higher than at 25°C [31].

During the drying process, anthocyanin content in roselle extract decreased with the increase of temperature (Table 6). In this case, by increasing drying temperatures 10°C

TABLE 7: The Arrhenius parameters for colour degradation.

T (°C)	Nonfoam				Foam			
	k_c (min^{-1})	R^2	$k_o \times 10^6$	Ea (kJ mol^{-1})	k_c (min^{-1})	R^2	$k_o \times 10^2$	Ea (kJ mol^{-1})
40	0.0012	0.82			0.0030	0.92		
50	0.0030	0.86	2.4	55.55	0.0048	0.93	4.4	30.86
60	0.0043	0.87			0.0061	0.87		

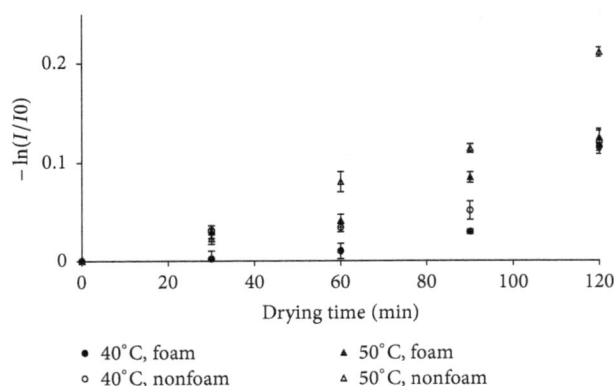

FIGURE 4: The antioxidant activity degradation at various drying conditions.

- 40°C, foam
- 40°C, nonfoam
- 50°C, foam
- 50°C, nonfoam

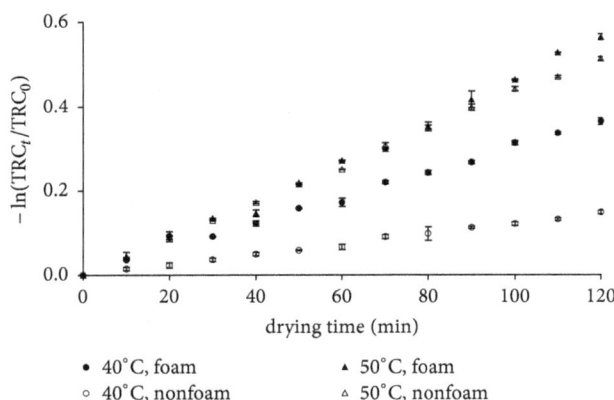

FIGURE 5: The colour degradation at various drying conditions.

- 40°C, foam
- 40°C, nonfoam
- 50°C, foam
- 50°C, nonfoam

4. Conclusion

Roselle extract drying has been successfully performed with ovalbumin and glycerol mono stearate as foaming agent and foam stabiliser at various drying air conditions. The experiment results showed that the drying time of roselle extract with foaming agent was three times shorter than that of conventional drying without foam. Furthermore, drying time became shorter at higher air temperature and lower air relative humidity. With this condition, the antioxidant activity and colour can be well retained. The mathematical models have been also developed to represent the foam drying kinetics as well as antioxidant and colour degradation kinetics of the roselle extract. These model parameters were validated by experimental data. Therefore, the models can be used to estimate drying time for different drying temperature.

Conflicts of Interest

The authors declare that they have no conflicts of interest.

Acknowledgments

This study was supported by Research and Community Service Centre, Diponegoro University, under Research and International Publication Grant 2016-2017.

References

[1] T. Chewonarin, T. Kinouchi, K. Kataoka et al., "Effects of roselle (Hibiscus sabdariffa Linn.), a Thai medicinal plant, on the mutagenicity of various known mutagens in Salmonella typhimurium and on formation of aberrant crypt foci induced by the colon carcinogens azoxymethane and 2-amino-1-methyl-6-phenylimidazo[4,5-b]pyridine in F344 rats," Food and Chemical Toxicology, vol. 37, no. 6, pp. 591–601, 1999.

[2] A. Castañeda-Ovando, M. de Lourdes Pacheco-Hernández, M. Elena Páez-Hernández, J. A. Rodríguez, and C. A. Galán-Vidal, "Chemical studies of anthocyanins: a review," Food Chemistry, vol. 113, pp. 859–871, 2009.

[3] V. Hirunpanich, A. Utaipat, N. P. Morales et al., "Hypocholesterolemic and antioxidant effects of aqueous extracts from the dried calyx of Hibiscus sabdariffa L. in hypercholesterolemic rats," Journal of Ethnopharmacology, vol. 103, no. 2, pp. 252–260, 2006.

[4] H.-H. Lin, J.-H. Chen, and C.-J. Wang, "Chemopreventive properties and molecular mechanisms of the bioactive compounds in hibiscus sabdariffa linne," Current Medicinal Chemistry, vol. 18, no. 8, pp. 1245–1254, 2011.

the anthocyanin content can decrease about 15% (58.71 ± 1.52 mg/100 gram at 40°C to 46.85 ± 1.25 mg/100 gram at 50°C). The decrease of anthocyanin content caused the colour to change significantly as formulated by Cao et al. [32].

The colour degradation rate constants k_c at various temperatures were calculated using (21), while their dependence on temperature was estimated using Arrhenius like correlation (analog with (7)). The Arrhenius-like constants obtained in this study were listed in Table 7. It was clear that the colour degradation was faster for foam drying due to dual degradation effect by caramelisation and Maillard reactions. Addition of ovalbumin as foaming agent, which was rich in amino acids in its protein, led to the occurrence of Maillard reaction.

[5] P.-J. Tsai, J. McIntosh, P. Pearce, B. Camden, and B. R. Jordan, "Anthocyanin and antioxidant capacity in Roselle (*Hibiscus sabdariffa* L.) extract," *Food Research International*, vol. 35, no. 4, pp. 351–356, 2002.

[6] S. S. Kumar, P. Manoj, N. P. Shetty, and P. Giridhar, "Effect of different drying methods on chlorophyll, ascorbic acid and antioxidant compounds retention of leaves of Hibiscus sabdariffa L.," *Journal of the Science of Food and Agriculture*, vol. 95, no. 9, pp. 1812–1820, 2015.

[7] P. Wong, S. Yusof, H. M. Ghazali, and Y. B. Che Man, "Physico-chemical characteristics of roselle *(Hibiscus sabdariffa L.)*," *Nutrition & Food Science*, vol. 32, no. 2, pp. 68–73, 2002.

[8] H. Wang, G. Cao, and R. L. Prior, "Oxygen radical absorbing capacity of anthocyanins," *Journal of Agricultural and Food Chemistry*, vol. 45, no. 2, pp. 304–309, 1997.

[9] H.-H. Chen, P.-J. Tsai, S.-H. Chen, Y.-M. Su, C.-C. Chung, and T.-C. Huang, "Grey relational analysis of dried roselle (Hibiscus sabdariffa L.)," *Journal of Food Processing and Preservation*, vol. 29, no. 3-4, pp. 228–245, 2005.

[10] G. Mazza and E. Miniati, *Anthocyanins in fruits, vegetables, and grains*, CRC Press, Boca Raton, Fl, USA, 1993.

[11] F. D. Sßnchez, E. M. S. L, S. F. Kerstupp et al., "Colorant extraction from a red prickly pear (*Opuntialasiacantha*) for food application," *Electronic Journal of Environmental, Agricultural and Food Chemistry*, vol. 5, no. 2, pp. 1330–1337, 2006.

[12] R. V. Tonon, C. Brabet, and M. D. Hubinger, "Influence of drying air temperature and carrier agent concentration on the physicochemical properties of açai juice powder," *Ciência e Tecnologia de Alimentos*, vol. 29, no. 2, pp. 444–450, 2009.

[13] M. Djaeni, A. Prasetyaningrum, S. B. Sasongko, W. Widayat, and C. L. Hii, "Application of foam-mat drying with egg white for carrageenan: drying rate and product quality aspects," *Journal of Food Science and Technology*, vol. 52, no. 2, pp. 1170–1175, 2013.

[14] K. Lomakina and K. Míková, "A study of the factors affecting the foaming properties of egg white - A review," *Czech Journal of Food Sciences*, vol. 24, no. 3, pp. 110–118, 2006.

[15] P. Kandasamy, N. Varadharaju, S. Kalemullah, and D. Maladhi, "Optimization of process parameters for foam-mat drying of papaya pulp," *Journal of Food Science and Technology*, vol. 51, no. 10, pp. 2526–2534, 2012.

[16] P. Chumsri, A. Sirichote, and A. Itharat, "Studies on the optimum conditions for the extraction and concentration of roselle *(Hibiscus sabdariffa Linn.)* extract," *Songklanakarin Journal of Science and Technology*, vol. 30, no. 1, pp. 133–139, 2008.

[17] AOAC, Official Methods of Analysis of AOAC International. AOAC International, USA, 2005.

[18] T. S. Franco, C. A. Perussello, L. D. S. N. Ellendersen, and M. L. Masson, "Foam mat drying of yacon juice: Experimental analysis and computer simulation," *Journal of Food Engineering*, vol. 158, pp. 48–57, 2015.

[19] J. Langová, D. Jaisut, R. Thuwapanichayanan et al., "Modelling the moisture sorption isotherms of roselle (hibiscus sabdariffa l.) in the temperature range of 5-35∘c," *Acta Universitatis Agriculturae et Silviculturae Mendelianae Brunensis*, vol. 61, no. 6, pp. 1769–1777, 2013.

[20] D. Christ, R. L. da Cunha, F. C. Menegalli, K. P. Takeuchi, S. R. M. Coelho, and L. H. P. Nóbrega, "Sorption isotherms of albumen dried in a spout fluidised bed," *Journal of Food, Agriculture and Environment (JFAE)*, vol. 10, no. 2, pp. 151–155, 2012.

[21] P. Rajkumar, R. Kailappan, R. Viswanathan, and G. S. V. Raghavan, "Drying characteristics of foamed alphonso mango pulp in a continuous type foam mat dryer," *Journal of Food Engineering*, vol. 79, no. 4, pp. 1452–1459, 2007.

[22] R. A. Wilson, D. M. Kadam, S. Chadha, and M. Sharma, "Foam Mat Drying Characteristics of Mango Pulp," *International Journal of Food Science and Nutrition Engineering*, vol. 2, no. 4, pp. 63–69, 2012.

[23] W. Brand-Williams, M. E. Cuvelier, and C. Berset, "Use of a free radical method to evaluate antioxidant activity," *LWT - Food Science and Technology*, vol. 28, no. 1, pp. 25–30, 1995.

[24] J. Ahmed, U. S. Shivhare, and G. S. V. Raghavan, "Thermal degradation kinetics of anthocyanin and visual colour of plum puree," *European Food Research and Technology*, vol. 218, no. 6, pp. 525–528, 2004.

[25] N. Anuar, A. F. Mohd Adnan, N. Saat, N. Aziz, and R. Mat Taha, "Optimization of extraction parameters by using response surface methodology, purification, and identification of anthocyanin pigments in melastoma malabathricum fruit," *The Scientific World Journal*, vol. 2013, Article ID 810547, 10 pages, 2013.

[26] M. Djaeni, S. B. Sasongko, A. Prasetyaningrum A, X. Jin, and A. J. Van Boxtel, "Carrageenan drying with dehumidified air: Drying characteristics and product quality," *International Journal of Food Engineering*, vol. 8, no. 3, 2012.

[27] S. A. Ambekar, S. V. Gokhale, and S. S. Lele, "Process optimization for foam mat-tray drying of passiflora edulis flavicarpa pulp and characterization of the dried powder," *International Journal of Food Engineering*, vol. 9, no. 4, pp. 433–443, 2013.

[28] E. Abbasi and M. Azizpour, "Evaluation of physicochemical properties of foam mat dried sour cherry powder," *LWT- Food Science and Technology*, vol. 68, pp. 105–110, 2016.

[29] T. Kudra and C. Ratti, "Foam-mat drying: Energy and cost analyses," *Canadian Biosystems Engineering*, vol. 48, pp. 27–32, 2006.

[30] I. Karabulut, A. Topcu, A. Duran, S. Turan, and B. Ozturk, "Effect of hot air drying and sun drying on color values and β-carotene content of apricot (Prunus armenica L.)," *LWT- Food Science and Technology*, vol. 40, no. 5, pp. 753–758, 2007.

[31] L. F. Reyes and L. Cisneros-Zevallos, "Degradation kinetics and colour of anthocyanins in aqueous extracts of purple- and red-flesh potatoes (Solanum tuberosum L.)," *Food Chemistry*, vol. 100, no. 3, pp. 885–894, 2007.

[32] S.-Q. Cao, L. Liu, and S.-Y. Pan, "Thermal degradation kinetics of anthocyanins and visual color of blood orange juice," *Agricultural Sciences in China*, vol. 10, no. 12, pp. 1992–1997, 2011.

21

Comparative Study between Ethanolic and β-Cyclodextrin Assisted Extraction of Polyphenols from Peach Pomace

Nada El Darra ⓘ,[1] Hiba N. Rajha,[2] Espérance Debs,[3] Fatima Saleh,[1] Iman El-Ghazzawi,[1] Nicolas Louka,[2] and Richard G. Maroun[2]

[1]Faculty of Health Sciences, Beirut Arab University, Tarik El Jedidah, Riad El Solh, P.O. Box 115020, Beirut 1107 2809, Lebanon
[2]Unité de Recherche Technologies et Valorisation Agro-Alimentaire, Centre d'Analyses et de Recherche, Faculté des Sciences, Université Saint-Joseph, Riad El Solh, BP 11-514, Beirut 1107 2050, Lebanon
[3]Faculty of Sciences, Department of Biology, University of Balamand, Koura, Lebanon

Correspondence should be addressed to Nada El Darra; n.aldarra@bau.edu.lb

Academic Editor: Salam A. Ibrahim

Peach byproducts are often regarded as food waste despite their high content in health-promoting components. Amongst the latter, polyphenols are bioactive molecules with significant health benefits. The present study investigated an eco-friendly and cost-effective method using a GRAS food additive, β-cyclodextrin (β-CD), for the recovery of polyphenols from peach pomace. β-CD assisted extraction of polyphenols was compared to that of conventional solvent (ethanol) extraction at the same concentrations (10 mg/mL, 20 mg/mL, 30 mg/mL, 40 mg/mL, and 50 mg/mL) in terms of quality (antiradical activity) and quantity. The extract obtained by 50 mg/mL β-CD assisted extraction showed the highest polyphenol (0.72 mg GAE/g DM) and flavonoid (0.35 mg catechin/g of DM) concentrations as maximal antiradical activity (6.82%) and a noted antibacterial activity. Our results showed the competitiveness of β-CD assisted extraction to recover a high quantity and quality of polyphenols from peach pomace suggesting β-CD as a green alternative method for phenolic extraction.

1. Introduction

Byproducts and waste obtained from food processing represent a major disposal problem for the food industry. A large number of these products are generated at different stages of the food supply chain of either vegetable or animal commodities [1]. Currently, waste management bodies are highly recommending that industrialists invest in new end-uses for such food byproducts. Valorisation of food waste sources is therefore becoming a prime interest, owing to their environmental and economic values. For instance, natural oils extracted from fruit seeds and kernels are used in cosmetics and pharmaceutical industry [2].

Peach (*Prunus persica* L.) is a fruit rich in polyphenols that are mainly localized in the pulp and peel tissues. Chlorogenic acid, catechin, epicatechin, rutin, and cyanidin-3-glucoside represent the main phenolic compounds of this fruit [3]. Although it is rich in ascorbic acid and carotenoids, it was found that the phenolic content of the peach is the major contributor to the observed antioxidative activity [4]. It is noteworthy to mention that, in China (mainland), 10^5 MT (Metric Ton) of peach pomace has been at least estimated to be produced annually from peach juice processing [5].

In the present study, we are interested in the recovery of polyphenols from the peach pomace, as the studies related to peach polyphenols are very limited in the literature. Adil and coworkers [6] have optimized the subcritical extraction of phenolic content from peach pomace which was performed by selecting the pressure between 20 and 60 MPa, temperature between 40 and 60°C, ethanol concentration at 14–20 wt.%, and extraction time from 10 to 40 min on subcritical (CO_2 + ethanol) extraction of polyphenol. The total phenolic content from peach pomace was 0.26 mg gallic acid equiv./g and the antiradical efficiency was 1.5 mg 1,1-diphenyl-2-picrylhydrazyl (DPPH)/mg [6].

Traditionally, extraction of phenolic compounds from natural resources is carried out using organic solvents, for example, methanol, ethanol, and acetone [7]. However, these extraction processes are quite laborious and involve large amounts of solvents associated with serious environmental issues, thus limiting their application. The need to develop new and cost-effective methods used to extract high levels of polyphenols with enhanced bioavailability from products such as peach turns out to be urgent. There is an increasing demand in recent years for cheaper, safer, and eco-friendly alternatives to organic solvents. Cyclodextrins (CDs) based extraction is an emerging "green" technology of great potential. CDs are naturally occurring cyclic oligosaccharides arising from the degradation of starch and are FDA "Food and Drug Administration" approved [8]. They appear as α-, β-, and γ-CDs, knowing that β-CDs are the least expensive and the most widely used [9]. β-CDs are increasingly employed as encapsulating agents for plants bioactive molecules such as polyphenols, hence preserving their biological properties, extending their shelf life, and protecting them away from environmental factors (light, temperature, oxidation, pH, and moisture) [9–11]. Compared to organic solvents, β-CDs assisted extraction is more economic, safe, and green [12].

A recent study conducted by Rajha and collaborators [13] has clearly shown the capacity of β-CD to extract polyphenols from vine shoots with a higher radical scavenging capacity compared to conventional ethanol extraction. Similarly, Ratnasooriya and Rupasinghe [14] demonstrated that the assisted recovery of total phenolic compounds from grape pomace slurry using β-CD at room temperature was significantly higher compared to water extraction [14]. Diamanti and coworkers [15] also reported that green extraction using β-CD enhanced the total phenolic content and the radical scavenging activity of whole pomegranate extracts [15].

To the best of our knowledge, there are no available studies on the use of CDs in recovering polyphenols from peach pomace. The aim of this study was to compare the efficiencies of polyphenols extraction using β-CD, aqueous and organic solvents. The extraction processes were optimized by monitoring solvent concentration, temperature, solvent volume to sample ratio, and extraction time. The resulting extracts were then analysed for total phenolic content, flavonoids, tannins, vitamin C, and carotenoids contents. In addition, these compounds were examined for their antioxidant and antimicrobial activity.

Finally, high-performance liquid chromatography (HPLC) was implemented to identify the phenolic components present in every extract.

2. Materials and Methods

2.1. Samples Preparation and Dry Matter Content.
Peach (*Prunus persica* L.) pomace was obtained from Conserves Modernes Chtaura (Chtaura, Lebanon) specialized in the production of jams and purees. The pomace consists of pressed skins and pulp residue. The dry matter content for the raw material was determined by weighing an appropriate amount of sample and drying it for 24 hours in a ventilated oven at 105°C [16].

2.2. Chemicals.
All chemicals were purchased from Fluka Chemie GmbH (Buchs, Switzerland) or from Sigma-Aldrich (Steinheim, Germany).

2.3. Bacterial Strains, Culture Media, and Growth Conditions.
Sixteen stock isolates of bacterial strains were used. They are clinical isolates obtained from previous research studies carried out in the Faculty of Medicine of Alexandria University, Egypt. The isolates included two strains of each of the following: Methicillin-resistant *Staphylococcus aureus* (MRSA), Methicillin-resistant *Staphylococcus epidermidis* (MRSE), coagulase-negative staphylococci, *Staphylococcus aureus*, high-level aminoglycoside-resistant enterococci (HLAR) (*Enterococcus faecalis* and *Enterococcus faecium* which were also resistant to vancomycin), *Pseudomonas aeruginosa*, and *Escherichia coli*. One strain of *Klebsiella pneumoniae* and *Acinetobacter baumannii* was also studied. The two strains of *Escherichia coli*, *Klebsiella pneumoniae*, and *Acinetobacter baumannii* were previously proved to be extended-spectrum β-lactamases (ESBL), by double-disc synergy and combined-disc tests.

2.4. Solid-Liquid Extraction Process.
The extraction process of polyphenols from peach pomace was performed with a solid-liquid ratio of 1 : 10 (w/v). After β-CD aqueous extraction at different concentrations (10, 20, 30, 40, and 50 mg/mL) and ethanol extraction at different concentrations (10, 20, 30, 40, 50, and 500 mg/mL), the extracts were centrifuged at 5000 rpm (rotation per minute) for 15 min. β-CD aqueous concentrations were prepared by dissolving the required weight of β-CD in the specific water volume in a 50°C water bath while stirring for 2 hours of diffusion time [13].

2.5. Total Phenolic Content Determination: Folin-Ciocalteu Method.
The total phenolic content was determined according to Folin-Ciocalteu (FC) method [17]. 0.2 mL of standard (gallic acid) or diluted sample, 1.0 mL of FC reagent, and 0.8 mL of Na_2CO_3 solution (7.5%) were mixed and allowed to stand for 2 hours at room temperature. Light absorption was measured at 750 nm by a spectrophotometer UV-VIS against a blank similarly prepared, but containing distilled water instead of extract. The total phenolic content (Y_e) was expressed in grams of gallic acid equivalent (GAE) per gram of dry matter (DM) (g GAE/g DM).

2.6. Determination of Tannin Concentration.
Total tannin content (g/L) and HCl index which represents the tannin polymerization degree were determined according to Ribérau-Gayon and collaborators [18]. Total tannin assay is based on the heating process of tannins in acidic medium leading to the formation of cyanidins. Two tubes were prepared, each containing 1 mL of diluted peach pomace extract, 0.5 mL of water, and 1.5 mL of 12 N HCl. The first tube was mixed and heated in a water bath at 100°C for 30 min. The second was kept at room temperature. Following the rapid cooling, 0.25 mL of ethanol was added to the mixture and the resulting absorbance was recorded at 520 nm.

The tannin concentration was calculated as follows:

$$\text{Tannin concentration (mg/L)} = 19.33 \times \Delta \text{ optical densities.} \tag{1}$$

2.7. Free Radical Scavenging Activity. The 1,1-diphenyl-2-picrylhydrazyl (DPPH) radical was used in the present study for the screening of the radical scavenging activity of the extracts [19]. The DPPH radical scavenging activity was measured using the spectrophotometer UV-VIS (Libra S32, Biochrom, France). The samples were tested at a concentration of 20 mg/mL and then mixed with 1000 μL of 0.1 mM DPPH-ethanol solution and 450 μL of 50 mM Tris-HCl buffer (pH 7.4). Methanol (50 μL) was used for blank measurements in this experiment. After 30 min of incubation at room temperature, the reduction of the DPPH free radical was measured by reading the absorbance at 517 nm. Butylhydroxytoluene (BHT) (a synthetic antioxidant) was used as a positive control. The inhibition ratio (percent) was calculated according to the following equation:

$$\% \text{ inhibition} = \left[\frac{(\text{absorbance of control} - \text{absorbance of test sample})}{\text{absorbance of control}} \right] \quad (2)$$
$$* 100.$$

2.8. Determination of Total Flavonoids (TF). The total flavonoids (TF) assay was conducted as previously described by Zhuang and coworkers [20] with some modification [20]. A volume of 1 mL of diluted extract or standard solution of catechin was placed in a 10 mL volumetric flask already containing 4 mL of H_2O. Five minutes later, 0.3 mL of $NaNO_2$ (5%) and 1.5 mL of $AlCl_3$ (2%) were added. The mixture was shaken for 5 min, then 2 mL of 1 M solution of NaOH was added, and the mixture was well shaken again. The absorbance was measured at 510 nm against the blank. The results were calculated according to the calibration curve for catechin ($R^2 = 0.99$). The content of TF was expressed as mg of catechin equivalent (CE) per g of dry matter content.

2.9. Vitamin C Analysis. Vitamin C estimation was done according to the Folin-Ciocalteu method [21]. Peach pomace extracts (0.2 mL) were added to 0.8 mL of 10% trichloroacetic acid and then well shaken. The mixtures were kept on ice for 5 min and then centrifuged at 3000g for 5 min. The extract was then diluted (1/10). Folin-Ciocalteu was diluted (1/10) and then 0.2 mL was added to the mixture and vigorously shaken. After 10 min, at room temperature, the absorbance was measured at 760 nm against distilled water as a blank and vitamin C was estimated through the calibration curve of the ascorbic acid standard.

2.10. β-Carotene Estimation. A simple UV spectrophotometric method was used for the analysis of β-carotene. The extraction of β-carotene was simply assessed on the liquid extracts by UV absorbance which was measured at 461 nm. The following equation was applied to determine the β-carotene concentration: $y = 0.1069x - 0.0057$, where x and y are, respectively, the β-carotene concentration in mg/L and UV absorption at 461 nm [22].

2.11. Determination of Minimum Inhibitory Concentration (MIC). The broth microdilution method was used in a sterile 96-well microtiter plate (U shaped base) [23].

2.11.1. Preparation of the Bacterial Inocula for MIC. Glycerol broth stocks were subcultured on a freshly prepared blood agar plate, incubated at 37°C overnight. Using a sterilized loop, five colonies of each strain were inoculated in 3 mL of cation adjusted Mueller Hinton broth, and the turbidity was compared to 0.5 McFarland standard. 1/100 dilution of this 0.5 McFarland was prepared to be used for the MIC.

2.11.2. Peach Phenolic Extracts Preparation for MIC Assessment. The peach ethanolic extracts at different concentrations (10, 20, 30, 40, 50, and 500 mg/mL) were subjected to drying by rotary evaporator, in order to remove the ethanol. Afterwards, five serial dilutions, in sterile distilled water, of each of the following extracts were prepared: aqueous extract, β-cyclodextrin extract (50 mg/mL), ethanolic extract (50 mg/mL), and ethanolic extract (500 mg/mL) until reaching the concentrations 53 μg/mL, 26 μg/mL, 13 μg/mL, 6 μg/mL, and 3 μg/mL. After the addition of equal volumes (100 μL) of each concentration of these extracts to the bacterial strains (100 μL) to be tested in each well of the plate, the final concentrations of the different phenolic extracts were reduced to 26 μg/mL, 13 μg/mL, 6.5 μg/mL, 3.25 μg/mL, and 1.6 μg/mL. Four microtiter plates were used in this experiment. After overnight incubation at 37°C, all the plates were examined for the MIC of each of the phenolic extracts concentration that inhibits the bacterial growth. All the peach phenolic extracts were filter sterilized using 0.4 μm disposable syringe filters prior to the assessment.

2.12. High-Performance Liquid Chromatography with Diode Array Detection (HPLC-DAD) Analysis. Polyphenol analyses were performed using a Jasco HPLC system (Japan) (PV-2089) equipped with an autosampler, an L-2130 pump, a Jetstream column oven, and an L-2450 diode array detector. The separation was carried out with a Column C18 25 × 0.46 mm (Teknokroma Professional Friendly Lichrospher 100 RP18 5 μM, 25 × 0.46, serial number NF-21378, Barcelona, Spain), using a gradient elution at a flow rate of 1 mL per min for 30 min. Trans-cinnamic acid, caffeic acid, hydroxybenzoic acid, chlorogenic acid, catechin, rutin, quercetin, protocatechin, gallic acid, epigallocatechin, kaempferol, catechin gallate, and myricetin standards were used for identification and quantification purposes with HPLC-DAD, respectively. The mobile phase consisted of acidified nanopure water at pH 2.3 with HCl (A) and acetonitrile HPLC grade (B). The elution was isocratic conditions from 0 to 5 min with 85% A and 15% B. Gradient from 5 to 30 min began with 85% A and 15% B and ended with 0% A and 100% B followed by isocratic conditions from 30 to 35 min with 0% A and 100% B to reequilibrate the column. The injection volume was 10 μL. The identification of peaks was based on retention time and the spectra of external standards. The concentration of phenolic compounds was determined from standard curves constructed for individual compounds by injecting different concentrations of corresponding standards [24].

2.13. Statistical Analysis. Each experiment was conducted in duplicate and analysis repeated twice. Means and standard deviations of data were calculated. The error bars in all

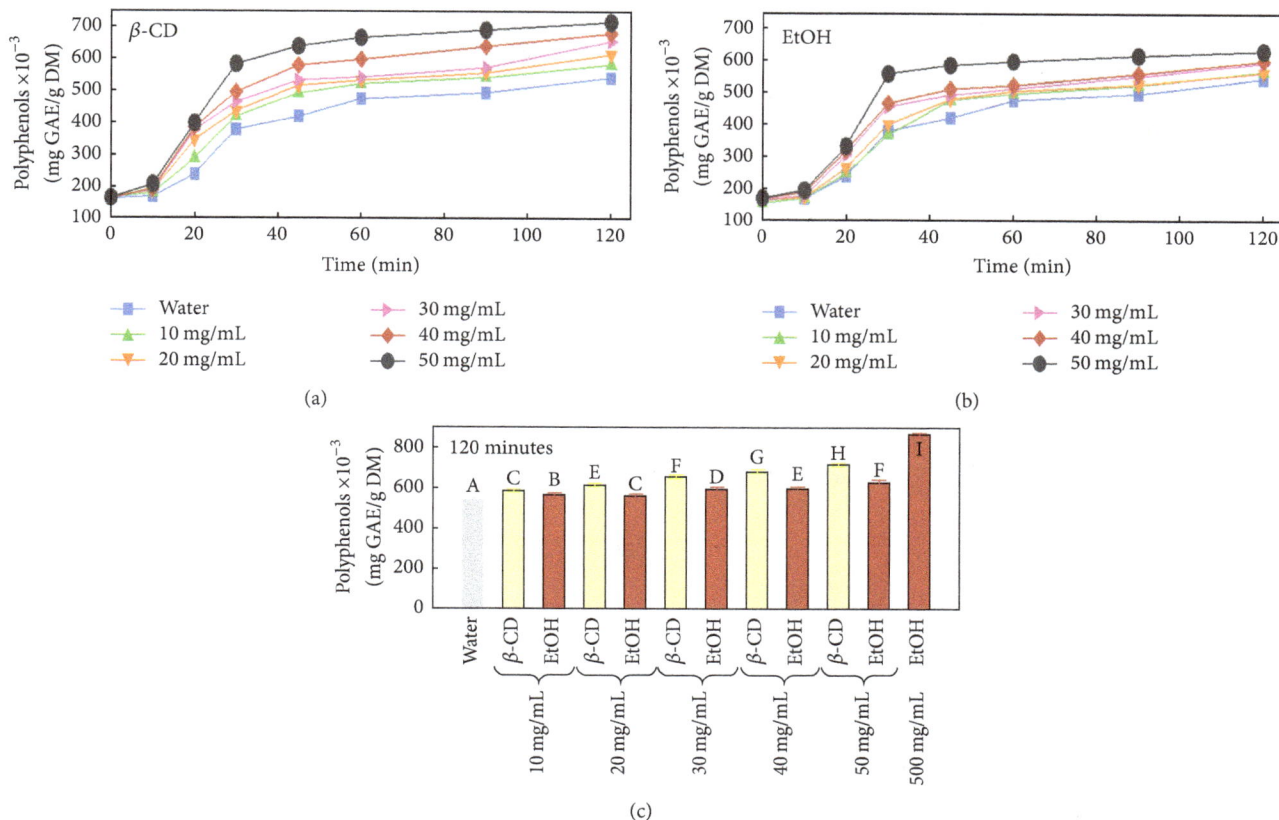

FIGURE 1: (a) Extraction kinetics of polyphenols with different concentrations (0, 10, 20, 30, 40, and 50 mg/mL) of beta-cyclodextrin (β-CD) and (b) different concentrations (0, 10, 20, 30, 40, and 50 mg/mL) of ethanol (EtOH). (c) Comparison of polyphenol recovery with both solvents (β-CD and EtOH) at the same concentrations after 120 minutes of extraction. Different superscript letters indicate significant statistical difference ($p < 0.05$).

figures correspond to the standard errors. Variance analyses (ANOVA) and Least Significant Difference (LSD) test were conducted to evaluate the significant differences between the results. STATGRAPHICS® Centurion XV (StatPoint Technologies, Inc.) was used to perform statistical analysis.

3. Results and Discussion

3.1. Effect of Ethanolic and Beta-Cyclodextrin Assisted Extractions on the Phenolic Content, Tannins, and Flavonoids of the Peach Pomace Extract. Figure 1 shows the recovery of polyphenols from peach pomace by solid-liquid extraction using β-CD or EtOH solvents. Many parameters affect the solid-liquid extraction process. One of these parameters is the extraction time that had to be studied before undertaking the trials. For this purpose, a systematic study, between 0 and 120 minutes, was conducted for the recovery of total phenolic content (TPC) using the three extraction media: water, beta-cyclodextrin, and ethanol at different concentrations. In Figures 1(a) and 1(b), we noticed that the longer the extraction time, the better the TPC yield. Moreover, a closer observation of the plots revealed a common kinetics pattern of extraction for the three solvents. All the kinetics followed a classical three-step model. A slight increase in the TPC yield

was observed during the first ten minutes of the extraction process. Less than 20% increase was recorded for the best datum. Then a second step occurred where a sharp rise of the extraction efficiency is clearly noticed. The TPC yield was around 3-fold higher at 30 min compared to 10 min. Finally, the slope decreased significantly between 30 and 120 min where the recovery capacity had plateaued out with barely 30% of improvement. To stretch out the argument, this model could strongly reflect what is happening at the microscopic level (Figure 1(c)). Step one marks the time needed by the solvent to diffuse inside the cellular structure and to prepare the biological material for the subsequent step. Step two represents the highest extraction efficiency of the cellular content in TPC. Step three witnesses an almost stabilization of the extraction yield, which means that the three solvents would have reached their maximum capacity of extraction under those conditions rather than unveiling the maximum cellular content in TPC.

On a different note, it is noteworthy to mention that the effect of the concentration of both solvents β-CD and ethanol was directly proportional to the TPC yield. The tested solvents were used over a concentration range of 10 to 50 mg/mL; the higher the concentration, the better the TPC yield. Water was significantly the least efficient in TPC extraction.

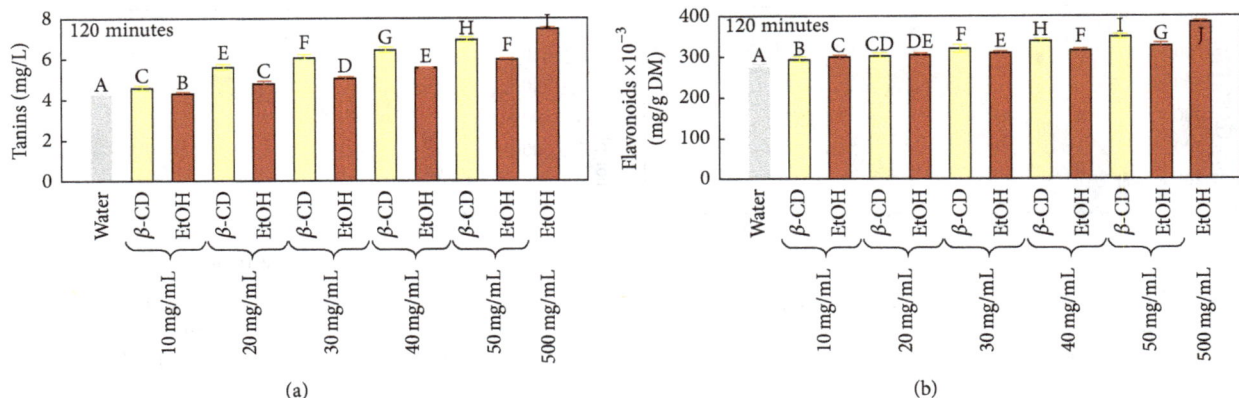

FIGURE 2: Comparison of (a) tannins and (b) flavonoids recovery using β-CD and EtOH at the same concentrations after 120 minutes of extraction. Different superscript letters indicate significant statistical difference ($p < 0.05$).

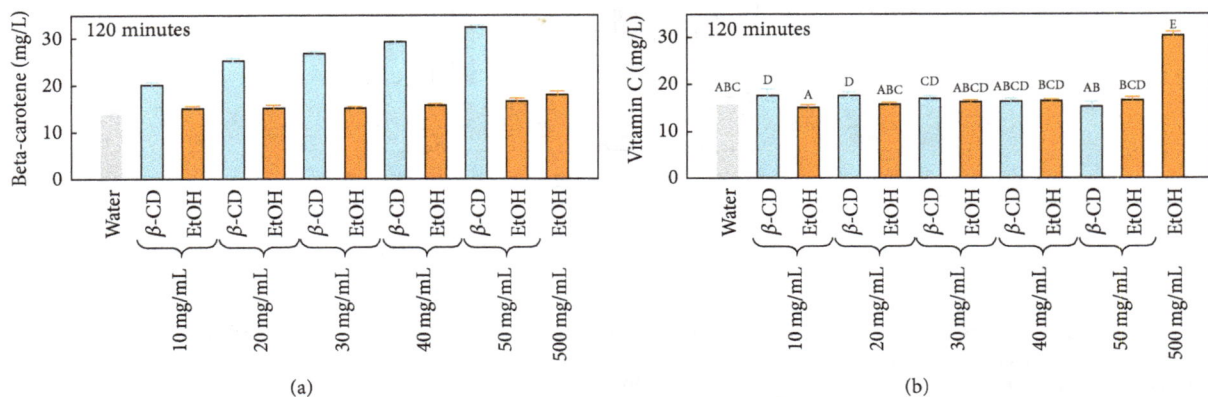

FIGURE 3: Comparison of (a) beta-carotene and (b) vitamin C recovery using β-CD and EtOH at the same concentrations after 120 minutes of extraction. Different superscript letters indicate significant statistical difference ($p < 0.05$).

The increase in both β-CD and EtOH concentrations up to 50 mg/mL significantly ($p < 0.05$) enhances polyphenol extraction compared to water solvent (Figures 1(a) and 1(b)). However, at the same concentrations, β-CD is more efficient than the organic solvent EtOH. For example, at 50 mg/mL of both solvents, polyphenol yields were 715 and 630 mg GAE/g DM for β-CD and EtOH, respectively. The highest polyphenol concentration was obtained with 500 mg/mL EtOH (865 mg GAE/g DM). However, compared to 50 mg/mL EtOH, the concentration was enhanced by only 1.2 times with an increase of tenfold EtOH concentration.

In concordance with total polyphenol extraction, tannins and flavonoids recovery from peach pomace was also enhanced by β-CD and EtOH compared to the aqueous extraction. At all the studied concentrations (10, 20, 30, 40, and 50 mg/mL), β-CD was more efficient than EtOH (Figures 2(a) and 2(b)). β-CD efficacy compared to water is likely due to the inclusion complexes it forms with bioactive molecules, which increases their solubility and therefore their recovery (Szente, L. et al., 2004, [9]). On the other hand, ethanol efficiency is rather related to the alteration it causes to the cellular membranes which increases their permeability and

therefore the diffusion process of intracellular components [25].

3.2. Effect of Ethanolic and Beta-Cyclodextrin Assisted Extraction on Beta-Carotene and Vitamin C Content of the Peach Pomace Extract. Figure 3 shows the recovery of beta-carotene and vitamin C from peach pomace with increasing β-CD and EtOH concentrations. β-CD ameliorates by far the extraction of beta-carotene compared to EtOH. For example, 50 mg/mL of β-CD permits the obtainment of a better beta-carotene yield (32 mg/L) than 500 mg/mL of EtOH (18 mg/L). In contrast, vitamin C extraction was not enhanced (16 mg/L) by β-CD or EtOH addition up to 50 mg/mL. This result is in agreement with the study of Navarro et al., 2011, who showed that the addition of β-CD produces low or null effect on the vitamin C content of pasteurized orange juice. The highest vitamin C yield (30 mg/L) was obtained with 500 mg/mL of EtOH.

3.3. Antiradical Activity of the Peach Pomace Extracted by Ethanolic and Beta-Cyclodextrin Assisted Extraction. Since no amelioration of vitamin C extraction was observed in

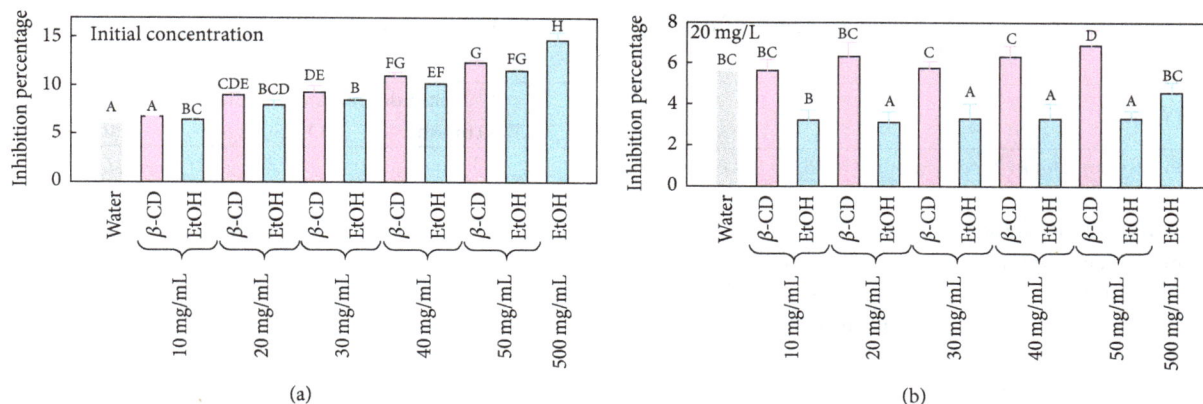

FIGURE 4: Comparison of the radical scavenging capacity of β-CD and EtOH extracts (a) at their initial concentration and (b) at 20 mg/mL of polyphenols. Different superscript letters indicate significant statistical difference ($p < 0.05$).

water, β-CD, and EtOH concentrations up to 50 mg/mL (Figure 3(b)), the radical scavenging capacity of those extracts was therefore attributed to their polyphenol and beta-carotene contents. This was in accordance with the study of Cantín et al., 2009, who showed no correlation between vitamin C and antiradical activity [26]. Figure 4(a) shows the inhibition percentage of the DPPH radical obtained with the different extracts at their initial polyphenol concentrations shown in Figure 1. The higher the polyphenol concentration in the extracts, the better their antiradical capacity was. Many authors showed the concentration-dependent radical scavenging activity of polyphenol extracts [27, 28]. At the same polyphenol concentration (Figure 4(b)), all β-CD extracts showed better antiradical capacity than those obtained with EtOH suggesting a better quality of the recovered molecules. The efficiency of β-CD in terms of polyphenol extraction and the enhancement of their radical scavenging capacity was also shown on vine shoots [13]. Similar study showed that microencapsulation of Mexican oregano essential oils with β-cyclodextrin enhanced their antiradical activity [29]. This is attributed to the protection effect of polyphenol encapsulation by β-CD since it is likely to preserve them from heat-degradation, UV light, and oxidation [30].

3.4. Antibacterial Activity of the Peach Pomace Extracted by Ethanolic and Beta-Cyclodextrin Assisted Extraction. The antimicrobial activities of the three peach pomace extracts (β-CD 50 mg/mL, EtOH 50 mg/mL, and EtOH 500 mg/mL) were tested against different Gram-positive and Gram-negative bacterial strains (Table 1) using MIC at different concentrations 1.6, 3.25, 6.5, 13, and 26 µg/mL. The ethanol extract (500 mg/mL) of peach pomace showed the highest inhibitory activity against the different tested Gram-positive and Gram-negative strains. Our results are in accordance with the study by Zarai et al. [31] who showed that ethanolic extracts were more effective against both Gram-positive and Gram-negative bacteria tested, which could be explained by a better extraction of phenolic and flavonoid components at ethanolic concentration of 500 mg/mL [31]. This could be probably due to the fact that ethanol at a high concentration

of 500 mg/mL is more selective in extracting polyphenols with higher biological activity. This was consistent with our radical scavenging activity and flavonoids findings, where ethanol 500 mg/mL showed the higher inhibitor activity of DPPH with high flavonoids content. Other extracts (β-CD 50 mg/mL, EtOH 50 mg/mL) were found to be active against all the tested species of Gram-positive bacteria whereas the Gram-negative bacteria remained unaffected. Our results are in agreement with the findings of Naz et al. [32] who showed that phenolic compounds have enhanced activity against Gram-positive strains compared to Gram-negative [32], due to the presence of an outer membrane in the cell wall of Gram-negative strains acting as permeability barrier and thus reducing the uptake [33]. Compared to EtOH 50 mg/mL, β-CD 50 mg/mL in water showed lower MIC for the different Gram-positive strains, which leads to a higher antibacterial activity. The ethanol was used to be compared at the same concentration as β-CD, since it is a solvent with relatively low toxic potential, and use of ethanol was permitted in the food industry [34]. The recovery of polyphenols possessing a higher antibacterial activity was noted for β-CD 50 mg/mL assisted extraction comparing to EtOH 50 mg/mL. This could be due to a better encapsulation and activity of polyphenols in β-CD 50 mg/mL, leading to a higher antibacterial activity.

3.5. Polyphenol Quantification by High-Performance Liquid Chromatography. Polyphenol quantity and diversity were also determined by HPLC on water, β-CD (50 mg/mL), and EtOH (50 and 500 mg/mL) extracts (Figure 5). β-CD selectively enhances the extraction of gallic and caffeic acids with yields equal to 220 and 328 µg/g DM, respectively. 500 mg/mL of EtOH was required to reach the same gallic and caffeic acids' yields. The higher concentrations of these phenolic acids in β-CD (50 mg/mL) and EtOH (500 mg/mL) could explain their high antibacterial activities against the bacterial strains stated above [35], as well as their high scavenging activity against the DPPH radical [36]. However, water seems to better extract protocatechin (7188 µg/g DM) compared to EtOH 50 mg/mL (3561 µg/g DM) and EtOH 500 mg/mL (4899 µg/g DM).

TABLE 1: Minimum inhibitory concentration (μg/mL) of different Gram + and Gram − bacteria obtained with β-CD 50 mg/mL, EtOH 50 mg/mL, and EtOH 500 mg/mL extracts.

Bacteria/POMs	Minimum inhibitory concentration (μg/mL)		
	β-CD 50 mg/mL	EtOH 50 mg/mL	EtOH 500 mg/mL
Methicillin-resistant *Staphylococcus aureus* (MRSA1) (Gram +)	13	26	13
Methicillin-resistant *Staphylococcus aureus* (MRSA2) (Gram +)	26	-	1.5
Methicillin-resistant *Staphylococcus epidermidis* MRSE 1297 (Gram +)	-	-	1.5
Methicillin-resistant *Staphylococcus epidermidis* MRSE 1296 (Gram +)	-	-	6.5
Coagulase-negative staphylococci 1664 (Gram +)	-	-	13
Coagulase-negative staphylococci 1530 (Gram +)	13	-	13
Staphylococcus aureus 1966 (Gram +)	6.5	13	13
Staphylococcus aureus 2030 (Gram +)	3	13	13
High-level aminoglycoside-resistance enterococci HLAR (Gram +)	13	-	13
Vancomycin-resistant enterococci VRE (Gram +)	13	-	1.5
Pseudomonas aeruginosa 27 (Gram −)	-	-	1.5
Pseudomonas aeruginosa 32 (Gram −)	-	-	1.5
Escherichia coli 1238 (Gram −)	-	-	13
Escherichia coli 1250 (Gram −)	-	-	13
Klebsiella pneumoniae 184 (Gram −)	-	-	6.5
Acinetobacter baumannii 1204 (Gram −)	-	-	1.5

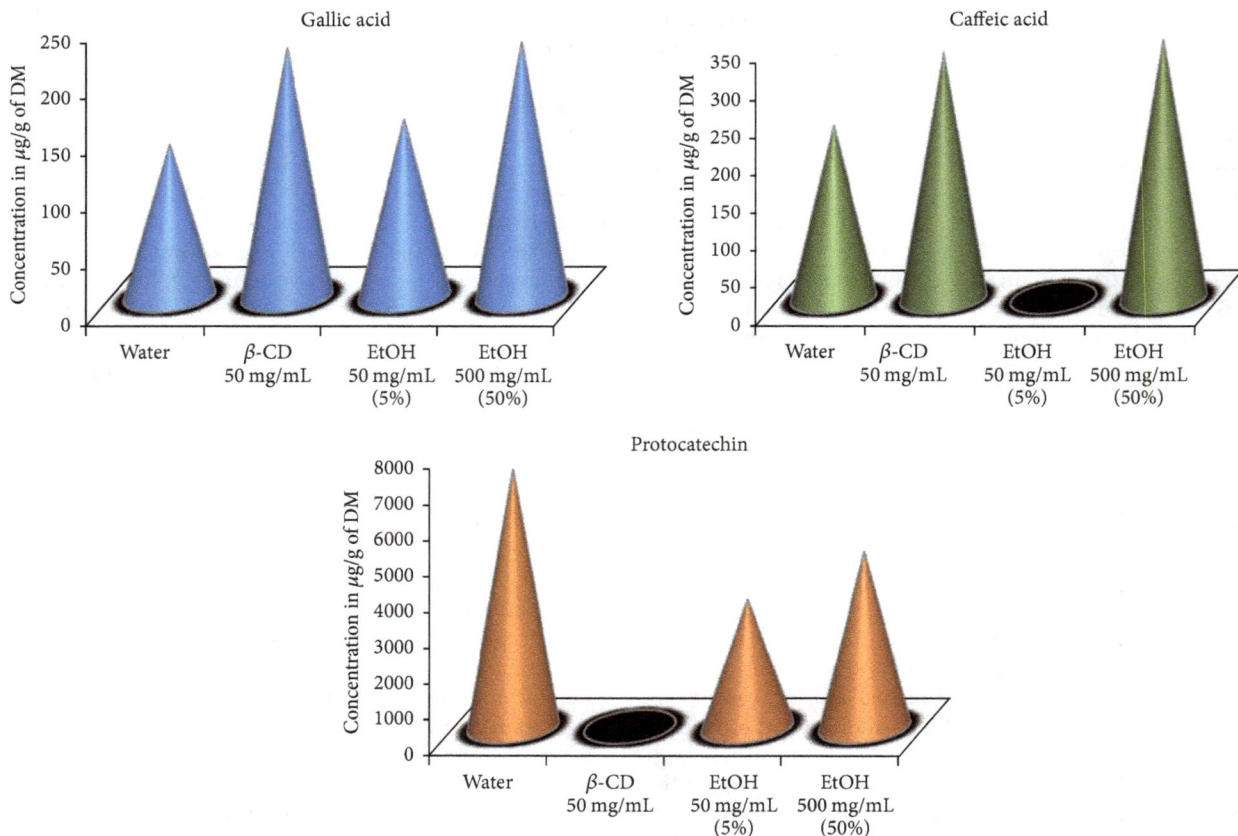

FIGURE 5: Gallic acid, caffeic acid, and protocatechin content of water, β-CD (50 mg/mL), EtOH (50 mg/mL), and EtOH (500 mg/mL) extracts.

4. Conclusion

Our study demonstrated that β-CD assisted extraction of peach pomace enhanced the phenolic content, tannins, flavonoids, carotenoids, antiradical, and antimicrobial activities compared to organic solvent extraction. Our study verified that encapsulation of peach pomace polyphenols in β-CD is efficient in terms of the quantity and quality of the extracted molecules. The use of β-CD in an assisted extraction process is a green technology for food waste recovery.

Disclosure

This work was presented as an oral presentation in "15th International Conference on Food Processing & Technology" in Italy 2016.

Conflicts of Interest

The authors declare that they have no conflicts of interest.

Authors' Contributions

Both Nada El Darra and Hiba N. Rajha equally contributed to this work.

References

[1] J. Gustavsson, C. Cederberg, U. Sonesson et al., "Global food losses and food waste: extent, causes and prevention," in *Proceedings of the International Congress: Save Food*, vol. 38, 2011.

[2] C. M. Galanakis, "Recovery of high added-value components from food wastes: conventional, emerging technologies and commercialized applications," *Trends in Food Science & Technology*, vol. 26, no. 2, pp. 68–87, 2012.

[3] C. Andreotti, D. Ravaglia, A. Ragaini, and G. Costa, "Phenolic compounds in peach (Prunus persica) cultivars at harvest and during fruit maturation," *Annals of Applied Biology*, vol. 153, no. 1, pp. 11–23, 2008.

[4] M. I. Gil, F. A. Tomás-Barberán, B. Hess-Pierce, and A. A. Kader, "Antioxidant capacities, phenolic compounds, carotenoids, and vitamin C contents of nectarine, peach, and plum cultivars from California," *Journal of Agricultural and Food Chemistry*, vol. 50, no. 17, pp. 4976–4982, 2002.

[5] H. Xu, Q. Jiao, F. Yuan, and Y. Gao, "Invitro binding capacities and physicochemical properties of soluble fiber prepared by microfluidization pretreatment and cellulase hydrolysis of peach pomace," *LWT- Food Science and Technology*, vol. 63, no. 1, pp. 677–684, 2015.

[6] I. H. Adil, H. I. Çetin, M. E. Yener, and A. Bayindirli, "Subcritical (carbon dioxide + ethanol) extraction of polyphenols from apple and peach pomaces, and determination of the antioxidant activities of the extracts," *The Journal of Supercritical Fluids*, vol. 43, no. 1, pp. 55–63, 2007.

[7] R. N. Mariano, D. Alberti, J. C. Cutrin, S. G. Crich, and S. Aime, "Design of PLGA based nanoparticles for imaging guided applications," *Molecular Pharmaceutics*, vol. 11, pp. 4100–4106, 2014.

[8] U.S. Food and Drug Administration, "Food additive status list," 2016, http://www.fda.gov/food/ingredientspackaginglabeling/foodadditivesingredients/ucm091048.

[9] E. Pinho, M. Grootveld, G. Soares, and M. Henriques, "Cyclodextrins as encapsulation agents for plant bioactive compounds," *Carbohydrate Polymers*, vol. 101, no. 1, pp. 121–135, 2014.

[10] N. Kalogeropoulos, K. Yannakopoulou, A. Gioxari, A. Chiou, and D. P. Makris, "Polyphenol characterization and encapsulation in β-cyclodextrin of a flavonoid-rich Hypericum perforatum (St John's wort) extract," *LWT - Food Science and Technology*, vol. 43, no. 6, pp. 882–889, 2010.

[11] M. Shulman, M. Cohen, A. Soto-Gutierrez et al., "Enhancement of naringenin bioavailability by complexation with hydroxypropoyl-β-cyclodextrin," *PLoS ONE*, vol. 6, no. 4, Article ID e18033, 2011.

[12] C. D. Santos, P. Buera, and F. Mazzobre, "Novel trends in cyclodextrins encapsulation, Applications in food science," *Current Opinion in Food Science*, 2017.

[13] H. N. Rajha, S. Chacar, C. Afif, E. Vorobiev, N. Louka, and R. G. Maroun, "β-cyclodextrin-assisted extraction of polyphenols from vine shoot cultivars," *Journal of Agricultural and Food Chemistry*, vol. 63, no. 13, pp. 3387–3393, 2015.

[14] C. C. Ratnasooriya and H. P. V. Rupasinghe, "Extraction of phenolic compounds from grapes and their pomace using β-cyclodextrin," *Food Chemistry*, vol. 134, no. 2, pp. 625–631, 2012.

[15] A. C. Diamanti, P. E. Igoumenidis, I. Mourtzinos, K. Yannakopoulou, and V. T. Karathanos, "Green extraction of polyphenols from whole pomegranate fruit using cyclodextrins," *Food Chemistry*, vol. 214, pp. 61–66, 2017.

[16] M. A. Madrau, A. Piscopo, A. M. Sanguinetti et al., "Effect of drying temperature on polyphenolic content and antioxidant activity of apricots," *European Food Research and Technology*, vol. 228, no. 3, pp. 441–448, 2009.

[17] K. Slinkard and V. L. Singleton, "Total phenol analysis: automation and comparison with manual methods," *American Journal of Enology and Viticulture*, vol. 28, no. 1, pp. 49–55, 1977.

[18] P. Ribéreau-Gayon, D. Dubourdieu, B. Donèche, and A. Lonvaud, Traité d'Œnologie, Tome 1: Microbiologie du Vin, Vinifications, Dunod, Paris, 1998.

[19] W. Brand-Williams, M. E. Cuvelier, and C. Berset, "Use of a free radical method to evaluate antioxidant activity," *LWT - Food Science and Technology*, vol. 28, no. 1, pp. 25–30, 1995.

[20] B. Zhuang, L.-L. Dou, P. Li, and E.-H. Liu, "Deep eutectic solvents as green media for extraction of flavonoid glycosides and aglycones from Platycladi Cacumen," *Journal of Pharmaceutical and Biomedical Analysis*, vol. 134, pp. 214–219, 2017.

[21] S. K. Jagota and H. M. Dani, "A new colorimetric technique for the estimation of vitamin C using folin phenol reagent," *Analytical Biochemistry*, vol. 127, no. 1, pp. 178–182, 1982.

[22] P. Karnjanawipagul, W. Nittayanuntawech, P. Rojsanga, and L. Suntornsuk, "Analysis of beta-carotene in Carrot by Spectrophotometer," *Mahidol University Journal of Pharmaceutical Science*, vol. 37, no. 8, 2010.

[23] J. M. Andrews, "Determination of minimum inhibitory concentrations," *Journal of Antimicrobial Chemotherapy*, vol. 48, no. 1, pp. 5–16, 2001.

[24] M. Vizzotto, W. Porter, D. Byrne, and L. Cisneros-Zevallos, "Polyphenols of selected peach and plum genotypes reduce cell viability and inhibit proliferation of breast cancer cells while not affecting normal cells," *Food Chemistry*, vol. 164, pp. 363–370, 2014.

[25] D. B. Goldstein and J. H. Chin, "Interaction of ethanol with biological membranes," *Federation Proceedings*, vol. 40, pp. 2073–2076, 1981.

[26] C. M. Cantín, M. A. Moreno, and Y. Gogorcena, "Evaluation of the antioxidant capacity, phenolic compounds, and vitamin C content of different peach and nectarine (Prunus persica (L.) Batsch) breeding progenies," *Journal of Agricultural and Food Chemistry*, vol. 57, no. 11, pp. 4586–4592, 2009.

[27] H. N. Rajha, N. Louka, N. Darra et al., "Multiple response optimization of high temperature, low time aqueous extraction process of phenolic compounds from grape byproducts," *Journal of Food and Nutrition Sciences*, vol. 5, pp. 351–360, 2014.

[28] M. N. Hasmida, A. R. N. Syukriah, M. S. Liza, and C. Y. M. Azizi, "Effect of different extraction techniques on total phenolic content and antioxidant activity of quercus infectoria galls," *International Food Research Journal*, vol. 21, no. 3, pp. 1039–1043, 2014.

[29] A. Arana-Sánchez, M. Estarrón-Espinosa, E. N. Obledo-Vázquez, E. Padilla-Camberos, R. Silva-Vázquez, and E. Lugo-Cervantes, "Antimicrobial and antioxidant activities of Mexican oregano essential oils (Lippia graveolens H. B. K.) with different composition when microencapsulated in β-cyclodextrin," *Letters in Applied Microbiology*, vol. 50, no. 6, pp. 585–590, 2010.

[30] Z. Fang and B. Bhandari, "Encapsulation of polyphenols—a review," *Trends in Food Science & Technology*, vol. 21, no. 10, pp. 510–523, 2010.

[31] Z. Zarai, E. Boujelbene, N. Ben Salem, Y. Gargouri, and A. Sayari, "Antioxidant and antimicrobial activities of various solvent extracts, piperine and piperic acid from Piper nigrum," *LWT- Food Science and Technology*, vol. 50, no. 2, pp. 634–641, 2013.

[32] S. Naz, S. Ahmad, S. A. Rasool, S. A. Sayeed, and R. Siddiqi, "Antibacterial activity directed isolation of compounds from Onosma hispidum," *Microbiological Research*, vol. 161, no. 1, pp. 43–48, 2006.

[33] A. D. Russell, "Similarities and differences in the responses of microorganisms to biocides," *Journal of Antimicrobial Chemotherapy*, vol. 52, no. 5, pp. 750–763, 2003.

[34] "Health Canada- Health Products and Food Branch," Lists of Permitted Food Additives, Government of Canada, 2016, http://www.hc-sc.gc.ca/fn-an/securit/addit/list/index-eng.php.

[35] M. J. R. Vaquero, M. R. Alberto, and M. C. M. de Nadra, "Antibacterial effect of phenolic compounds from different wines," *Food Control*, vol. 18, no. 2, pp. 93–101, 2007.

[36] M. Karamac, A. Kosińska, and R. B. Pegg, "Comparison of radical-scavenging activities for selected phenolic acids," *Polish Journal of Food and Nutrition Science*, vol. 14, pp. 165–169, 2005.

PERMISSIONS

LIST OF CONTRIBUTORS

Tesfaye Wolde
Applied Biology Department, Wolkite University, Wolkite, Ethiopia
Biology Department, Jimma University, Jimma, Ethiopia

Ketema Bacha
Biology Department, Jimma University, Jimma, Ethiopia

Leonardo Gomes Braga Ferreira
Laboratory of Inflammation, Oswaldo Cruz Foundation, Av. Brazil, 4365 Rio de Janeiro, RJ, Brazil

Robson Xavier Faria
Laboratory of Toxoplasmosis, Oswaldo Cruz Foundation, Av. Brazil, 4365 Rio de Janeiro, RJ, Brazil

Natiele Carla da Silva Ferreira and Rômulo José Soares-Bezerra
Laboratory of Cellular Communication, Oswaldo Cruz Foundation, Av. Brazil, 4365 Rio de Janeiro, RJ, Brazil

W. H. Karoki, D. N. Karanja, L. C. Bebora and L. W. Njagi
Department of Veterinary Pathology, Microbiology and Parasitology, University of Nairobi, Nairobi, Kenya

Séraphin C. Atidégla, Euloge K. Agbossou and Romain Glèlè Kakai
Faculté des Sciences Agronomiques, Université d'Abomey-Calavi, 01 BP 526 Cotonou, Benin

Joël Huat
CIRAD UPR HortSys, 34498 Montpellier Cedex 05, France

Hervé Saint-Macary
CIRAD, UPR Recyclage et Risque, 34398 Montpellier Cedex 05, France

Aline P. Oliveira, Geyssa Ferreira Andrade, Bianca S. O. Mateó and Juliana Naozuka
Departamento de Química, Universidade Federal de São Paulo, Diadema, SP, Brazil

Amani H. Al-Jahani
Nutrition and Food Science Department, College of Home Economics, Princess Nourah Bint Abdulrahman University, Riyadh, Saudi Arabia

Marta del Puerto
Department of Animal Production & Pastures, Nutrition and Food Quality Laboratory, Faculty of Agronomy, University of the Republic (UDELAR), E. Garzón 809, Montevideo, Uruguay

M. Cristina Cabrera
Department of Animal Production & Pastures, Nutrition and Food Quality Laboratory, Faculty of Agronomy, University of the Republic (UDELAR), E. Garzón 809, Montevideo, Uruguay
Physiology & Nutrition, Faculty of Sciences, University of the Republic (UDELAR), Iguá 4225, Montevideo, Uruguay

Ali Saadoun
Physiology & Nutrition, Faculty of Sciences, University of the Republic (UDELAR), Iguá 4225, Montevideo, Uruguay

Lamye Glory Moh and Kuiate Jules-Roger
Department of Biochemistry, University of Dschang, Dschang, Cameroon

Lunga Paul Keilah
Department of Biochemistry, University of Yaounde 1, Yaounde, Cameroon

Pamo Tedonkeng Etienne
Department of Animal Production, University of Dschang, Dschang, Cameroon

Andinet Abera Hailu and Getachew Addis
Ethiopian Public Health Institute, Addis Ababa, Ethiopia

M. A. Valdivia-López, A. Tecante, S. Granados-Navarrete and C. Martínez-García
Departamento de Alimentos y Biotecnología, Facultad de Química, Universidad Nacional Autónoma de México, 04510 Ciudad de México, Mexico

Ernest Ekow Abano
Department of Agricultural Engineering, School of Agriculture, College of Agricultural and Natural Sciences, University of Cape Coast, Cape Coast, Ghana

Vimbai R. Hamandishe, Petronella T. Saidi and Venancio E. Imbayarwo-Chikosi
Department of Animal Science, Faculty of Agriculture, University of Zimbabwe, Mt Pleasant Harare, Zimbabwe

Tamuka Nhiwatiwa
Department of Biological Sciences, Faculty of Science, University of Zimbabwe, Mt Pleasant Harare, Zimbabwe

Melkamnesh Azage and Mulugeta Kibret
Department of Biology, Science College, Bahir Dar University, Bahir Dar, Ethiopia

Purabi R. Ghosh, Derek Fawcett and Gerrard Eddy Jai Poinern
Murdoch Applied Nanotechnology Research Group, Department of Physics, Energy Studies and Nanotechnology, School of Engineering and Energy, Murdoch University, Murdoch, WA 6150, Australia

Shashi B. Sharma
Department of Agriculture and Food, 3 Baron-Hay Court, South Perth, WA 6151, Australia

Dayanne Vigo Miranda and Sandra Pagador
School of Agroindustrial Engineering, Universidad César Vallejo (UCV), AV. Víctor Larco Herrera 13009, Trujillo, Peru

Meliza Lindsay Rojas
Department of Agri-food Industry, Food and Nutrition (LAN), Luiz de Queiroz College of Agriculture (ESALQ), University of São Paulo (USP), AV. Padua dias 11, Piracicaba, SP, Brazil

Leslie Lescano, Jesús Sanchez-Gonzalez and Guillermo Linares
School of Agroindustrial Engineering, Universidad Nacional de Trujillo (UNT), Av. Juan Pablo II s/n, Trujillo, Peru

Anoma Chandrasekara and Thamilini Josheph Kumar
Department of Applied Nutrition, Wayamba University of Sri Lanka, Makandura, 60170 Gonawila, Sri Lanka

H. A. Rathnayake and S. B. Navaratne
Department of Food Science and Technology, Faculty of Applied Sciences, University of Sri Jayewardenepura, Gangodawila, Nugegoda, Sri Lanka

C. M. Navaratne
Department of Agricultural Engineering, Faculty of Agriculture, University of Ruhuna, Mapalana, Kamburupitiya, Sri Lanka

A. R. Sloan, M. L. Dunn, L. K. Jefferies, O. A. Pike, Sarah E. Nielsen Barrows and F. M. Steele
Department of Nutrition, Dietetics and Food Science, Brigham Young University, S-221 ESC, Provo, UT 84602, USA

Ruvini Liyanage, Chathuni Jayathilake and Rizliya Visvanathan
Division of Nutritional Biochemistry, National Institute of Fundamental Studies, Hantana Road, Kandy, Sri Lanka

Harshani Nadeeshani and Swarna Wimalasiri
Department of Food Science and Technology, Faculty of Agriculture, University of Peradeniya, Peradeniya, Sri Lanka

Mohamad Djaeni, Andri Cahyo Kumoro, Setia Budi Sasongko and Febiani Dwi Utari
Department of Chemical Engineering, Faculty of Engineering, Diponegoro University, Jl. Prof. H. Soedarto, SH, Tembalang, Semarang 50275, Indonesia

Nada El Darra, Fatima Saleh and Iman El-Ghazzawi
Faculty of Health Sciences, Beirut Arab University, Tarik El Jedidah, Riad El Solh, Beirut 1107 2809, Lebanon

Hiba N. Rajha, Nicolas Louka and Richard G. Maroun
Unité de Recherche Technologies et Valorisation Agro-Alimentaire, Centre d'Analyses et de Recherche, Faculté des Sciences, Université Saint-Joseph, Riad El Solh, BP 11-514, Beirut 1107 2050, Lebanon

Espérance Debs
Faculty of Sciences, Department of Biology, University of Balamand, Koura, Lebanon

Index